Advanced Textbooks in Control and Signal Processing

Springer
London
Berlin
Heidelberg
New York
Barcelona
Hong Kong
Milan
Paris
Santa Clara
Singapore
Tokyo

Series Editors

Professor Michael J. Grimble, Professor of Industrial Systems and Director
Professor Michael A. Johnson, Professor of Control Systems and Deputy Director
Industrial Control Centre, Department of Electronic and Electrical Engineering,
University of Strathclyde, Graham Hills Building, 50 George Street, Glasgow G1 1QE, U.K.

Other titles published in this series:

Genetic Algorithms: Concepts and Designs
K.F. Man, K.S. Tang and S. Kwong

Model Predictive Control
E. F. Camacho and C. Bordons

Discrete-Time Signal Processing
D. Williamson
Publication Due September 1999

E. W. Kamen and J. K. Su

Introduction to Optimal Estimation

With 43 Figures

Springer

E. W. Kamen, PhD
School of Electrical and Computer Engineering, Georgia Institute of Technology, Atlanta, GA 30332-0250

J. K. Su, PhD
Telecommunications Laboratory, University of Erlangen-Nurnberg, Cauerstrasse 7, D-91058 Erlangen, Germany

ISBN 1-85233-133-X Springer-Verlag London Berlin Heidelberg

British Library Cataloguing in Publication Data
Kamen, Edward
Introduction to optimal estimation. - (control and signal processing)
1.Signal processing - Digital techniques 2.Estimation theory
I.Title II.Su, Jonathan
621.3'822
ISBN 185233133X

Library of Congress Cataloging-in-Publication Data
Kamen, Edward W.
Introduction to optimal estimation / Edward Kamen and Jonathan Su.
p. cm. -- (Advanced textbooks in control and signal processing)
Includes bibliographical references (p.).
ISBN 1-85233-133-X (alk. paper)
1. Signal processing. 2. Estimation theory. 3. Mathematical optimaization. I. Su, Jonathan, 1969- . II. Title. III. Series.
TK5102.9.K36 1999
621.382'2--dc21 99-13005

To R. E. Kalman
— EWK

To Jennifer and Kendall
— JKS

Series Editors' Foreword

The topics of control engineering and signal processing continue to flourish and develop. In common with general scientific investigation, new ideas, concepts and interpretations emerge quite spontaneously and these are then discussed, used, discarded or subsumed into the prevailing subject paradigm. Sometimes these innovative concepts coalesce into a new sub-discipline within the broad subject tapestry of control and signal processing. This preliminary battle between old and new usually takes place at conferences, through the Internet and in the journals of the discipline. After a little more maturity has been acquired has been acquired by the new concepts then archival publication as a scientific or engineering monograph may occur.

A new concept in control and signal processing is known to have arrived when sufficient material has developed for the topic to be taught as a specialised tutorial workshop or as a course to undergraduates, graduates or industrial engineers. The *Advanced Textbooks in Control and Signal Processing* Series is designed as a vehicle for the systematic presentation of course material for both popular and innovative topics in the discipline. It is hoped that prospective authors will welcome the opportunity to publish a structured presentation of either existing subject areas or some of the newer emerging control and signal processing technologies.

Out of the 1940's came Wiener filtering and from the 1960's, and Kalman, emerged the state-space system description, the Kalman filter and the basis of the optimal linear gaussian regulator. This new technology dominated the control research activities of the 1960's and 1970's. Many of the control achievements of that era pervade the control curriculum today. The discrete Kalman filter was a major achievement for the field of optimal estimation. It is right therefore that this is the centrepiece of the new textbook Introduction *to Optimal Estimation* by Edward Kamen, and Jonathan Su of the Georgia Institute of Technology, U.S.A. In this textbook there is an introductory trio of chapters covering the basics of optimal estimation, an allegro of a chapter on Wiener filtering and a solid concluding quartet of chapters on the theoretical and applications aspects of the Kalman filter. The thorough and complete development presented is adaptable for graduate courses, self-study or can even be used as a good reference text.

M.J. Grimble and M.A. Johnson
Industrial Control Centre
Glasgow, Scotland, U.K.
June, 1999

Preface

This book began as a set of lecture notes prepared by the first-named author for a senior elective on estimation taught at the University of Florida some years ago. The notes were then expanded with a substantial amount of material added by the second-named author and used for a first-year graduate course on estimation taught in the School of Electrical Engineering at the Georgia Institute of Technology. Over the past few years, we have continued to develop and refine the notes based in part on several teachings of the estimation course at Georgia Tech, with the result being the present version of the text. We have also developed a number of examples in the book using MATLAB, and some of the homework problems require the use of MATLAB.

The primary objective in writing this book is to provide an introductory, yet comprehensive, treatment of both Wiener and Kalman filtering along with a development of least-squares estimation, maximum likelihood estimation, and maximum *a posteriori* estimation based on discrete-time measurements. Although this is a fairly broad range of estimation techniques, it is possible to cover all of them in some depth in a single textbook, which is precisely what we have attempted to do here. We have also placed a good deal of emphasis on showing how these different approaches to estimation fit together to form a systematic development of optimal estimation. It is possible to cover the bulk of material in the book in a one-semester course, and in fact, the book has been written to be used in a single course on estimation for seniors or first-year graduate students. The book can also be used for a one-quarter course, although in this case, some material must be omitted due to the shorter time period.

The background required for reading this book consists of a standard course on probability and random variables and one or more courses on signals and systems including a development of the state space theory of linear systems. It is helpful, but not necessary, to have had some exposure to random signals and the study of deterministic systems driven by random signals with random initial conditions. A summary treatment of this material which is needed in the book is given in Chapter 2. In teachings of the course based on the text material at Georgia Tech, we typically devote four or five 50-

minute lectures to the material in Chapter 2, so the students in the class are on somewhat the same level of proficiency in working with random variables and signals. In this chapter and in other parts of the book we emphasize the difference between formulations based on sample realizations of random signals and formulations based on random signals. This brings out the difference between the issue of actually computing estimates versus the issue of characterizing the properties of estimates viewed as random variables.

The book begins in Chapter 1 with the description of the estimation problem in a deterministic framework. Signal estimation is illustrated using a frequency-domain approach and state estimation is approached using the least squares methodlogy. Then the treatment of estimation in a stochastic framework begins in Chapter 2 with a summary of the theory of random variables, random signals, and systems driven by random signals. In Chapter 3, different versions of the optimal signal estimation problem are studied, with maximum likelihood (ML), maximum *a posteriori* (MAP), and minimum mean square error (MMSE) estimation covered. The case of linear MMSE estimation leads to the Wiener filter, which is developed in Chapter 4. The finite impulse response (FIR) Wiener filter, the noncausal infinite impulse response (IIR) Wiener filter, and the causal IIR Wiener filter are all derived in Chapter 4.

Chapter 5 begins the development of the Kalman filter for estimating the state of a linear system specified by a state model. The filter is derived using the orthogonality principle. The innovations approach to the derivation of the Kalman filter is given in Chapter 6. Chapter 6 also contains results on the time-varying case, robustness of the filter to model errors, the Kalman predictor, and the Kalman smoother. Applications of the Kalman filter to target tracking, system identification, and the case of nonwhite noise are considered in Chapter 7. The last chapter focuses on the case when the system state model is nonlinear, beginning with the derivation of the extended Kalman filter (EKF). A new measurement update, which is more accurate in general than the EKF measurement update, is derived using the Levenburg-Marquardt (LM) algorithm. Then applications of nonlinear filtering are considered including the identification of nonlinear systems modeled by neural networks, FM demodulation, target tracking based on polar-coordinate measurements, and multiple target tracking. The book also contains appendices on the state model formulation, the z-transform, and expanded developments of the properties of the Kalman filter.

The authors wish to thank the following individuals for their suggestions and comments on various drafts of the text: Louis Bellaire, Yong Lee, Brent Romine, Chellury Sastry, Jeff Schodorf, and Jim Sills. Thanks also go to the many students who have taken the course at Georgia Tech based on the material in the book and who have offered helpful comments.

EWK, JKS

Contents

Chapter 1

Introduction

One of the most common problems in science and engineering is the estimation of various quantities based on a collection of measurements. This includes the estimation of a signal based on measurements that relate to the signal, the estimation of the state of a system based on noisy measurements of the state, and the estimation of parameters in some functional relationship. The use of estimation techniques occurs in a very wide range of technology areas such as aerospace systems, communications, manufacturing, and biomedical engineering. Specific examples include the estimation of an aircraft's or spacecraft's position and velocity based on radar measurements of position, the estimation of congestion in a computer communications network, the estimation of process parameters in a manufacturing production system, and the estimation of the heath of a person's heart based on an electrocardiogram (ECG).

In this chapter we provide an introduction to the estimation problem, beginning in Section 1.1 with the estimation of a real-valued signal from a collection of measurements. Then in Section 1.2 we consider state estimation, and in the last section of the chapter, we present a deterministic approach to estimation based on the least squares method.

1.1 Signal Estimation

Consider a signal $s(t)$ which is a real-valued function of the continuous-time variable t. Suppose that there is another signal $z(t)$ that is generated from $s(t)$; that is, in mathematical terms

$$z(t) = g(s(t), v(t), t), \tag{1.1}$$

where $v(t)$ is a noise or disturbance term and g is a function that represents the degradation that occurs in the generation of $z(t)$ from $s(t)$. In general,

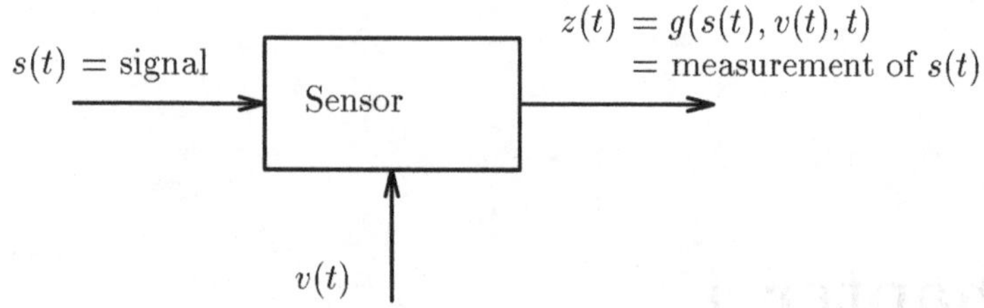

Figure 1.1. Sensor which provides a measurement $z(t) = g(s(t), v(t), t)$ of the signal $s(t)$.

the function g may depend on time t, which explains the appearance of t in the argument of g in (1.1). The type of relationship given by (1.1) arises in many applications. For example, in a communications system $s(t)$ may be a transmitted signal and $z(t)$ is the received signal which is a distorted version of $s(t)$. In other applications, $z(t)$ may be a measurement of the signal $s(t)$ obtained from a sensor as illustrated in Figure 1.1. The signal $s(t)$ may be the output $y(t)$ of a process or system and $z(t)$ is a measurement of $y(t)$.

In many applications, the measurement $z(t) = g(s(t), v(t), t)$ can be expressed in the signal-plus-noise form

$$z(t) = s(t) + v(t). \tag{1.2}$$

In this case, $z(t)$ is simply a sum of the signal $s(t)$ and *additive noise* $v(t)$. When $z(t)$ is a measurement provided by a sensor, the noise term $v(t)$ could be due to physical phenomena occurring within the sensor such as thermal noise and/or it could be a result of external effects such as gravity acting on a force sensor.

The particular form of the measurement equation (1.2) arises in target tracking where $z(t)$ is a noisy measurement of a target's position provided by a radar. There are many other applications where the signal-plus-noise form (1.2) arises, such as in the measurement of bioelectric signals including electrocardiogram (ECG) and electroencephalogram (EEG) signals.

In some applications, the signal-plus-noise form (1.2) is not valid; for example, it is possible that $z(t)$ could be given in terms of *multiplicative noise*; that is

$$z(t) = s(t)v(t).$$

Many other types of signal degradation can occur in practice, but we shall not pursue these. Often, the additive noise model (1.2) is used in practice since it is a simplifying assumption that makes the analysis tractable.

Given $z(t) = g(s(t), v(t), t)$, one of the fundamental problems in science and engineering is the reconstruction of $s(t)$ from $z(t)$. The determination of

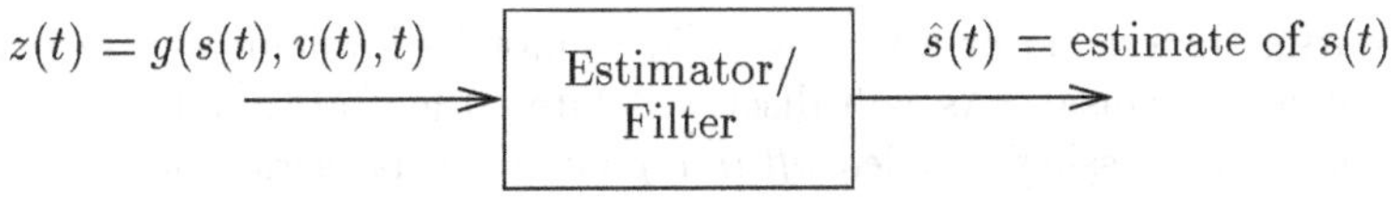

Figure 1.2. Signal reconstruction using an estimator/filter.

$s(t)$ from $z(t)$ is a type of filtering or estimation problem. In particular, a device which produces an estimate $\hat{s}(n)$ of $s(t)$ is called an *estimator* or *filter*. Such a device is illustrated in Figure 1.2. In this book, the term "estimator" will usually be used in preference to the term "filter."

In general, an estimator based on the measurements $z(t) = g(s(t), v(t), t)$ is a dynamical system; that is, the estimate $\hat{s}(t)$ at time t generated by the estimator is not simply a function of $z(t)$ at time t. Rather, the estimate $\hat{s}(t)$ will depend in general on $z(\tau)$ for some range of values of the variable τ. A general mathematical form for $\hat{s}(t)$ is

$$\hat{s}(t) = \alpha\left(\{z(\tau) : -\infty < \tau \leq t\}, t\right), \tag{1.3}$$

where α is a function of time t and the set of measurements $\{z(\tau) : -\infty < \tau \leq t\}$. The estimator defined by (1.3) is causal since $z(\tau)$ depends on $z(\tau)$ for $-\infty < \tau \leq t$. If the estimator is allowed to be noncausal, then the estimate $\hat{s}(t)$ at time t may depend on $z(\tau)$ for $\tau > t$. For example, if $\hat{s}(t)$ is generated from $z(\tau)$ for $-\infty < \tau \leq t + a$, where a is some positive number, then the input/output relationship of the estimator is given by

$$\hat{s}(t) = \alpha\left(\{z(\tau) : -\infty < \tau \leq t + a\}, t\right).$$

Linear Estimators

The causal estimator defined by (1.3) is linear if and only if the function α is linear, in which case the estimate $\hat{s}(t)$ becomes

$$\hat{s}(t) = \int_{-\infty}^{t} h(t, \tau) z(\tau)\, d\tau, \tag{1.4}$$

where $h(t, \tau)$ is the impulse response function of the estimator; that is, $h(t, \tau)$ is the output of the estimator at time t when the unit impulse $\delta(t)$ is applied at time $t = \tau$. If the estimator defined by (1.4) is allowed to be noncausal, so that the estimate $\hat{s}(t)$ depends on $z(\tau)$ for $-\infty < \tau \leq t + a$, where a is some positive number, then the input/output relationship of the estimator $\hat{s}(t)$ becomes

$$\hat{s}(t) = \int_{-\infty}^{t+a} h(t, \tau) z(\tau)\, d\tau, \tag{1.5}$$

Of course, the estimator given by (1.5) cannot be implemented in real time since it is noncausal. As will discussed later, there are applications where "noncausal processing" (called *off-line processing*) is carried out.

The estimator given by (1.4) or (1.5) is time varying in general. It is time invariant if and only if $h(t,\tau)$ is a function of the difference $t-\tau$, in which case we write

$$h(t,\tau) = h(t-\tau), \tag{1.6}$$

where $h(t)$ is the response of the estimator at time t when the impulse $\delta(t)$ is applied at time $t=0$. Hence, if the estimator is linear, causal, and time invariant, the input/output relationship is given by

$$\hat{s}(t) = \int_{-\infty}^{t} h(t-\tau)z(\tau)\,d\tau. \tag{1.7}$$

The right-hand side of (1.7) is the convolution of $h(t)$ with $z(t)$ which is denoted by $h(t) * z(t)$, and thus

$$\hat{s}(t) = h(t) * z(t). \tag{1.8}$$

A frequency domain representation of the estimator given by (1.8) can be generated by taking the Fourier transform of both sides of (1.8). First, recall that the Fourier transform $S(\Omega)$ of a continuous-time signal $s(t)$ is defined by

$$S(\Omega) = \int_{-\infty}^{\infty} s(t)e^{-j\Omega t}\,dt,$$

where Ω is the frequency variable, which is given in terms of radians per second (rad/sec). In general, the transform $S(\Omega)$ is a complex-valued function of Ω, and thus $S(\Omega)$ is often given in polar form

$$S(\Omega) = |S(\Omega)| \exp\left[j\angle S(\Omega)\right],$$

where $|S(\Omega)|$ is the *amplitude spectrum* of the signal $s(t)$ and the angle $\angle S(\Omega)$ is the *phase spectrum* of $s(t)$. The amplitude spectrum $|S(\Omega)|$ and the phase spectrum $\angle S(\Omega)$ display the magnitude and phase angles, respectively, of the frequency components comprising the signal $s(t)$.

Now taking the Fourier transform of both sides of (1.8) results in the frequency domain representation:

$$\hat{S}(\Omega) = H(\Omega)Z(\Omega), \tag{1.9}$$

where $\hat{S}(\Omega)$, $H(\Omega)$, and $Z(\Omega)$ are the Fourier transforms of $\hat{s}(t)$, $h(t)$, and $z(t)$, respectively. The Fourier transform $H(\Omega)$ of the impulse response $h(t)$ is referred to as the *frequency response function* of the estimator given by (1.8) or (1.9).

Estimator Design

Now given $z(t) = g(s(t), v(t), t)$, suppose that in the problem of estimating $s(t)$ we restrict attention to the case of a linear estimator given by (1.4) or (1.5). Then the problem of determining the estimator boils down to determining the function $h(t, \tau)$ of the estimator. If the estimator is further restricted to be time invariant, the problem is to determine the impulse response $h(t)$ or the frequency response function $H(\Omega)$ of the estimator.

Suppose that the measurement $z(t)$ is equal to signal-plus-noise $s(t)+v(t)$. In this case, the input/output relationship (1.4) of the estimator becomes

$$\begin{aligned}\hat{s}(t) &= \int_{-\infty}^{t} h(t,\tau)\,[s(\tau)+v(\tau)]\,d\tau \\ &= \int_{-\infty}^{t} h(t,\tau)s(\tau)\,d\tau + \int_{-\infty}^{t} h(t,\tau)v(\tau)\,d\tau, \end{aligned} \tag{1.10}$$

or if the estimator is time invariant, in the frequency domain we have

$$\hat{S}(\Omega) = H(\Omega)[S(\Omega)+V(\Omega)] = H(\Omega)S(\Omega) + H(\Omega)V(\Omega), \tag{1.11}$$

where in (1.11), $S(\Omega)$ and $V(\Omega)$ are the Fourier transforms of the signal $s(t)$ and noise $v(t)$, respectively.

From (1.10) or (1.11), it is clear that the design of the estimator should be carried out so that it "rejects" $v(t)$ and "passes" $s(t)$. In the time-invariant case, this means that we want to design the frequency response function $H(\Omega)$ so that

$$H(\Omega)S(\Omega) = S(\Omega) \tag{1.12}$$

$$H(\Omega)V(\Omega) = 0. \tag{1.13}$$

If (1.12) and (1.13) are satisfied, from (1.11), $\hat{S}(\Omega) = S(\Omega)$, and thus $\hat{s}(t) = s(t)$. Hence, perfect reconstruction of $s(t)$ is possible if (1.12) and (1.13) hold.

Unfortunately, (1.12) and (1.13) can never be exactly satisfied in practice, but it is possible to come close to achieving (1.12) and (1.13) in the case when $S(\Omega)$ and $V(\Omega)$ do not overlap in frequency; that is, when the frequency components in the signal $s(t)$ do not overlap with the frequency components in the noise $v(t)$. To put this in mathematical terms, suppose that there are non-overlapping frequency intervals $[\Omega_1, \Omega_2]$ and $[\Omega_3, \Omega_4]$ such that the magnitude $|S(\Omega)|$ is nonzero only for $\Omega \in [\Omega_1, \Omega_2]$, and the magnitude $|V(\Omega)|$ is nonzero only for $\Omega \in [\Omega_3, \Omega_4]$. Then (1.12) and (1.13) can be satisfied approximately if the frequency function $H(\Omega)$ of the estimator is chosen so that

$$H(\Omega) \approx \begin{cases} 1, & \Omega \in [\Omega_1, \Omega_2]; \\ 0, & \Omega \in [\Omega_3, \Omega_4]; \end{cases} \tag{1.14}$$

where "$\approx$" means approximately. If the frequency components of $s(t)$ and $v(t)$ do overlap, the design of an estimator based on (1.14) may still yield good results if a relatively small portion of the spectral content of $v(t)$ overlaps with the spectral content of $s(t)$. This includes the situation when the noise $v(t)$ is "wideband" and the signal $s(t)$ is "narrowband." In other words, the frequency components of the noise are distributed over a wide range of frequencies, whereas the frequency components of the signal are contained within a narrow range of frequencies.

However, there are many applications where the design of an estimator cannot be based on a simple frequency domain analysis of the signal $s(t)$ and noise $v(t)$. In such cases, a more sophisticated approach is needed, which leads us to Wiener and Kalman filtering. In both these approaches to estimation, information on the signal $s(t)$ and the noise $v(t)$ is exploited to generate the estimate $\hat{s}(t)$. In particular, Wiener filtering is based on a frequency domain approach given in terms of the power spectrum of the signal $s(t)$; whereas, Kalman filtering is based on a time domain approach given in terms of a state model for $s(t)$.

Estimation Based on Discrete Measurements

In this book the estimation problem will be given in terms of a discrete-time formulation since signal measurements are almost always taken at discrete points in time and signals are processed using digital computers. To set up the discrete-time version of signal estimation, we begin by considering the general setup where $z(t) = g(s(t), v(t), t)$, and the problem is again to estimate $s(t)$ from knowledge of $z(t)$. In practice, $z(t)$ is usually specified for t equal to discrete time points $t = t_n$ where n is an integer-valued time index. The time points t_n are usually taken to be T seconds apart, in which case $t_n = nT$. Here T is some fixed positive real number equal to the sampling interval.

The values of $z(t)$, $s(t)$, and $v(t)$ at $t = nT$ will be denoted by $z(n)$, $s(n)$, and $v(n)$, respectively. Hence, when $z(t) = g(s(t), v(t), t)$, we have $z(n) = g(s(n), v(n), nT)$. Then the problem of interest is the generation of an estimate $\hat{s}(n)$ of $s(n)$ based on the measurements $z(i)$ for some range of integer values of i. In the causal case, the general form of the estimate is

$$\hat{s}(n) = \alpha\left(\{z(i) : -\infty < i \leq n\}\right),$$

where α is a function of the time index n and the set of measurement values $\{z(i) : -\infty < i \leq n\}$. If the estimator is causal and linear, the estimate $\hat{s}(n)$ is given by

$$\hat{s}(n) = \sum_{i=-\infty}^{n} h(n,i)z(i), \tag{1.15}$$

where $h(n,i)$ is the output of the estimator at time n when the unit pulse $\delta(n)$ is applied at time i. (Recall that $\delta(0) = 0$ and $\delta(n) = 0$ for all $n \neq 0$.) Note that the estimator given by (1.15) is the "discrete-time counterpart" to the estimator given by (1.4). We also have a discrete-time counterpart of the noncausal estimator (1.5) given by

$$\hat{s}(n) = \sum_{i=-\infty}^{n+q} h(n,i)z(i), \tag{1.16}$$

where q is a positive integer.

The estimator given by (1.15) is time invariant if and only if $h(n,i)$ is a function of the difference $n - i$, in which case we write $h(n,i) = h(n-i)$, where $h(n)$ is the impulse response of the estimator. Thus, if the estimator is causal, linear, and time invariant, the estimate is given by

$$\hat{s}(n) = \sum_{i=-\infty}^{n} h(n-i)z(i). \tag{1.17}$$

The right-hand side of (1.17) is the discrete-time convolution of $h(n)$ and $z(n)$ which is denoted by $h(n) * z(n)$, and therefore,

$$\hat{s}(n) = h(n) * z(n). \tag{1.18}$$

Example 1.1 Mean Filter

Suppose that $s(n)$ is the constant signal $s(n) = s$ for all n, and we have measurements of $s(n)$ given by $z(n) = s + v(n)$ for $n = 1, 2, \ldots$. In this case we can estimate $s(n)$ using a *mean filter* where the estimate $\hat{s}(n)$ at time nT is equal to the mean value of the measurements up to time nT; that is,

$$\hat{s}(n) = \frac{1}{n}\left[z(1) + z(2) + \cdots + z(n)\right]. \tag{1.19}$$

Equation (1.19) can be expressed in the form

$$\hat{s}(n) = \sum_{i=1}^{n} h(n,i)z(i),$$

where $h(n,i) = 1/n$ for $i = 1, 2, \ldots, n$, and thus the mean filter is linear. However, it is not time invariant since $h(n,i)$ cannot be expressed as a function of the difference $n - i$. Hence, the mean filter is time varying. Inserting $z(i) = s + v(i)$ into (1.19) gives

$$\hat{s}(n) = s + \frac{1}{n}\left[v(1) + v(2) + \cdots + v(n)\right].$$

If $v(n)$ is "randomly varying" about zero, we would expect $(1/n)[v(1) + v(2) + \cdots + v(n)]$ to converge to zero as $n \to \infty$, and thus the mean filter should yield an accurate estimate of s if a sufficient number of measurements are available.

Frequency Domain Representation

The frequency domain representation of the estimator given by (1.18) can be generated by taking the discrete-time Fourier transform (DTFT) of both sides of (1.18). Recall that the DTFT $S(e^{j\omega})$ of a discrete-time signal $s(n)$ is defined by

$$S(e^{j\omega}) = \sum_{n=-\infty}^{\infty} s(n)e^{-j\omega},$$

where ω is the frequency variable with values ranging over a 2π interval (e.g., $-\pi \leq \omega < \pi$). The polar form of $S(e^{j\omega})$ is

$$S(e^{j\omega}) = \left|S(e^{j\omega})\right| \exp\left[j\angle S(e^{j\omega})\right],$$

where $\left|S(e^{j\omega})\right|$ is the amplitude spectrum of the discrete-time signal $s(n)$ and the angle $\angle S(e^{j\omega})$ is the phase spectrum of $s(n)$. It should be noted that the transform $S(e^{j\omega})$ is usually computed for discrete values of ω using the Fast Fourier transform (FFT). This is not considered here.

Then taking the DTFT of both sides of (1.18) yields the frequency domain representation

$$\hat{S}(e^{j\omega}) = H(e^{j\omega})Z(ejw),$$

where $\hat{S}(e^{j\omega})$, $H(e^{j\omega})$, and $Z(e^{j\omega})$ are the DTFTs of $\hat{s}(n)$, $h(n)$, and $z(n)$, respectively. The DTFT $H(e^{j\omega})$ of the impulse response $h(n)$ is the frequency response function of the estimator defined by (1.18).

If the estimator is constrained to be linear and time invariant, then the determination of the estimator reduces to finding either the impulse response $h(n)$ or the frequency response function $H(e^{j\omega})$ of the estimator. As in the continuous-time case, in the Kalman filtering approach it is necessary to have a state model for the signal $s(n)$. This is pursued in the next section.

Estimation of Signal Parameters

In many applications, a discrete-time signal $s(n)$ can be expressed in the form

$$s(n) = \sum_{j=1}^{q} \theta_j \gamma_j(n), \tag{1.20}$$

where $\theta_1, \theta_2, \ldots, \theta_q$ are constants referred to as the *parameters* of the signal, and $\gamma_1(n), \gamma_2(n), \ldots, \gamma_q(n)$ are known functions of n. For example, if $s(n)$ is the constant signal $s(n) = s$ for all n, then $s(n)$ can be expressed in the form (1.20) with the single parameter $\theta = s$ and the function $\gamma(n) = 1$ for all n.

More generally, the functions $\gamma_j(n)$ may be powers of n; that is $\gamma_j(n) = n^{j-1}$ for $j = 1, 2, \ldots, q$, in which case, the signal is given by

$$s(n) = \sum_{j=1}^{q} \theta_j n^{j-1}.$$

The form given by (1.20) is said to be *linear in the parameters* since it is a linear combination of $\theta_1, \theta_2, \ldots, \theta_q$. For a signal specified in the form (1.20), a major problem of interest is estimating the parameters $\theta_1, \theta_2, \ldots, \theta_q$, based on measurements $z(n) = g(s(n), v(n), nT)$ of the signal. This is a parameter estimation problem that we will solve in Section 1.3 using the least squares method.

1.2 State Estimation

Estimation is often carried in a state model framework defined in terms of state variables $x_1(t)$, $x_2(t)$, $\ldots$, $x_N(t)$, where N is the number of state variables. (See Appendix A for a brief review of the state model.) In the continuous-time case, the general form of a linear time-invariant state model used in estimation is as follows:

$$\frac{dx(t)}{dt} = Ax(t) + Bw(t) \tag{1.21}$$

$$z(t) = Cx(t) + v(t), \tag{1.22}$$

where $x(t)$ is the N-element state vector defined by

$$x(t) = \begin{bmatrix} x_1(t) \\ x_2(t) \\ \vdots \\ x_N(t) \end{bmatrix}.$$

In (1.21), $w(t)$ is the *process noise* which is an m-element column vector, A is a $N \times N$ matrix, and B is a $N \times m$ matrix. In (1.22), $z(t)$ is the measurement which is assumed to be a p-vector, C is a $p \times N$ matrix, and $v(t)$ is the measurement noise which is also a p-vector. Equations (1.21) and (1.22) comprise the state model; (1.22) is referred to as the *measurement equation.*

The problem is to generate an estimate $\hat{x}(t)$ of the state $x(t)$ at time t using the measurements $z(\tau)$ for $0 \leq \tau \leq t$, assuming measurements begin at time $t = 0$. The Kalman-Bucy filter for estimating $x(t)$ is given directly in terms of the matrices A and C in the state model for $x(t)$. The general form of the Kalman-Bucy filter is

$$\frac{d\hat{x}(t)}{dt} = A\hat{x}(t) + K(t)\left[z(t) - C\hat{x}(t)\right],$$

where $K(t)$ is a $N \times p$ matrix with time-varying entries, called the gain matrix of the estimator. The derivation of the "optimal gain" $K(t)$ requires an extensive development given in terms of a random signal formulation. This will be carried out in Chapter 5 for the discrete-time version of the filter (called the Kalman filter).

Example 1.2 Tracking an Object

As an illustration of the state formulation, we consider the problem of tracking the position and velocity of an object moving in three-dimensional space. It is assumed that the object has constant velocity and thus its motion can be decoupled into separate x, y, z coordinates. We shall consider only the x-coordinate motion with the x-coordinate position of the object denoted by $x_1(t)$. It is assumed that a radar provides a noisy measurement $z(t)$ of $x_1(t)$ with $z(t)$ given by

$$z(t) = x_1(t) + v(t), \tag{1.23}$$

where $v(t)$ is the measurement noise. Equation (1.23) is the model for the measurement which is an approximation to the actual measurement process in a radar. However, it is known that the additive noise form (1.23) does lead to good results in practice. To construct a state model, we let $x_1(t)$ and the velocity $\dot{x}_1(t)$ be the state variables. Then since the object is moving with constant velocity, the acceleration in the x-direction must be zero, and thus the derivative of $\dot{x}_1(t)$ must be zero. Hence, the state model (1.21)–(1.22) in this case is given by

$$\frac{dx(t)}{dt} = \begin{bmatrix} 0 & 1 \\ 0 & 0 \end{bmatrix} x(t) \tag{1.24}$$

$$z(t) = \begin{bmatrix} 1 & 0 \end{bmatrix} x(t) + v(t), \tag{1.25}$$

where

$$x(t) = \begin{bmatrix} x_1(t) \\ \dot{x}_1(t) \end{bmatrix}.$$

Note that in this example, there is no process noise. The problem is then to estimate $x(t)$ based on the measurements $z(\tau)$ for $0 \le \tau \le t$. Since the first entry of $x(t)$ is the position and the second entry is the velocity, the estimate of $x(t)$ will yield estimates of both the position and velocity of the object. The estimation of $x(t)$ will be carried out in Chapter 7 using the discrete-time Kalman filter, which is based on a discretization-in-time of the state model given by (1.24) and (1.25). The discretization is considered below.

Estimation Based on a Discrete-Time State Model

The discrete-time version of the state model (1.21)–(1.22) is given by

$$x(n+1) = \Phi x(n) + \Gamma w(n) \tag{1.26}$$

$$z(n) = Cx(n) + v(n), \tag{1.27}$$

where $x(n)$, $w(n)$, and $v(n)$ are sampled versions of the state, process noise, and measurement noise. From the results in Appendix A, we have that the coefficient matrices Φ and Γ in (1.26) are given by

$$\Phi = e^{AT}, \qquad \text{and} \qquad \Gamma = \int_0^T e^{A\tau} B\, d\tau,$$

where T is the sampling interval.

Example 1.3 State Model for Tracking an Object

Again consider the tracking of the object defined in Example 1.2. In this case,

$$e^{AT} = \begin{bmatrix} 1 & T \\ 1 & 1 \end{bmatrix}.$$

The above matrix exponential was computed using the result that e^{At} is equal to the inverse Laplace transform of $(sI - A)^{-1}$ where I is the $N \times N$ identity matrix. The discrete-time state model for tracking the object is then given by

$$x(n+1) = \begin{bmatrix} 1 & T \\ 1 & 1 \end{bmatrix} x(n)$$
$$z(n) = \begin{bmatrix} 1 & 0 \end{bmatrix} x(n) + v(n).$$

This state model will be utilized in Chapter 7 to generate the Kalman filter for estimating the position and velocity of the object from the noisy radar measurements $z(i)$ for $i \leq n$.

Estimation of Signals Using a State Model

For a large class of signals, the problem of signal estimation can be approached using a state model. In particular, this is the case for any continuous-time signal $s(t)$ that satisfies an Nth-order differential equation:

$$s^{(N)}(t) + a_{N-1}s^{(N-1)}(t) + \ldots + a_1 s^{(1)}(t) + a_0 s(t) = 0, \tag{1.28}$$

where $s^{(i)}(t)$ is the ith derivative of $s(t)$ and the a_i are known constants. Defining the state variables

$$x_1(t) = s(t)$$
$$x_2(t) = s^{(1)}(t)$$
$$\vdots$$
$$x_N(t) = s^{(N-1)}(t),$$

we can express (1.28) in the state model form:

$$\frac{dx(t)}{dt} = Ax(t), \qquad s(t) = Cx(t),$$

where $x(t)$ is the N-element state vector whose ith entry is equal to $x_i(t)$, A is the $N \times N$ matrix defined by

$$A = \begin{bmatrix} 0 & 1 & 0 & \cdots & 0 \\ 0 & 0 & 1 & & 0 \\ \vdots & \vdots & & \ddots & \vdots \\ 0 & 0 & 0 & \cdots & 1 \\ -a_0 & -a_1 & -a_2 & \cdots & -a_{N-1} \end{bmatrix},$$

and C is the N-element row vector given by

$$C = \begin{bmatrix} 1 & 0 & \cdots & 0 \end{bmatrix}.$$

Then given the measurements

$$z(t) = s(t) + v(t) = Cx(t) + v(t)$$

since the first entry of $x(t)$ is equal to $s(t)$, the signal $s(t)$ can be estimated by estimating the state $x(t)$. In particular, the estimate $\hat{s}(t)$ of $s(t)$ is equal to $C\hat{x}(t)$ where $\hat{x}(t)$ is the estimate of $x(t)$. The discrete-time version of the signal estimator can be constructed by using the discretized state model:

$$x(n+1) = e^{AT}x(n)$$
$$z(n) = Cx(n) + v(n).$$

Parameter Estimation Using a State Model

The problem of estimating the parameters of a signal as discussed above can also be approached using a state model. To show this, we first repeat the expression (1.20) for a signal with parameters:

$$s(n) = \sum_{j=1}^{q} \theta_j \gamma_j(n). \tag{1.29}$$

We then construct a state model with the state variables defined by

$$x_i(n) = \theta_i, \quad \text{for } i = 1, 2, \ldots, q.$$

Since the state variables are constants (equal to the parameters of the signal), we have

$$x(n+1) = \Phi x(n), \tag{1.30}$$

where $x(n)$ is the q-element state vector whose ith entry is equal to θ_i and Φ is the $q \times q$ identity matrix. In addition, by (1.29) we have

$$s(n) = \gamma(n)x(n),$$

where $\gamma(n)$ is the q-element row vector defined by

$$\gamma(n) = \begin{bmatrix}\gamma_1(n) & \gamma_2(n) & \cdots & \gamma_q(n)\end{bmatrix}.$$

Finally, with the measurements given by

$$z(n) = s(n) + v(n) = \gamma(n)x(n) + v(n), \tag{1.31}$$

we can proceed to estimate the state $x(n)$ using the state model (1.30)-(1.31), which will result in an estimate of the signal parameters. Note that in this case, the coefficient $\gamma(n)$ of $x(n)$ in the measurement equation (1.31) is in general a function of the time index n. It turns out that the Kalman filter can easily handle this time-varying case.

1.3 Least Squares Estimation

A very powerful approach to estimation is based on the method of least squares, which is presented in this section. The least squares method is a deterministic approach to estimation; in other words, the measurements and the quantities being estimated are not modeled as random variables or random signals. We begin by considering least squares estimation of the parameters of a signal and then we consider least squares estimation of the state in a state variable framework.

Least Squares Estimation of Signal Parameters

Consider the discrete-time signal $s(n)$ given by

$$s(n) = \sum_{j=1}^{q} \theta_j \gamma_j(n), \tag{1.32}$$

where $\theta_1, \theta_2, \ldots, \theta_q$ are unknown parameters and $\gamma_1(n), \gamma_2(n), \ldots, \gamma_q(n)$ are known functions of n. Measurements $z(n)$ of the signal are given by

$$z(n) = s(n) + v(n).$$

With θ equal to the q-vector

$$\theta = \begin{bmatrix}\theta_1 \\ \theta_2 \\ \vdots \\ \theta_q\end{bmatrix},$$

and $\gamma(n)$ equal to the q-element row vector

$$\gamma(n) = \begin{bmatrix}\gamma_1(n) & \gamma_2(n) & \cdots & \gamma_q(n)\end{bmatrix},$$

we can rewrite (1.32) in the form

$$s(n) = \gamma(n)\theta. \tag{1.33}$$

The objective is to generate an estimate $\hat{\theta}(n)$ of θ at time nT using the measurements $z(1)$, $z(2)$, ..., $z(n)$. Given an estimate $\hat{\theta}(n)$, from (1.33) we see that we can take the estimate $\hat{s}(n)$ of the signal $s(n)$ at time nT to be

$$\hat{s}(n) = \gamma(n)\hat{\theta}(n). \tag{1.34}$$

We can also take the estimate of $s(i)$ at time iT for $i < n$ to be

$$\hat{s}(i) = \gamma(i)\hat{\theta}(n), \quad i < n. \tag{1.35}$$

Now with the estimate $\hat{s}(i)$ defined by (1.34)–(1.35), consider the sum of the squares

$$[z(1) - \hat{s}(1)]^2 + [z(2) - \hat{s}(2)]^2 + \cdots + [z(n) - \hat{s}(n)]^2. \tag{1.36}$$

If the estimate $\hat{\theta}(n)$ is equal to θ and $v(i) = 0$ for $i = 1, 2, \ldots, n$, then the sum of squares in (1.36) is equal to zero. If the estimate $\hat{\theta}(n)$ is not equal to θ, the sum of squares will equal some positive number whose magnitude depends on how far off the estimate is. Thus the sum of squares can be viewed as an error function (for the estimation of θ) that we want to minimize. The estimate $\hat{\theta}(n)$ that minimizes (1.36) is the least squares (LS) estimate of θ at time nT.

To compute the LS estimate, we first need to express (1.36) in vector form: Let Z_n denote the vector of measurements given by

$$Z_n = \begin{bmatrix} z(1) \\ z(2) \\ \vdots \\ z(n) \end{bmatrix},$$

and let Γ_n denote the $n \times q$ matrix whose ith row is equal to $\gamma(i)$; that is,

$$\Gamma_n = \begin{bmatrix} \gamma(1) \\ \gamma(2) \\ \vdots \\ \gamma(n) \end{bmatrix}.$$

Then (1.36) can be written in the vector form

$$\left[Z_n - \Gamma_n \hat{\theta}(n)\right]^{\mathrm{T}} \left[Z_n - \Gamma_n \hat{\theta}(n)\right], \tag{1.37}$$

where the superscript "T" denotes transposition. The reader should verify that (1.37) is equivalent to the sum of squares in (1.36).

To compute the LS estimate, we first take the partial derivative of (1.37) with respect to $\hat{\theta}(n)$:

$$\begin{aligned} &\frac{\partial}{\partial \hat{\theta}(n)} \left[Z_n - \Gamma_n \hat{\theta}(n)\right]^{\mathrm{T}} \left[Z_n - \Gamma_n \hat{\theta}(n)\right] \\ &= 2\left\{\frac{\partial}{\partial \hat{\theta}(n)} \left[Z_n - \Gamma_n \hat{\theta}(n)\right]^{\mathrm{T}}\right\} \left[Z_n - \Gamma_n \hat{\theta}(n)\right] = -2\Gamma_n^{\mathrm{T}} \left[Z_n - \Gamma_n \hat{\theta}(n)\right]. \end{aligned}$$

Setting the derivative to zero gives

$$\Gamma_n^{\mathrm{T}} \Gamma_n \hat{\theta}(n) = \Gamma_n^{\mathrm{T}} Z_n. \tag{1.38}$$

A sufficient condition for (1.38) to have a solution for $\hat{\theta}(n)$ is that the rank of Γ_n be equal to q, in which case the $q \times q$ matrix $\Gamma_n^{\mathrm{T}} \Gamma_n$ is invertible. The condition that the rank of Γ_n be equal to q places a constraint on the time functions $\gamma_j(n)$ comprising the representation of the signal given by (1.32). Note that in order for Γ_n to have rank q, it is necessary that $n \geq q$.

If the rank of Γ_n is equal to q so that $\Gamma_n^{\mathrm{T}} \Gamma_n$ is invertible, (1.38) has a unique solution given by

$$\hat{\theta}(n) = \left[\Gamma_n^{\mathrm{T}} \Gamma_n\right]^{-1} \Gamma_n^{\mathrm{T}} Z_n. \tag{1.39}$$

This is the LS estimate of θ at time nT. It is said to be in *batch form* since $\hat{\theta}(n)$ is specified in terms of all the measurements $z(1)$, $z(2)$, ... , $z(n)$. Due to numerical complexity, the batch form estimate may be difficult to compute when n is a very large number. Fortunately, it turns out that the LS estimate can be expressed in a recursive form that is derived next.

Recursive Form of LS

To derive the recursive form of the LS estimate, we first note that

$$\Gamma_{n+1}^{\mathrm{T}} Z_{n+1} = \Gamma_n^{\mathrm{T}} Z_n + \gamma^{\mathrm{T}}(n+1) z(n+1), \tag{1.40}$$

$$\Gamma_{n+1}^{\mathrm{T}} \Gamma_{n+1} = \Gamma_n^{\mathrm{T}} \Gamma_n + \gamma^{\mathrm{T}}(n+1) \gamma(n+1). \tag{1.41}$$

Applying the matrix inversion lemma to (1.41) gives

$$\left[\Gamma_{n+1}^{\mathrm{T}} \Gamma_{n+1}\right]^{-1} = \left[\Gamma_n^{\mathrm{T}} \Gamma_n\right]^{-1} - \frac{\left(\Gamma_n^{\mathrm{T}} \Gamma_n\right)^{-1} \gamma^{\mathrm{T}}(n+1) \gamma(n+1) \left(\Gamma_n^{\mathrm{T}} \Gamma_n\right)^{-1}}{1 + \gamma(n+1) \left(\Gamma_n^{\mathrm{T}} \Gamma_n\right)^{-1} \gamma^{\mathrm{T}}(n+1)}. \tag{1.42}$$

Then replacing n by $n+1$ in (1.39) yields

$$\hat{\theta}(n+1) = \left[\Gamma_{n+1}^{\mathrm{T}}\Gamma_{n+1}\right]^{-1}\Gamma_{n+1}^{\mathrm{T}}Z_{n+1}. \tag{1.43}$$

Inserting (1.40) and (1.42) into (1.43) and using (1.39) results in the following recursive form

$$\hat{\theta}(n+1) = \hat{\theta}(n) + K(n)\left[z(n+1) - \gamma(n+1)\hat{\theta}(n)\right], \tag{1.44}$$

where $K(n)$ is a q-element row vector defined by

$$K(n) = \frac{\left(\Gamma_n^{\mathrm{T}}\Gamma_n\right)^{-1}\gamma^{\mathrm{T}}(n+1)}{1+\gamma(n+1)\left(\Gamma_n^{\mathrm{T}}\Gamma_n\right)^{-1}\gamma^{\mathrm{T}}(n+1)}. \tag{1.45}$$

The row vector $K(n)$ and the $q \times q$ matrix $\left[\Gamma_n^{\mathrm{T}}\Gamma_n\right]^{-1}$ can also be computed recursively: Using (1.45) in (1.42) yields

$$\begin{aligned}\left[\Gamma_{n+1}^{\mathrm{T}}\Gamma_{n+1}\right]^{-1} &= \left[\Gamma_n^{\mathrm{T}}\Gamma_n\right]^{-1} - K(n)\gamma(n+1)\left[\Gamma_n^{\mathrm{T}}\Gamma_n\right]^{-1} \\ &= \left[I - K(n)\gamma(n+1)\right]\left[\Gamma_n^{\mathrm{T}}\Gamma_n\right]^{-1}.\end{aligned} \tag{1.46}$$

Then letting $P_n = \left[\Gamma_n^{\mathrm{T}}\Gamma_n\right]^{-1}$, from (1.45) and (1.46) we have

$$K(n) = \frac{P_n\gamma^{\mathrm{T}}(n+1)}{1+\gamma(n+1)P_n\gamma^{\mathrm{T}}(n+1)}, \tag{1.47}$$

$$P_{n+1} = \left[I - K(n)\gamma(n+1)\right]P_n. \tag{1.48}$$

Thus $K(n)$ can be computed from P_n using (1.47) and P_{n+1} can be computed from P_n and $K(n)$ using (1.48).

The recursive form given by (1.44), (1.47), and (1.48) is much more efficient than the batch form given by (1.39) in terms of the computations required to compute estimates. However, the recursive form does require that an initial estimate of $\hat{\theta}$ be specified. The initial estimate can be given at time rT, where r is the smallest positive integer for which $\Gamma_r^{\mathrm{T}}\Gamma_r$ is invertible. If there is no *a priori* knowledge of the parameter vector θ, the initial estimate $\hat{\theta}(r)$ can be taken to be the zero vector. It is also possible to evaluate the recursive form by taking the initial estimate to be at time 0 and by initializing (1.48) with $P_0 = aI$ where $a > 0$. The details of this evaluation are very similar to the generation of estimates using the Kalman filter, so we shall not pursue this here.

Example 1.4 LS Estimation of a Constant Signal

Suppose that $s(n)$ is the constant signal $s(n) = s$ for all n, and we have measurements of $s(n)$ given by $z(n) = s + v(n)$ for $n = 1, 2, \ldots$. Then $s(n)$ can be expressed in the form (1.33) with the single parameter $\theta = s$ and the function $\gamma(n) = 1$ for all n. Thus, Γ_n is the column vector

$$\Gamma_n = \begin{bmatrix} 1 \\ 1 \\ \vdots \\ 1 \end{bmatrix},$$

and $\Gamma_n^{\mathrm{T}}\Gamma_n = n$. Then using (1.39), we have that the batch form of the LS estimate is

$$\hat{\theta}(n) = \frac{1}{n}\sum_{i=1}^{n} z(i).$$

Thus, the LS estimate $\hat{\theta}(n)$ is equal to the mean value of the measurements. We can determine the recursive form of the LS estimate by using (1.44) and (1.45). First, evaluating (1.45) gives

$$K(n) = \frac{1}{n}\left[1 + \frac{1}{n}\right]^{-1} = \frac{1}{n+1}.$$

Hence, the recursive form of the LS estimate is

$$\hat{\theta}(n+1) = \hat{\theta}(n) + \frac{1}{n+1}\left[z(n+1) - \hat{\theta}(n)\right]. \tag{1.49}$$

Since the LS estimate is equal to the mean value of the measurements up to time nT, (1.49) is the recursive form of the mean filter.

Weighted Least Squares

In applications, the least-squares function (1.37) is often modified to include a $n \times n$ weighting matrix W_n which is assumed to be positive definite; that is, W_n is symmetric and all eigenvalues are strictly positive. In this case, (1.37) becomes

$$\left[Z_n - \Gamma_n\hat{\theta}(n)\right]^{\mathrm{T}} W_n \left[Z_n - \Gamma_n\hat{\theta}(n)\right], \tag{1.50}$$

If the weighting matrix W_n is diagonal with positive elements w_1, w_2, $\ldots$, w_n on the diagonal, (1.50) is equivalent to the sum of squares criterion

$$\sum_{i=1}^{n} w_i \left[z(i) - \hat{s}(i)\right]^2.$$

This result shows that the w_i can be chosen to provide "forgetting of data," so that "old data" can be discounted in generating the estimate; that is, we can take $w_n = 1$ with $w_i \to 0$ as $i \to 1$.

For the weighted least-squares function (1.50), it is easy to show that the batch form of the least-squares estimate (1.39) becomes

$$\hat{\theta}(n) = \left[\Gamma_n^{\mathrm{T}} W_n \Gamma_n\right]^{-1} \Gamma_n^{\mathrm{T}} W_n Z_n. \tag{1.51}$$

If W_n is a diagonal matrix with $\alpha^{n-1}, \alpha^{n-2}, \ldots, \alpha, 1$ on the diagonal where α is the *forgetting factor* with $0 < \alpha < 1$, then there still is a recursive form given by (1.44), but the expression for $K(n)$ changes to

$$K(n) = \frac{\left(\Gamma_n^{\mathrm{T}} W_n \Gamma_n\right)^{-1} \gamma^{\mathrm{T}}(n+1)}{\alpha + \gamma(n+1)\left(\Gamma_n^{\mathrm{T}} W_n \Gamma_n\right)^{-1} \gamma^{\mathrm{T}}(n+1)}. \tag{1.52}$$

We also have the counterpart to (1.46):

$$\left[\Gamma_{n+1}^{\mathrm{T}} \Gamma_{n+1}\right]^{-1} = \left[I - K(n)\gamma(n+1)\right]\left[\Gamma_n^{\mathrm{T}} W_n \Gamma_n\right]^{-1}. \tag{1.53}$$

The verification of both the batch form (1.51) and the recursive form given by (1.44), (1.52), and (1.53) is left to Problem 1.15.

LS State Estimation

Consider the discrete-time state model

$$x(n+1) = \Phi x(n), \tag{1.54}$$

$$z(n) = Cx(n) + v(n), \tag{1.55}$$

where the state $x(n)$ is an N-vector, Φ is a $N \times N$ matrix, and C is a N-element row vector. As noted previously, the objective is to estimate the state $x(n)$ based on the measurements $z(1), z(2), \ldots, z(n)$. In the deterministic case (as given above), a viable approach for estimating $x(n)$ from the measurements $z(1), z(2), \ldots, z(n)$, is to use the method of least-squares, which is presented below.

We assume that the matrix Φ is invertible. (This constraint will be removed later.) Then (1.54) can be solved "backwards" starting with $x(n)$. This gives

$$x(i) = \Phi^{-n+i} x(n), \quad i = 1, 2, \ldots, n. \tag{1.56}$$

Then using (1.54)–(1.56), we have the vector equation

$$\begin{bmatrix} z(1) \\ z(2) \\ \vdots \\ z(n) \end{bmatrix} = \begin{bmatrix} Cx(1) \\ Cx(2) \\ \vdots \\ Cx(n) \end{bmatrix} + \begin{bmatrix} v(1) \\ v(2) \\ \vdots \\ v(n) \end{bmatrix} = \begin{bmatrix} C\Phi^{-n+1} \\ C\Phi^{-n+2} \\ \vdots \\ C \end{bmatrix} x(n) + \begin{bmatrix} v(1) \\ v(2) \\ \vdots \\ v(n) \end{bmatrix}. \tag{1.57}$$

Defining

$$Z_n = \begin{bmatrix} z(1) \\ z(2) \\ \vdots \\ z(n) \end{bmatrix}, \qquad V_n = \begin{bmatrix} v(1) \\ v(2) \\ \vdots \\ v(n) \end{bmatrix},$$

and letting U_n denote the $n \times N$ matrix

$$U_n = \begin{bmatrix} C\Phi^{-n+1} \\ C\Phi^{-n+2} \\ \vdots \\ C \end{bmatrix},$$

we can write (1.57) in the form

$$Z_n = U_n x(n) + V_n. \tag{1.58}$$

Then the LS estimate is the value of $x(n)$ that minimizes the scalar function

$$[Z_n - U_n x(n)]^{\mathrm{T}} [Z_n - U_n x(n)]. \tag{1.59}$$

Note that (1.59) is equal to the sum of the squares of the components of the vector $Z_n - U_n x(n)$. To compute the least-squares estimate, we take the partial derivative of (1.59) with respect to $x(n)$ and set the result equal to 0. This gives

$$U_n^{\mathrm{T}} U_n x(n) = U_n^{\mathrm{T}} Z_n. \tag{1.60}$$

Assuming that U_n has rank N so that $U_n^{\mathrm{T}} U_n$ is invertible, we can solve (1.60) for $x(n)$, which yields the LS estimate:

$$\hat{x}(n) = \left[U_n^{\mathrm{T}} U_n\right]^{-1} U_n^{\mathrm{T}} Z_n. \tag{1.61}$$

We can express the LS estimate (1.61) in terms of the n-step observability matrix O_n (see Appendix A) defined by

$$O_n = \begin{bmatrix} C \\ C\Phi \\ \vdots \\ C\Phi^{n-1} \end{bmatrix}. \tag{1.62}$$

First note that

$$U_n = O_n \Phi^{1-n}. \tag{1.63}$$

Then

$$U_n^{\mathrm{T}} = \left(\Phi^{\mathrm{T}}\right)^{1-n} O_n^{\mathrm{T}}, \tag{1.64}$$

and thus

$$U_n^{\mathrm{T}} U_n = \left(\Phi^{\mathrm{T}}\right)^{1-n} O_n^{\mathrm{T}} O_n \Phi^{1-n}. \tag{1.65}$$

Inverting the right-hand side of (1.65) yields

$$\left[U_n^{\mathrm{T}} U_n\right]^{-1} = \Phi^{n-1} \left[O_n^{\mathrm{T}} O_n\right]^{-1} \left(\Phi^{\mathrm{T}}\right)^{n-1}. \tag{1.66}$$

Finally, inserting (1.64) and (1.66) into (1.61) gives

$$\begin{aligned} \hat{x}(n) &= \Phi^{n-1} \left[O_n^{\mathrm{T}} O_n\right]^{-1} \left(\Phi^{\mathrm{T}}\right)^{n-1} \left(\Phi^{\mathrm{T}}\right)^{1-n} O_n^{\mathrm{T}} Z_n \\ &= \Phi^{n-1} \left[O_n^{\mathrm{T}} O_n\right]^{-1} O_n^{\mathrm{T}} Z_n. \end{aligned} \tag{1.67}$$

From the expression (1.67) for the LS estimate, we see that the inverse of Φ does not appear, and thus the LS estimate exists without requiring Φ to be invertible. Also note that the inverse $\left[O_n^{\mathrm{T}} O_n\right]^{-1}$ exists if and only if the rank of the n-step observability matrix O_n is equal to N, which is equivalent to requiring that the system defined by the state model (1.54)–(1.55) be observable.

The LS estimate given by (1.67) is the batch form. The recursive form is given by

$$\hat{x}(n+1) = \Phi\hat{x}(n) + K(n)\left[z(n+1) - C\Phi\hat{x}(n)\right], \tag{1.68}$$

where

$$K(n) = \frac{P_n C^{\mathrm{T}}}{1 + C P_n C^{\mathrm{T}}}, \tag{1.69}$$

$$P_{n+1} = \Phi\left[I - K(n)C\right] P_n \Phi^{\mathrm{T}}, \tag{1.70}$$

$$P_n = \Phi^n \left[O_n^{\mathrm{T}} O_n\right]^{-1} \left(\Phi^{\mathrm{T}}\right)^n. \tag{1.71}$$

Example 1.5 One-Dimensional Case

Suppose that the state model is $x(n+1) = \phi x(n)$, $z(n) = cx(n) + v(n)$, where ϕ and c are nonzero real numbers. In this case,

$$O_n^{\mathrm{T}} O_n = \sum_{i=1}^{n} c^2 \phi^{2(i-1)},$$

and using (1.67) yields the LS estimate:

$$\hat{x}(n) = \left[\frac{\psi^{n-1}}{\sum_{i=1}^{n} c^2 \phi^{2(i-1)}} \right] \sum_{i=1}^{n} c^2 \phi^{i-1} z(i).$$

When $\phi \neq 1$,

$$\sum_{i=1}^{n} c^2 \phi^{2(i-1)} = \frac{c^2 \left(1 - \phi^{2n}\right)}{1 - \phi^2},$$

and thus when $\phi \neq 1$, the LS estimate is

$$\hat{x}(n) = \left[\frac{\phi^{n-1} \left(1 - \phi^2\right)}{1 - \phi^{2n}} \right] \sum_{i=1}^{n} \phi^{i-1} z(i).$$

From (1.68) the recursive form of the estimate is

$$\hat{x}(n+1) = \phi \hat{x}(n) + K(n) \left[z(n+1) - c\phi \hat{x}(n) \right].$$

Evaluating (1.69) and (1.70), we have that $K(n)$ and P_n are given recursively by

$$K(n) = \frac{P_n c}{1 + c^2 P_n},$$
$$P_{n+1} = \phi^2 \left[1 - K(n)c\right] P_n.$$

The recursive form can be evaluated by initializing with $\hat{x}(1) = 0$ and $P_1 = \phi^2/c^2$, where P_1 was evaluated using (1.71). The reader should verify that when $\phi = c = 1$, the recursive form reduces to (1.49) which is the recursive form for the mean filter.

Estimation Error

For the LS estimate given by the batch form (1.67), we shall compute the estimation error $\tilde{x}(n)$ defined by

$$\tilde{x}(n) = x(n) - \hat{x}(n). \tag{1.72}$$

First, from (1.58) we have

$$Z_n = U_n x(n) + V_n. \tag{1.73}$$

But by (1.63), $U_n = O_n \Phi^{1-n}$, and inserting this into (1.73) gives

$$Z_n = O_n \Phi^{1-n} x(n) + V_n. \tag{1.74}$$

Using (1.74) in (1.67) gives

$$\begin{aligned} \hat{x}(n) &= \Phi^{n-1} \left[O_n^{\mathrm{T}} O_n\right]^{-1} O_n^{\mathrm{T}} \left[O_n \Phi^{1-n} x(n) + V_n\right] \\ &= x(n) + \Phi^{n-1} \left[O_n^{\mathrm{T}} O_n\right]^{-1} O_n^{\mathrm{T}} V_n. \end{aligned} \tag{1.75}$$

Then inserting (1.75) into (1.72), we have

$$\tilde{x}(n) = -\Phi^{n-1}\left[O_n^{\mathrm{T}}O_n\right]^{-1}O_n^{\mathrm{T}}V_n. \tag{1.76}$$

This result shows that the magnitude of the estimation error is directly proportional to the magnitude of the noise $v(n)$. In particular, if $v(n) = 0$ for all n (so that $V_n = 0$), then $\tilde{x} = 0$ and $\hat{x}(n) = x(n)$. When $v(n)$ is not zero, to improve on the estimate given by (1.67) it is necessary to incorporate information on $v(n)$ into the estimation process. For example, if $v(n)$ is randomly varying, this information can be used to achieve a much better estimate of $x(n)$ than that given by (1.67). To consider this, it is necessary to have a stochastic formulation of the estimation problem given in terms of random signals and noise. The study of random discrete-time signals and discrete-time systems with random inputs is covered in the next chapter.

Problems

1.1. Given the measurement $z(t) = s(t) + v(t)$, a simple type of estimator is the *finite-window mean filter*, which is defined as follows: Given a positive number $\Delta > 0$, the estimate $\hat{s}(t)$ is defined by

$$\hat{s}(t) = \frac{1}{\Delta}\int_{t-\Delta}^{t} z(\tau)\,d\tau.$$

Here Δ is the length of the time window over which the mean is computed.

(a) Show that this filter is causal, linear, and time invariant by expressing $\hat{s}(t)$ in the form

$$\hat{s}(t) = \int_{-\infty}^{t} h(t-\tau)z(\tau)\,d\tau.$$

Express your answer by giving an explicit mathematical expression for the impulse response $h(t)$ of the filter.

(b) Suppose that $s(t) = s$ for all t where s is a constant and $v(t) = B\cos(Ct+\theta)$ for all t where B and C are constants. Show that for specific values of Δ, the mean-filter estimate $\hat{s}(t)$ is exactly equal to $s(t)$. Give all values of Δ for which this is the case.

(c) Give conditions on the time behaviors of the signal $s(t)$ and the disturbance/noise $v(t)$ which insure that the mean-filter estimate $\hat{s}(t)$ is "close" to the actual signal $s(t)$.

(d) Compute the Fourier transform $H(\Omega)$ of the mean filter's impulse response found in Part (a).

(e) Using MATLAB, sketch the magnitude function $|H(\Omega)|$ when the window length $\Delta = 1$, $\Delta = 0.1$, and $\Delta = 0.01$.

(f) Based on your results in Part (e), give conditions on the amplitude spectrum $|S(\Omega)|$ of the signal $s(t)$ and the amplitude spectrum $|V(\Omega)|$ of the disturbance/noise $v(t)$ which insure that the mean-filter estimate $\hat{s}(t)$ is "close" to the actual signal $s(t)$.

1.2. The discrete-time version of the finite-window mean filter defined in Problem 1.1 is the *q-step mean filter* given by

$$\hat{s}(n) = \frac{1}{q}\left[\sum_{i=n-q+1}^{n} z(i)\right],$$

where $z(n) = s(n) + v(n)$ and q is a positive integer.

(a) Show that the q-step mean filter is causal, linear, and time invariant by expressing $\hat{s}(n)$ in the form

$$\hat{s}(n) = \sum_{i=-\infty}^{n} h(n-i)z(i).$$

Express your answer by giving an explicit mathematical expression for the impulse response $h(n)$ of the filter.

(b) Suppose that $s(n) = s$ for all n where s is a constant and $v(n) = B\cos(Cn + \theta)$ for all n where B and C are constants. Show that for specific values of q and C, the q-step mean-filter estimate $\hat{s}(n)$ is exactly equal to $s(n)$. Give all values of C and q for which this is the case.

(c) Give conditions on the time behaviors of the signal $s(n)$ and the disturbance/noise $v(n)$ which insure that the q-step mean-filter estimate $\hat{s}(n)$ is "close" to the actual signal $s(n)$.

(d) Compute the discrete-time Fourier transform $H(e^{j\omega})$ of the q-step mean filter's impulse response found in Part (a).

(e) Using MATLAB, sketch the magnitude function $|H(e^{j\omega})|$ when $q = 2$, $q = 4$, and $q = 10$.

(f) Based on your results in Part (e), give conditions on the amplitude spectrum $|S(e^{j\omega})|$ of the signal $s(n)$ and the amplitude spectrum $|V(e^{j\omega})|$ of the disturbance/noise $v(n)$ which insure that the q-step mean-filter estimate $\hat{s}(n)$ is "close" to the actual signal $s(n)$.

1.3. For each of the following continuous-time signals $s(t)$, generate a state model given by $\dot{x}(t) = Ax(t)$, $s(t) = Cx(t)$. In each case, the coefficients A and C must be known and the dimension of the state model must be as small as possible.

(a) $s(t) = \alpha + \beta t$, where α and β are unknown constants.

(b) $s(t) = \alpha + \beta t + \gamma t^2$, where α, β, and γ are unknown constants.

(c) $s(t) = \alpha + \beta e^{\gamma t}$, where α and β are unknown constants and γ is a known constant.

(d) $s(t) = \alpha + \beta t + \gamma e^{\mu t}$, where α, β, and γ are unknown constants and μ is a known constant.

(e) $s(t) = \alpha \cos(\beta t + \theta)$ where α and θ are unknown constants and β is a known constant.

1.4. For each of the signal state models found in Problem 1.3, compute the discretized state model given by $x(n+1) = \Phi x(n)$, $s(n) = Cx(n)$ with the sampling interval equal to an arbitrary positive number T (so that $s(n) = s(t)|_{t=nT}$). Express your answer by giving Φ for each case.

1.5. A signal $s(n)$ is given by $s(n) = \alpha n$ where α is an unknown parameter. Measurements of $s(n)$ are given by $z(n) = s(n) + v(n)$, $n = 1, 2, \ldots$, where $v(n)$ is a measurement error term. Using the measurements $z(1), z(2), \ldots, z(n)$, generate an estimate $\hat{\alpha}(n)$ of α at time n that minimizes the sum of the squares

$$[z(1) - \hat{z}(1)]^2 + [z(2) - \hat{z}(2)]^2 + \cdots + [z(n) - \hat{z}(n)]^2,$$

where $\hat{z}(i) = [\hat{\alpha}(i)]\, i$ for $i = 1, 2, \ldots, n$. Express $\hat{\alpha}(n)$ in terms of $z(1), z(2), \ldots, z(n)$.

1.6. The signal $s(n)$ is given by

$$s(n) = \begin{cases} 1, & \text{when } n \text{ is even;} \\ 2, & \text{when } n \text{ is odd.} \end{cases}$$

(a) Derive an expression for the batch form of the LS estimate of $s(n)$ based on the measurements $z(1), z(2), \ldots, z(n)$, where $z(n) = s(n) + v(n)$. The coefficients of the $z(i)$ in the expression for $\hat{s}(n)$ must be completely evaluated.

(b) Give the recursive form of the LS estimate of $s(n)$. Specify an initial value of P_n for the recursion.

1.7. A deterministic signal $s(t)$ is given by

$$s(t) = a \cos\left(\frac{\pi}{2} t + \theta\right),$$

where a and θ are unknown constants. Derive the batch form of the LS estimate of $s(3)$ based on the measurements $z(1)$, $z(2)$, and $z(3)$, where $z(n) = s(n) + v(n)$.

1.8. Consider a constant-velocity object moving in three-dimensional space with x-coordinate position $s(t)$ (see Examples 1.2 and 1.3).

(a) Derive an expression for the batch form of the LS estimate of the velocity $\dot{s}(n) = \dot{s}(t)|_{t=nT}$ with the estimate given in terms of the measured position values $z(1), z(2), \ldots, z(n)$, where $z(n) = s(n) + v(n)$. Express your answer in terms of the sampling interval T and the measurements $z(1), z(2), \ldots, z(n)$.

(b) Assume that $T = 1$. For $n = 2$ and $n = 3$, express the LS estimate of the velocity $\dot{s}(n)$ in terms of $z(1), z(2), \ldots, z(n)$. All coefficients in these expressions must be evaluated.

(c) Are the results obtained in Part (b) expected? Explain.

(d) Give the recursive form of the LS estimate of the velocity $\dot{s}(n)$. Specify an initial value of P_n for the recursion.

1.9. Again consider the state model $x(n+1) = \Phi x(n)$, $s(n) = Cx(n)$ with the measurements $z(n) = s(n) + v(n)$. In some cases, the noise/disturbance term $v(n)$ is given by a state model of the form

$$\gamma(n+1) = \Psi\gamma(n)$$
$$v(n) = D\gamma(n) + \varepsilon(n),$$

where the state $\gamma(n)$ is a p-element vector, Ψ and D are known, and $\varepsilon(n)$ is an unknown signal. It is possible to use this state model for $v(n)$ to generate a much better estimate of x(n) than that which can be obtained using the LS procedure derived in Chapter 1. The basic idea is to *augment* the state model to include the state model for $v(n)$. More precisely, we begin by defining a new $(N+p)$-dimensional state vector given by

$$\begin{bmatrix} x(n) \\ \gamma(n) \end{bmatrix}.$$

Then an augmented state model is generated in terms of this state vector, and the LS approach is applied to the resulting model. This results in a type of *generalized least-squares* (LS) *estimation.*

(a) Derive an expression for the generalized LS estimate of $x(n)$ using this approach.

(b) What condition is needed to insure the existence and uniqueness of the generalized LS estimate?

(c) Using your result in Part (a), derive an expression for the estimate of $s(n)$.

1.10. Consider the state model $x(n+1) = \Phi x(n)$, $s(n) = Cx(n)$ with the measurements $z(n) = s(n) + v(n)$. It is also known that

$$v(n) = \alpha + b(n),$$

where α is an unknown constant and $b(n)$ is an unknown time-varying signal.

(a) Using the approach in Problem 1.9, derive an expression for the generalized LS estimate of $x(n)$.

(b) What condition is needed to insure the existence and uniqueness of the LS estimate found in Part (a)?

1.11. Repeat Problem 1.9 for the case when the least-squares criterion includes a weighting matrix W_n.

1.12. Given the state model $x(n+1) = \Phi x(n)$, $s(n) = Cx(n)$, derive an expression for the LS estimate of $x(n)$ based on the measurements $z(n)$, $z(n-1)$, $\ldots$, $z(n-q)$, where $z(n) = s(n) + v(n)$ and q is a positive integer. This is called the *finite-window* (or *finite horizon*) LS estimate of $x(n)$. Assume that the signal state model is observable.

1.13. Derive an expression for the two-step ($q = 2$) finite-window LS estimate of $x(n)$ in the case when $\Phi = a$ and $C = 1$, where a is a nonzero constant.

1.14. Repeat Problem 1.13 for the case when

$$\Phi = \begin{bmatrix} 1 & 0 \\ 1 & -1 \end{bmatrix}, \qquad C = \begin{bmatrix} 1 & 1 \end{bmatrix}.$$

The coefficients in your expression for $\hat{x}(n)$ must be completely evaluated.

1.15. Show that the batch form of the weighted least squares estimate is given by (1.51) and that the recursive form is given by (1.44), (1.52), and (1.53).

1.16. Show that the LS estimate $\hat{x}(n)$ given by the batch form (1.67) can be expressed in the recursive form given by (1.68)–(1.71). **Hint:** Use the relationships

$$O_{n+1}^{\mathrm{T}} O_{n+1} = O_n^{\mathrm{T}} O_n + \left(\Phi^{\mathrm{T}}\right)^n c^{\mathrm{T}} c \Phi^n$$

$$O_{n+1}^{\mathrm{T}} Z_{n+1} = O_n^{\mathrm{T}} Z_n + \left(\Phi^{\mathrm{T}}\right)^n c^{\mathrm{T}} z(n+1).$$

Chapter 2

Random Signals and Systems with Random Inputs

In the problem of estimating a signal $s(n)$ from the measurements $z(n) = g(s(n), v(n), n)$, the noise term $v(n)$ usually varies "randomly," and thus modeling $v(n)$ requires that we use a random signal formulation. The signal $s(n)$ may also include some random variation, and thus it too must be modeled in general as a random signal. The random signal formulation is generated by taking $v(n)$ and $s(n)$ to be random variables for each value of the time index n. We begin by presenting the fundamentals of random variables in Section 2.1, and then in Section 2.2 we consider random discrete-time signals. In the last section of the chapter, we study linear time-varying and time-invariant discrete-time systems driven by random signal inputs. The treatment of random signals and systems with random inputs given in this chapter is presented in sufficient depth to allow the reader to then follow the development of optimal filtering given in this text.

2.1 Random Variables

A random variable is defined with respect to an experiment with a set S of possible outcomes of the experiment. A single performance of the experiment is called a *trial*. At each trial, an outcome α belonging to the set S is observed. An *event* A is a subset of S, denoted by $A \subset S$. An event A is said to occur during a trial if the outcome of the trial is an element of A. Given an event A, at each trial either the event A occurs or the event $\bar{A}$ occurs, where $\bar{A}$ is the complement of A. The *certain event* is the event $A = S$, and the *impossible event* is the event $A = \emptyset$, where $\emptyset$ is the empty set.

A random variable (RV) $\boldsymbol{x}$ is a function from S into the set of real numbers. That is, for any outcome $\alpha \in S$, the value $\boldsymbol{x}(\alpha)$ of $\boldsymbol{x}$ at α is a real number. It is important to stress that a RV $\boldsymbol{x}$ is a function, not a real number. In particular, note that a RV $\boldsymbol{x}$ cannot be equal to zero, but $\boldsymbol{x}$ could be the zero function defined by $\boldsymbol{x}(\alpha) = 0$ for all $\alpha \in S$. . In this book, random variables will always be denoted by a boldface symbol to emphasize the fact that they are functions.

By definition, a RV $\boldsymbol{x}$ is a real-valued function of the outcomes α belonging to S. But since the outcomes arise as a result of performing trials of the experiment, it is also possible to view $\boldsymbol{x}$ as a function $\boldsymbol{x}(i)$ of the performance of trials $i = 1, 2, \ldots$. To make this precise, suppose that for $i = 1, 2, \ldots$, the ith trial of the experiment is performed and the outcome is α_i. Then the value $\boldsymbol{x}(i)$ of the RV $\boldsymbol{x}$ at the ith trial can be defined to be $\boldsymbol{x}(i) = \boldsymbol{x}(\alpha_i)$. The values $\boldsymbol{x}(i) = \boldsymbol{x}(\alpha_i)$ for $i = 1, 2, \ldots$, are called *sample realizations* or *sample values* of the RV $\boldsymbol{x}$.

A RV $\boldsymbol{x}$ is defined in terms of the probability of the events $\{\alpha \in S : \boldsymbol{x}(\alpha) \leq x\}$, where x ranges over the set of real numbers. This requires that a positive real number $P(A)$, called the probability of the event A, be assigned to events $A \subset S$. The number $P(A)$ is the probability that the outcome of a trial is an element of A; in other words, $P(A)$ is the probability that the event A occurred. The probabilities of the events $A \subset S$ must be assigned so that the following axioms are satisfied

$$0 \leq P(A) \leq 1, \tag{2.1}$$

$$P(S) = 1, \tag{2.2}$$

$$\text{If } A \cap B = \emptyset, \text{ then } P(A \cup B) = P(A) + P(B), \tag{2.3}$$

where $A \cup B$ is the union of A and B. It follows from (2.1)–(2.3) that $P(\emptyset) = 0$, $P(\bar{A}) = 1 - P(A)$, and for any $A, B \subset S$,

$$P(A \cup B) = P(A) + P(B) - P(A \cap B),$$

where $A \cap B$ is the intersection of A and B. Once the assignment of probabilities satisfying the axioms (2.1)–(2.3) is made, S is called a *probability space*.

A RV $\boldsymbol{x}$ defined on a probability space S is specified in terms of its *probability distribution function* $F_{\boldsymbol{x}}(x)$ given by

$$F_{\boldsymbol{x}}(x) = P\{\alpha \in S : \boldsymbol{x}(\alpha) \leq x\}.$$

Note that the values of $F_{\boldsymbol{x}}(x)$ belong to the interval [0,1]. Also note that for a given number x, $F_{\boldsymbol{x}}(x)$ is the probability that the value of the RV $\boldsymbol{x}$ is less than or equal to x.

As a consequence of the axioms (2.1)–(2.3) of a probability space, the probability distribution function $F_{\boldsymbol{x}}(x)$ has the following properties:

$$F_{\boldsymbol{x}}(-\infty) = 0 \text{ and } F_{\boldsymbol{x}}(\infty) = 1, \tag{2.4}$$

$$0 \le F_{\boldsymbol{x}}(x) \le 1, \text{for all } x, \tag{2.5}$$

$$\text{If } x1 < x2, \text{ then } F_{\boldsymbol{x}}(x_1) \le F_{\boldsymbol{x}}(x_2). \tag{2.6}$$

Note that by (2.6), $F_{\boldsymbol{x}}(x)$ is a nondecreasing function of the variable x.

A RV $\boldsymbol{x}$ can also be specified by its *probability density function* $f_{\boldsymbol{x}}(x)$, which is equal to the derivative of $F_{\boldsymbol{x}}(x)$; that is,

$$f_{\boldsymbol{x}}(x) = \frac{d}{dx} F_{\boldsymbol{x}}(x).$$

If the distribution function $F_{\boldsymbol{x}}(x)$ is discontinuous (i.e., jumps) at a point $x = x_1$, then the derivative will contain a term of the form $c\delta(x - x_1)$, where $\delta(x - x_1)$ is the impulse function located at the point $x = x_1$ and c is the amount of the jump. The reader should recall that the impulse function $\delta(x - x_1)$ is defined to be zero for all $x \neq x_1$, and for any real number $a > 0$,

$$\int_{x_1 - a}^{x_1 + a} \delta(x - x_1)\, dx = 1. \tag{2.7}$$

By (2.7), the area under the impulse $\delta(x - x_1)$ is equal to 1.

Given a RV $\boldsymbol{x}$ with probability distribution function $F_{\boldsymbol{x}}(x)$, since the probability density function $f_{\boldsymbol{x}}(x)$ is the derivative of $F_{\boldsymbol{x}}(x)$, $F_{\boldsymbol{x}}(x)$ is equal to the integral of $f_{\boldsymbol{x}}(x)$; that is,

$$F_{\boldsymbol{x}}(x) = \int_{-\infty}^{x} f_{\boldsymbol{x}}(x)\, dx. \tag{2.8}$$

Note that since $F_{\boldsymbol{x}}(\infty) = 1$, from (2.8)

$$\int_{-\infty}^{\infty} f_{\boldsymbol{x}}(x)\, dx = 1.$$

Hence the total area under the density function is always equal to one. It also follows from (2.8) that for any real numbers x_1 and x_2 with $x_1 < x_2$,

$$F_{\boldsymbol{x}}(x_2) - F_{\boldsymbol{x}}(x_1) = \int_{x_1}^{x_2} f_{\boldsymbol{x}}(x)\, dx.$$

And by definition of $F_{\boldsymbol{x}}(x)$, we have

$$P\{x_1 < \boldsymbol{x} \le x2\} = \int_{x_1}^{x_2} f_{\boldsymbol{x}}(x)\, dx.$$

Therefore, the probability that the value of $\boldsymbol{x}$ lies between x_1 and x_2 is equal to the area under the probability density function from $x = x_1$ to $x = x_2$.

Given a small positive number ε, the probability that $\boldsymbol{x}$ lies in the interval $(x - \varepsilon, x + \varepsilon)$ for any real number x is equal to

$$\int_{x-\varepsilon}^{x+\varepsilon} f_{\boldsymbol{x}}(x)\, dx. \tag{2.9}$$

If $f_{\boldsymbol{x}}(x)$ is *unimodal*; that is, $f_{\boldsymbol{x}}(x)$ has a single peak, then the *most likely value* of the RV $\boldsymbol{x}$ is the value of x for which the integral in (2.9) is maximum. Clearly, the most likely value of $\boldsymbol{x}$ is equal to the location of the peak of $f_{\boldsymbol{x}}(x)$, which is the value of x for which $df_{\boldsymbol{x}}(x)/dx = 0$.

Three examples of random variables are given next.

Example 2.1 A Discrete RV

Let the experiment be the operation of a machine with the set of outcomes $S = \{\text{working normally}, \text{not working normally}\}$. The assignment of probabilities is given by

$$P\{\text{working normally}\} = 0.9, \quad P\{\text{not working normally}\} = 0.1.$$

We define the RV $\boldsymbol{x}$ by $\boldsymbol{x}(\text{working normally}) = 0$ and $\boldsymbol{x}(\text{not working normally}) = 1$. The probability distribution function for this RV is given by

$$F_{\boldsymbol{x}}(x) = P\{\alpha \in S : \boldsymbol{x}(\alpha) \le x\} = \begin{cases} 0, & x < 0; \\ 0.9, & 0 < x < 1; \\ 1, & x > 1. \end{cases}$$

And the probability density function is given by

$$f_{\boldsymbol{x}}(x) = 0.9\delta(x) + 0.1\delta(x - 1).$$

This RV is referred to as a *discrete RV* since there are only a finite number of different values of $\boldsymbol{x}$.

Example 2.2 Uniformly-Distributed RV

Let the experiment be taking a measurement with the set S of outcomes equal to any number between -1 and 1. We define the RV $\boldsymbol{z}$ by $\boldsymbol{z}(\alpha) = \alpha$; that is, when a measurement is taken (the experiment is performed), the value of z is equal to the value of the measurement. We assume that the distribution function is given by

$$F_{\boldsymbol{z}}(z) = \begin{cases} 0, & z < -1; \\ 0.5(z + 1), & -1 < z < 1; \\ 1, & z > 1. \end{cases}$$

Then the density function is

$$f_z(z) = \begin{cases} 0.5, & -1 < z < 1; \\ 0, & \text{otherwise.} \end{cases}$$

Since the density function is constant over the range of possible measurement values, the measurement values are *equally likely*. This RV is said to be *uniformly distributed from* -1 *to* 1 since its density function is a constant over the range from -1 to 1 where it is nonzero. The RV $\boldsymbol{z}$ is also said to be a *continuous random variable* since its distribution function $F_{\boldsymbol{z}}(z)$ is a continuous function of the variable z. This is equivalent to the probability density function $f_{\boldsymbol{z}}(z)$ not having any impulse functions.

Example 2.3 Gaussian RV

Again let the experiment be taking a measurement with the set S of outcomes equal to any number between -1 and 1, and let $\boldsymbol{z}$ be the RV with the value of z equal to the value of the measurement. In contrast to the RV in Example 2.2, we assume that the density function is

$$f_{\boldsymbol{z}}(z) = \frac{1}{\sqrt{2\pi}\sigma} e^{-\frac{(z-\eta)^2}{2\sigma^2}}, \tag{2.10}$$

where η is a real number and σ is a positive number. Note that $f_{\boldsymbol{z}}(z)$ is completely determined by the numbers η and σ. A plot of $f_{\boldsymbol{z}}(z)$ with $\eta = 2$ and $\sigma = 1$ is given in Figure 2.1. From the plot we see that the most likely value of $\boldsymbol{z}$ is $z = 2$. A RV given by (2.10) is said to be a *Gaussian* or *normal* RV. A Gaussian RV $\boldsymbol{z}$ with the density function (2.10) will sometimes be denoted by $\boldsymbol{z} \sim \mathcal{N}(\eta, \sigma^2)$, where "$\mathcal{N}$" stands for "normal."

Conditional Distributions and Densities

Let $\boldsymbol{x}$ be a RV defined on the probability space S and let $A \subset S$ be an event. Given a number x, the intersection $\{\boldsymbol{x} \le x\} \cap A$ is the event consisting of all outcomes $\alpha \in S$ such that $\boldsymbol{x}(\alpha) \le x$ and $\alpha \in A$. The *conditional distribution function* $F_{\boldsymbol{x}}(x|A)$ of the RV $\boldsymbol{x}$ given that the event A occurred is defined to be the conditional probability of the event $\{\alpha \in S : \boldsymbol{x}(\alpha) \le x\}$ given A. That is,

$$F_{\boldsymbol{x}}(x|A) = P(\{\alpha \in S : \boldsymbol{x}(\alpha) \le x\} | A) = \frac{P(\{\alpha \in S : \boldsymbol{x}(\alpha) \le x\} \cap A)}{P(A)}. \tag{2.11}$$

The *conditional density function* $f_{\boldsymbol{x}}(x|A)$ is equal to the derivative with respect to x of $F_{\boldsymbol{x}}(x|A)$.

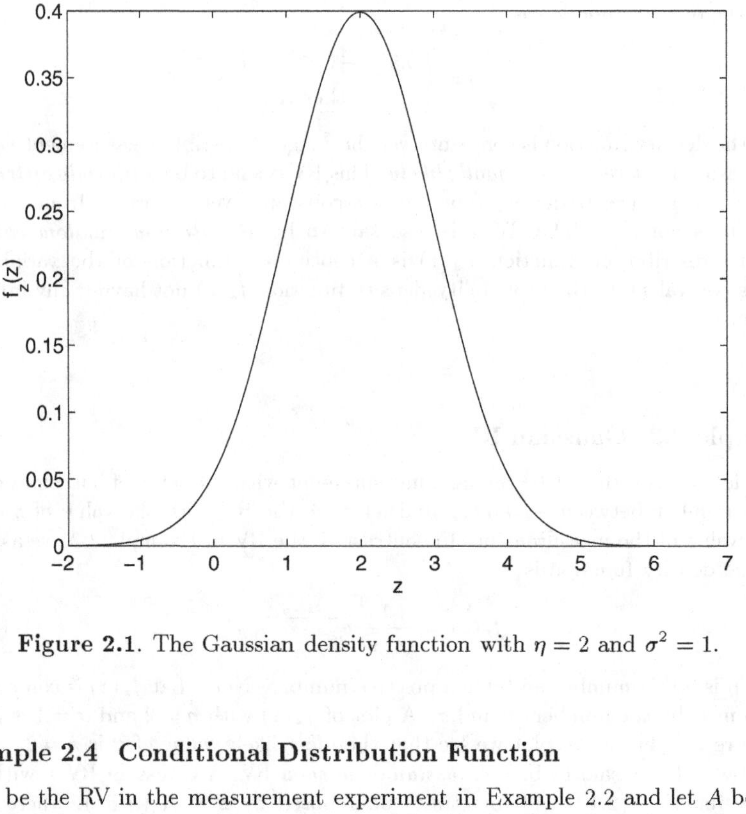

Figure 2.1. The Gaussian density function with $\eta = 2$ and $\sigma^2 = 1$.

Example 2.4 Conditional Distribution Function

Let $\boldsymbol{z}$ be the RV in the measurement experiment in Example 2.2 and let A be the event that the measurement is non-negative; that is, $A = \{z \in S : 0 \leq z \leq 1\}$. Then $P(A) = 0.5$ and

$$P\left(\{\alpha \in S : \boldsymbol{x}(\alpha) \leq x\} \cap A\right) = \begin{cases} 0, & z < 0; \\ 0.5z, & 0 \leq z \leq 1; \\ 0.5, & z > 1. \end{cases}$$

Hence, by (2.11) the conditional distribution $F_{\boldsymbol{z}}(z|A)$ of $\boldsymbol{z}$ given A is:

$$F_{\boldsymbol{z}}(z|A) = \begin{cases} 0, & z < 0; \\ z, & 0 \leq z \leq 1; \\ 1, & z > 1. \end{cases}$$

Functions of a Random Variable

Let Ψ be a real-valued function defined on the set of real numbers; that is, for any real number b, the value $\Psi(b)$ of Ψ at b is a real number. Then given any random variable $\boldsymbol{x}$ defined with respect to the probability space S, we can define a new random variable $\boldsymbol{y}$ to be the function of $\boldsymbol{x}$ given by

$\boldsymbol{y} = \Psi(\boldsymbol{x})$. Here the value $\boldsymbol{y}(\alpha)$ of the random variable $\boldsymbol{y}$ at $\alpha \in S$ is defined by $\boldsymbol{y}(\alpha) = \Psi(\boldsymbol{x}(\alpha))$.

Example 2.5 Scalar Multiple

Given a fixed nonzero real number a, let Ψ be the function defined by $\Psi(b) = ab$. Then given any RV $\boldsymbol{x}$ defined with respect to the probability space S, one can define the RV $\boldsymbol{y} = \Psi(\boldsymbol{x})$, which is denoted by $\boldsymbol{y} = a\boldsymbol{x}$. The RV $\boldsymbol{y}$ is said to be a "scalar multiple" of the RV $\boldsymbol{x}$. By definition of the RV $\boldsymbol{y}$, $\boldsymbol{y}(\alpha) = a\boldsymbol{x}(\alpha)$ for all $\alpha \in S$. If $a > 0$, the probability distribution function $F_{\boldsymbol{y}}(y)$ for the RV $\boldsymbol{y} = a\boldsymbol{x}$ is equal to $F_{\boldsymbol{x}}(y/a)$, where $F_{\boldsymbol{x}}(x)$ is the distribution function for $\boldsymbol{x}$. To show this, first note that

$$F_{\boldsymbol{y}}(y) = P\left\{\alpha \in S : \boldsymbol{y}(\alpha) \leq y\right\}.$$

But $\boldsymbol{y}(\alpha) = a\boldsymbol{x}(\alpha)$, and thus

$$F_{\boldsymbol{y}}(y) = P\left\{\alpha \in S : a\boldsymbol{x}(\alpha) \leq y\right\},$$

and since a is strictly positive,

$$F_{\boldsymbol{y}}(y) = P\left\{\alpha \in S : \boldsymbol{x}(\alpha) \leq y/a\right\} = F_{\boldsymbol{x}}(y/a).$$

Example 2.6 Translation

Given a fixed real number a, let Ψ be the function defined by $\Psi(b) = b + a$. Then given any RV $\boldsymbol{x}$, the RV $\boldsymbol{y} = \Psi(\boldsymbol{x})$ is a *translation* of $\boldsymbol{x}$, and is denoted by $\boldsymbol{y} = \boldsymbol{x} + a$. Here $\boldsymbol{y}(\alpha) = \boldsymbol{x}(\alpha) + a$. The distribution function $F_{\boldsymbol{y}}(y)$ is given by

$$\begin{aligned} F_{\boldsymbol{y}}(y) &= P\left\{\alpha \in S : \boldsymbol{y}(\alpha) \leq y\right\} \\ &= P\left\{\alpha \in S : \boldsymbol{x}(\alpha) + a \leq y\right\} \\ &= P\left\{\alpha \in S : \boldsymbol{x}(\alpha) \leq y - a\right\} \\ &= F_{\boldsymbol{x}}(y - a), \end{aligned}$$

where $F_{\boldsymbol{x}}(x)$ is the probability distribution function of the RV $\boldsymbol{x}$. Hence if $a > 0$, $F_{\boldsymbol{y}}(y)$ is equal to a shift of $F_{\boldsymbol{x}}(x)$ to the right by amount a, and if $a < 0$, $F_{\boldsymbol{y}}(y)$ is equal to shift of $F_{\boldsymbol{x}}(x)$ to the left by amount $-a$.

Example 2.7 Square of a Translation

Given a fixed real number a, let Ψ be the function defined by $\Psi(b) = (b - a)^2$. Then the RV $\boldsymbol{y} = \Psi(\boldsymbol{x})$ is the "square of $\boldsymbol{x}$ minus a," which we denote by $\boldsymbol{y} = (\boldsymbol{x} - a)^2$. The derivation of the distribution function $F_{\boldsymbol{y}}(y)$ is left to the reader.

The Moments of a Random Variable

Let $\boldsymbol{x}$ be a RV with probability density function $f_{\boldsymbol{x}}(x)$. Given a positive integer i, the *ith moment* of $\boldsymbol{x}$ is denoted by $\mathrm{E}\left[\boldsymbol{x}^i\right]$ and is defined by

$$\mathrm{E}\left[\boldsymbol{x}^i\right] = \int_{-\infty}^{\infty} x^i f_{\boldsymbol{x}}(x)\, dx. \tag{2.12}$$

Note that the moments of a RV are real numbers.

The first moment $\mathrm{E}[\boldsymbol{x}]$ is also called the *mean* or *expected value* of $\boldsymbol{x}$. Setting $i = 1$ in (2.12), we have that the mean $\mathrm{E}[\boldsymbol{x}]$ is given by

$$\mathrm{E}\left[\boldsymbol{x}\right] = \int_{-\infty}^{\infty} x f_{\boldsymbol{x}}(x)\, dx. \tag{2.13}$$

By definition, the mean is the *center of gravity* of the probability density function.

The mean of a RV $\boldsymbol{x}$ is the expected value of $\boldsymbol{x}$ in the following sense. Suppose that after having performed N trials, the values $x_1, x_2, \ldots, x_N$ of $\boldsymbol{x}$ were observed. As noted previously, the x_i are said to be *sample values* of the RV $\boldsymbol{x}$. Now it can be shown that as $N \to \infty$,

$$\frac{1}{N}\sum_{i=1}^{N} x_i \to \mathrm{E}[\boldsymbol{x}]. \tag{2.14}$$

In other words, the average of the sample values of $\boldsymbol{x}$ converges to the mean of $\boldsymbol{x}$ as the number of samples approaches infinity. The proof of (2.14) is omitted (see [1, p. 104]). Hence, the mean $\mathrm{E}[\boldsymbol{x}]$ can be viewed as the *average of x* taken over a collection of sample values of $\boldsymbol{x}$.

Example 2.8 Mean Value of a Uniformly-Distributed RV

As first noted in Example 2.2, a RV $\boldsymbol{x}$ is uniformly distributed if its probability density function $f_{\boldsymbol{x}}(x)$ is equal to $1/(x_2 - x_1)$ over the interval $x_1 \le x \le x_2$ for some real numbers x_1, x_2 with $x_1 < x_2$, and $f_{\boldsymbol{x}}(x)$ is zero for all other x. By evaluating (2.13), it can be shown that the mean $\mathrm{E}[\boldsymbol{x}]$ is equal to the *midpoint* of the density function $f_{\boldsymbol{x}}(x)$; that is

$$\mathrm{E}[\boldsymbol{x}] = \frac{x_1 + x_2}{2}. \tag{2.15}$$

Example 2.9 Mean Value of a Gaussian RV

As defined in Example 2.3, a Gaussian RV $\boldsymbol{x} \sim \mathcal{N}\left(\eta, \sigma^2\right)$ has the density function

$$f_{\boldsymbol{x}}(x) = \frac{1}{\sqrt{2\pi}\sigma} e^{-\frac{(x-\eta)^2}{2\sigma^2}}.$$

Inserting $f_{\boldsymbol{x}}(x)$ into (2.13) and performing the integration gives $\mathrm{E}[\boldsymbol{x}] = \eta$. Note that the mean $\mathrm{E}[\boldsymbol{x}]$ is equal to the peak value of $f_{\boldsymbol{x}}(x)$, which is the most likely value of $\boldsymbol{x}$.

Given a RV $\boldsymbol{x}$ with density function $f_{\boldsymbol{x}}(x)$, and given a real-valued function Ψ defined on the set of real numbers, consider the RV $\boldsymbol{y} = \Psi(\boldsymbol{x})$. It turns out that the mean $\mathrm{E}[\boldsymbol{y}]$ of the RV $\boldsymbol{y}$ can be computed by using the expression

$$\mathrm{E}[\boldsymbol{y}] = \int_{-\infty}^{\infty} \Psi(x) f_{\boldsymbol{x}}(x)\, dx. \tag{2.16}$$

It is important to note that by using (2.16), the mean $\mathrm{E}[\boldsymbol{y}]$ can be computed without having to determine the density function $F_{\boldsymbol{y}}(y)$ of the RV $\boldsymbol{y}$. For example, if $\boldsymbol{y} = a\boldsymbol{x}$ for some real number a, it follows directly from (2.16) that $\mathrm{E}[\boldsymbol{y}] = a\mathrm{E}[\boldsymbol{x}]$. For a proof verifying (2.16), see [1, p. 105].

Given a RV $\boldsymbol{x}$, the second-order moment $\mathrm{E}\left[\boldsymbol{x}^2\right]$ is also called the *mean of the square*, or the *mean square*. Setting $i = 2$ in (2.12), we have that the mean square is given by

$$\mathrm{E}\left[\boldsymbol{x}^2\right] = \int_{-\infty}^{\infty} x^2 f_{\boldsymbol{x}}(x)\, dx. \tag{2.17}$$

It follows from the properties of $f_{\boldsymbol{x}}(x)$ that $\mathrm{E}\left[\boldsymbol{x}^2\right] > 0$ for any RV $\boldsymbol{x}$. If x_1, x_2, ..., x_N are samples values of $\boldsymbol{x}$, $\mathrm{E}\left[\boldsymbol{x}^2\right]$ can be approximated by the sum:

$$\mathrm{E}\left[\boldsymbol{x}^2\right] = \frac{1}{N}\sum_{i=1}^{N} x_i^2.$$

Closely connected to the mean square of a RV $\boldsymbol{x}$ is the *variance* $\mathrm{Var}\left[\boldsymbol{x}\right]$ of $\boldsymbol{x}$, which is the expected value of the RV $\boldsymbol{y} = (\boldsymbol{x} - \eta)^2$, where η is the mean of $\boldsymbol{x}$. Also, the *standard deviation* σ of $\boldsymbol{x}$ is defined to be the square root of the variance $\mathrm{Var}\left[\boldsymbol{x}\right]$. Hence, in mathematical notation,

$$\mathrm{Var}\left[\boldsymbol{x}\right] = \mathrm{E}\left[(\boldsymbol{x} - \eta)^2\right],$$
$$\sigma = \sqrt{\mathrm{Var}\left[\boldsymbol{x}\right]}.$$

By (2.16), the variance $\mathrm{Var}\left[\boldsymbol{x}\right]$ can be written in the form

$$\mathrm{Var}\left[\boldsymbol{x}\right] = \int_{-\infty}^{\infty} (x - \eta)^2 f_{\boldsymbol{x}}(x)\, dx. \tag{2.18}$$

It follows from (2.18) that for any scalar a,

$$\mathrm{Var}[a\boldsymbol{x}] = a^2\mathrm{Var}[\boldsymbol{x}].$$

The variance $\mathrm{Var}[\boldsymbol{x}]$ or the standard deviation σ of a RV $\boldsymbol{x}$ gives a measure of the spread of the values of $\boldsymbol{x}$ from the mean. (By the values of $\boldsymbol{x}$, we mean the sample values resulting from the performance of trials). This can be seen from *Tchebycheff's inequality* given by

$$P\{|\boldsymbol{x}-\eta| \geq c\} \leq \frac{\sigma^2}{c^2}, \tag{2.19}$$

where c is any positive real number and $\eta = \mathrm{E}[\boldsymbol{x}]$. By the inequality (2.19), the probability that the value of the RV $\boldsymbol{x}-\eta$ lies outside of the interval $[-c, c]$ is bounded above by σ^2/c^2. Hence, for a small value of c, the smaller σ is, the higher the probability is that the value of $\boldsymbol{x}-\eta$ lies in the interval $[-c, c]$. Thus, if σ is sufficiently small, the values of the RV $\boldsymbol{x}$ will be concentrated near the mean of $\boldsymbol{x}$.

Note that if the mean η of $\boldsymbol{x}$ is zero, then $\mathrm{Var}[\boldsymbol{x}] = \mathrm{E}[\boldsymbol{x}^2]$. Hence, the variance and the mean square are the same quantity in the case when the mean of $\boldsymbol{x}$ is zero. When $\eta \neq 0$, expanding the square in the integrand of (2.18) and using linearity of the integration operation gives

$$\begin{aligned}\mathrm{Var}[\boldsymbol{x}] &= \int_{-\infty}^{\infty} x^2 f_{\boldsymbol{x}}(x)\,dx - 2\eta\int_{-\infty}^{\infty} x f_{\boldsymbol{x}}(x)\,dx + \int_{-\infty}^{\infty} \eta^2 f_{\boldsymbol{x}}(x)\,dx \\ &= \mathrm{E}[\boldsymbol{x}^2] - 2\eta\mathrm{E}[\boldsymbol{x}] + \eta^2.\end{aligned}$$

But $\eta = \mathrm{E}[\boldsymbol{x}]$, and thus

$$\mathrm{Var}[\boldsymbol{x}] = \mathrm{E}[\boldsymbol{x}^2] - (\mathrm{E}[\boldsymbol{x}])^2. \tag{2.20}$$

Thus, in the general case (when $\mathrm{E}[\boldsymbol{x}] \neq 0$) the variance is equal to the mean square minus the square of the mean.

Example 2.10 Mean Square of a Uniformly-Distributed RV

Again let $\boldsymbol{x}$ be a uniformly distributed RV with probability density function $f_{\boldsymbol{x}}(x)$ equal to $1/(x_2 - x_1)$ over the interval $x_1 \leq x \leq x_2$ for some real numbers x_1, x_2 with $x_1 < x_2$. By evaluating (2.17), it can be shown that the mean square $\mathrm{E}[\boldsymbol{x}^2]$ is given by

$$\mathrm{E}[\boldsymbol{x}^2] = \frac{(x_2 - x_1)^2}{12}. \tag{2.21}$$

Example 2.11 Variance of a Gaussian RV

Consider the RV $\boldsymbol{x} \sim \mathcal{N}\left(\eta, \sigma^2\right)$ with the density function

$$f_{\boldsymbol{x}}(x) = \frac{1}{\sqrt{2\pi}\sigma} e^{-\frac{(x-\eta)^2}{2\sigma^2}}.$$

To compute the variance of $\boldsymbol{x}$, first note that since the area under $f_{\boldsymbol{x}}(x)$ is equal to 1, we have

$$\int_{-\infty}^{\infty} e^{-\frac{(x-\eta)^2}{2\sigma^2}}\, dx = \sqrt{2\pi}\sigma. \tag{2.22}$$

Differentiating both sides of (2.22) with respect to σ gives

$$\int_{-\infty}^{\infty} \frac{(x-\eta)^2}{\sigma^3} e^{-\frac{(x-\eta)^2}{2\sigma^2}}\, dx = \sqrt{2\pi}. \tag{2.23}$$

Multiplying both sides of (2.23) by $\frac{\sigma^2}{\sqrt{2\pi}}$ reveals that $\mathrm{Var}[\boldsymbol{x}] = \sigma^2$.

In applications, it is often the case that the only knowledge one has of a RV $\boldsymbol{x}$ are the first-order and second-order statistics of $\boldsymbol{x}$ given by the mean $\mathrm{E}[\boldsymbol{x}]$ and the mean square $\mathrm{E}\left[\boldsymbol{x}^2\right]$ or the variance $\mathrm{Var}[\boldsymbol{x}] = \mathrm{E}\left[(\boldsymbol{x}-\eta)^2\right]$, where $\eta = \mathrm{E}[\boldsymbol{x}]$. In some instances, one may also know the higher-order moments (or higher-order statistics) of $\boldsymbol{x}$ given by $\mathrm{E}\left[\boldsymbol{x}^i\right]$ for $i = 3, 4, 5, \ldots$. For random variables that are uniformly distributed or Gaussian distributed, knowledge of the mean and variance is equivalent to knowing the distribution or density function. However, in general, knowledge of the mean and variance is not sufficient for determining the distribution or density function.

Two Random Variables

Two random variables $\boldsymbol{x}$ and $\boldsymbol{y}$ are said to be *jointly distributed* if they are defined on the same probability space S. In this case, for any real numbers x, y, we let $\{\boldsymbol{x} \le x, \boldsymbol{y} \le y\}$ denote the event consisting of all $\alpha \in S$ such that $\boldsymbol{x}(\alpha) \le x$ and $\boldsymbol{y}(\alpha) \le y$. We can then define the *joint distribution function* $F_{\boldsymbol{x},\boldsymbol{y}}(x, y)$ by

$$F_{\boldsymbol{x},\boldsymbol{y}}(x, y) = P\left(\{\boldsymbol{x} \le x, \boldsymbol{y} \le y\}\right). \tag{2.24}$$

The *joint density function* $f_{\boldsymbol{x},\boldsymbol{y}}(x, y)$ is equal to the partial derivative of $F_{\boldsymbol{x},\boldsymbol{y}}(x, y)$ with respect to x and y:

$$f_{\boldsymbol{x},\boldsymbol{y}}(x, y) = \frac{\partial^2 F_{\boldsymbol{x},\boldsymbol{y}}(x, y)}{\partial x \partial y}. \tag{2.25}$$

We have

$$P\left(\{\boldsymbol{x} \le x, \boldsymbol{y} \le y\}\right) = \int_{-\infty}^{x} \int_{-\infty}^{y} f_{\boldsymbol{x},\boldsymbol{y}}(x, y)\, dx\, dy.$$

Integrating $f_{\boldsymbol{x},\boldsymbol{y}}(x,y)$ with respect to y, or with respect x, results in the *marginal densities* (the density functions of $\boldsymbol{x}$ and $\boldsymbol{y}$, respectively):

$$f_{\boldsymbol{x}}(x) = \int_{-\infty}^{\infty} f_{\boldsymbol{x},\boldsymbol{y}}(x,y)\,dy, \tag{2.26}$$

$$f_{\boldsymbol{y}}(y) = \int_{-\infty}^{\infty} f_{\boldsymbol{x},\boldsymbol{y}}(x,y)\,dx. \tag{2.27}$$

Two jointly distributed RVs $\boldsymbol{x}$ and $\boldsymbol{y}$ are said to be *uncorrelated* if

$$\mathrm{E}[\boldsymbol{x}\boldsymbol{y}] = \mathrm{E}[\boldsymbol{x}]\mathrm{E}[\boldsymbol{y}],$$

where

$$\mathrm{E}[\boldsymbol{x}\boldsymbol{y}] = \int_{-\infty}^{\infty}\int_{-\infty}^{\infty} xy f_{\boldsymbol{x},\boldsymbol{y}}(x,y)\,dx\,dy.$$

The covariance of $\boldsymbol{x}$ and $\boldsymbol{y}$ is denoted by $\mathrm{Cov}[\boldsymbol{x},\boldsymbol{y}]$, and is defined by

$$\mathrm{Cov}[\boldsymbol{x},\boldsymbol{y}] = \mathrm{E}\left[(\boldsymbol{x} - \mathrm{E}[\boldsymbol{x}])\,(y - \mathrm{E}[\boldsymbol{y}])\right]. \tag{2.28}$$

It follows that $\boldsymbol{x}$ and $\boldsymbol{y}$ are uncorrelated if and only if $\mathrm{Cov}[\boldsymbol{x},\boldsymbol{y}] = 0$.

The RVs $\boldsymbol{x}$ and $\boldsymbol{y}$ are said to be *independent* if the joint distribution function $F_{\boldsymbol{x},\boldsymbol{y}}(x,y)$ is equal to the product of the marginals; that is,

$$F_{\boldsymbol{x},\boldsymbol{y}}(x,y) = F_{\boldsymbol{x}}(x)F_{\boldsymbol{y}}(y). \tag{2.29}$$

The condition (2.29) is equivalent to requiring that the joint density function $f_{\boldsymbol{x},\boldsymbol{y}}(x,y)$ be expressable in the form

$$f_{\boldsymbol{x},\boldsymbol{y}}(x,y) = f_{\boldsymbol{x}}(x)f_{\boldsymbol{y}}(y), \tag{2.30}$$

and thus $\boldsymbol{x}$ and $\boldsymbol{y}$ are independent if and only if (2.30) holds. If $\boldsymbol{x}$ and $\boldsymbol{y}$ are independent, it follows that they are uncorrelated.

Given two jointly distributed RVs $\boldsymbol{x}$, $\boldsymbol{y}$ defined on the probability space S and a function $\Psi(x,y)$, we can define a new random variable $\boldsymbol{z}$ to be the function of $\boldsymbol{x}$ and $\boldsymbol{y}$ defined by

$$\boldsymbol{z}(\alpha) = \Psi(\boldsymbol{x}(\alpha),\boldsymbol{y}(\alpha)), \quad \text{for all } \alpha \in S. \tag{2.31}$$

The RV $\boldsymbol{z}$ defined by (2.31) is denoted by $\boldsymbol{z} = \Psi(\boldsymbol{x},\boldsymbol{y})$. The expected value $\mathrm{E}[\boldsymbol{z}]$ of $\boldsymbol{z}$ is given by

$$\mathrm{E}[\boldsymbol{z}] = \int_{-\infty}^{\infty}\int_{-\infty}^{\infty} \Psi(x,y)f_{\boldsymbol{x},\boldsymbol{y}}(x,y)\,dx\,dy. \tag{2.32}$$

Example 2.12 Sum of Two RVs

Suppose that $\boldsymbol{z} = \boldsymbol{x} + \boldsymbol{y}$. Then using (2.26) and (2.27), (2.32), it follows that

$$\mathrm{E}[\boldsymbol{z}] = \mathrm{E}[\boldsymbol{x}] + \mathrm{E}[\boldsymbol{y}]. \tag{2.33}$$

If $\boldsymbol{x}$ and $\boldsymbol{y}$ are uncorrelated, the variance $\mathrm{Var}[\boldsymbol{z}]$ is given by

$$\mathrm{Var}[\boldsymbol{z}] = \mathrm{Var}[\boldsymbol{x}] + \mathrm{Var}[\boldsymbol{y}]. \tag{2.34}$$

If $\boldsymbol{x}$ and $\boldsymbol{y}$ are independent, the density function $f_{\boldsymbol{z}}(z)$ is given by

$$f_{\boldsymbol{z}}(z) = \int_{-\infty}^{\infty} f_{\boldsymbol{x}}(z - y) f_{\boldsymbol{y}}(y)\, dy = f_{\boldsymbol{x}}(z) * f_{\boldsymbol{y}}(z). \tag{2.35}$$

The proofs of (2.34) and (2.35) are left to the reader.

Example 2.13 Sum of Two Gaussian RVs

Suppose that $\boldsymbol{z} = \boldsymbol{x} + \boldsymbol{y}$, where $\boldsymbol{x} \sim \mathcal{N}(\eta_x, \sigma_x^2)$ and $\boldsymbol{y} \sim \mathcal{N}(\eta_y, \sigma_y^2)$ with $\boldsymbol{x}$ and $\boldsymbol{y}$ independent. It follows from (2.37) that $\boldsymbol{z}$ is also Gaussian, and by (2.33) and (2.34), the mean of $\boldsymbol{z}$ is equal to $\eta_x + \eta_y$ and the variance of $\boldsymbol{z}$ is equal to $\sigma_x^2 + \sigma_y^2$.

Conditional Density Functions

Again let $\boldsymbol{x}$ and $\boldsymbol{y}$ be jointly distributed RVs. For any two real numbers x_1 and x_2 with $x_1 < x_2$, the conditional distribution function $F_{\boldsymbol{y}}(y|x_1 < \boldsymbol{x} \le x_2)$ of $\boldsymbol{y}$ given that $x_1 < \boldsymbol{x} \le x_2$ occurred is defined to be the conditional probability of the event $\{\boldsymbol{y} \le y\}$ given $x_1 < \boldsymbol{x} \le x_2$. That is,

$$F_{\boldsymbol{y}}(y|x_1 < \boldsymbol{x} \le x_2) = P(\{\boldsymbol{y} \le y\}|x_1 < \boldsymbol{x} \le x_2) = \frac{P\{\boldsymbol{y} \le y, x_1 < \boldsymbol{x} \le x_2\}}{P\{x_1 < \boldsymbol{x} \le x_2\}}. \tag{2.36}$$

By definition of the joint density function,

$$P\{\boldsymbol{y} \le y, x_1 < \boldsymbol{x} \le x_2\} = \int_{x_1}^{x_2} \int_{-\infty}^{y} f_{\boldsymbol{x},\boldsymbol{y}}(x, y)\, dx\, dy, \tag{2.37}$$

and by definition of $f_{\boldsymbol{x}}(x)$,

$$P\{x_1 < \boldsymbol{x} \le x_2\} = \int_{x_1}^{x_2} f_{\boldsymbol{x}}(x)\, dx. \tag{2.38}$$

Using (2.37) and (2.38) in (2.36) gives

$$F_{\boldsymbol{y}}(y|x_1 < \boldsymbol{x} \le x_2) = \frac{\displaystyle\int_{x_1}^{x_2} \int_{-\infty}^{y} f_{\boldsymbol{x},\boldsymbol{y}}(x, y)\, dx\, dy}{\displaystyle\int_{x_1}^{x_2} f_{\boldsymbol{x}}(x)\, dx}. \tag{2.39}$$

Now

$$f_{\boldsymbol{y}}(y|x_1 < \boldsymbol{x} \leq x_2) = \frac{\partial}{\partial y} F_{\boldsymbol{y}}(y|x_1 < \boldsymbol{x} \leq x_2). \tag{2.40}$$

Inserting the right-hand side of (2.39) into (2.40) gives

$$f_{\boldsymbol{y}}(y|x_1 < \boldsymbol{x} \leq x_2) = \frac{\displaystyle\int_{x_1}^{x_2} f_{\boldsymbol{x},\boldsymbol{y}}(x, y)\, dx}{\displaystyle\int_{x_1}^{x_2} f_{\boldsymbol{x}}(x)\, dx}. \tag{2.41}$$

Then setting $x_1 = x$ and $x_2 = x + \Delta x$ in (2.41), we have

$$f_{\boldsymbol{y}}(y|x < \boldsymbol{x} \leq x + \Delta x) = \frac{\displaystyle\int_{x}^{x+\Delta x} f_{\boldsymbol{x},\boldsymbol{y}}(x, y)\, dx}{\displaystyle\int_{x}^{x+\Delta x} f_{\boldsymbol{x}}(x)\, dx} \approx \frac{f_{\boldsymbol{x},\boldsymbol{y}}(x, y)\Delta x}{f_{\boldsymbol{x}}(x)\Delta x}.$$

Finally, we can define $f_{\boldsymbol{y}}(y|\boldsymbol{x} = x)$ as the limit:

$$f_{\boldsymbol{y}}(y|\boldsymbol{x} = x) = \lim_{\Delta x \to 0} f_{\boldsymbol{y}}(y|x < \boldsymbol{x} \leq x + \Delta x),$$

and thus, taking the limit in (2.41) gives

$$f_{\boldsymbol{y}}(y|\boldsymbol{x} = x) = \frac{f_{\boldsymbol{x},\boldsymbol{y}}(x, y)}{f_{\boldsymbol{x}}(x)}. \tag{2.42}$$

Similarly, we have

$$f_{\boldsymbol{x}}(x|\boldsymbol{y} = y) = \frac{f_{\boldsymbol{x},\boldsymbol{y}}(x, y)}{f_{\boldsymbol{y}}(y)}. \tag{2.43}$$

By (2.43),

$$f_{\boldsymbol{x},\boldsymbol{y}}(x, y) = f_{\boldsymbol{x}}(x|\boldsymbol{y} = y) f_{\boldsymbol{y}}(y),$$

and inserting this into (2.42) results in Bayes' formula:

$$f_{\boldsymbol{y}}(y|\boldsymbol{x} = x) = \frac{f_{\boldsymbol{x}}(x|\boldsymbol{y} = y) f_{\boldsymbol{y}}(y)}{f_{\boldsymbol{x}}(x)}. \tag{2.44}$$

Example 2.14 Conditional Density Function

Suppose that $\boldsymbol{z} = \boldsymbol{x} + \boldsymbol{y}$ where $\boldsymbol{x}$ and $\boldsymbol{y}$ are independent and $\boldsymbol{y} \sim \mathcal{N}\left(0, \sigma_y^2\right)$. The goal is to determine the conditional density function $f_{\boldsymbol{z}}(z|\boldsymbol{x} = x)$. First note that

$$\begin{aligned} P\{\boldsymbol{z} \leq z, \boldsymbol{x} = x\} &= P\{\boldsymbol{x} + \boldsymbol{y}|z, \boldsymbol{x} = x\} \\ &= P\{\boldsymbol{y} \leq z - \boldsymbol{x}, \boldsymbol{x} = x\} = P\{\boldsymbol{y} \leq z - x, \boldsymbol{x} = x\}, \end{aligned}$$

and since $\boldsymbol{x}$ and $\boldsymbol{y}$ are independent

$$P\{\boldsymbol{y} \leq z - x, \boldsymbol{x} = x\} = P\{\boldsymbol{y} \leq z - x\} P\{\boldsymbol{x} = x\}.$$

It follows that $f_{\boldsymbol{z}}(z|\boldsymbol{x} = x) = f_{\boldsymbol{y}}(z - x)$, and since $\boldsymbol{y} \sim \mathcal{N}\left(0, \sigma_y^2\right)$, $f_{\boldsymbol{z}}(z|\boldsymbol{x} = x)$ is Gaussian with mean x and variance σ_y^2; that is,

$$f_{\boldsymbol{z}}(z|\boldsymbol{x} = x) = \frac{1}{\sqrt{2\pi}\sigma_y} \exp\left[-\frac{(z - x)^2}{2\sigma_y^2}\right]. \tag{2.45}$$

Example 2.15 Use of Bayes' Formula

Suppose that $\boldsymbol{z} = \boldsymbol{x} + \boldsymbol{y}$ where $\boldsymbol{x} \sim \mathcal{N}\left(\eta_x, \sigma_x^2\right)$, and $\boldsymbol{y} \sim \mathcal{N}\left(0, \sigma_y^2\right)$ with $\boldsymbol{x}$ and $\boldsymbol{y}$ independent. Now the objective is to determine the conditional density function $f_{\boldsymbol{x}}(x|\boldsymbol{z} = z)$. By Bayes' formula (2.44)

$$f_{\boldsymbol{x}}(x|\boldsymbol{z} = z) = \frac{f_{\boldsymbol{z}}(z|\boldsymbol{x} = x) f_{\boldsymbol{x}}(x)}{f_{\boldsymbol{z}}(z)}. \tag{2.46}$$

From the result in Example 2.13, $\boldsymbol{z}$ is Gaussian with mean η_x and variance $\sigma_x^2 + \sigma_y^2$, and thus

$$f_{\boldsymbol{z}}(z) = \frac{1}{\sqrt{2\pi\left(\sigma_x^2 + \sigma_y^2\right)}} \exp\left[-\frac{(z - \eta_x)^2}{2\left(\sigma_x^2 + \sigma_y^2\right)}\right]. \tag{2.47}$$

In addition,

$$f_{\boldsymbol{x}}(x) = \frac{1}{\sqrt{2\pi}\sigma_x} \exp\left[-\frac{(x - \eta_x)^2}{2\sigma_x^2}\right], \tag{2.48}$$

and $f_{\boldsymbol{z}}(z|\boldsymbol{x} = x)$ is given by (2.45). Then inserting (2.45), (2.47), and (2.48) into (2.46) gives

$$f_{\boldsymbol{x}}(x|\boldsymbol{z} = z) = \frac{\sqrt{\sigma_x^2 + \sigma_y^2}}{\sqrt{2\pi}\sigma_x\sigma_y} \exp\left[-\frac{(z - x)^2}{2\sigma_y^2} - \frac{(z - \eta_x)^2}{2\sigma_x^2} + \frac{(z - \eta_x)^2}{2\left(\sigma_x^2 + \sigma_y^2\right)}\right]. \tag{2.49}$$

It can be shown that the conditional density function $f_{\boldsymbol{x}}(x|\boldsymbol{z} = z)$ given by (2.49) is Gaussian with mean

$$\eta_x + \frac{\sigma_x^2}{\sigma_x^2 + \sigma_y^2}(z - \eta_x), \tag{2.50}$$

and variance

$$\frac{\sigma_x^2 \sigma_y^2}{\sigma_x^2 + \sigma_y^2}.$$

The verification of these expressions is left to a homework problem.

Given jointly distributed RVs $\boldsymbol{x}$ and $\boldsymbol{y}$ with conditional density function $f_{\boldsymbol{y}}(y|\boldsymbol{x} = x)$, the conditional expectation of $\boldsymbol{y}$ given $\boldsymbol{x} = x$, denoted by $E[\boldsymbol{y}|\boldsymbol{x} = x]$, is defined by

$$\mathrm{E}[\boldsymbol{y}|\boldsymbol{x} = x] = \int_{-\infty}^{\infty} y f_{\boldsymbol{y}}(y|\boldsymbol{x} = x)\, dy. \tag{2.51}$$

Let Ψ denote the real-valued function given by

$$\Psi(x) = \int_{-\infty}^{\infty} y f_{\boldsymbol{y}}(y|\boldsymbol{x} = x)\, dy. \tag{2.52}$$

Then we can define the RV

$$\Psi(\boldsymbol{x}) = \mathrm{E}[\boldsymbol{y}|\boldsymbol{x}], \tag{2.53}$$

and thus the conditional expectation can be viewed as a random variable. It follows from the definition (2.52) of $\Psi(x)$ that

$$\mathrm{E}\left[\mathrm{E}\left[\boldsymbol{y}|\boldsymbol{x}\right]\right] = \mathrm{E}[\boldsymbol{y}]. \tag{2.54}$$

The verification of (2.54) is left to the reader.

Example 2.16 Conditional Expectation

Let $\boldsymbol{z} = \boldsymbol{x} + \boldsymbol{y}$ be as defined in Example 2.15. Since $f_{\boldsymbol{x}}(x|\boldsymbol{z} = z)$ is Gaussian with mean given by (2.50), it follows that

$$\mathrm{E}[\boldsymbol{x}|\boldsymbol{z} = z] = \eta_x + \frac{\sigma_x^2}{\sigma_x^2 + \sigma_y^2}(z - \eta_x). \tag{2.55}$$

Then $\mathrm{E}[\boldsymbol{x}|\boldsymbol{z}]$ is the RV defined by

$$\mathrm{E}[\boldsymbol{x}|\boldsymbol{z}] = \eta_x + \frac{\sigma_x^2}{\sigma_x^2 + \sigma_y^2}(\boldsymbol{z} - \eta_x). \tag{2.56}$$

Vector Random Variables

For some positive integer N, let $\boldsymbol{x}_1, \boldsymbol{x}_2, \ldots, \boldsymbol{x}_N$ be N jointly distributed RVs defined on the probability space S. Then we can define the vector random variable $\boldsymbol{x}$ to be the N-element column vector consisting of the $\boldsymbol{x}_i$; that is,

$$\boldsymbol{x} = \begin{bmatrix} \boldsymbol{x}_1 \\ \boldsymbol{x}_2 \\ \vdots \\ \boldsymbol{x}_N \end{bmatrix}.$$

Note that for any $\alpha \in S$, the value $\boldsymbol{x}(\alpha)$ of $\boldsymbol{x}$ at α is an N-vector given by

$$\boldsymbol{x}(\alpha) = \begin{bmatrix} \boldsymbol{x}_1(\alpha) \\ \boldsymbol{x}_2(\alpha) \\ \vdots \\ \boldsymbol{x}_N(\alpha) \end{bmatrix}.$$

The mean $\mathrm{E}[\boldsymbol{x}]$ of $\boldsymbol{x}$ is the N-vector defined by

$$\mathrm{E}[\boldsymbol{x}] = \begin{bmatrix} \mathrm{E}[\boldsymbol{x}_1] \\ \mathrm{E}[\boldsymbol{x}_2] \\ \vdots \\ \mathrm{E}[\boldsymbol{x}_N] \end{bmatrix}.$$

The *covariance* of $\boldsymbol{x}$ is the $N \times N$ matrix defined by

$$\mathrm{Cov}\,[\boldsymbol{x}] = \mathrm{E}\left[(\boldsymbol{x} - \mathrm{E}[\boldsymbol{x}])\,(\boldsymbol{x} - \mathrm{E}[\boldsymbol{x}])^{\mathrm{T}}\right], \tag{2.57}$$

where superscript "T" denotes matrix transposition. Note that when $N = 1$, so that $\boldsymbol{x}$ is scalar-valued, $\mathrm{Cov}\,[\boldsymbol{x}] = \mathrm{Var}\,[\boldsymbol{x}]$. It can shown that for any N-vector random variable $\boldsymbol{x}$, $\mathrm{Cov}\,[\boldsymbol{x}]$ is always positive definite; that is, $\mathrm{Cov}\,[\boldsymbol{x}]$ is symmetric and the eigenvalues are always strictly positive. As a result, $\mathrm{Cov}\,[\boldsymbol{x}]$ is always invertible. For any $N \times N$ matrix M, it can also be shown that

$$\mathrm{Cov}\,[M\boldsymbol{x}] = M(\mathrm{Cov}\,[\boldsymbol{x}])M^{\mathrm{T}}. \tag{2.58}$$

The distribution function $F_{\boldsymbol{x}}(x)$ and density function $f_{\boldsymbol{x}}(x)$ of the vector RV $\boldsymbol{x}$ are defined by

$$F_{\boldsymbol{x}}(x) = P\left\{\boldsymbol{x}_1 \le x_1, \boldsymbol{x}_2 \le x_2, \ldots, \boldsymbol{x}_N \le x_N\right\},$$
$$f_{\boldsymbol{x}}(x) = \frac{\partial^N}{\partial x_1\, \partial x_2 \cdots \partial x_N} F_{\boldsymbol{x}}(x),$$

where x is an N-element column vector whose entries are real numbers equal to $x_1, x_2, \ldots, x_N$. Note that both $F_{\boldsymbol{x}}(x)$ and $f_{\boldsymbol{x}}(x)$ are real-valued functions of the N-vector x.

Example 2.17 Linear Combination of RVs

Given N jointly distributed normal RVs $\boldsymbol{x}_1, \boldsymbol{x}_2, \ldots, \boldsymbol{x}_N$, the $\boldsymbol{x}_i$ are said to be *jointly normal* if every linear combination $a_1\boldsymbol{x}_1 + a_2\boldsymbol{x}_2 + \cdots + a_N\boldsymbol{x}_N$ is a normal RV, where the a_i are arbitrary real numbers. If the $\boldsymbol{x}_i$ are independent, they can be shown to be jointly normal. If the $\boldsymbol{x}_i$ are jointly normal, it can be shown that the density function $f_{\boldsymbol{x}}(x)$ of the N-element Gaussian vector random variable $\boldsymbol{x} = \begin{bmatrix}\boldsymbol{x}_1 & \boldsymbol{x}_2 & \cdots & \boldsymbol{x}_N\end{bmatrix}^{\mathrm{T}}$ is given by

$$f_{\boldsymbol{x}}(x) = \frac{1}{(2\pi)^{N/2}|P|^{1/2}} \exp\left[-\frac{1}{2}\left(x - \mathrm{E}[\boldsymbol{x}]\right)^{\mathrm{T}} P^{-1}\left(x - \mathrm{E}[\boldsymbol{x}]\right)\right], \qquad (2.59)$$

where P is the covariance of $\boldsymbol{x}$ defined by (2.57), $|P|$ is the determinant of P, and $\mathrm{E}[\boldsymbol{x}]$ is the mean of $\boldsymbol{x}$. The function $f_{\boldsymbol{x}}(x)$ given by (2.59) is called the *N-variate Gaussian density function*. The Gaussian vector random variable $\boldsymbol{x}$ with the density function (2.59) will sometimes be denoted by $\boldsymbol{x} \sim \mathcal{N}(\mathrm{E}[\boldsymbol{x}], P)$.

2.2 Random Discrete-Time Signals

A random discrete-time signal $\boldsymbol{x}(n)$ is a sequence

$$\ldots, \boldsymbol{x}(-2),\ x(-1), \boldsymbol{x}(0), \boldsymbol{x}(1), \boldsymbol{x}(2), \ldots$$

of jointly distributed random variables defined on a probability space S. Here it is assumed that the discrete-time index n begins at $n = -\infty$ and ends at $n = \infty$, so that $\boldsymbol{x}(n)$ is a "two sided" random signal. An example of a one-sided random signal $\boldsymbol{x}(n)$ is the sequence

$$\boldsymbol{x}(0), \boldsymbol{x}(1), \boldsymbol{x}(2), \ldots.$$

Note that in this case, $\boldsymbol{x}(i)$ is defined for $i \geq 0$ only.

A two-sided random signal $\boldsymbol{x}(n)$ can be viewed as function from S into the set of all two-sided sequences of real numbers. More precisely, for any $\alpha \in S$, the value $[\boldsymbol{x}(n)](\alpha)$ of $\boldsymbol{x}(n)$ at α is the sequence of real numbers:

$$\ldots, [\boldsymbol{x}(-2)](\alpha), [\boldsymbol{x}(-1)](\alpha), [\boldsymbol{x}(0)](\alpha), [\boldsymbol{x}(1)](\alpha), [\boldsymbol{x}(2)](\alpha), \ldots \qquad (2.60)$$

where $[\boldsymbol{x}(i)](\alpha)$ is the value of the RV $\boldsymbol{x}(i)$ at α. The sequence (2.60) is called a *sample realization* of the random signal $\boldsymbol{x}(n)$ and is denoted by $x(n)$.

A random discrete-time signal $\boldsymbol{x}(n)$ may be viewed as a sampled random continuous-time signal $\boldsymbol{x}(t)$. In other words,

$$\boldsymbol{x}(i) = \boldsymbol{x}(iT), \quad i = 0, \pm 1, \pm 2, \ldots,$$

where T is the sampling interval.

In dealing with a random signal $\boldsymbol{x}(n)$, we will usually denote the RVs comprising $\boldsymbol{x}(n)$ by the same symbol; that is, the RV at time instant n will also be denoted by $\boldsymbol{x}(n)$. It will always be clear from the context as to whether $\boldsymbol{x}(n)$ refers to the random signal, or to the RV at time instant n.

A random discrete-time signal $\boldsymbol{x}(n)$ can be characterized in terms of its *autocorrelation function*, defined by

$$\begin{aligned} R_x(i,j) &= \mathrm{E}[\boldsymbol{x}(i)\boldsymbol{x}(j)] \\ &= \int_{-\infty}^{\infty}\int_{-\infty}^{\infty} x(i)x(j)f_{\boldsymbol{x}(i),\boldsymbol{x}(j)}(x(i),x(j))\,dx(i)\,dx(j), \\ &\qquad \text{for all integers } i,j, \end{aligned} \tag{2.61}$$

where $f_{\boldsymbol{x}(i),\boldsymbol{x}(j)}(x(i),x(j))$ is the joint density function of the random variables $\boldsymbol{x}(i)$ and $x(j)$. Note that the autocorrelation is a function of two integer variables i and j, with $-\infty < i < \infty$ and $-\infty < j < \infty$. The autocorrelation function $R_x(i,j)$ measures the correlation between signal samples. Two examples are given below.

Example 2.18 Autocorrelation Function

Suppose that $x(n+1) = a\boldsymbol{x}(n)$, where a is a nonzero real number. Then

$$\boldsymbol{x}(n) = a^n\boldsymbol{x}(0) \quad \text{for } n = 0, \pm1, \pm2, \ldots,$$

and the autocorrelation function is

$$R_x(i,j) = \mathrm{E}[\boldsymbol{x}(i)\boldsymbol{x}(j)] = \mathrm{E}\left[a^i\boldsymbol{x}(0)a^j\boldsymbol{x}(0)\right] = a^{i+j}\mathrm{E}\left[(\boldsymbol{x}(0))^2\right]. \tag{2.62}$$

Note that if a is negative, $R_x(i,j)$ is negative for all integers i and j such that $i+j$ is odd. Hence, it is possible that the RVs comprising a random signal $\boldsymbol{x}(n)$ may be "negatively corrrelated."

Example 2.19 Purely Random Signal

Suppose that the random variables comprising the random signal $\boldsymbol{x}(n)$ are independent and all have zero mean. Then

$$R_x(i,j) = \mathrm{E}[\boldsymbol{x}(i)\boldsymbol{x}(j)] = \begin{cases} \mathrm{E}\left[\boldsymbol{x}_i^2\right], & \text{when } i = j; \\ \mathrm{E}\left[\boldsymbol{x}_i\right]\mathrm{E}\left[\boldsymbol{x}_j\right], & \text{when } i \neq j. \end{cases}$$

In this case, since $R_x(i,j) = 0$ when $i \neq j$, there is no correlation between the signal samples at different times. Such a random signal is sometimes said to be *purely random.*

Suppose that the random variables comprising a random signal $\boldsymbol{x}(n)$ are independent and have means $\mathrm{E}[\boldsymbol{x}(n)] = c_n$, where the c_n are arbitrary nonzero real numbers. Then $\mathrm{E}[\boldsymbol{x}(i)\boldsymbol{x}(j)] = c_i c_j \neq 0$ for $i \neq j$, and thus even though $\boldsymbol{x}(i)$ and $\boldsymbol{x}(j)$ are independent (for $i \neq j$), and therefore are uncorrelated, $\mathrm{E}[\boldsymbol{x}(i)\boldsymbol{x}(j)]$ is nonzero. So there appears to be "correlation" between $\boldsymbol{x}(i)$ and $\boldsymbol{x}(j)$. This is a result of the nonzero mean; in fact, a nonzero mean can be interpreted as the existence of a *deterministic part* of the random signal $\boldsymbol{x}(n)$.

If random signal $\boldsymbol{x}(n)$ has a nonzero mean $\mathrm{E}[\boldsymbol{x}(n)] = c_n$ corresponding to a deterministic part, we can remove this part from $\boldsymbol{x}(n)$ by subtracting the mean from $\boldsymbol{x}(n)$. In other words, we can form the random signal $\boldsymbol{y}(n) = \boldsymbol{x}(n) - c_n$. Then of course $\boldsymbol{y}(n)$ has zero mean, and so there is no deterministic part in $\boldsymbol{y}(n)$.

In contrast to "subtracting out a deterministic part," we can also "add a deterministic part" as follows. Let $s(n)$ be a deterministic signal; that is, for any integer value of n, $s(n)$ is a real number. (Note that we are using nonbold notation for a deterministic signal.) Then given a random discrete-time signal $\boldsymbol{x}(n)$, we can add $s(n)$ to $\boldsymbol{x}(n)$ by defining the random signal $\boldsymbol{y}(n) = s(n) + \boldsymbol{x}(n)$. Note that if $\mathrm{E}[\boldsymbol{x}(n)] = 0$, then $\mathrm{E}[\boldsymbol{y}(n)] = s(n)$. Also note that if $\boldsymbol{x}(n)$ is a "noise signal," then the random signal $\boldsymbol{y}(n) = s(n) + \boldsymbol{x}(n)$ can be interpreted as signal plus noise.

Wide-Sense Stationary Signals

A random discrete-time signal $\boldsymbol{x}(n)$ is said to be *wide-sense stationary* (WSS) if the following two conditions are satisfied:

$$\text{(1)} \quad \mathrm{E}[\boldsymbol{x}(n)] = c(\text{a constant}), \quad \text{for all integers } n,$$
$$\text{(2)} \mathrm{E}[\boldsymbol{x}(i)\boldsymbol{x}(j)] = \mathrm{E}[\boldsymbol{x}(i+k)\boldsymbol{x}(j+k)], \quad \text{for all integers } i, j, k.$$

It is very important to note that the concept of wide-sense stationarity applies only to random signals $\boldsymbol{x}(n)$ that are defined for n ranging from $-\infty$ to ∞. In particular, a random signal $\boldsymbol{x}(n)$ that is defined for $n \geq 0$ **only** cannot be WSS.

Taking $j = i$ in Condition (2), we see that the mean square $\mathrm{E}\left[(\boldsymbol{x}(n))^2\right]$ must be a constant for all integers n, and thus the variance $\mathrm{Var}\left[\boldsymbol{x}(n)\right]$ must be a constant for all n. Hence, in order for a random signal $\boldsymbol{x}(n)$ to be WSS, it is necessary that both the mean $\mathrm{E}[\boldsymbol{x}(n)]$ and the variance $\mathrm{Var}\left[\boldsymbol{x}(n)\right]$ be constant for all n. As a result, if the random variables comprising $\boldsymbol{x}(n)$ are Gaussian distributed or uniformly distributed, it turns out that $\boldsymbol{x}(n)$ is WSS if and only if both the mean and the variance are constant. Unfortunately, this result is not valid in general; that is, a random signal $\boldsymbol{x}(n)$ may not be WSS even though the mean and variance are constant.

If $\boldsymbol{x}(n)$ is a WSS random signal, the autocorrelation function $R_x(i,j) = \mathrm{E}[\boldsymbol{x}(i)\boldsymbol{x}(j)]$ is a function of the difference $i - j$. To see this, set $k = -i$ in

Condition (2) given above. Then

$$R_x(i,j) = \mathrm{E}[\boldsymbol{x}(i)\boldsymbol{x}(j)] = \mathrm{E}[\boldsymbol{x}(0)\boldsymbol{x}(j-i)] = R_x(0, j-i).$$

Setting $k = j - i$, we see that the autocorrelation can be viewed as a function of the single integer variable k. Therefore, defining

$$R_x(k) = R_x(0,k), \quad k = 0, \pm 1, \pm 2, \ldots,$$

we can take $R_x(k)$ to be the autocorrelation function in the case when $\boldsymbol{x}(n)$ is WSS. Again from Condition(2) above, we have that

$$R_x(k) = \mathrm{E}[\boldsymbol{x}(n)\boldsymbol{x}(n+k)], \quad \text{for any integer } n. \tag{2.63}$$

Note that $R_x(0)$ is equal to the mean square $\mathrm{E}\left[(\boldsymbol{x}(n))^2\right]$, and thus $R_x(0)$ is always a strictly positive real number. Also note that since $R_x(-k) = \mathrm{E}[\boldsymbol{x}(n)\boldsymbol{x}(n-k)]$, replacing n by $n+k$ we have that

$$R_x(-k) = \mathrm{E}[\boldsymbol{x}(n)\boldsymbol{x}(n+k)] = R_x(k).$$

Hence, the autocorrelation is an *even function* of k.

Example 2.20 Nonstationary Signal

Consider the random signal $\boldsymbol{x}(n)$ in Example 2.18 given by $\boldsymbol{x}(n+1) = a\boldsymbol{x}(n)$, where a is a nonzero real number. From (2.62), it is clear that $R_x(i,j)$ is not a function of the difference $i - j$, and thus $\boldsymbol{x}(n)$ is not WSS.

Example 2.21 White Noise

Suppose that the random variables comprising the random signal $\boldsymbol{x}(n)$ are independent and all have zero mean. From the result in Example 2.19, we see that $\boldsymbol{x}(n)$ is WSS if and only if the mean square $\mathrm{E}\left[(\boldsymbol{x}(n))^2\right]$, which is equal to the variance $\mathrm{Var}[\boldsymbol{x}(n)]$ in this case, is a constant for all integers n. If $\mathrm{Var}[\boldsymbol{x}(n)] = \sigma^2$ for all n, then the autocorrelation function $R_x(k)$ is equal to $\sigma^2\delta(k)$, where $\delta(k)$ is the impulse (unit pulse) concentrated at $k = 0$ (i.e., $\delta(0) = 1$ and $\delta(k) = 0$ for $k \neq 0$). A random signal with this particular autocorrelation function is often referred to as *white noise*. Later, we will see the justification for this terminology.

Example 2.22 WSS Signal

Suppose that $\boldsymbol{x}(n) = \boldsymbol{w}(n) + \boldsymbol{w}(n-1)$, where $\boldsymbol{w}(n)$ is zero-mean white noise with $\mathrm{E}\left[(\boldsymbol{w}(n))^2\right] = \sigma^2$ for all n. From the results of the previous example, $R_w(k) = \sigma^2\delta(k)$. Using additivity of the expectation operator [see (2.33)], we have

$$\mathrm{E}[\boldsymbol{x}(n)] = \mathrm{E}[\boldsymbol{w}(n)] + \mathrm{E}[\boldsymbol{w}(n-1)] = 0, \quad \text{for all } n,$$

and thus Condition (1) is satisfied. Now

$$\begin{aligned}\mathrm{E}[\boldsymbol{x}(i)\boldsymbol{x}(j)] &= \mathrm{E}\left[[\boldsymbol{w}(i)+\boldsymbol{w}(i-1)][\boldsymbol{w}(j)+\boldsymbol{w}(j-1)]\right] \\ &= \mathrm{E}[\boldsymbol{w}(i)\boldsymbol{w}(j)+\boldsymbol{w}(i)\boldsymbol{w}(j-1)+\boldsymbol{w}(i-1)\boldsymbol{w}(j)+\boldsymbol{w}(i-1)\boldsymbol{w}(j-1)] \\ &= \mathrm{E}[\boldsymbol{w}(i)\boldsymbol{w}(j)]+\mathrm{E}[\boldsymbol{w}(i)\boldsymbol{w}(j-1)] \\ &\quad +\mathrm{E}[\boldsymbol{w}(i-1)\boldsymbol{w}(j)]+\mathrm{E}[\boldsymbol{w}(i-1)\boldsymbol{w}(j-1)]. \end{aligned} \tag{2.64}$$

Since $\boldsymbol{w}(n)$ is WSS, we can replace i by $i+k$ and j by $j+k$ in the left-hand side of (2.64), and thus Condition (2) is satisfied. Hence $\boldsymbol{x}(n)$ is WSS, and

$$\begin{aligned}R_x(k) = \mathrm{E}[x(n)x(n+k)] &= \mathrm{E}\left[[\boldsymbol{w}(n)+\boldsymbol{w}(n-1)][\boldsymbol{w}(n+k)+\boldsymbol{w}(n+k-1)]\right] \\ &= \mathrm{E}[\boldsymbol{w}(n)\boldsymbol{w}(n+k)]+\mathrm{E}[\boldsymbol{w}(n)\boldsymbol{w}(n+k-1)] \\ &\quad +\mathrm{E}[\boldsymbol{w}(n-1)\boldsymbol{w}(n+k)]+\mathrm{E}[\boldsymbol{w}(n-1)\boldsymbol{w}(n+k-1)].\end{aligned}$$

But

$$R_w(k) = \mathrm{E}[\boldsymbol{w}(n)\boldsymbol{w}(n+k)], \quad \text{for all } n,$$

and thus

$$\begin{aligned}R_x(k) &= R_w(k)+R_w(k-1)+R_w(k+1)+R_w(k) \\ &= 2R_w(k)+R_w(k-1)+R_w(k+1).\end{aligned}$$

Finally, since $R_w(k) = \sigma^2\delta(k)$, we have

$$R_x(k) = \sigma^2\left[2\delta(k)+\delta(k-1)+\delta(k+1)\right].$$

This result shows that the RVs $\boldsymbol{x}(i)$ and $\boldsymbol{x}(j)$ comprising $\boldsymbol{x}(n)$ are correlated when $j = i+1$, but $\boldsymbol{x}(i)$ and $\boldsymbol{x}(j)$ are not correlated when $j = k+i$ for $k = \pm 2, \pm 3, \ldots$.

Estimation of the Autocorrelation Function

Let $\boldsymbol{x}(n)$ be a zero-mean WSS random signal, and let $x(1)$, $x(2)$, ..., $x(N)$ be sample values of the RVs $\boldsymbol{x}(1)$, $\boldsymbol{x}(2)$, ..., $\boldsymbol{x}(N)$ comprising $\boldsymbol{x}(n)$. We can then generate an estimate $\hat{R}_x(k)$ of the autocorrelation function $R_x(k)$ defined by

$$\hat{R}_x(k) = \frac{1}{N-1}\sum_{i=0}^{N-k-1} x(i)x(i+k), \quad k = 0,1,2,\ldots,N-1. \tag{2.65}$$

For negative values of k, $\hat{R}_x(k)$ is defined by

$$\hat{R}_x(k) = \hat{R}_x(-k), \quad k = -1,-2,\ldots,-N+1. \tag{2.66}$$

The function $\hat{R}_x(k)$ defined by (2.65) and (2.66) is called the *sample autocrrelation.*

To determine how good of an estimate $\hat{R}_x(k)$ is, define the random signal

$$\hat{\boldsymbol{R}}_x(k) = \frac{1}{N-1}\sum_{i=0}^{N-k-1} \boldsymbol{x}(i)\boldsymbol{x}(i+k), \quad k = 0,1,2,\ldots,N-1.$$

Then

$$\mathrm{E}\left[\hat{\boldsymbol{R}}_x(k)\right] = \frac{1}{N-1}\sum_{i=0}^{N-k-1} \mathrm{E}\left[\boldsymbol{x}(i)\boldsymbol{x}(i+k)\right],$$

and by definition of $R_x(k)$,

$$\mathrm{E}\left[\hat{\boldsymbol{R}}_x(k)\right] = \frac{N-k}{N-1}R_x(k).$$

This result shows that $\hat{R}_x(k)$ is a "biased estimate" of $R_x(k)$. In other words, $\mathrm{E}\left[\hat{\boldsymbol{R}}_x(k)\right] \neq R_x(k)$ in general. But if the number N of sample points is large, the estimate $\hat{R}_x(k)$ is "good on the average" for values of k that are small relative to N.

Power Spectrum

Let $\boldsymbol{x}(n)$ be a zero-mean WSS random signal with autocorrelation function $R_x(k)$. The *power spectral density* or *power spectrum* of $\boldsymbol{x}(n)$, which is denoted by $S_x\left(e^{j\omega}\right)$, is defined to be the discrete-time Fourier transform of $R_x(k)$; that is

$$S_x\left(e^{j\omega}\right) = \sum_{k=-\infty}^{\infty} R_x(k)e^{-j\omega k}. \tag{2.67}$$

The function $S_x\left(e^{j\omega}\right)$ represents the distribution of power with respect to frequency (where ω is the frequency variable).

Since $R_x(k)$ is an even function of k, it follows from (2.67) that $S_x\left(e^{j\omega}\right)$ is real valued. Furthermore, it can be shown that $S_x\left(e^{j\omega}\right) \geq 0$ for all ω, so $S_x\left(e^{j\omega}\right)$ is always positive or zero. It also follows from (2.67) that $S_x\left(e^{j\omega}\right)$ is a periodic function of ω with period 2π and is symmetric about $\omega = 0$. Usually, the magnitude $\left|S_x\left(e^{j\omega}\right)\right|$ and the angle $\angle S_x\left(e^{j\omega}\right)$ are plotted from $-\pi$ to π or from 0 to 2π.

Example 2.23 White Noise Spectrum

Let $\boldsymbol{x}(n)$ be the zero-mean WSS random signal (white noise) with autocorrelation function $R_x(k) = \sigma^2\delta(k)$. Then $S_x\left(e^{j\omega}\right) = \sigma^2$ for all ω. This random signal has a flat spectrum; that is, it has spectral components of the same magnitude at all frequencies, which is the reason why it is referred to as white noise (in analogy with white light, which contains all colors of light.) The constant σ^2 is the *signal power.*

In the definition of $S_x(e^{j\omega})$ given by (2.67), it is assumed that $R_x(k)$ is constrained so that the summation in (2.67) is well defined (i.e., converges) for all values of ω. A sufficient condition for convergence of the summation is that $R_x(k)$ be absolutely summable, which means that

$$\sum_{k=-\infty}^{\infty} |R_x(k)| < \infty. \tag{2.68}$$

The condition (2.68) can be checked by determining the two-sided z-transform $S_x(z)$ of $R_x(k)$ defined by (see Appendix B)

$$S_x(z) = \sum_{k=-\infty}^{\infty} R_x(k) z^{-k}. \tag{2.69}$$

If $S_x(z)$ is a rational function of z; that is, $S_x(z) = N(z)/D(z)$, where $N(z)$ and $D(z)$ are polynomials in positive powers of z, the condition (2.68) is satisfied if and only if $S_x(z)$ has no poles on the unit circle, which is the set of all z such that $|z| = 1$. If this is the case, then setting $z = e^{j\omega}$ in $S_x(z)$ gives $S_x(e^{j\omega})$.

Two Random Signals

Two random signals $\boldsymbol{x}(n)$ and $\boldsymbol{y}(n)$ are said to be *jointly distributed* if they are defined on the same probability space S. Given jointly distributed RVs $\boldsymbol{x}(n)$ and $\boldsymbol{y}(n)$, the *cross correlation function* $R_{x,y}(i,j)$ is defined by

$$R_{x,y}(i,j) = \mathrm{E}[x(i)y(j)], \quad \text{for all integers } i, j. \tag{2.70}$$

If both $\boldsymbol{x}(n)$ and $\boldsymbol{y}(n)$ are WSS, the cross correlation $R_{x,y}(i,j)$ is a function of the difference $i-j$, in which case the cross correlation is defined by

$$R_{x,y}(i-j) = R_{x,y}(k) = \mathrm{E}[\boldsymbol{x}(n)\boldsymbol{y}(n+k)], \quad \text{for all integers } n \text{ and } k. \tag{2.71}$$

Given jointly distributed RVs $\boldsymbol{x}(n)$ and $\boldsymbol{y}(n)$ and a real-valued function $\Psi(x,y)$ of two real variables x any y, we can define a new random signal $\boldsymbol{z}(n)$ to be the function of $\boldsymbol{x}(n)$ and $\boldsymbol{y}(n)$ defined by

$$[\boldsymbol{z}(n)](\alpha) = \Psi\left([\boldsymbol{x}(n)](\alpha), [\boldsymbol{y}(n)](\alpha)\right), \quad \alpha \in S. \tag{2.72}$$

The random signal $\boldsymbol{z}(n)$ defined by (2.72) is denoted by $\boldsymbol{z}(n) = \Psi(\boldsymbol{x}(n), \boldsymbol{y}(n))$. For example, $\boldsymbol{z}(n)$ may be the sum or product of $\boldsymbol{x}(n)$ and $\boldsymbol{y}(n)$, denoted by $\boldsymbol{x}(n) + \boldsymbol{y}(n)$ and $\boldsymbol{x}(n)\boldsymbol{y}(n)$.

2.3 Discrete-Time Systems with Random Inputs

Let's first consider a single-input single-output deterministic discrete-time system with deterministic input signal $w(n)$ and output signal $y(n)$. Here again we are using the nonbold notation for deterministic signals. If the system is linear, the output $y(n)$ resulting from input $w(n)$ with no initial energy at time $-\infty$ is given by

$$y(n) = \sum_{i=-\infty}^{\infty} h(n,i)w(i). \tag{2.73}$$

In (2.73), $h(n,i)$ is the output response at time n due to the application of the unit pulse (impulse) at time i with no initial energy in the system prior to the application of the unit pulse. More precisely, if $w(n) = \delta(n-i)$ = unit pulse applied at time $n = i$, then inserting this into (2.73) gives $y(n) = h(n,i)$. If the system is also time invariant, (2.73) reduces to the standard convolution expression

$$y(n) = h(n) * w(n) = \sum_{i=-\infty}^{\infty} h(n-i)w(i), \tag{2.74}$$

where $h(n)$ is the impulse response; that is, $h(n)$ is the output response at time n due to the unit pulse $\delta(n)$ applied at time 0 with zero initial energy in the system prior to the application of $\delta(n)$.

There is an important technical question with respect to (2.73) and (2.74), and that is the issue as to when the infinite summations in (2.73) and (2.74) are well defined. If $w(n) = 0$ for $n = -1, -2, \ldots,$ (i.e., $w(n)$ is one-sided) and the system is causal, it turns out that the summations in (2.73) and (2.74) are actually over a finite range of integers, and thus in this case the summations always exist. But if the system is not causal or $w(n)$ is nonzero for negative values of n back to $-\infty$, in general the summations in (2.73) and (2.74) do not exist. In the time-invariant case, a sufficient condition for the summation in (2.74) to exist is that $w(n)$ be bounded as a function of n and $h(n)$ be absolutely summable. In this case, the response $y(n)$ will be bounded and the system is said to be *bounded-input bounded-output* (BIBO) *stable*. If the transfer function $H(z)$ of the system defined by (2.74) is equal to a ratio $N(z)/D(z)$ of polynomials in z, the system is BIBO stable if and only if all the poles of $H(z)$ lie within the open unit disk of the complex plane. (Recall that $H(z)$ is equal to the z-transform of the impulse response $h(n)$.)

Let us return to the general input/output relationship (2.73). If the input is now a random discrete-time signal $\boldsymbol{w}(n)$ defined on the probability space

S, we define the resulting output random signal by

$$\boldsymbol{y}(n) = \sum_{i=-\infty}^{\infty} h(n,i)\boldsymbol{w}(i). \tag{2.75}$$

Note that $\boldsymbol{y}(n)$ is also defined on the probability space S. In the time-invariant case, (2.75) reduces to

$$\boldsymbol{y}(n) = \sum_{i=-\infty}^{\infty} h(n-i)\boldsymbol{w}(i). \tag{2.76}$$

The expression (2.76) looks like a convolution relationship, but it is not, since the $\boldsymbol{w}(i)$ in (2.76) are random variables. In other words, since $\boldsymbol{w}(i)$ is a function (not a number) for any value of i, there is no way to evaluate (2.76). Hence the random output signal cannot be directly "computed" from (2.76), or (2.75) in the time-varying case.

However, given a sample realization $w(n)$ of the random input signal $\boldsymbol{w}(n)$, in the time-varying case the corresponding sample realization $y(n)$ of the random output signal $\boldsymbol{y}(n)$ is given by

$$y(n) = \sum_{i=-\infty}^{\infty} h(n,i)\boldsymbol{w}(i),$$

and in the time-invariant case, the corresponding sample realization $y(n)$ of the random output signal $\boldsymbol{y}(n)$ is given by

$$y(n) = \sum_{i=-\infty}^{\infty} h(n-i)\boldsymbol{w}(i).$$

Thus, sample realizations of the output can be readily computed from sample realizations of the input, but we would also like to be able to determine the statistical properties of the output in terms of the statistical properties of the input. Unfortunately, in general it is not possible to compute the distribution or density functions of the random variables comprising the output random signal $\boldsymbol{y}(n)$ in terms of the distribution or density functions of the random variables comprising the input random signal $\boldsymbol{w}(n)$. One exception is the case when $\boldsymbol{w}(n)$ is Gaussian distributed: Since any linear combination of independent Gaussian-distributed random variables is Gaussian distributed, the output $\boldsymbol{y}(n)$ of the system defined by (2.75) or (2.76) will also be Gaussian distributed. So in this case the output random signal is completely specified once the mean and variance have been determined. This remarkable property is true whether the system is time invariant or time varying.

Autocorrelation Function of the Output

It turns out that it is possible to derive an expression for the autocorelation function of the output $\boldsymbol{y}(n)$ in terms of the autocorrelation function of the input $\boldsymbol{w}(n)$. We carry out this derivation below in the linear time-invariant case with the input signal $\boldsymbol{w}(n)$ assumed to be WSS.

Given the linear time-invariant system defined by (2.76), we assume that $h(n)$ is absolutely summable (so that the system is BIBO stable) and the input random signal $\boldsymbol{w}(n)$ is WSS. We will first show that the resulting random output $\boldsymbol{y}(n)$ must then be WSS.

Taking the mean of both sides of (2.76), we obtain

$$\mathrm{E}[\boldsymbol{y}(n)] = \mathrm{E}\left[\sum_{i=-\infty}^{\infty} h(n-i)\boldsymbol{w}(i)\right] = \mathrm{E}\left[\sum_{i=-\infty}^{\infty} h(n-i)\right]\eta, \tag{2.77}$$

where $\eta = \mathrm{E}[\boldsymbol{w}(n)]$. Now

$$\sum_{i=-\infty}^{\infty} h(n-i) = \sum_{i=-\infty}^{\infty} h(i) = A, \tag{2.78}$$

where A is a finite constant (recall that $h(n)$ is absolutely summable). Then combining (2.77) and (2.78) gives

$$\mathrm{E}[\boldsymbol{y}(n)] = A\eta, \quad \text{for all integers } n,$$

and thus the mean of the output signal is also constant.

Now we have

$$\begin{aligned}\mathrm{E}[\boldsymbol{y}(i)\boldsymbol{y}(j)] &= \mathrm{E}\left[\left(\sum_{r=-\infty}^{\infty} h(i-r)\boldsymbol{w}(r)\right)\left(\sum_{\ell=-\infty}^{\infty} h(j-\ell)\boldsymbol{w}(\ell)\right)\right] \\ &= \sum_{r=-\infty}^{\infty}\sum_{\ell=-\infty}^{\infty} h(i-r)h(j-\ell)\mathrm{E}[\boldsymbol{w}(r)\boldsymbol{w}(\ell)] \\ &= \sum_{r=-\infty}^{\infty}\sum_{\ell=-\infty}^{\infty} h(i-r)h(j-\ell)\mathrm{E}[\boldsymbol{w}(r+k)\boldsymbol{w}(\ell+k)], \quad \text{for any } k.\end{aligned}$$

With the change of indices $\bar{r} = r + k$ and $\bar{\ell} = \ell + k$, we obtain

$$\begin{aligned}\mathrm{E}[\boldsymbol{y}(i)\boldsymbol{y}(j)] &= \sum_{\bar{r}=-\infty}^{\infty}\sum_{\bar{\ell}=-\infty}^{\infty} h\left(i-\bar{r}+k\right)h\left(j-\bar{\ell}+k\right)\mathrm{E}\left[\boldsymbol{w}\left(\bar{r}\right)\boldsymbol{w}\left(\bar{\ell}\right)\right] \\ &= \mathrm{E}[\boldsymbol{y}(i+k)\boldsymbol{y}(j+k)].\end{aligned}$$

Thus the second condition in the definition of wide-sense stationarity is also satisfied, which proves that $\boldsymbol{y}(n)$ is wide-sense stationary.

The autocorrelation function of the output $\boldsymbol{y}(n)$ is

$$
\begin{aligned}
R_y(k) &= \mathrm{E}[\boldsymbol{y}(n)\boldsymbol{y}(n+k)] = \mathrm{E}[\boldsymbol{y}(0)\boldsymbol{y}(k)] \\
&= \mathrm{E}\left[\left(\sum_{r=-\infty}^{\infty} h(-r)\boldsymbol{w}(r)\right)\left(\sum_{\ell=-\infty}^{\infty} h(k-\ell)\boldsymbol{w}(\ell)\right)\right] \\
&= \sum_{r=-\infty}^{\infty}\sum_{\ell=-\infty}^{\infty} h(-r)h(k-\ell)\mathrm{E}[\boldsymbol{w}(r)\boldsymbol{w}(\ell)]. \qquad (2.79)
\end{aligned}
$$

Then since

$$
\mathrm{E}[\boldsymbol{w}(r)\boldsymbol{w}(\ell)] = \mathrm{E}[\boldsymbol{w}(0)\boldsymbol{w}(\ell-r)]R_w(\ell-r),
$$

from (2.79) we have

$$
R_y(k) = \sum_{r=-\infty}^{\infty}\sum_{\ell=-\infty}^{\infty} h(-r)h(k-\ell)R_w(\ell-r), \qquad (2.80)
$$

and (2.80) can be expressed in the form

$$
R_y(k) = h(k) * h(-k) * R_w(k), \qquad (2.81)
$$

where $*$ denotes the convolution operation. The expression (2.80) or (2.81) can be viewed as a type of "input/output relationship" for a linear time-invariant system whose input is a WSS random signal.

It should be stressed that if the input is not WSS, in general the resulting output is not WSS, and thus the expression (2.80) or (2.81) is not valid. Also, if the given system is time varying so that the system is given by (2.75), in general the output will not be WSS even if the input is WSS.

Example 2.24 Output Autocorelation Function

Consider the causal linear time-invariant system with $h(n) = a^n$ for $n \geq 0$ and $h(n) = 0$ for $n < 0$, where a is a real number with $|a| < 1$. Then

$$
\sum_{n=-\infty}^{\infty} h(n) = \sum_{n=0}^{\infty} a^n = \frac{1}{1-a},
$$

and thus,

$$
\mathrm{E}[\boldsymbol{y}(n)] = \left(\frac{1}{1-a}\right)\mathrm{E}[\boldsymbol{w}(n)].
$$

From (2.80),

$$
R_y(k) = \sum_{r=-\infty}^{\infty}\sum_{\ell=-\infty}^{\infty} a^{k-r-\ell}R_w(\ell-r). \qquad (2.82)
$$

If the input $\boldsymbol{w}(n)$ is zero-mean white noise with $R_w(k) = \sigma^2 \delta(k)$, (2.82) becomes

$$R_y(k) = \sum_{r=-\infty}^{0} a^{k-2r}\sigma^2 = \sigma^2 a^k \sum_{r=0}^{\infty} a^{2r} = \left(\frac{\sigma^2}{1-a^2}\right) a^k, \quad \text{for } k \geq 0. \tag{2.83}$$

Given the linear time-invariant system defined by (2.76), we can also derive a useful expression for the cross correlation function

$$R_{wy}(k) = \mathrm{E}[\boldsymbol{w}(n)\boldsymbol{y}(n+k)].$$

First, setting $n = 0$ in the expression for $R_{wy}(k)$ gives

$$R_{wy}(k) = \mathrm{E}[\boldsymbol{w}(0)\boldsymbol{y}(k)]. \tag{2.84}$$

Inserting the expression (2.76) for $\boldsymbol{y}(n)$ with $n = k$ into (2.84) yields

$$\begin{aligned} R_{wy}(k) &= \mathrm{E}\left[\boldsymbol{w}(0) \sum_{i=-\infty}^{\infty} h(k-i)\boldsymbol{w}(i)\right] = \sum_{i=-\infty}^{\infty} h(k-i)\mathrm{E}[\boldsymbol{w}(0)\boldsymbol{w}(i)] \\ &= \sum_{i=-\infty}^{\infty} h(k-i) R_w(i). \end{aligned} \tag{2.85}$$

The right-hand side of (2.85) is equal to the convolution of $h(k)$ and $R_w(k)$, and thus

$$R_{wy}(k) = h(k) * R_w(k). \tag{2.86}$$

Power Spectral Density of the Output

For a linear time-invariant system, the power spectral density or power spectrum $S_y\left(e^{j\omega}\right)$ of the output $\boldsymbol{y}(n)$ can be computed by taking the discrete-time Fourier transform of both sides of (2.81). This gives

$$S_y\left(e^{j\omega}\right) = H\left(e^{j\omega}\right) H\left(e^{-j\omega}\right) S_w\left(e^{j\omega}\right), \tag{2.87}$$

where $H\left(e^{j\omega}\right)$ is the discrete-time Fourier transform of $h(n)$ given by

$$H\left(e^{j\omega}\right) = \sum_{n=-\infty}^{\infty} h(n) e^{-j\omega n}.$$

Now $H\left(e^{-j\omega}\right)$ = complex congugate of $H\left(e^{j\omega}\right)$ and

$$H\left(e^{j\omega}\right) H\left(e^{-j\omega}\right) = \left|H\left(e^{j\omega}\right)\right|^2.$$

Hence, (2.87) can be written in the form

$$S_y\left(e^{j\omega}\right) = \left|H\left(e^{j\omega}\right)\right|^2 S_w\left(e^{j\omega}\right). \tag{2.88}$$

Note that if $w(n)$ is zero-mean white noise with variance σ^2, then $S_w(e^{j\omega}) = \sigma^2$, and thus

$$S_y(e^{j\omega}) = \left|H(e^{j\omega})\right|^2 \sigma^2.$$

Hence, it is possible to determine $\left|H(e^{j\omega})\right|^2$ by measuring the power spectrum $S_y(e^{j\omega})$ of the response due to a white noise input.

Example 2.25 Output Power Spectral Density

Again consider the causal linear time-invariant system in Example 2.24 with $h(n) = a^n$ for $n \geq 0$ and $h(n) = 0$ for $n < 0$, where a is a real number with $|a| < 1$. Then

$$H\left(e^{j\omega}\right) = \sum_{n=-\infty}^{\infty} a^n e^{-j\omega n},$$

which can be written in the closed form

$$H\left(e^{j\omega}\right) = \frac{1}{1 - ae^{-j\omega}} = \frac{1}{1 - a\cos\omega + ja\sin\omega}.$$

Thus

$$\left|H\left(e^{j\omega}\right)\right| = \frac{1}{\sqrt{1 + a - 2a\cos\omega}},$$

and

$$\left|H\left(e^{j\omega}\right)\right|^2 = \frac{1}{1 + a - 2a\cos\omega}.$$

Suppose that $x(n)$ is zero-mean white noise with variance σ^2. Then $S_w(e^{j\omega}) = \sigma^2$, and using (2.88), we have

$$S_y\left(e^{j\omega}\right) = \frac{\sigma^2}{1 + a - 2a\cos\omega}.$$

Using table lookup, the inverse transform of $S_y(e^{j\omega})$, which is equal to the autocorrelation function $R_y(k)$, is given by

$$R_y(k) = \left(\frac{\sigma^2}{1 - a^2}\right) a^k, \quad \text{for } k \geq 0.$$

Note that this corresponds with the result given in Example 2.24 [see (2.83)].

The relationship (2.81) for the autocorrelation $R_y(k)$ and the relationship (2.86) for the cross correlation $R_{wy}(k)$ can be expressed in the z-transform domain by taking the z-transform of both sides of (2.81) and (2.86). This gives

$$S_y(z) = H(z)H(z^{-1})S_w(z), \tag{2.89}$$

$$S_{wy}(z) = H(z)S_w(z), \tag{2.90}$$

where $H(z)$ is the transfer function of the system and $S_y(z)$, $S_w(z)$, and $S_{wy}(z)$ are the z-transforms of $R_y(k)$, $R_w(k)$, and $R_{wy}(k)$, respectively.

Input/Output Difference Equation Representation

A linear time-invariant discrete-time system with random input $\boldsymbol{w}(n)$ defined on the probability space S is sometimes specified by an input/output *difference equation* given by

$$\boldsymbol{y}(n) = \sum_{i=1}^{N} a_i \boldsymbol{y}(n-i) + \sum_{i=0}^{M} b_i \boldsymbol{w}(n-i). \tag{2.91}$$

In (2.91), N is a positive integer, M is a nonnegative integer, the a_i and b_i are real numbers, and $\boldsymbol{w}(n-i)$ and $\boldsymbol{y}(n-i)$ are i-step right shifts of the random signals $\boldsymbol{w}(n)$ and $\boldsymbol{y}(n)$, respectively. Note that for the system defined by (2.91), the transfer function $H(z)$ is given by

$$H(z) = \frac{b_0 + b_1 z^{-1} + \cdots + b_{M-1} z^{-M+1} + b_M z^{-M}}{1 - a_1 z^{-1} + \cdots + a_{N-1} z^{-N+1} + a_N z^{-N}}. \tag{2.92}$$

As noted previously, the system is BIBO stable if and only if the poles of $H(z)$ lie within the unit disk of the complex plane.

It is important to stress that (2.91) is a recursion of random variables, not a recursion of numbers, and thus (2.91) cannot be "evaluated." However, if $w(n)$ is a sample realization of $\boldsymbol{w}(n)$, the corresponding sample realization $y(n)$ of the output $\boldsymbol{y}(n)$ is given by

$$y(n) = \sum_{i=1}^{N} a_i y(n-i) + \sum_{i=0}^{M} b_i w(n-i). \tag{2.93}$$

If we start the recursion (2.91) at $n = 0$, we need to specify the initial conditions $\boldsymbol{y}(-1)$, $\boldsymbol{y}(-2)$, ..., $\boldsymbol{y}(-N)$, where the initial conditions are random variables defined on the probability space S. To specify the initial conditions, in general we usually give the mean and variance of the random variables.

Note that the system defined by (2.91) is causal. To consider the noncausal case, we must add left shifts to the right-hand side of (2.91). Note also that the representation (2.91) is always well-defined whether or not the system's unit-pulse response $h(n)$ is absolutely summable. This is in contrast to the representation (2.74), which is well defined in general only if $h(n)$ is absolutely summable.

Example 2.26 Mean of the Output Response

Consider the linear time-invariant system defined by the first-order recursion

$$\boldsymbol{y}(n) = a\boldsymbol{y}(n-1) + \boldsymbol{w}(n) \tag{2.94}$$

where a is a real number. The impulse response $h(n)$ of the system is equal to a^n for $n \geq 0$ and is equal to 0 for $n < 0$. Taking the expected value of both sides of (2.94), we have that $\mathrm{E}[y(n)]$ satisfies the recursion

$$\mathrm{E}[y(n)] = a\mathrm{E}[y(n-1)] + \mathrm{E}[w(n)], \tag{2.95}$$

with initial condition $\mathrm{E}[y(-1)]$. Since the mean is a real number, (2.95) is a recursion of real numbers, and thus it can be solved by iteration, which gives

$$\mathrm{E}[y(n+1)] = a^{n+1}\mathrm{E}[y(n-1)] + \sum_{i=0}^{N} a^{n-i}\mathrm{E}[w(i)]. \tag{2.96}$$

Note that if $E[w(n)] = \eta$ (a constant) for all $n \geq 0$, from (2.96) we see that $\mathrm{E}[y(n)]$ is not necessarily constant for $n \geq 0$. This result may appear to conflict with the results obtained above; that is, should not the mean of the output be constant since the mean of the input is constant and the system is time invariant? The answer is "no," and in fact there is no conflict, since here the time index n begins at $n = 0$. Recall that the above result on wide-sense stationarity of the output given wide-sense stationarity of the input was derived for the case of two-sided signals (where n ranges from $-\infty$ to ∞).

Example 2.27 Output Autocorrelation Function

Again consider the linear time-invariant system in Example 2.26 defined by (2.94), but now suppose that the random input $w(n)$ is applied to the system at time $n = -\infty$. To insure that the response $y(n)$is well defined (i.e., exists), we must require that the unit-pulse response $h(n)$ be absolutely summable, which in this case means that $|a| < 1$. Now suppose that $w(n)$ has constant mean equal to η. Then the output $y(n)$ must have constant mean, and therefore, $\mathrm{E}[y(n)] = \mathrm{E}[y(n-1)]$. Hence from (2.94), we have

$$(1-a)\mathrm{E}[y(n)] = \mathrm{E}[w(n)] = \eta,$$

or

$$\mathrm{E}[y(n)] = \frac{\eta}{1-a}, \quad \text{for } n \geq 0.$$

Note that this is consistent with the result in Example 2.24.

Given a system specified by the input/output difference equation (2.91), the autocorrelation function of the output $y(n)$ can be directly determined from (2.91). The computation is illustrated by the following example.

Example 2.28 Direct Computation of the Output Autocorrelation Function

Again consider the system given by (2.94) and suppose that the random input $\boldsymbol{w}(n)$ is WSS with autocorrelation function $R_w(k)$. Then

$$R_y(k) = \mathrm{E}[\boldsymbol{y}(n)\boldsymbol{y}(n+k)] = \mathrm{E}[\boldsymbol{y}(n)(a\boldsymbol{y}(n+k-1) + \boldsymbol{w}(n+k))]. \qquad (2.97)$$

Now $\mathrm{E}[\boldsymbol{y}(n)\boldsymbol{w}(n+k)] = \mathrm{E}[\boldsymbol{y}(k)]\mathrm{E}[\boldsymbol{w}(n+k)]$ for $k \geq 1$ since $\boldsymbol{y}(n)$ does not depend on $\boldsymbol{w}(j)$ for $j > n$. Using this in (2.97) gives

$$R_y(k) = a\mathrm{E}[\boldsymbol{y}(n)\boldsymbol{y}(n+k-1)], \quad k \geq 1.$$

Thus,

$$R_y(k) = a^k R_y(0), \quad k \geq 1. \qquad (2.98)$$

We also have

$$R_y(0) = \mathrm{E}\left[\boldsymbol{y}^2(n)\right] = \mathrm{E}\left[(a\boldsymbol{y}(n-1) + \boldsymbol{w}(n))^2\right] = a^2 R_y(0) + R_w(0).$$

Therefore,

$$R_y(0) = \left(\frac{1}{1-a^2}\right) R_w(0),$$

and using (2.98) gives

$$R_y(k) = \left(\frac{a^k}{1-a^2}\right) R_w(0), \quad k \geq 0.$$

Note that if $R_w(0) = \sigma^2$, we obtain the result in Example2.24.

State Model

An m-input, p-output linear time-invariant N-dimensional system with input $w(n)$ and output $y(n)$ may be specified by a state model given by

$$x(n+1) = \Phi x(n) + \Gamma w(n),$$
$$y(n) = Cx(n),$$

where $x(n)$ is the N-dimensional state vector, Φ is a $N \times N$ matrix, Γ is a $N \times m$ matrix, and C is a $p \times N$ matrix. For a brief treatment of the state model, see Appendix A.

If the input is now an m-element random vector signal $\boldsymbol{w}(n)$ defined on the probability space S and the initial state $\boldsymbol{x}(0)$ is defined on S, the state model becomes

$$\boldsymbol{x}(n+1) = \Phi\boldsymbol{x}(n) + \Gamma\boldsymbol{w}(n), \qquad (2.99)$$
$$\boldsymbol{y}(n) = C\boldsymbol{x}(n). \qquad (2.100)$$

In (2.99)–(2.100), the state $\boldsymbol{x}(n)$ at each time n is an N-element random vector defined on S, and the output $\boldsymbol{y}(n)$ at each time n is a p-element random vector defined on S.

The system given by (2.99)-(2.100) is often characterized in terms of the *propagation of the mean and covariance* of $\boldsymbol{x}(n)$ and $\boldsymbol{y}(n)$, starting from the initial state $\boldsymbol{x}(0)$. To determine the propagation of the mean, first take the expected value of both sides of (2.99). This gives

$$\mathrm{E}[\boldsymbol{x}(n+1)] = \Phi\mathrm{E}[\boldsymbol{x}(n)] + \Gamma\mathrm{E}[\boldsymbol{w}(n)]. \tag{2.101}$$

Solving (2.101) with initial condition $\mathrm{E}[\boldsymbol{x}(0)] = \boldsymbol{x}_0$, we have that the propagation of the mean $\mathrm{E}[\boldsymbol{x}(n)]$ is given by

$$\mathrm{E}[\boldsymbol{x}(n)] = \Phi^n\boldsymbol{x}_0 + \sum_{i=0}^{n-1} \Phi^{n-i-1}\Gamma\mathrm{E}[\boldsymbol{w}(n)], \quad \text{for } n = 1, 2, \ldots . \tag{2.102}$$

From (2.100) and (2.102), it follows easily that the propagation of the mean $\mathrm{E}[\boldsymbol{y}(n)]$ is given by

$$\mathrm{E}[\boldsymbol{y}(n)] = C\Phi^n\boldsymbol{x}_0 + \sum_{i=0}^{n-1} C\Phi^{n-i-1}\Gamma\mathrm{E}[\boldsymbol{w}(n)], \quad \text{for } n = 1, 2, \ldots . \tag{2.103}$$

If $\mathrm{E}[\boldsymbol{w}(n)] = 0$ for all n, (2.102) and (2.103) reduce to

$$\mathrm{E}[\boldsymbol{x}(n)] = \Phi^n\boldsymbol{x}_0,$$
$$\mathrm{E}[\boldsymbol{y}(n)] = C\Phi^n\boldsymbol{x}_0.$$

To determine the propagation of the covariance of $\boldsymbol{x}(n)$ and $\boldsymbol{y}(n)$, first let $P(n) = \mathrm{Cov}\,[\boldsymbol{x}(n)]$, so that by definition of the covariance [see (2.57)],

$$P(n) = \mathrm{E}\left[(\boldsymbol{x}(n) - \mathrm{E}[\boldsymbol{x}(n)])\,(\boldsymbol{x}(n) - \mathrm{E}[\boldsymbol{x}(n)])^{\mathrm{T}}\right].$$

If it is assumed that $\boldsymbol{w}(n)$ and $\boldsymbol{x}(n)$ are independent each value of n, it follows that

$$\mathrm{Cov}\,[\Phi\boldsymbol{x}(n) + \Gamma\boldsymbol{w}(n)] = \mathrm{Cov}\,[\Phi\boldsymbol{x}(n)] + \mathrm{Cov}\,[\Gamma\boldsymbol{w}(n)].$$

In addition, by (2.58)

$$\mathrm{Cov}\,[\Phi\boldsymbol{x}(n)] = \Phi\mathrm{Cov}\,[\boldsymbol{x}(n)]\,\Phi^{\mathrm{T}}, \quad \text{and} \quad \mathrm{Cov}\,[\Gamma\boldsymbol{w}(n)] = \Gamma\mathrm{Cov}\,[\boldsymbol{w}(n)]\,\Gamma^{\mathrm{T}},$$

and thus

$$\mathrm{Cov}\,[\Phi\boldsymbol{x}(n) + \Gamma\boldsymbol{w}(n)] = \Phi\mathrm{Cov}\,[\boldsymbol{x}(n)]\,\Phi^{\mathrm{T}} + \Gamma\mathrm{Cov}\,[\boldsymbol{w}(n)]\,\Gamma^{\mathrm{T}}. \tag{2.104}$$

Then taking the covariance of both sides of (2.99) and using (2.104), we have

$$\mathrm{Cov}\left[\boldsymbol{x}(n+1)\right] = \Phi \mathrm{Cov}\left[\boldsymbol{x}(n)\right]\Phi^{\mathrm{T}} + \Gamma \mathrm{Cov}\left[\boldsymbol{w}(n)\right]\Gamma^{\mathrm{T}},$$

or

$$P(n+1) = \Phi P(n)\Phi^{\mathrm{T}} + \Gamma \mathrm{Cov}\left[\boldsymbol{w}(n)\right]\Gamma^{\mathrm{T}}. \tag{2.105}$$

The propagation of $P(n)$ can then be determined by solving (2.105) with initial condition $P(0) = P_0$. This gives

$$P(n) = \Phi^n P_0 \left(\Phi^{\mathrm{T}}\right)^n + \sum_{i=0}^{n-1} \Phi^i \Gamma \mathrm{Cov}\left[\boldsymbol{w}(i)\right]\Gamma^{\mathrm{T}} \left(\Phi^{\mathrm{T}}\right)^i, \quad \text{for } n = 1,\, 2,\, \ldots\,. \tag{2.106}$$

Using (2.100) and (2.106), we have that the propagation of $\mathrm{Cov}\left[\boldsymbol{y}(n)\right]$ is given by

$$\mathrm{Cov}\left[\boldsymbol{y}(n)\right] = C\Phi^n P_0 \left(\Phi^{\mathrm{T}}\right)^n C^{\mathrm{T}} + \sum_{i=0}^{n-1} C\Phi^i \Gamma \mathrm{Cov}\left[\boldsymbol{w}(i)\right]\Gamma^{\mathrm{T}} \left(\Phi^{\mathrm{T}}\right)^i C^{\mathrm{T}}, \quad \text{for } n = 1,\, 2,\, \ldots\,. \tag{2.107}$$

To conclude, suppose that the output equation (2.100) is modified to include an additive noise term $\boldsymbol{v}(n)$ defined on the probability space S; that is, (2.100) is replaced by

$$\boldsymbol{z}(n) = C\boldsymbol{x}(n) + \boldsymbol{v}(n).$$

Then if $\boldsymbol{x}(n)$ and $\boldsymbol{v}(n)$ are independent for each value of n, using (2.107) we have

$$\mathrm{Cov}\left[\boldsymbol{z}(n)\right] = C\Phi^n P_0 \left(\Phi^{\mathrm{T}}\right)^n C^{\mathrm{T}} + \sum_{i=0}^{n-1} C\Phi^i \Gamma \mathrm{Cov}\left[\boldsymbol{w}(i)\right]\Gamma^{\mathrm{T}} \left(\Phi^{\mathrm{T}}\right)^i C^{\mathrm{T}} + \mathrm{Cov}\left[\boldsymbol{v}(n)\right]. \tag{2.108}$$

Problems

2.1. This problem examines the distribution of sequences produced by MATLAB's random number generator. Consult the MATLAB documentation on the functions `rand` and `randn`, or run MATLAB and enter `help rand` and `help randn` to obtain a description of the random number generator.

(a) Use `randn` to create a 250-element sample realization of white Gaussian noise with zero-mean and unit variance. Call the realization `w` and plot it.

(b) The following MATLAB code plots a histogram of the values in w and superimposes the PDF of a Gaussian.

```
Nbins = 50;                             % Number of
bins = linspace(-4, 4, Nbins);          %  histogram bins.
[n,x] = hist(w, bins);                  % Make histogram.
i = min(find(bins>=0)) + [-1 0 1];      % Get center bins.
hold off                                % Plot histogram
bar(x, n/mean(n(i))*0.3989)             %  & compare with
hold on                                 %  a Gaussian.
y = linspace(-4, 4, 100);               % Generate pdf for
f = exp(-(y.^2)/2)/sqrt(2*pi);          %  Gaussian(0,1).
plot(y, f, '--')                        % Plot pdf.
hold off
```

Enter this code directly or write it as a MATLAB script and try it for the sample realization w. How well does MATLAB simulate the desired Gaussian process?

(c) Next use rand to create a 250-element sample realization of a random process uniformly distributed from 0 to 1. Call this realization v and plot it.

(d) The following MATLAB code plots a histogram and the PDF of the desired uniform random process.

```
Nbins = 20;                             % Number of
bins = linspace(0, 1, Nbins);           %  histogram bins.
[n,x] = hist(v, bins);
hold off                                % Plot normalized
axis([-0.2 1.2 0 1.1*max(n)/(2*mean(n))])
bar(x, n/(2*mean(n)))                   %  histogram.
hold on                                 % Plot unif[0,1]
plot([0 0 1 1], [0 0.5 0.5 0], '--')%  pdf.
hold off
```

Try this code with v. How well does the sample realization approximate the uniform random process?

2.2. This problem looks for correlation between element of a sample realization created by randn. Use randn to make x, a 50-element sample realization of a zero-mean, unit-variance Gaussian random process.

(a) Use the function mean to estimate the mean of x. Note that this mean is an ensemble average,

$$\text{estimated } \mathrm{E}\left[x(n)\right] = \hat{\eta}_x = \frac{1}{N}\sum_{n=0}^{N-1} x(n),$$

where N is the number of samples in the sample realization. (Since the $x(n)$ are elements of a sample realization, they are no longer random and are not boldfaced.)

(b) Estimate the variance of $x(n)$ from the ensemble average

$$\text{estimated } \sigma_x^2 = \hat{\sigma}_x^2 = \frac{1}{N}\sum_{n=0}^{N} [x(n) - \hat{\eta}_x]^2 .$$

(c) Use the MATLAB function **cov** to estimate the variance of **x**. Note that **cov** uses the alternative estimate,

$$\text{estimated } \sigma_x^2 = \hat{\sigma}_x'^2 = \frac{1}{N-1}\sum_{n=0}^{N-1} [x(n) - \hat{\eta}_x]^2 .$$

2.3. MATLAB's **randn** and **rand** functions simulate Gaussian and uniform random processes, with fixed means and variances. By shifting and scaling these processes, we can generate a Gaussian or uniform random process with arbitrary parameters.

(a) Use **randn** to generate **w**, a 50-element sample realization of a Gaussian random process with mean 8 and variance 30. Plot **w** and check the ensemble average and variance.

(b) Use **rand** to create a 50-element sample realization **v** of a random process uniformly distributed from -6 to 10. Plot **v** and check its mean, range, and variance.

2.4. Use **randn** to make **x**, a 50-element sample realization of a zero-mean, unit-variance Gaussian random process.

(a) The autocorrelation function of $x(n)$ may be approximated using the ensemble average

$$\text{estimated } R_x(k) = \frac{1}{N}\sum_{i=0}^{N-1} x(i)x(i-k) = \frac{1}{N}x(k) * x(-k).$$

Consult the MATLAB documentation for information on the functions **conv**, **fliplr**, and **flipud**. Use these functions to estimate the autocorrelation function $R_x(k)$ and call the estimate **Rx**. Plot **Rx** versus the time lag k, which ranges from $-(N-1)$ to $N-1$. What is the variance σ_x^2 estimated from **Rx**?

(b) Write a MATLAB function called **crosscorr** that computes the cross-correlation between random processes $x(n)$ and $y(n)$ from two sample realizations **x** and **y**. The function header for this function is given below.

```
function C = crosscorr(x, y)
% CROSSCORR  Estimate cross-correlation function using
%            ensemble averages.
% C = crosscorr(x) returns the autocorrelation
%         function for the sample realization vector x.
% C = crosscorr(x, y) returns the cross-correlation
%         function for the vectors x and y.
```

If only one vector is given, crosscorr should return an estimate of the autocorrelation function. (The MATLAB variable nargin indicates the number of arguments passed to a function.)

(c) The following function produces an estimate of the power density spectrum $S_x(e^{j\omega})$ from Rx.

```
function [W, S] = ccorr2pspec(R)
% CCORR2PSPEC  Convert cross-correlation to power
%              density spectrum
% ccorr2pspec(R) does not return anything but plots
%      the spectrum.
% S = ccorr2pspec(R) returns the spectrum.
% [W, S] = ccorr2pspec(R) returns a frequency vector
%      and the spectrum.  Use plot(W,S) to view the
%      spectrum.
% JKS 19 Sep 1993
L = length(R);
inW = 2*pi*[0:(L-1)]/L;
inS = real(fft(rotate(R, (L+1)/2)));
if nargout == 0,
   plot(inW, inS)
   xlabel('frequency w (radians)')
   ylabel('S(w)')
   title('Estimated power density spectrum')
elseif nargout == 1,
   W = inW;
else
   W = inW;
   S = inS;
end
```

Enter this function and use it to obtain W and S. Plot the estimated spectrum. What is the average value of S? Based on the results of this problem, how closely does MATLAB simulate zero-mean white noise?

2.5. This problem demonstrates how MATLAB can be used to create colored processes from white noise. Let $s(n)$ be a random process described by the difference equation

$$s(n) = 0.8s(n-1) - 0.5s(n-2) + y(n) + 1.25y(n-1),$$

where $y(n)$ is zero-mean white Gaussian noise with variance $\sigma_y^2 = 8$.

(a) Find the system function $H(z) = S(z)/Y(z)$ and the power density spectrum $S_s(e^{j\omega})$.

(b) Use randn to generate a 50-element sample realization y of $y(n)$.

(c) Estimate the mean, variance, and autocorrelation function $R_y(k)$ from y. Let Ry be the estimate of $R_y(k)$. [See Problems 2.2 and 2.4 for more details.]

(d) Use `ccorr2pspec` [see p. 64] to estimate the power density spectrum $S_y(e^{j\omega})$. Call the estimate `Sy`.

(e) Use the MATLAB function `filter` to simulate the system described by $H(z)$ with `y` as the input. Let `s` be the output of $H(z)$.

(f) Estimate the mean, variance, and $R_s(k)$ from `s`. Also estimate `Ss`, the power density spectrum $S_s(e^{j\omega})$. Plot the actual power density spectrum against `Ss`. How well does MATLAB simulate the desired process $s(n)$?

2.6. Given a probability space S and a "deterministic number" c (i.e., c is a real number), it is possible to embed c into a stochastic setting by defining the RV $\boldsymbol{c}(\alpha) = c$ for all $\alpha \in S$.

(a) Determine the probability density function $f_{\boldsymbol{c}}(c)$ and the probability distribution function $F_{\boldsymbol{c}}(c)$ of the RV $\boldsymbol{c}$.

(b) Compute $\mathrm{E}[\boldsymbol{c}]$, $\mathrm{Var}[\boldsymbol{c}]$, and $\mathrm{E}\left[\boldsymbol{c}^2\right]$.

(c) Now define the RV $\boldsymbol{y} = \boldsymbol{c} - c$.

(i) Determine $f_{\boldsymbol{y}}(c)$ and $F_{\boldsymbol{y}}(c)$.

(ii) Compute $\mathrm{E}[\boldsymbol{y}]$, $\mathrm{Var}[\boldsymbol{y}]$, and $\mathrm{E}\left[\boldsymbol{y}^2\right]$.

(iii) Does the RV $\boldsymbol{y}$ correspond to a deterministic number? If so, what number?

2.7. Consider the experiment of measuring a voltage with $S = \{v : 0 \leq v \leq 1\}$. Define the probabilities of the events by $P\{v_1 \leq v \leq v_2\} = (v_2 - v_1)(v_1 + v_2)$.

(a) Verify that S is a probability space with the above definition of probabilities.

(b) For the probability space defined above, let $\boldsymbol{x}$ denote the RV defined by $\boldsymbol{x}(v) = v$ for all $v \in S$.

(i) Compute $\mathrm{E}[\boldsymbol{x}]$, $\mathrm{Var}[\boldsymbol{x}]$, and $\mathrm{E}\left[\boldsymbol{x}^2\right]$.

(ii) Determine the probability density function $f_{\boldsymbol{x}}(x)$.

2.8. A RV $\boldsymbol{x}$ is uniformly distributed between a and b, where $a < b$.

(a) Show that $\mathrm{E}[\boldsymbol{x}] = (b - a)/2$.

(b) Show that $E\left[\boldsymbol{x}^2\right] = (b - a)^2/12$.

2.9. A RV $\boldsymbol{x}$ has the probability density function

$$f_{\boldsymbol{x}}(x) = \begin{cases} Ae^{-2x}, & x \geq 0; \\ 0, & x < 0. \end{cases}$$

(a) Compute A.

(b) Compute $\mathrm{E}[\boldsymbol{x}]$, $\mathrm{Var}[\boldsymbol{x}]$, and $\mathrm{E}\left[\boldsymbol{x}^2\right]$.

(c) Compute $P\{1 < \boldsymbol{x} \leq 3\}$.

2.10. A RV $\boldsymbol{x}$ has the sample values 1, 2, 4, 8, 1, 3, 6, 1, 2, 5, 1, 2.

(a) Determine approximate values for $\mathrm{E}[\boldsymbol{x}]$, $\mathrm{Var}[\boldsymbol{x}]$, and $\mathrm{E}\left[\boldsymbol{x}^2\right]$.

(b) Determine an upper bound on the probability that

(i) $\boldsymbol{x} \geq 9$ or $\boldsymbol{x} \leq -9$;

(ii) $\boldsymbol{x} \geq 5$ or $\boldsymbol{x} \leq -5$.

2.11. The RV $\boldsymbol{z}$ is given by $\boldsymbol{z} = \boldsymbol{x} + \boldsymbol{v}$ where $\boldsymbol{x}$ is a RV that is uniformly distributed between 0 and 4, $\boldsymbol{v}$ is a RV that is uniformly distributed from -1 to 1, and $\boldsymbol{x}$ and $\boldsymbol{v}$ are independent, jointly distributed RVs.

(a) Compute $\mathrm{E}[\boldsymbol{z}]$, $\mathrm{Var}[\boldsymbol{z}]$, and $\mathrm{E}\left[\boldsymbol{z}^2\right]$.

(b) Determine the probability density function $f_{\boldsymbol{z}}(z)$. Express $f_{\boldsymbol{z}}(z)$ in mathematical form and sketch it.

2.12. Repeat Problem 2.11 but now suppose that $\boldsymbol{x}$ and $\boldsymbol{v}$ are independent, jointly distributed Gaussian RVs, where $\boldsymbol{x}$ has mean 1 and variance 2, and $\boldsymbol{v}$ has mean 0 and variance 1.

2.13. The RV $\boldsymbol{z}$ is given by $\boldsymbol{z} = \boldsymbol{x} + \boldsymbol{v}$ where $\boldsymbol{x}$ is uniformly distributed from 0 to 2, $\boldsymbol{v}$ is Gaussian with mean 0 and variance 1, and $\boldsymbol{x}$ and $\boldsymbol{v}$ are independent, jointly distributed RVs.

(a) Compute $P\{\boldsymbol{x} \leq 0.5\}$, $P\{\boldsymbol{v} \leq 0.5\}$, $P\{\boldsymbol{z} \leq 0.5 | \boldsymbol{x} = 1\}$, and $P\{\boldsymbol{z} \leq 0.5 | \boldsymbol{v} = 1\}$.

(b) Compute $\mathrm{E}[\boldsymbol{z}]$, $\mathrm{Var}[\boldsymbol{z}]$, and $\mathrm{E}\left[\boldsymbol{z}^2\right]$.

2.14. Consider the signal plus noise $\boldsymbol{z} = \boldsymbol{s} + \boldsymbol{v}$, where $\boldsymbol{s} \sim \mathcal{N}\left(\eta, \sigma^2\right)$ and $\boldsymbol{v}$ is the RV with the exponential density function $f_{\boldsymbol{v}}(v) = 5e^{-5v}1(v)$, where $1(v) = 1$ for $v \geq 0$ and $1(v) = 0$ for $v < 0$. Determine the conditional density function $f_{\boldsymbol{z}}(z|\boldsymbol{s} = s)$. Express your answer in mathematical form.

2.15. Suppose that $\boldsymbol{z} = \boldsymbol{s} + \boldsymbol{v}$, where $\boldsymbol{s}$ and $\boldsymbol{v}$ are independent, jointly distributed RVs with $\boldsymbol{s} \sim \mathcal{N}\left(\eta, \sigma^2\right)$ and $\boldsymbol{v} \sim \mathcal{N}\left(0, V^2\right)$.

(a) Derive an expression for $\mathrm{E}\left[\boldsymbol{s} | \boldsymbol{z} = z\right]$.

(b) Derive an expression for $\mathrm{E}\left[\boldsymbol{s}^2 | \boldsymbol{z} = z\right]$.

2.16. Any deterministic signal $x(n)$ can be embedded into a random signal framework by defining the random signal $\boldsymbol{x}(n)$ by $[\boldsymbol{x}(n)](\alpha) = x(n)$ for all $\alpha \in S$, where S is some probability space.

(a) For each integer value of i, determine the density function and distribution function of the RV $\boldsymbol{x}(i)$ comprising the random signal $\boldsymbol{x}(n)$.

(b) For each integer value of i, compute $\mathrm{E}[\boldsymbol{x}(i)]$, $\mathrm{E}\left[(\boldsymbol{x}(i))^2\right]$, and $\mathrm{Var}\left[\boldsymbol{x}(i)\right]$.

(c) Determine the joint distribution function for the RVs $\boldsymbol{x}(i)$ and $\boldsymbol{x}(j)$ comprising $\boldsymbol{x}(n)$ for all integer values of i and j with $i \neq j$. Express your answer in mathematical form.

(d) Based on your result in Part (c), are the RVs $\boldsymbol{x}(i)$ and $\boldsymbol{x}(j)$ independent for all $i \neq j$? Are they uncorrelated? Justify your answers.

(e) Compute the autocorrelation function $R_x(i,j)$ for all integers i, j. Express your answer in mathematical form.

2.17. Consider the signal plus noise $z(n) = s + v(n)$, where s is a RV with $\text{E}[s] = 1$, $\text{E}[s^2] = 2$, and for each value of n, $v(n) \sim \mathcal{N}(0,1)$. It is known that $\text{E}[sw(i)] = 1$ for all i, and $w(i)$ is independent of $w(j)$ for all $i \neq j$.

(a) Compute the autocorrelation function $R_z(i,j)$ of $z(n)$ for all integers i, j.

(b) Is $z(n)$ WSS? If so, derive a mathematical expression for $R_z(k)$.

2.18. Apply the random signal $z(n)$ defined in Problem 2.17 to a filter given by the input/output difference equation

$$\beta(n) = z(n+1) + z(n),$$

where $\beta(n)$ is the output of the filter.

(a) Compute the autocorrelation function $R_\beta(i,j)$ of the output $\beta(n)$ for all integers i, j.

(b) If $\beta(n)$ is WSS, derive a mathematical express for $R_\beta(k)$.

2.19. Let $w(n)$ be zero-mean white noise with variance σ^2, and suppose that $w(n)$ is applied to the q-step mean filter whose output $y(n)$ is given by

$$y(n) = \frac{1}{q} \sum_{i=n-q+1}^{n} w(i).$$

(a) Derive an expression for $E\left[(y(n))^2\right]$ for all n.

(b) What does your result in Part (a) imply about the filter? Explain.

(c) Determine the autocorrelation function $R_y(i,j)$. If $y(n)$ is WSS, determine $R_y(k)$.

2.20. A WSS random signal $x(n)$ is applied to a causal, linear time-invariant, discrete-time system with impulse response $h(n) = (0.25)^n$ for $n \geq 0$, $h(n) = 0$ for $n < 0$. The autocorrelation function $R_x(k)$ of $x(n)$ is given by $R_x(k) = 1 + (0.5)^{|k|}$ for $k = 0, \pm 1, \pm 2, \ldots$. Let $y(n)$ denote the output resulting from $x(n)$ with no initial energy.

(a) Compute $\text{E}[y(n)]$ for all n.

(b) Compute $\text{E}\left[(y(n))^2\right]$ for all n.

2.21. A zero-mean white noise signal $w(n)$ with variance 1 is applied to each of the systems defined below. For each part given below, determine the autocorrelation function $R_y(k)$ of the output $y(n)$. Express $R_y(k)$ in mathematical form and sketch the result.

(a) $y(n) = -0.25y(n-1) + w(n)$.

(b) $y(n) = -0.5y(n-1) + w(n) - w(n-1)$.

(c) $y(n) = -y(n-1) - 0.25y(n-2) + w(n)$.

2.22. For each of the parts in Problem 2.21, determine the power spectral density $S_y(z)$ of the output $y(n)$. Express $S_y(z)$ in mathematical form.

2.23. Zero-mean white noise $w(n)$ with variance 1 is applied to a causal, linear time-invariant discrete-time system. The power spectral density $S_y\left(e^{j\omega}\right)$ of the resulting output $y(n)$ is given by

$$S_y\left(e^{j\omega}\right) = \frac{9 + 9\cos\omega}{1.25 + \cos 4\omega}.$$

Determine the input/output difference equation of the system. All coefficients in the input/output difference equation must be completely evaluated (in numerical form).

2.24. A WSS random signal $w(n)$ with autocorrelation function

$$R_w(k) = (0.5)^{|k|}, \quad \text{for } k = 0, \pm 1, \pm 2, \dots,$$

is applied to a causal linear time-invariant discrete-time system. The z-transform $R_y(z)$ of the autocorrelation function $R_y(k)$ of the output $y(n)$ is given by

$$R_y(z) = \frac{0.75}{(1 - 0.25z^{-1})(1 - 0.5z)(1 - 0.25z)(1 - 0.5z^{-1})}.$$

Determine the input/output difference equation of the system. All coefficients in the input/output difference equation must be completely evaluated (in numerical form).

2.25. A zero-mean white noise signal $w(n)$ with variance 1 is applied to each of the systems defined below. For each of the parts below, derive an expression for $\mathrm{E}[y(n)]$ and $\mathrm{E}\left[(y(n))^2\right]$ for all integers $n \geq 0$.

(a) $y(n) = -0.75y(n-1) + w(n)$, $\mathrm{E}[y(-1)] = c$, and $E\left[(y(-1))^2\right] = b$.

(b) $y(n) = -0.5y(n-2) + w(n) + w(n-1)$, $\mathrm{E}[y(-2)] = 0$, $\mathrm{E}[y(-1)] = c$, $\mathrm{E}\left[(y(-2))^2\right] = b_2$, and $\mathrm{E}\left[(y(-1))^2\right] = b_1$.

2.26. A linear time-invariant discrete-time system has the state model

$$\begin{aligned} x_1(n+1) &= -0.5x_2(n) + w_1(n), \\ x_2(n+1) &= 0.75x_1(n) + w_2(n), \\ y(n) &= x_1(n) + x_2(n), \end{aligned}$$

where $w_1(n)$ and $w_2(n)$ are independent of $x_1(0)$ and $x_2(0)$ for all $n > 0$, $w_1(n)$ and $w_2(n)$ are white noises with $\mathrm{E}[w_1(n)] = \mathrm{E}[w_2(n)] = 2$ for $n \geq 0$, and $\mathrm{E}\left[(w_1(-1))^2\right] = \mathrm{E}\left[(w_2(-1))^2\right] = 5$ for $n \geq 0$.

(a) Assuming that $\mathrm{E}[x_1(0)] = \mathrm{E}[x_2(0)] = 1$, compute $E[x_1(n)]$ and $\mathrm{E}[x_2(n)]$ for $n = 1$ and 2. Give your answers in numerical form (all constants must be evaluated).

(b) Compute $\mathrm{Var}[x_1(n)]$ and $\mathrm{Var}[x_2(n)]$ for $n = 1$ and 2 assuming that $\mathrm{E}[x_1(0)] = \mathrm{E}[x_2(0)] = 1$, and $\mathrm{Var}[x_1(0)] = \mathrm{Var}[x_2(0)] = 2$. Give your answers in numerical form (all constants must be evaluated).

Chapter 3

Optimal Estimation

The preceding chapters provide the background necessary to introduce the optimal estimation problem. An "optimal estimate" is a best guess. However, we may express the "goodness" of an estimate in different ways, depending upon the particular engineering problem. After presenting the basic optimal estimation problem and some desirable properties of an estimate, we introduce three commonly-used optimality criterion: the maximum-likelihood, maximum *a posteriori* , and minimum mean-square error criteria. Each leads to a different estimate and a different form for the estimator. The estimators we discuss are typically implemented in digital systems, so we restrict ourselves to discrete-time signals and systems. Finally, we compare and contrast the different approaches.

3.1 Formulating the Problem

With $s(n)$ equal to a sampled version of a signal $s(t)$, as first considered in Chapter 1, suppose that we have measurements $z(n)$ of $s(n)$ given by

$$z(n) = g(s(n), v(n), n),$$

where $v(n)$ is a noise signal and g is a function that represents the degradation of $s(n)$ in generating the measurement $z(n)$. The goal is to compute the "optimal estimate" of $s(n)$ at time n based on the measurements or observations $z(1)$, $z(2)$, ... , $z(n)$. Denoting the estimate by $\hat{s}(n)$, we can express $\hat{s}(n)$ in the form

$$\hat{s}(n) = \alpha_n(z(1), z(2), \ldots, z(n)) \tag{3.1}$$

where α_n is some function that generally depends on n. The dependence of α on n implies that the estimator defined by (3.1) may be time varying. The estimator may also be nonlinear if α_n is a nonlinear function of $z(1)$, $z(2)$, $\ldots$, $z(n)$.

To handle the uncertainty that is present and to be able to define an appropriate optimality criterion, we model $s(n)$ and $v(n)$ as random signals $\boldsymbol{s}(n)$ and $\boldsymbol{v}(n)$, respectively, defined on a probability space S, so that for each n, $\boldsymbol{s}(n)$ and $\boldsymbol{v}(n)$ are jointly-distributed RVs. Then the measurement $z(n)$ becomes a random signal $\boldsymbol{z}(n)$ defined on S and given by

$$\boldsymbol{z}(n) = g(\boldsymbol{s}(n), \boldsymbol{v}(n), n),$$

and the estimate $\hat{s}(n)$ becomes a random signal $\hat{\boldsymbol{s}}(n)$ defined on S given by

$$\hat{\boldsymbol{s}}(n) = \alpha_n(\boldsymbol{z}(1), \boldsymbol{z}(2), \ldots, \boldsymbol{z}(n)). \tag{3.2}$$

Given sample realizations $z(i)$ of $\boldsymbol{z}(i)$ for $i = 1, 2, \ldots, n$, the estimate $\hat{s}(n)$ of the signal realization $s(n)$ at time n is defined by (3.1). Note that we use non-bold notation to indicate sample values.

The random signal representation of the estimate given by (3.2) allows us to define various optimality criteria, which are presented later. Below we formalize the problem we wish to solve in terms of the random signal framework.

Optimal Estimation Problem

> *Given the measurements $\boldsymbol{z}(1)$, $\boldsymbol{z}(2)$, $\ldots$, $\boldsymbol{z}(n)$, the corruption function g, and an optimality criterion, design an estimator that generates an optimal estimate $\hat{\boldsymbol{s}}(n)$ of $\boldsymbol{s}(n)$ given by*
>
> $$\hat{\boldsymbol{s}}(n) = \alpha_n(\boldsymbol{z}(1), \boldsymbol{z}(2), \ldots, \boldsymbol{z}(n)),$$
>
> *for some function α_n.*

Prediction, Filtering, and Smoothing

So far, we have been concerned with estimating the signal process $\boldsymbol{s}(n)$ at time n based on the measurements $\boldsymbol{z}(1)$, $\boldsymbol{z}(2)$, $\ldots$, $\boldsymbol{z}(n)$. However, we may want to estimate $\boldsymbol{s}(n+1)$, based on the same set of measurements; that is, we may want to *predict* the next value of the signal. Or perhaps we want to estimate $\boldsymbol{s}(n-1)$ from this set of measurements, which is known as *smoothing.* In each case we are still estimating the signal based on the measurements $\boldsymbol{z}(1)$, $\boldsymbol{z}(2)$, $\ldots$, $\boldsymbol{z}(n)$, but the time index of the estimate has changed. We now summarize the terminology that describes some common types of estimates.

- *Filtering:* In this case we estimate the signal process $\boldsymbol{s}(n)$ at time n based on $\boldsymbol{z}(1)$, $\boldsymbol{z}(2)$, $\ldots$, $\boldsymbol{z}(n)$.

ℓ	Estimate	Estimation Problem
n	$\hat{s}(n)$	filtering
$n+1$	$\hat{s}(n+1)$	one-step prediction
$n+m, m > 0$	$\hat{s}(n+m)$	m-step prediction
$n-1$	$\hat{s}(n-1)$	smoothing with lag 1
$n-m, m > 0$	$\hat{s}(n-m)$	smoothing with lag m
m constant	$\hat{s}(m)$	fixed-point smoothing

Table 3.1. Types of Estimation Problems

- *Prediction:* In this case we estimate the signal process $s(n)$ at a future time point, beyond the time frame of the observations $z(1)$, $z(2)$, ..., $z(n)$. For example, we estimate $s(n+1)$ or $s(n+20)$.
- *Smoothing:* Here we estimate $s(n)$ at a time point before n and use our subsequent observations up to time n to "smooth" the estimation error. For example, we estimate $s(n-1)$ or $s(n-16)$.

We may generalize the estimation problem to include filtering, prediction, and smoothing in the following way: "Given observations $\{z(1), \ldots, z(n)\}$ make the best guess of the value of $s(\ell)$," where we are free to choose ℓ to fit the type of estimation problem. Several common choices for ℓ are given in Table 3.1.

Properties of Estimates

To evaluate an estimator, we now present a few properties that describe how an estimator may behave. One of the most intuitively appealing properties of $\hat{s}(n)$ is that the mean of $\hat{s}(n)$ should equal the true mean of $s(n)$. That is,

$$\mathrm{E}\left[\hat{s}(n)\right] = \mathrm{E}\left[s(n)\right]. \tag{3.3}$$

If we denote the *estimation error* by

$$\tilde{s}(n) = s(n) - \hat{s}(n), \tag{3.4}$$

this property becomes

$$\mathrm{E}\left[\tilde{s}(n)\right] = 0. \tag{3.5}$$

In other words, we would like the expected value of the estimation error to be zero. If our estimate satisfies Condition (3.3) or (3.5), we say that $\hat{s}(n)$ is an *unbiased estimate*, and the estimator α is an *unbiased estimator*.

For estimation conducted over a period of time, we say the estimates are *asymptotically unbiased* if

$$\lim_{n\to\infty} \mathrm{E}\left[\hat{s}(n)\right] = \mathrm{E}\left[s(n)\right]. \tag{3.6}$$

That is, the estimate becomes unbiased in the limit, a weaker property than if it were simply unbiased.

We may also examine the mean square error (MSE), given by $\mathrm{E}\left[\tilde{s}^2(n)\right]$. If the MSE satisfies

$$\lim_{n\to\infty} \mathrm{E}\left[\tilde{s}^2(n)\right] = 0, \tag{3.7}$$

then we say that the estimator that generates $\hat{s}(n)$ is *consistent*.

Finally, observe that if an estimator is both (asymptotically) unbiased and consistent, then $\mathrm{E}\left[\tilde{s}^2\right] \to 0$ as $n \to \infty$. Both the average error and the MSE tend to zero as n becomes large. In this sense, we have a perfect estimate of $s(n)$ as $n \to \infty$.

Example 3.1 The Mean Filter

Consider the case of a constant signal s in zero-mean, additive noise $v(n)$. Hence, $s(n) = s$ and $z(n) = s + v(n)$. Also assume that s and $v(n)$ are independent. Next consider the mean filter, or averaging filter:

$$\hat{s}(n) = \frac{1}{n}\sum_{j=1}^{n} z(j).$$

We show that this estimator is unbiased and consistent.

First, to see that it is unbiased, write

$$\begin{aligned}
\mathrm{E}\left[\hat{s}(n)\right] &= \frac{1}{n}\sum_{j-1}^{n} \mathrm{E}\left[z(j)\right] \\
&= \frac{1}{n}\sum_{j-1}^{n} \left(\mathrm{E}\left[s\right] + \mathrm{E}\left[v(n)\right]\right) \\
&= \frac{1}{n}\sum_{j-1}^{n} \mathrm{E}\left[s\right] \\
&= \mathrm{E}\left[s\right].
\end{aligned}$$

Second to show that it is consistent, we have

$$\begin{aligned}
\mathrm{E}\left[\tilde{s}^2(n)\right] &= \mathrm{E}\left[(s - \hat{s}(n))^2\right] \\
&= \mathrm{E}\left[s^2 - 2s\hat{s}(n) + \hat{s}^2(n)\right].
\end{aligned}$$

Now

$$\mathrm{E}\left[\boldsymbol{s}\hat{\boldsymbol{s}}(n)\right] = \frac{1}{n}\sum_{j=1}^{n} \mathrm{E}\left[\boldsymbol{s}\boldsymbol{z}(j)\right] = \frac{1}{n}\sum_{j=1}^{n} \mathrm{E}\left[\boldsymbol{s}^2 | \boldsymbol{s}\boldsymbol{v}(j)\right] = \mathrm{E}\left[\boldsymbol{s}^2\right].$$

Also,

$$\begin{aligned}\mathrm{E}\left[\hat{\boldsymbol{s}}^2(n)\right] &= \frac{1}{n^2}\mathrm{E}\left[\left(\boldsymbol{z}(1) + \boldsymbol{z}(2) + \ldots \boldsymbol{z}(n)\right)\left(\boldsymbol{z}(1) + \boldsymbol{z}(2) + \ldots \boldsymbol{z}(n)\right)\right] \\ &= \frac{1}{n^2}\left[\left(n\mathrm{E}\left[\boldsymbol{s}^2\right] + \sigma_v^2\right) n\right].\end{aligned}$$

Therefore, we obtain

$$\mathrm{E}\left[\tilde{\boldsymbol{s}}^2(n)\right] = \mathrm{E}\left[\boldsymbol{s}^2\right] - 2\mathrm{E}\left[\boldsymbol{s}^2\right] + \mathrm{E}\left[\boldsymbol{s}^2\right] + \frac{\sigma_v^2}{n} = \frac{\sigma_v^2}{n},$$

and clearly the right-hand side approaches zero as n becomes large.

Recalling the optimal estimation problem stated above, we have not yet specified the measure of optimality, or "goodness." This generality is intentional because optimality may be specified in a variety of ways. We present several common optimality criteria in the following sections. Depending upon the application, one criterion may be preferred over the others.

3.2 Maximum Likelihood and Maximum *a posteriori* Estimation

Maximum Likelihood Estimation

Given a RV $\boldsymbol{x}$ with a unimodal[1] probability density function (PDF) $f_{\boldsymbol{x}}(x)$, we recall from Chapter 2 that the most-probable or "most-likely" value of $\boldsymbol{x}$ corresponds to the peak of $f_{\boldsymbol{x}}(x)$. That is,

$$\text{most-likely value of } \boldsymbol{x} = \text{value of } x \text{ that maximizes } f_{\boldsymbol{x}}(x).$$

By the same reasoning, suppose we have a single measurement z, a sample realization from $\boldsymbol{z} = g(\boldsymbol{s}, \boldsymbol{v})$. Then it is natural to estimate $\boldsymbol{s}$ by finding the value of s that is most likely to have produced z. We treat the conditional density $f_{\boldsymbol{z}}(z|\boldsymbol{s} = s)$ as a function of s and call it the *likelihood function*. We then seek the value of s that maximizes the likelihood function, and this estimation method is known as *maximum likelihood* (ML) estimation. We denote the ML estimate by $\hat{s}_{\mathrm{ML}}$:

$$\hat{s}_{\mathrm{ML}} = \text{value of } s \text{ that maximizes } f_{\boldsymbol{z}}(z|\boldsymbol{s} = s). \tag{3.8}$$

[1] That is, the PDF has a unique maximum.

Let us assume that the likelihood function is differentiable with a unique maximum in the interior of its domain. Then

$$\hat{s}_{\text{ML}} = \text{value of } s \text{ for which } \frac{\partial f_{\boldsymbol{z}}(z|\boldsymbol{s}=s)}{\partial s} = 0, \tag{3.9}$$

where z is the sample realization of $\boldsymbol{z}$. Since the natural logarithm is a monotonically increasing function, we may equivalently maximize $\ln(f_{\boldsymbol{z}}(z|\boldsymbol{s}=s))$, which is called the log-likelihood function. In this case we have

$$\hat{s}_{\text{ML}} = \text{value of } s \text{ for which } \frac{\partial \ln f_{\boldsymbol{z}}(z|\boldsymbol{s}=s)}{\partial s} = 0, \tag{3.10}$$

For either case, we can write $\hat{s}_{\text{ML}}$ in the form

$$\hat{s}_{\text{ML}} = \alpha(z) \tag{3.11}$$

for some function $\alpha(z)$.

We should emphasize that in ML estimation, $f_{\boldsymbol{z}}(z|\boldsymbol{s}=s)$ is *not* a density. Normally the density $f_{\boldsymbol{z}}(z|\boldsymbol{s}=s)$ is regarded as a function of z. Since we are now considering $f_{\boldsymbol{z}}(z|\boldsymbol{s}=s)$ as a function of s rather than z, strictly speaking, $f_{\boldsymbol{z}}(z|\boldsymbol{s}=s)$ is no longer a density. Hence the term "likelihood function."

Example 3.2 ML Estimation

Suppose $\boldsymbol{s}$ and $\boldsymbol{z}$ are random variables with joint PDF

$$f_{\boldsymbol{s},\boldsymbol{z}}(s,z) = \begin{cases} \frac{1}{12}(s+z)e^{-z}, & 0 \le s \le 4, \quad 0 \le z < \infty; \\ 0, & \text{otherwise}; \end{cases}$$

and the goal is to compute the ML estimate of $\boldsymbol{s}$ based on $\boldsymbol{z}$.

To find the likelihood function, we use the relationship

$$f_{\boldsymbol{z}}(z|\boldsymbol{s}=s) = \frac{f_{\boldsymbol{s},\boldsymbol{z}}(s,z)}{f_{\boldsymbol{s}}(s)},$$

and recall that

$$\begin{aligned} f_{\boldsymbol{s}}(s) &= \int_{-\infty}^{\infty} f_{\boldsymbol{s},\boldsymbol{z}}(s,z)\,dz = \int_0^{\infty} \tfrac{1}{12}(s+z)e^{-z}\,dz \\ &= \tfrac{1}{12}\left[-se^{-z} + (-z-1)e^{-z}\right]_{z=0}^{\infty} = \frac{1}{12}(s+1), \quad 0 \le s \le 4. \end{aligned}$$

Hence, the likelihood function is

$$f_{\boldsymbol{z}}(z|\boldsymbol{s}=s) = \frac{s+z}{s+1}e^{-z}, \quad 0 \le s \le 4, \quad 0 \le z < \infty.$$

Now we seek the value of s that maximizes $f_z(z|s = s)$. In this example, $f_z(z|s = s)$ does not have a maximum within the interior of its domain. However, it is differentiable, and we have

$$\frac{\partial f_z(z|s=s)}{\partial s} = \frac{\partial}{\partial s}\frac{s+z}{s+1}e^{-z} = \frac{1-z}{(s+1)^2}e^{-z}.$$

From this expression we see that, for $z > 1$, $f_z(z|s = s)$ is strictly decreasing with respect to s. Hence its maximum occurs at the smallest value of s, namely $s = 0$. If $0 \leq z < 1$, then $f_z(z|s = s)$ is strictly increasing, and its maximum occurs at $s = 4$.

Finally, if $z = 1$, $f_z(z|s = s) = e^{-1}$ for all s between 0 and 4, so the maximum of $f_z(z|s = s)$ is not generated by a unique value of s. In other words, any value of s is equally likely to produce the observation $z = 1$. We arbitrarily choose $\hat{s} = 22/9$, the mean of s.

Overall, the maximum-likelihood estimate of s given the observation z is

$$\hat{s}_{\text{ML}} = \begin{cases} 4, & 0 \leq z < 1; \\ 22/9, & z = 1; \\ 0, & z > 1. \end{cases}$$

Example 3.3 ML Estimation with Gaussian Noise

Suppose that $z = s + v$, where s and v are independent and $v \sim \mathcal{N}(0, \sigma^2)$. Then

$$f_v(v) = \frac{1}{\sqrt{2\pi}\sigma}e^{-v^2/2\sigma^2}.$$

From Example 2.14, given the sample realization z, the likelihood function is

$$f_z(z|s = s) = f_v(v)\big|_{v=z-s} = \frac{1}{\sqrt{2\pi}\sigma}e^{-(z-s)^2/2\sigma^2}.$$

Clearly, the peak of the likelihood function occurs when $s = z$, and thus the ML estimate is $\hat{s}_{\text{ML}} = z = \alpha(z)$. When z is viewed as a RV z, the ML estimate is $\hat{s}_{\text{ML}} = z = \alpha(z)$.

Example 3.4 ML Application to Signal Detection

Consider the single measurement

$$z = \begin{cases} s + v, & \text{when the signal } s \text{ is present;} \\ v, & \text{when the signal } s \text{ is not present;} \end{cases} \tag{3.12}$$

where s and v are independently jointly distributed RVs defined on the probability space $\mathcal{S}_1$. Based on the measurement $z = z$, the goal is to determine whether or not the signal s is present, or to detect the presence (or absence) of s. We can approach this problem by using ML estimation as follows.

First, let c denote the discrete RV defined on the probability space $\mathcal{S}_2$ = {signal is present, signal is absent} with

$$c = \begin{cases} 1, & \text{when the signal is present;} \\ 0, & \text{when the signal is absent.} \end{cases}$$

Then we can rewrite (3.12) in the form

$$z = cs + v. \tag{3.13}$$

Note that the RV z given by (3.13) is defined on the product space $\mathcal{S}_1 \times \mathcal{S}_2$; that is, we have a "hybrid" formulation, where "hybrid" means that we have a combination of discrete and continuous RVs.

Now the detection problem defined above is equivalent to the problem of estimating the value of c based on the measurement $z = z$. We can take the estimate to be the ML estimate; that is,

$$\hat{c}_{\text{ML}} = \text{value of } i \text{ that maximizes the likelihood function } f_z(z|c = i),$$

where i can be 0 or 1. Since $f_z(z|c = i)$ as a function of i has only two values, we see that

$$\hat{c}_{\text{ML}} = \begin{cases} 1, & \text{when } f_z(z|c = 1) > f_z(z|c = 0); \\ 0, & \text{otherwise.} \end{cases} \tag{3.14}$$

Observe that this result does not require any *a priori* information on c. As discussed in Example 3.7, if the density $f_c(c)$ is known (which is equivalent to knowing the probability that $c = 1$), then we can generate a much better estimate than the one given by (3.14).

Example 3.5 ML with Multiple Measurements

We now extend Example 3.3 to the case of multiple measurements. Suppose $z(n) = g(s(n), v(n))$. To compute the ML estimate $\hat{s}_{\text{ML}}(n)$ of $s(n)$, it is necessary to know how the signal terms $s(n)$ are related to each other. For simplicity, let us assume that $s(n)$ is constant, i.e.,

$$s(n) = s \quad \text{for all } n,$$

and assume that the measurements $z(n)$ are given by

$$z(n) = s + v(n),$$

where s and $v(n)$ are independent for all n.

Let $\boldsymbol{Z}_n$ denote the n-vector of measurements,

$$\boldsymbol{Z}_n = \begin{bmatrix} z(1) \\ z(2) \\ \vdots \\ z(n) \end{bmatrix}.$$

Then the ML estimate $\hat{s}_{\mathrm{ML}}(n)$ is the value of s that maximizes the likelihood function $f_{\boldsymbol{Z}_n}(Z_n|\boldsymbol{s} = s)$. To compute $\hat{s}_{\mathrm{ML}}(n)$, let $\boldsymbol{V}_n$ denote the n-vector of noise terms,

$$\boldsymbol{V}_n = \begin{bmatrix} v(1) \\ v(2) \\ \vdots \\ v(n) \end{bmatrix}.$$

Assuming that each $v(n) \sim \mathcal{N}\left(0, \sigma^2\right)$, $\boldsymbol{V}_n$ is a n-variate Gaussian with PDF

$$f_{\boldsymbol{V}_n}(V_n) = \frac{1}{(2\pi)^{n/2}|P_n|^{1/2}} \exp\left[-\tfrac{1}{2}V_n^{\mathrm{T}} P_n^{-1} V_n\right],$$

where $P_n = \mathrm{Cov}\left[\boldsymbol{V}_n\right]$ and $|P_n|$ is the determinant of P_n.

Now we may write

$$\boldsymbol{Z}_n = \beta \boldsymbol{s} + \boldsymbol{V}_n,$$

where β is the n-vector $\beta = \begin{bmatrix}1 & 1 & \cdots & 1\end{bmatrix}^{\mathrm{T}}$. Then the likelihood function becomes

$$\begin{aligned} f_{\boldsymbol{Z}_n}(Z_n|\boldsymbol{s} = s) &= f_{\boldsymbol{V}_n}(V_n)|_{V_n = Z_n - \beta s} \\ &= \frac{1}{(2\pi)^{n/2}|P_n|^{1/2}} \exp\left[-\tfrac{1}{2}(Z_n - \beta s)^{\mathrm{T}} P_n^{-1}(Z_n - \beta s)\right]. \end{aligned}$$

Now $\hat{s}_{\mathrm{ML}}$ is the value of s that maximizes $f_{\boldsymbol{Z}_n}(Z_n|\boldsymbol{s} = s)$, or equivalently, that minimizes $(Z_n - \beta s)^{\mathrm{T}} P_n^{-1}(Z_n - \beta s)$. Solving the latter is a weighted least squares problem, which has been addressed in Section 1.3, and we have that

$$\hat{s}_{\mathrm{ML}} = \left[\beta^{\mathrm{T}} P_n^{-1} \beta\right]^{-1} \beta^{\mathrm{T}} P_n^{-1} Z_n.$$

If P_n is diagonal with $P_n = \sigma^2 I$, then $\beta^{\mathrm{T}} P_n^{-1} \beta = n/\sigma^2$, and

$$\beta^{\mathrm{T}} P_n^{-1} = \frac{1}{\sigma^2}\begin{bmatrix}1 & 1 & \cdots & 1\end{bmatrix},$$

and the ML estimate for observations $\{z(1), \ldots, z(n)\}$ is

$$\hat{s}_{\mathrm{ML}} = \frac{1}{n}\sum_{j=1}^{n} z(j) = \alpha(z(1), z(2), \ldots, z(n)).$$

Hence, the ML estimate is simply the mean of the measurements in this case.

We make an additional important note about ML estimation. Observe that in (3.8), there is no dependence upon the density $f_{\boldsymbol{s}}(s)$. As shown later, some other optimality criteria depend upon $f_{\boldsymbol{s}}(s)$ through Bayes' formula (2.42) and are therefore known as Bayesian approaches. In contrast, ML estimation is sometimes said to be a *non-Bayesian approach.*

Since the ML criterion does not require $f_{\boldsymbol{s}}(s)$, it does not require any knowledge about the signal itself. This feature makes the ML criterion the most general optimality criterion. Many other optimality criteria can be interpreted as special cases of the ML criterion when additional knowledge about the signal is available. However, the determination of the likelihood function presents a nontrivial task, often more complicated than solving either (3.9) or (3.10). In addition, the choice of the likelihood function is often subjective. These difficulties suggest the adoption of alternative optimality criterion in some cases.

Maximum *a posteriori* Estimation

We now consider another optimality criterion. Suppose that we have $\boldsymbol{z} = g(\boldsymbol{s}, \boldsymbol{v})$. Given the observation z, the most-likely value of $\boldsymbol{s}$ to have occurred is the value of s that maximizes the conditional density $f_{\boldsymbol{s}}(s|\boldsymbol{z} = z)$. This density is known as the *a posteriori* density since it is the density after the measurement z has become available. This estimate is called the *maximum* a posteriori (MAP) estimate, denoted by $\hat{s}_{\text{MAP}}$, and is given by

$$\hat{s}_{\text{MAP}} = \text{value of } s \text{ that maximizes } f_{\boldsymbol{s}}(s|\boldsymbol{z} = z) \tag{3.15}$$

Assuming $f_{\boldsymbol{s}}(s|\boldsymbol{z} = z)$ is differentiable and has a unique maximum in the interior of its domain, we have

$$\hat{s}_{\text{MAP}} = \text{value of } s \text{ for which } \frac{\partial f_{\boldsymbol{s}}(s|\boldsymbol{z} = z)}{\partial s} = 0, \tag{3.16}$$

By Bayes' formula (2.42),

$$f_{\boldsymbol{s}}(s|\boldsymbol{z} = z) = \frac{f_{\boldsymbol{z}}(z|\boldsymbol{s} = s) f_{\boldsymbol{s}}(s)}{f_{\boldsymbol{z}}(z)}.$$

Since z is given, $f_{\boldsymbol{z}}(z)$ is constant and thus independent of s, so we may neglect the denominator and express $\hat{s}_{\text{MAP}}$ as

$$\hat{s}_{\text{MAP}} = \text{value of } s \text{ that maximizes } f_{\boldsymbol{z}}(z|\boldsymbol{s} = s) f_{\boldsymbol{s}}(s). \tag{3.17}$$

Observe that in this case, the density $f_{\boldsymbol{s}}(s)$ of $\boldsymbol{s}$ must be known, which is in contrast to the ML estimate. Due to the use of Bayes' formula, MAP estimation is a form of *Bayesian estimation.*

Example 3.6 MAP Estimation with Gaussian Noise

Again consider the additive-noise case $\boldsymbol{z} = \boldsymbol{s} + \boldsymbol{v}$, where $\boldsymbol{v} \sim \mathcal{N}(0, \sigma_v^2)$. MAP estimation requires some knowledge about the density of $\boldsymbol{s}$, so suppose that $\boldsymbol{s} \sim \mathcal{N}(\eta_s, \sigma_s^2)$. Then

$$f_{\boldsymbol{s}}(s) = \frac{1}{\sqrt{2\pi}\sigma_s} e^{-(s-\eta_s)^2/2\sigma_s^2},$$

and from Example 2.14,

$$f_{\boldsymbol{z}}(z|\boldsymbol{s} = s) = \frac{1}{\sqrt{2\pi}\sigma_v} e^{-(z-s)^2/2\sigma_v^2}.$$

Hence

$$f_{\boldsymbol{z}}(z|\boldsymbol{s} = s) f_{\boldsymbol{s}}(s) = \frac{1}{2\pi\sigma_s\sigma_v} \exp\left[-\frac{(z-s)^2}{2\sigma_v^2} - \frac{(s-\eta_s)^2}{2\sigma_s^2}\right].$$

The MAP estimate maximizes the above expression, which amounts to minimizing the term $\left[(z-s)^2/2\sigma_v^2 + (s-\eta_s)^2/2\sigma_s^2\right]$. Differentiating with respect to s and setting the partial derivative to zero gives

$$\frac{1}{2\sigma_v^2}[2(z-s)(-1)] + \frac{1}{2\sigma_s^2}[2(s-\eta_s^2(1)] = 0.$$

Solving for s yields the MAP estimate

$$\hat{s}_{\text{MAP}} = \frac{\sigma_v^2}{\sigma_v^2 + \sigma_s^2}\eta_s + \frac{\sigma_s^2}{\sigma_v^2 + \sigma_s^2} z,$$

which can be rewritten in the form

$$\hat{s}_{\text{MAP}} = \eta_s + \frac{\sigma_s^2}{\sigma_v^2 + \sigma_s^2}(z - \eta_s). \tag{3.18}$$

Note that both the mean η_s and variance σ_s^2 of $\boldsymbol{s}$ must be known before we can compute $\hat{s}_{\text{MAP}}$. Also, observe that when $\sigma_v^2 \ll \sigma_s^2$ (the noise power is much less than the signal power), the MAP estimate becomes

$$\hat{s}_{\text{MAP}} = \eta_s + \frac{\sigma_s^2}{\sigma_s^2}(z - \eta_s) = z = \hat{s}_{\text{ML}}.$$

In other words, the MAP estimate is equal to the ML estimate. This result makes sense because, as $\sigma_s^2 \to \infty$, the signal's Gaussian density $f_{\boldsymbol{s}}(s)$ approaches a uniform density, which implies that there is no *a priori* information about $\boldsymbol{s}$.

Example 3.7 MAP Application to Signal Detection

Consider the single measurement

$$\boldsymbol{z} = \boldsymbol{c}\boldsymbol{s} + \boldsymbol{v},$$

where $\boldsymbol{c}$, $\boldsymbol{s}$, and $\boldsymbol{v}$ are defined as in Example 3.4, and the goal is to determine the value of $\boldsymbol{c}$ based on the measurement $\boldsymbol{z} = z$. If we know the density $f_{\boldsymbol{c}}(c)$, then we can take the MAP estimate, rather than the ML estimate (as in Example 3.4). Then

$$\hat{c}_{\mathrm{MAP}} = \text{value of } i \text{ that maximizes } f_{\boldsymbol{z}}(z|\boldsymbol{c}=i)f_{\boldsymbol{c}}(i).$$

Since $\boldsymbol{c}$ is a discrete RV with two possible values, let P denote the probability that $\boldsymbol{c}=1$. Then $(1-P)$ is the probability that $\boldsymbol{c}=0$. It follows that

$$f_{\boldsymbol{z}}(z|\boldsymbol{c}=i)f_{\boldsymbol{c}}(i) = \begin{cases} f_{\boldsymbol{z}}(z|\boldsymbol{c}=1)P, & \text{if } \boldsymbol{c}=1; \\ f_{\boldsymbol{z}}(z|\boldsymbol{c}=0)(1-P), & \text{if } \boldsymbol{c}=0. \end{cases}$$

Therefore, $\hat{c}_{\mathrm{MAP}}$ is given by

$$\hat{c}_{\mathrm{MAP}} = \begin{cases} 1, & \text{if } f_{\boldsymbol{z}}(z|\boldsymbol{c}=1)P > f_{\boldsymbol{z}}(z|\boldsymbol{c}=0)(1-P); \\ 0, & \text{otherwise.} \end{cases}$$

This expression can be rewritten as

$$\hat{c}_{\mathrm{MAP}} = \begin{cases} 1, & \text{if } \frac{f_{\boldsymbol{z}}(z|\boldsymbol{c}=1)}{f_{\boldsymbol{z}}(z|\boldsymbol{c}=0)} > \frac{1-P}{P}; \\ 0, & \text{otherwise.} \end{cases} \tag{3.19}$$

The ratio $f_{\boldsymbol{z}}(z|\boldsymbol{c}=1)/f_{\boldsymbol{z}}(z|\boldsymbol{c}=0)$ is known as the *likelihood ratio*, which appears in many detection and hypothesis testing problems.

In general, the estimate given by (3.19) is better than the ML estimate of (3.14) since (3.19) incorporates the *a priori* information given by the probability P that $\boldsymbol{c}=1$. However, P may not be known. In this case, we can take $P=1/2$, in which case the MAP estimate of $\boldsymbol{c}$ reduces to the ML estimate given by (3.14).

3.3 Minimum Mean-Square Error Estimation

Let us continue to consider to the single-measurement case $\boldsymbol{z} = g(\boldsymbol{s}, \boldsymbol{v})$ and present another common measure of optimality. For the estimate $\hat{\boldsymbol{s}}$ of $\boldsymbol{s}$, the *estimation error* is defined to be the difference between the signal $\boldsymbol{s}$ and the estimate $\hat{\boldsymbol{s}}$. Denoting the estimation error by $\tilde{\boldsymbol{s}}$, we have

$$\tilde{\boldsymbol{s}} = \boldsymbol{s} - \hat{\boldsymbol{s}}. \tag{3.20}$$

The MSE is then

$$\text{MSE} = \text{E}\left[\text{E}\left[\tilde{\boldsymbol{s}}^2|\boldsymbol{z}\right]\right] = \text{E}\left[\text{E}\left[(\boldsymbol{s}-\hat{\boldsymbol{s}})^2|\boldsymbol{z}\right]\right] = \text{E}\left[(\boldsymbol{s}-\hat{\boldsymbol{s}})^2\right]. \tag{3.21}$$

The MSE gives the average power of the error. It is natural to try and minimize the average error power and find the *minimum mean-square error* (MMSE) estimate of $\boldsymbol{s}$; we denote this estimate by $\hat{\boldsymbol{s}}_{\text{MMSE}}$.

We next show that, given the RV $\boldsymbol{z}$, the MMSE estimate $\hat{\boldsymbol{s}}_{\text{MMSE}}$ is the conditional expectation $\text{E}[\boldsymbol{s}|\boldsymbol{z}]$.

Theorem 3.1 *Given the RV* $\boldsymbol{z}$, *the MMSE estimate* $\hat{\boldsymbol{s}}_{MMSE}$ *of* $\boldsymbol{s}$ *is the conditional expectation*

$$\hat{\boldsymbol{s}}_{\text{MMSE}} = E[\boldsymbol{s}|\boldsymbol{z}]. \tag{3.22}$$

Proof. Let $\boldsymbol{s}$ and $\boldsymbol{z}$ be jointly distributed RVs with joint PDF $f_{\boldsymbol{s},\boldsymbol{z}}(s,z)$, conditional PDF $f_{\boldsymbol{s}}(s|\boldsymbol{z}=z)$, and marginal PDF $f_{\boldsymbol{z}}(z)$. The estimate is $\hat{\boldsymbol{s}} = \alpha(\boldsymbol{z})$, where α is to be determined.

The MSE may be written as

$$\begin{aligned}
\text{MSE} &= \text{E}\left[(\boldsymbol{s}-\hat{\boldsymbol{s}})^2\right] \\
&= \text{E}\left[(\boldsymbol{s}-\alpha(\boldsymbol{z}))^2\right] \\
&= \int_{-\infty}^{\infty}\int_{-\infty}^{\infty} (s-\alpha(z))^2 f_{\boldsymbol{s},\boldsymbol{z}}(s,z)\,ds\,dz \\
&= \int_{-\infty}^{\infty}\int_{-\infty}^{\infty} (s-\alpha(z))^2 f_{\boldsymbol{s}}(s|\boldsymbol{z}=z)\,f_{\boldsymbol{z}}(z)\,ds\,dz. \\
&= \int_{-\infty}^{\infty}\int_{-\infty}^{\infty} \left[s^2 - 2s\alpha(z) + \alpha^2(z)\right] f_{\boldsymbol{s}}(s|\boldsymbol{z}=z)\,ds\,f_{\boldsymbol{z}}(z)\,dz.
\end{aligned}$$

The integral of a PDF is nonnegative and $f_{\boldsymbol{s}}(s|\boldsymbol{z}=z) \geq 0$ for all s, so to minimize the MSE we only need to minimize the inner integral in the above expression. That is, we should minimize

$$\int_{-\infty}^{\infty} \left[s^2 - 2s\alpha(z) + \alpha^2(z)\right] f_{\boldsymbol{s}}(s|\boldsymbol{z}=z)\,ds. \tag{3.23}$$

The estimator α is found by taking the partial derivative of (3.23) with respect to α and setting it equal to zero. Doing so produces

$$\int_{-\infty}^{\infty} \left[-2s + 2\alpha(z)\right] f_{\boldsymbol{s}}(s|\boldsymbol{z}=z)\,ds = 0.$$

Then

$$\alpha(z)\int_{-\infty}^{\infty} f_{\boldsymbol{s}}(s|\boldsymbol{z}=z)\,ds = \alpha(z)\times 1 = \int_{-\infty}^{\infty} s f_{\boldsymbol{s}}(s|\boldsymbol{z}=z)\,ds = \text{E}[\boldsymbol{s}|\boldsymbol{z}=z], \tag{3.24}$$

which means that for the RV $\boldsymbol{z}$,

$$\hat{\boldsymbol{s}}_{\text{MMSE}} = \alpha(\boldsymbol{z}) = \text{E}[\boldsymbol{s}|\boldsymbol{z}]. \tag{3.25}$$

Q.E.D. ■

The proof also establishes that the MMSE estimate $\hat{s}_{\text{MMSE}}$ is unique. Intuitively, this property makes sense because MSE is a quadratic expression in $\hat{s}$ (see (3.21)), and therefore it has a unique minimum. Also note that, from (3.24) and Bayes' formula (2.42), MMSE estimation requires information about $\boldsymbol{s}$, much like the MAP approach. Hence MMSE estimation is another type of *Bayesian estimation.*

Note that the estimate $\hat{\boldsymbol{s}}_{\text{MMSE}}$ is itself a RV because $\boldsymbol{z}$ is random. When we have a realization z of $\boldsymbol{z}$, the MMSE estimate takes on a specific value, given by

$$\hat{s}_{\text{MMSE}} = \alpha(z) = \mathrm{E}\left[\boldsymbol{s}|\boldsymbol{z}=z\right] = \int_{-\infty}^{\infty} s f_{\boldsymbol{s}}(s|\boldsymbol{z}=z)\, ds. \tag{3.26}$$

where α is the function defined by

$$\alpha(z) = \int_{-\infty}^{\infty} s f_{\boldsymbol{s}}(s|\boldsymbol{z}=z) ds.$$

One important property of the MMSE estimate $\hat{\boldsymbol{s}}_{\text{MMSE}}$ is that it is unbiased. From (2.52),

$$\mathrm{E}\left[\mathrm{E}\left[\boldsymbol{s}|\boldsymbol{z}\right]\right] = \mathrm{E}\left[\boldsymbol{s}\right],$$

and thus defining the estimation error

$$\tilde{\boldsymbol{s}} = \boldsymbol{s} - \hat{\boldsymbol{s}}_{\text{MMSE}},$$

we have that

$$\begin{aligned} \mathrm{E}\left[\tilde{\boldsymbol{s}}\right] &= \mathrm{E}\left[\boldsymbol{s} - \hat{\boldsymbol{s}}_{\text{MMSE}}\right] = \mathrm{E}\left[\boldsymbol{s} - \mathrm{E}\left[\boldsymbol{s}|\boldsymbol{z}\right]\right] \\ &= \mathrm{E}\left[\boldsymbol{s}\right] - \mathrm{E}\left[\mathrm{E}\left[\boldsymbol{s}|\boldsymbol{z}\right]\right] = \mathrm{E}\left[\boldsymbol{s}\right] - \mathrm{E}\left[\boldsymbol{s}\right] \\ &= 0. \end{aligned} \tag{3.27}$$

Now (3.27) shows that the MMSE estimate $\hat{\boldsymbol{s}}_{\text{MMSE}}$ is unbiased.

The above results extend to the case of MMSE estimation of a constant signal $\hat{\boldsymbol{s}}$ when a finite number of measurements $\boldsymbol{z}(1), \boldsymbol{z}(2), \ldots, \boldsymbol{z}(n)$ is available. In this case,

$$\hat{\boldsymbol{s}}_{\text{MMSE}} = \mathrm{E}\left[\boldsymbol{s}|\boldsymbol{z}(1), \boldsymbol{z}(2), \ldots, \boldsymbol{z}(n)\right], \tag{3.28}$$

and

$$\mathrm{E}\left[\boldsymbol{s} - \hat{\boldsymbol{s}}_{\text{MMSE}}\right] = 0. \tag{3.29}$$

Example 3.8 MMSE Estimation with Gaussian Noise

Again consider the additive-noise case

$$\boldsymbol{z} = \boldsymbol{s} + \boldsymbol{v},$$

with $\boldsymbol{v} \sim \mathcal{N}(0, \sigma_v^2)$. We require knowledge of $\boldsymbol{s}$, so assume $\boldsymbol{s} \sim \mathcal{N}(\bar{s}, \sigma_s^2)$. Also assume that $\boldsymbol{s}$ and $\boldsymbol{v}$ are uncorrelated, so that $\mathrm{E}[\boldsymbol{z}] = \mathrm{E}[\boldsymbol{s}] = \bar{s}$ and

$$\mathrm{Var}[\boldsymbol{z}] = \mathrm{Var}[\boldsymbol{s}] + \mathrm{Var}[\boldsymbol{v}] = \sigma_s^2 + \sigma_v^2.$$

Then the PDF of $\boldsymbol{z}$ is

$$f_{\boldsymbol{z}}(z) = \frac{1}{\sqrt{2\pi}\sqrt{\sigma_s^2 + \sigma_v^2}} \exp\left[-\frac{(z - \bar{s})^2}{2(\sigma_s^2 + \sigma_v^2)}\right].$$

Using Bayes' formula (2.42) and the conditional density $f_{\boldsymbol{z}}(z|\boldsymbol{s} = s)$ given in Example 3.3, we have that the conditional density is

$$f_{\boldsymbol{s}}(s|\boldsymbol{z} = z) = \frac{1}{2\pi\sigma_s\sigma_v f_{\boldsymbol{z}}(z)} \exp\left\{-\left[\frac{(z - s)^2}{2\sigma_v^2} + \frac{(s - \bar{s})^2}{2\sigma_s^2}\right]\right\}.$$

When we substitute the expression for $f_{\boldsymbol{z}}(z)$ into this equation, we find

$$f_{\boldsymbol{s}}(s|\boldsymbol{z} = z) = \frac{1}{\sqrt{2\pi}\sqrt{\frac{\sigma_s^2\sigma_v^2}{\sigma_s^2+\sigma_v^2}}} \exp\left\{-\left[-\frac{(z - \bar{s})^2}{2(\sigma_s^2 + \sigma_v^2)} + \frac{(z - s)^2}{2\sigma_v^2} + \frac{(s - \bar{s})^2}{2\sigma_s^2}\right]\right\}.$$

Now

$$-\frac{(z - \bar{s})^2}{2(\sigma_s^2 + \sigma_v^2)} + \frac{(z - s)^2}{2\sigma_v^2} + \frac{(s - \bar{s})^2}{2\sigma_s^2} = \frac{(s - \hat{s}_{\mathrm{MAP}})^2}{2\frac{\sigma_s^2\sigma_v^2}{\sigma_s^2+\sigma_v^2}},$$

where $\hat{s}_{\mathrm{MAP}}$ is the MAP estimate from Example 3.6 (3.18). The reader may verify this result. The conditional density becomes

$$f_{\boldsymbol{s}}(s|\boldsymbol{z} = z) = \frac{1}{\sqrt{2\pi}\sqrt{\frac{\sigma_s^2\sigma_v^2}{\sigma_s^2+\sigma_v^2}}} \exp\left[-\frac{(s - \hat{s}_{\mathrm{MAP}})^2}{2\frac{\sigma_s^2\sigma_v^2}{\sigma_s^2+\sigma_v^2}}\right],$$

which shows that $f_{\boldsymbol{s}}(s|\boldsymbol{z} = z)$ is a Gaussian RV with mean $\hat{s}_{\mathrm{MAP}}$ and variance $\sigma_s^2\sigma_v^2/(\sigma_s^2 + \sigma_v^2)$. (Compare this result with Example 2.15.)

Since $\hat{s}_{\mathrm{MMSE}}$ is the conditional expectation $\mathrm{E}[\boldsymbol{s}|\boldsymbol{z} = z]$, we have

$$\hat{s}_{\mathrm{MMSE}} = \mathrm{E}[\boldsymbol{s}|\boldsymbol{z} = z] = \hat{s}_{\mathrm{MAP}} = \bar{s} + \frac{\sigma_s^2}{\sigma_v^2 + \sigma_s^2}(z - \bar{s}).$$

In other words, if $\boldsymbol{s}$ and $\boldsymbol{v}$ are uncorrelated, then the MMSE estimate of $\boldsymbol{s}$ is identical to the MAP estimate.

The Orthogonality Principle

An important property of the MMSE estimate is that the estimation error $s - \mathrm{E}[s|z]$ is orthogonal to every function g of the observation z. This property is known as the *orthogonality principle* and is stated below.

Theorem 3.2 (Orthogonality Principle) *The error $s - E[s|z]$ is orthogonal to every function $\gamma(z)$, i.e.,*

$$E[(s - E[s|z])\,\gamma(z)] = 0. \tag{3.30}$$

Proof.

$$\mathrm{E}[\gamma(z)\,(s - \mathrm{E}[s|z])] = \mathrm{E}[\gamma(z)s] - \mathrm{E}[\gamma(z)\mathrm{E}[s|z]]\,. \tag{3.31}$$

Now consider $\mathrm{E}[\gamma(z)s|z]$. Once z is given, $\gamma(z)$ behaves like a constant. Hence,

$$\mathrm{E}[\gamma(z)s|z] = \gamma(z)\mathrm{E}[s|z]\,.$$

Taking the expected value of both sides of this expression gives

$$\mathrm{E}[\mathrm{E}[\gamma(z)s|z]] = \mathrm{E}[\gamma(z)\mathrm{E}[s|z]]\,,$$

and from (2.52), we know that $\mathrm{E}[\gamma(z)s] = \mathrm{E}[\mathrm{E}[\gamma(z)s|z]]$, so

$$\mathrm{E}[\gamma(z)s] = \mathrm{E}[\gamma(z)\mathrm{E}[s|z]]\,. \tag{3.32}$$

Therefore, when we insert (3.32) into (3.31), we have

$$\begin{aligned}
\mathrm{E}[\gamma(z)\,(s - \mathrm{E}[s|z])] &= \mathrm{E}[\gamma(z)s] - \mathrm{E}[\mathrm{E}[\gamma(z)s|z]] \\
&= \mathrm{E}[\gamma(z)s] - \mathrm{E}[\gamma(z)s] \\
&= 0.
\end{aligned}$$

Q.E.D. ∎

Proof. (Alternate method.)

$$\begin{aligned}
\mathrm{E}[(s - \mathrm{E}[s|z])\,\gamma(z)] &= \mathrm{E}[\mathrm{E}[(s - \mathrm{E}[s|z])\,\gamma(z)|\,z]] \\
&= \mathrm{E}[\mathrm{E}[(s - \mathrm{E}[s|z])\,|\,z]\,\gamma(z)] \\
&= \mathrm{E}[(\mathrm{E}[s|z] - \mathrm{E}[\mathrm{E}[s|z]|\,z])\,\gamma(z)] \\
&= \mathrm{E}[(\mathrm{E}[s|z] - \mathrm{E}[s|z])\,\gamma(z)] \\
&= 0.
\end{aligned}$$

Q.E.D. ∎

We now present another important property related to the orthogonality principle.

Theorem 3.3 *The estimate given by* $\hat{s} = \alpha(z)$ *is the MMSE estimate of* s *given* z *if and only if the error* $s - \alpha(z)$ *is orthogonal to every function* $\gamma(z)$*; that is,*

$$E[(s - \alpha(z))\,\gamma(z)] = 0. \tag{3.33}$$

Proof. (Sufficiency.) We will prove sufficiency by contradiction. Assume that $\alpha_1(z)$ is a MMSE estimate and $\gamma(z)$ is a function that is *not* orthogonal to the error. That is,

$$\mathrm{E}[(s - \alpha_1(z))\,\gamma(z)] \neq 0. \tag{3.34}$$

Define a new estimate α_2 by

$$\alpha_2(z) = \alpha_1(z) + c\gamma(z),$$

where

$$c = \frac{\mathrm{E}[(s - \alpha_1(z))\,\gamma(z)]}{\mathrm{E}[\gamma^2(z)]}.$$

Clearly $\gamma(z) \neq 0$ (otherwise (3.34) would not hold), so c is well-defined and nonzero.

The MSE associated with $\alpha_2(z)$ is

$$\begin{aligned}
\mathrm{MSE}(\alpha_2(z)) &= \mathrm{E}\left[(s - \alpha_1(z) - c\gamma(z))^2\right] \\
&= \mathrm{E}\left[(s - \alpha_1(z))^2\right] - 2c\mathrm{E}[(s - \alpha_1(z))\,\gamma(z)] \\
&\quad + c^2\mathrm{E}\left[\gamma^2(z)\right] \\
&= \mathrm{MSE}(\alpha_1(z)) - 2\frac{\mathrm{E}[(s - \alpha_1(z))\,\gamma(z)]}{\mathrm{E}[\gamma^2(z)]}\mathrm{E}[(s - \alpha_1(z))\,\gamma(z)] \\
&\quad + \frac{(\mathrm{E}[(s - \alpha_1(z))\,\gamma(z)])^2}{(\mathrm{E}[\gamma^2(z)])^2}\mathrm{E}\left[\gamma^2(z)\right] \\
&= \mathrm{MSE}(\alpha_1(z)) - \frac{(\mathrm{E}[(s - \alpha_1(z))\,\gamma(z)])^2}{\mathrm{E}[\gamma^2(z)]} \\
&< \mathrm{MSE}(\alpha_1(z)).
\end{aligned}$$

This result contradicts the assumption that $\alpha_1(z)$ is a MMSE estimate.

(Necessity.) Now assume that $\alpha_1(z)$ is an estimate of s with the property that for all functions $\gamma(z)$,

$$\mathrm{E}[(s - \alpha_1(z))\,\gamma(z)] = 0.$$

Let $\alpha_2(z)$ be any other estimate of s. Then

$$\begin{aligned}
\mathrm{MSE}(\alpha_2(z)) &= \mathrm{E}\left[(s - \alpha_2(z))^2\right] \\
&= \mathrm{E}\left[(s - \alpha_1(z) + \alpha_1(z) - \alpha_2(z))^2\right] \\
&= \mathrm{E}\left[(s - \alpha_1(z))^2\right] + 2\mathrm{E}[(s - \alpha_1(z))\,(\alpha_1(z) - \alpha_2(z))] \\
&\quad + \mathrm{E}\left[(\alpha_1(z) - \alpha_2(z))^2\right].
\end{aligned}$$

As a result of the orthogonality assumption, the middle term in the above expression is zero, so

$$\begin{aligned}\mathrm{MSE}(\alpha_2(\boldsymbol{z})) &= \mathrm{E}\left[(\boldsymbol{s}-\alpha_1(\boldsymbol{z}))^2\right] + \mathrm{E}\left[(\alpha_1(\boldsymbol{z})-\alpha_2(\boldsymbol{z}))^2\right] \\ &\geq \mathrm{E}\left[(\boldsymbol{s}-\alpha_1(\boldsymbol{z}))^2\right] \\ &\geq \mathrm{MSE}(\alpha_1(\boldsymbol{z})).\end{aligned}$$

Hence $\alpha_1(\boldsymbol{z})$ is the MMSE estimate of $\boldsymbol{s}$ given $\boldsymbol{z}$. **Q.E.D.** ∎

At first glance Theorems 3.2 and 3.3 may appear to be equivalent; however, the latter states that orthogonality is not simply a property of the MMSE estimate, but that only the MMSE estimate has the orthogonality property. Theorem 3.2 simply states a property of the error associated with the MMSE estimate: $\boldsymbol{s} - \mathrm{E}[\boldsymbol{s}|\boldsymbol{z}]$ is orthogonal to $\gamma(\boldsymbol{z})$ for all functions γ. Theorem 3.3 provides necessary and sufficient conditions for the MMSE estimator. It says that if an estimator α can be found such that $\boldsymbol{s} - \alpha(\boldsymbol{z})$ is orthogonal to every function $\gamma(\boldsymbol{z})$, then α must be the MMSE estimator.

MMSE Estimation in the General Case

Suppose that we wish to estimate the random *signal* $\boldsymbol{s}(n)$ given a set of measurements $\{\boldsymbol{z}(n) : n_1 \leq n \leq n_2\}$. Recall that a random signal evaluated at a particular time index may be considered a RV. Then the result of Equation (3.28) applies, provided n_1 and n_2 are finite. That is, the MMSE estimate is

$$\hat{\boldsymbol{s}}(n) = \mathrm{E}\left[\boldsymbol{s}(n)|\boldsymbol{z}(n_1), \ldots, \boldsymbol{z}(n_2)\right], \quad -\infty < n_1 \leq n_2 < \infty. \tag{3.35}$$

One might expect (3.35) to extend to include the case of an infinite number of observations. In fact, this conclusion is correct. However, a proof similar to that given for Theorem 3.1 introduces some mathematical intricacies that would take us too far afield. Instead, we note that the conditional expectation is guaranteed to exist and present a theorem that includes the estimator (3.25), (3.28), and (3.35) as special cases. This theorem applies to both finite and infinite sets of measurements.

Theorem 3.4 *Let* $\boldsymbol{Z}$ *denote the (possibly infinite) set of observations,* $\boldsymbol{Z} = \{\boldsymbol{z}(n) : n_1 \leq n \leq n_2\}$,[2] *and let* $E[\boldsymbol{s}(n)|\boldsymbol{Z}]$ *denote the conditional expectation of* $\boldsymbol{s}(n)$ *given* $\boldsymbol{Z}$. *Then the MMSE estimate of* $\boldsymbol{s}(n)$ *given* $\boldsymbol{Z}$ *is* $E[\boldsymbol{s}(n)|\boldsymbol{Z}]$; *that is,*

$$\hat{\boldsymbol{s}}(n) = E[\boldsymbol{s}(n)|\boldsymbol{Z}]. \tag{3.36}$$

[2] The set of measurements $\boldsymbol{Z}$ should not be confused with the set of integers $\mathbb{Z} = \{\ \ldots, -2, -1, 0, 1, 2, \ldots\ \}$.

Proof. Let $\boldsymbol{Z}$ and $\mathrm{E}[\boldsymbol{s}(n)|\boldsymbol{Z}]$ be as hypothesized above. Let $\hat{\boldsymbol{s}}(n) = \alpha(\boldsymbol{Z})$ be the optimal estimate. Then we may write the conditional MSE as

$$\begin{aligned}\mathrm{E}\left[(\boldsymbol{s}(n)-\alpha(\boldsymbol{Z}))^2\middle|\,\boldsymbol{Z}\right] &= \mathrm{E}\left[(\boldsymbol{s}(n)-\mathrm{E}[\boldsymbol{s}(n)|\boldsymbol{Z}]+\mathrm{E}[\boldsymbol{s}(n)|\boldsymbol{Z}]-\alpha(\boldsymbol{Z}))^2\middle|\,\boldsymbol{Z}\right]\\ &= \mathrm{E}\left[(\boldsymbol{s}(n)-\mathrm{E}[\boldsymbol{s}(n)|\boldsymbol{Z}])^2\middle|\,\boldsymbol{Z}\right]\\ &\quad + 2\mathrm{E}\left[(\boldsymbol{s}(n)-\mathrm{E}[\boldsymbol{s}(n)|\boldsymbol{Z}])(\mathrm{E}[\boldsymbol{s}(n)|\boldsymbol{Z}]-\alpha(\boldsymbol{Z}))|\,\boldsymbol{Z}\right]\\ &\quad + \mathrm{E}\left[(\mathrm{E}[\boldsymbol{s}(n)|\boldsymbol{Z}]-\alpha(\boldsymbol{Z}))^2\middle|\,\boldsymbol{Z}\right]. \end{aligned} \tag{3.37}$$

Because $\boldsymbol{Z}$ is given, $(\mathrm{E}[\boldsymbol{s}(n)|\boldsymbol{Z}]-\alpha(\boldsymbol{Z}))$ may be treated as a constant. Then the middle term in (3.37) becomes

$$\begin{aligned}&\mathrm{E}\left[(\boldsymbol{s}(n)-\mathrm{E}[\boldsymbol{s}(n)|\boldsymbol{Z}])(\mathrm{E}[\boldsymbol{s}(n)|\boldsymbol{Z}]-\alpha(\boldsymbol{Z}))|\,\boldsymbol{Z}\right]\\ &\quad = \{\mathrm{E}\left[(\boldsymbol{s}(n)-\mathrm{E}[\boldsymbol{s}(n)|\boldsymbol{Z}])\mid \boldsymbol{Z}\right]\}\{\mathrm{E}[\boldsymbol{s}(n)|\boldsymbol{Z}]-\alpha(\boldsymbol{Z})\}\\ &\quad = \{\mathrm{E}[\boldsymbol{s}(n)|\boldsymbol{Z}]-\mathrm{E}[\mathrm{E}[\boldsymbol{s}(n)|\boldsymbol{Z}]\,|\boldsymbol{Z}]\}\{\mathrm{E}[\boldsymbol{s}(n)|\boldsymbol{Z}]-\alpha(\boldsymbol{Z}))\\ &\quad = \{\mathrm{E}[\boldsymbol{s}(n)|\boldsymbol{Z}]-\mathrm{E}[\boldsymbol{s}(n)|\boldsymbol{Z}]\}\{\mathrm{E}[\boldsymbol{s}(n)|\boldsymbol{Z}]-\alpha(\boldsymbol{Z})\}\\ &\quad = 0. \end{aligned} \tag{3.38}$$

So (3.37) becomes

$$\begin{aligned}\mathrm{E}\left[(\boldsymbol{s}(n)-\alpha(\boldsymbol{Z}))^2\middle|\,\boldsymbol{Z}\right] &= \mathrm{E}\left[(\boldsymbol{s}(n)-\mathrm{E}[\boldsymbol{s}(n)|\boldsymbol{Z}])^2\middle|\,\boldsymbol{Z}\right]\\ &\quad + \mathrm{E}\left[(\mathrm{E}[\boldsymbol{s}(n)|\boldsymbol{Z}]-\alpha(\boldsymbol{Z}))^2\middle|\,\boldsymbol{Z}\right]\\ &\geq \mathrm{E}\left[(\boldsymbol{s}(n)-\mathrm{E}[\boldsymbol{s}(n)|\boldsymbol{Z}])^2\middle|\,\boldsymbol{Z}\right]. \end{aligned} \tag{3.39}$$

We take the expected value of both sides of (3.39) and find

$$\mathrm{MSE} = \mathrm{E}\left[(\boldsymbol{s}(n)-\alpha(\boldsymbol{Z}))^2\right] \geq \mathrm{E}\left[(\boldsymbol{s}(n)-\mathrm{E}[\boldsymbol{s}(n)|\boldsymbol{Z}])^2\right].$$

Clearly, the MSE will be minimized if and only if

$$\hat{\boldsymbol{s}}(n) = \alpha(\boldsymbol{Z}) = \mathrm{E}[\boldsymbol{s}(n)|\boldsymbol{Z}],$$

Q.E.D. ∎

Note that this proof also shows that the MMSE estimate $\hat{\boldsymbol{s}}(n)$ is unique.

3.4 Linear MMSE Estimation

We have determined the MMSE estimate of a signal $\boldsymbol{s}$ given the measurement $\boldsymbol{z}$ and seen that the MMSE estimate is a conditional expectation. Unfortunately, it is often difficult to determine the conditional expectation in practice because the relationship $f_{\boldsymbol{s}}(s|\boldsymbol{z}=z)$ may be difficult to find. (Note that this problem can also make MAP estimation difficult.)

Much of the difficulty lies in the fact that the MMSE estimator is very general; there are no restrictions on its form. As a result, the function α in the MMSE estimate $\hat{\boldsymbol{s}} = \alpha(\boldsymbol{z})$ can be very complicated, and it may be impossible to find a practical solution. However, suppose we constrain α to belong to a class of functions that we can implement. In other words, we restrict the estimation problem to produce a *tractable* solution for α.

Usually, the estimate $\hat{\boldsymbol{s}} = \alpha(\boldsymbol{z})$ will be *suboptimal*; that is, $\alpha(\boldsymbol{z})$ will not be the optimal MMSE estimate, but it will be optimal within the class of functions we have chosen. We are trading overall optimality for tractability.

One particularly powerful class of functions is the class of linear functions. We restrict $\hat{\boldsymbol{s}} = \alpha(\boldsymbol{z})$ to be a *linear estimate*, so we seek optimal *linear MMSE* (LMMSE) estimate. Often we can determine the LMMSE estimate much more easily than the overall MMSE estimate without a significant loss of performance. Now the estimate has the form

$$\hat{\boldsymbol{s}} = \lambda \boldsymbol{z}, \tag{3.40}$$

for some constant λ. Then the problem becomes that of determining λ.

We can find λ by direct minimization of the MSE:

$$\text{MSE} = \mathrm{E}\left[(\boldsymbol{s} - \lambda\boldsymbol{z})^2\right] = \mathrm{E}\left[\boldsymbol{s}^2 - 2\lambda\boldsymbol{s}\boldsymbol{z} + \lambda^2\boldsymbol{z}^2\right].$$

Taking the partial derivative with respect to λ and setting the result equal to zero gives

$$-2\mathrm{E}\left[\boldsymbol{s}\boldsymbol{z}\right] + 2\lambda\mathrm{E}\left[\boldsymbol{z}^2\right] = 0, \tag{3.41}$$

so that

$$\lambda = \frac{\mathrm{E}\left[\boldsymbol{s}\boldsymbol{z}\right]}{\mathrm{E}\left[\boldsymbol{z}^2\right]}. \tag{3.42}$$

The LMMSE estimate is given by

$$\hat{s}_{\text{LMMSE}} = \alpha(\boldsymbol{z}) = \left(\frac{\mathrm{E}\left[\boldsymbol{s}\boldsymbol{z}\right]}{\mathrm{E}\left[\boldsymbol{z}^2\right]}\right)\boldsymbol{z}. \tag{3.43}$$

Observe that we do *not* require knowledge about any likelihood function or densities. Instead, we only need the second-order moments $\mathrm{E}\left[\boldsymbol{s}\boldsymbol{z}\right]$ and $\mathrm{E}\left[\boldsymbol{z}^2\right]$. Typicaly, we can estimate $\mathrm{E}\left[\boldsymbol{s}\boldsymbol{z}\right]$ and $\mathrm{E}\left[\boldsymbol{z}^2\right]$ from experimental training data. For example, in a controlled environment we can generate or accurately measure sample realizations $s_1, s_2, \ldots, s_M$ of $\boldsymbol{s}$ and obtain corresponding measurements $z_1, z_2, \ldots, z_M$. Then

$$\mathrm{E}\left[\boldsymbol{s}\boldsymbol{z}\right] \approx \frac{1}{M}\sum_{m=1}^{M} s_m z_m,$$

and

$$\mathrm{E}\left[\boldsymbol{z}^2\right] \approx \frac{1}{M}\sum_{m=1}^{M} z_m^2,$$

Also observe that by imposing the linear form (3.40) on the estimator, the equation that must be solved to find λ is a linear equation (3.42), which is easy to solve and which has a unique solution.

Orthogonality Principle for LMMSE Estimation

A form of the orthogonality principle (Theorem 3.2) also exists for linear estimators.

Theorem 3.5 (Orthogonality Principle) *Let $\alpha(\boldsymbol{z})$ be the LMMSE estimate of $\boldsymbol{s}$ given $\boldsymbol{z}$. Then the error $\boldsymbol{s} - \alpha(\boldsymbol{z})$ is orthogonal to every linear function $\gamma(\boldsymbol{z})$, i.e.,*

$$E[(\boldsymbol{s} - \alpha(\boldsymbol{z}))\,\gamma(\boldsymbol{z})] = 0. \tag{3.44}$$

Proof. $\alpha(\boldsymbol{z})$ is given by (3.43), and $\gamma(\boldsymbol{z})$, being linear, can be written as $\gamma(\boldsymbol{z}) = \beta\boldsymbol{z}$ for some β. Then

$$\begin{aligned}
\mathrm{E}[(\boldsymbol{s} - \alpha(\boldsymbol{z}))\,\gamma(\boldsymbol{z})] &= \mathrm{E}\left[\left[\boldsymbol{s} - \left(\mathrm{E}[\boldsymbol{s}\boldsymbol{z}]/\mathrm{E}\left[\boldsymbol{z}^2\right]\right)\boldsymbol{z}\right]\beta\boldsymbol{z}\right] \\
&= \beta\mathrm{E}[\boldsymbol{s}\boldsymbol{z}] - \mathrm{E}\left[\beta\left(\mathrm{E}[\boldsymbol{s}\boldsymbol{z}]/\mathrm{E}\left[\boldsymbol{z}^2\right]\right)\boldsymbol{z}^2\right] \\
&= \beta\mathrm{E}[\boldsymbol{s}\boldsymbol{z}] - \beta\frac{\mathrm{E}[\boldsymbol{s}\boldsymbol{z}]}{\mathrm{E}[\boldsymbol{z}^2]}\mathrm{E}\left[\boldsymbol{z}^2\right] \\
&= \beta\mathrm{E}[\boldsymbol{s}\boldsymbol{z}] - \beta\mathrm{E}[\boldsymbol{s}\boldsymbol{z}] \\
&= 0.
\end{aligned}$$

Q.E.D. ■

Orthogonality Principle for Vector RVs

We will often employ vector RVs in this book. We therefore require the orthogonality principle in a form suitable for vector RVs. Let $\boldsymbol{s}$ and $\boldsymbol{z}$ be jointly distributed random vectors of lengths m and q, respectively. Let us find the LMMSE estimate of $\boldsymbol{s}$ given $\boldsymbol{z}$, so $\hat{\boldsymbol{s}} = M\boldsymbol{z}$, where M is an $m \times q$ matrix to be determined. Let

$$P = \mathrm{E}\left[(\boldsymbol{s} - \hat{\boldsymbol{s}})(\boldsymbol{s} - \hat{\boldsymbol{s}})^{\mathrm{T}}\right].$$

Recall that the *trace* of a matrix is the sum of its diagonal elements. That is, for an $m \times m$ matrix A,

$$A = \begin{bmatrix} a_{11} & a_{12} & \cdots & a_{1m} \\ a_{21} & a_{22} & \cdots & a_{2m} \\ \vdots & \vdots & \ddots & \vdots \\ a_{m1} & a_{m2} & \cdots & a_{mm} \end{bmatrix},$$

the trace of A is denoted by $\operatorname{tr}(A)$ and defined by

$$\operatorname{tr}(A) = \sum_{i=1}^{m} a_{ii}. \tag{3.45}$$

Then the MSE can be related to P via

$$\text{MSE} = \operatorname{tr}(P) = \mathrm{E}\left[(\boldsymbol{s} - \hat{\boldsymbol{s}})^{\mathrm{T}}(\boldsymbol{s} - \hat{\boldsymbol{s}})\right].$$

Because MSE $= \operatorname{tr}(P)$, we can find M by taking the partial derivative of $\operatorname{tr}(P)$ with respect to M, setting the result equal to the $m \times q$ zero matrix.

We expand P as follows:

$$\begin{aligned} P &= \mathrm{E}\left[(\boldsymbol{s} - \hat{\boldsymbol{s}})(\boldsymbol{s} - \hat{\boldsymbol{s}})^{\mathrm{T}}\right] \\ &= \mathrm{E}\left[\boldsymbol{s}\boldsymbol{s}^{\mathrm{T}}\right] - M\mathrm{E}\left[\boldsymbol{z}\boldsymbol{s}^{\mathrm{T}}\right] - \mathrm{E}\left[\boldsymbol{s}\boldsymbol{z}^{\mathrm{T}}\right] M^{\mathrm{T}} + M\mathrm{E}\left[\boldsymbol{z}\boldsymbol{z}^{\mathrm{T}}\right] M^{\mathrm{T}}. \end{aligned}$$

Note that if A and B are $m \times m$ matrices,

$$\operatorname{tr}(A + B) = \operatorname{tr}(A) + \operatorname{tr}(B), \tag{3.46}$$

so that

$$\begin{aligned} \operatorname{tr}(P) = {} & \operatorname{tr}\left(\mathrm{E}\left[\boldsymbol{s}\boldsymbol{s}^{\mathrm{T}}\right]\right) - \operatorname{tr}\left(M\mathrm{E}\left[\boldsymbol{z}\boldsymbol{s}^{\mathrm{T}}\right]\right) \\ & - \operatorname{tr}\left(\mathrm{E}\left[\boldsymbol{s}\boldsymbol{z}^{\mathrm{T}}\right] M^{\mathrm{T}}\right) + \operatorname{tr}\left(M\mathrm{E}\left[\boldsymbol{z}\boldsymbol{z}^{\mathrm{T}}\right] M^{\mathrm{T}}\right). \end{aligned} \tag{3.47}$$

Since we must differentiate $\operatorname{tr}(P)$ with respect to a matrix, we require some results from matrix calculus, namely:

$$\frac{\partial}{\partial A}\left[\operatorname{tr}\left(ABA^{\mathrm{T}}\right)\right] = 2AB, \tag{3.48}$$

and

$$\frac{\partial}{\partial A}\left[\operatorname{tr}(AB)\right] = \frac{\partial}{\partial A}\left[\operatorname{tr}\left(B^{\mathrm{T}}A^{\mathrm{T}}\right)\right] = B^{\mathrm{T}}, \quad \text{if } AB \text{ is a square matrix.} \tag{3.49}$$

Differentiating (3.47) yields

$$\frac{\partial \operatorname{tr}(P)}{\partial M} = -2\mathrm{E}\left[\boldsymbol{s}\boldsymbol{z}^{\mathrm{T}}\right] + 2M\mathrm{E}\left[\boldsymbol{z}\boldsymbol{z}^{\mathrm{T}}\right],$$

and setting this result equal to the zero matrix and solving for M gives

$$M = \mathrm{E}\left[\boldsymbol{s}\boldsymbol{z}^{\mathrm{T}}\right]\left(\mathrm{E}\left[\boldsymbol{z}\boldsymbol{z}^{\mathrm{T}}\right]\right)^{-1}. \tag{3.50}$$

Thus the LMMSE estimate of $\hat{\boldsymbol{s}}$ given $\boldsymbol{z}$ is

$$\hat{\boldsymbol{s}} = \mathrm{E}\left[\boldsymbol{s}\boldsymbol{z}^{\mathrm{T}}\right]\left(\mathrm{E}\left[\boldsymbol{z}\boldsymbol{z}^{\mathrm{T}}\right]\right)^{-1}\boldsymbol{z}. \tag{3.51}$$

In the random vector case, just as in the scalar RV case, the LMMSE estimate $\hat{\boldsymbol{s}}$ is unique.

We now have the following useful theorem:

Theorem 3.6 (Orthogonality Principle for Vector RVs) *Let $\boldsymbol{s}$ and $\boldsymbol{z}$ be jointly distributed random vectors, where $\boldsymbol{s}$ is an m-vector and $\boldsymbol{z}$ is a q-vector, and let $\hat{\boldsymbol{s}} = \alpha(\boldsymbol{z})$ be the LMMSE estimate of $\boldsymbol{s}$ given $\boldsymbol{z}$. Then the estimation error $\boldsymbol{s} - \hat{\boldsymbol{s}}$ is orthogonal to $\boldsymbol{z}$, i.e.,*

$$E\left[(\boldsymbol{s} - \hat{\boldsymbol{s}})\boldsymbol{z}^T\right] = \mathbf{0}, \tag{3.52}$$

where $\mathbf{0}$ is the $m \times q$ zero matrix.

Proof. Observe that $\hat{\boldsymbol{s}}$ is given by (3.51). Then

$$\mathrm{E}\left[(\boldsymbol{s} - \hat{\boldsymbol{s}})\boldsymbol{z}^{\mathrm{T}}\right] = \mathrm{E}\left[\left\{\boldsymbol{s} - \mathrm{E}\left[\boldsymbol{s}\boldsymbol{z}^{\mathrm{T}}\right]\left(\mathrm{E}\left[\boldsymbol{z}\boldsymbol{z}^{\mathrm{T}}\right]\right)^{-1}\boldsymbol{z}\right\}\boldsymbol{z}^{\mathrm{T}}\right],$$

and since $\mathrm{E}\left[\boldsymbol{s}\boldsymbol{z}^{\mathrm{T}}\right]$ and $\left(\mathrm{E}\left[\boldsymbol{z}\boldsymbol{z}^{\mathrm{T}}\right]\right)^{-1}$ are constant matrices, it follows that

$$\begin{aligned}\mathrm{E}\left[(\boldsymbol{s} - \hat{\boldsymbol{s}})\boldsymbol{z}^{\mathrm{T}}\right] &= \mathrm{E}\left[\boldsymbol{s}\boldsymbol{z}^{\mathrm{T}}\right] - \mathrm{E}\left[\boldsymbol{s}\boldsymbol{z}^{\mathrm{T}}\right]\left(\mathrm{E}\left[\boldsymbol{z}\boldsymbol{z}^{\mathrm{T}}\right]\right)^{-1}\mathrm{E}\left[\boldsymbol{z}\boldsymbol{z}^{\mathrm{T}}\right] \\ &= \mathrm{E}\left[\boldsymbol{s}\boldsymbol{z}^{\mathrm{T}}\right] - \mathrm{E}\left[\boldsymbol{s}\boldsymbol{z}^{\mathrm{T}}\right] \\ &= \mathbf{0}.\end{aligned}$$

Q.E.D. ■

Another theorem, analogous to Theorem 3.3, suggests that the orthogonality principle can be used to find the LMMSE estimate:

Theorem 3.7 *Let $\alpha(\boldsymbol{z})$ be a linear estimate of $\boldsymbol{s}$ given $\boldsymbol{z}$. Then α minimizes the MSE if and only if the error $\boldsymbol{s} - \alpha(\boldsymbol{z})$ is orthogonal to the measurement $\boldsymbol{z}$,*

$$E\left[(\boldsymbol{s} - \alpha(\boldsymbol{z}))\,\boldsymbol{z}^T\right] = \mathbf{0}. \tag{3.53}$$

This result implies that if we can find the linear estimate $\alpha(\boldsymbol{z})$ that satisfies (3.53), then $\hat{\boldsymbol{s}} = \alpha(\boldsymbol{z})$ is the LMMSE estimate. Compare this corollary with Theorems 3.2 and 3.3, the orthogonality principle for unconstrained, generally nonlinear, estimators. The latter theorems imply that we must design α such that the error $\boldsymbol{s} - \alpha(\boldsymbol{z})$ is orthogonal to *all* functions $\gamma(\boldsymbol{z})$. Since the form of $\gamma(\boldsymbol{z})$ is unconstrained, it is virtually impossible to consider every possible form of $\gamma(\boldsymbol{z})$. Often the only way to find the optimum nonlinear estimator is to find the conditional expectation $\mathrm{E}\left[\boldsymbol{s}|\boldsymbol{z}\right]$, which is usually very difficult.

However, when α is limited to the linear case, Theorem 3.7 states that we only need to design α such that the error is orthogonal to $\boldsymbol{z}$. This property makes it possible to design α using the orthogonality principle in the LTI case.

For example, we can find λ in (3.40) by applying this theorem. From (3.53) the value of λ that produces the MMSE estimate must satisfy

$$\mathrm{E}\left[(\boldsymbol{s}-\lambda\boldsymbol{z})\beta\boldsymbol{z}\right]=0,$$

so that

$$\beta\mathrm{E}\left[\boldsymbol{s}\boldsymbol{z}\right]=\beta\lambda\mathrm{E}\left[\boldsymbol{z}^2\right]$$

Hence,

$$\lambda=\mathrm{E}\left[\boldsymbol{s}\boldsymbol{z}\right]/\mathrm{E}\left[\boldsymbol{z}^2\right].$$

the same result as (3.42).

Let us examine the linear estimator in greater detail and assume that $\boldsymbol{s}$ and $\boldsymbol{z}$ are both zero-mean. In this case we may write λ as

$$\lambda=\frac{\mathrm{E}\left[\boldsymbol{s}\boldsymbol{z}\right]-\mathrm{E}\left[\boldsymbol{s}\right]\mathrm{E}\left[\boldsymbol{z}\right]}{\mathrm{E}\left[\boldsymbol{z}^2\right]-\left(\mathrm{E}\left[\boldsymbol{z}\right]\right)^2}=\frac{\mathrm{Cov}\left[\boldsymbol{s},\boldsymbol{z}\right]}{\sigma_z^2}.$$

Defining the *correlation coefficient* by

$$\rho_{s,z}=\mathrm{Cov}\left[\boldsymbol{s},\boldsymbol{z}\right]/\sqrt{\sigma_s^2\sigma_z^2}, \tag{3.54}$$

we have that

$$\lambda=\frac{\sigma_s}{\sigma_z}\rho_{s,z}.$$

Then the LMMSE estimate is

$$\hat{\boldsymbol{s}}_{\mathrm{LMMSE}}=\rho_{s,z}\frac{\sigma_s}{\sigma_z}\boldsymbol{z}. \tag{3.55}$$

The weighting term in (3.55) balances the signal and noise. σ_s^2 provides a measure of the signal power in $\boldsymbol{z}$, so a large value of σ_s^2 increases the weight on $\boldsymbol{z}$. Conversely, σ_z^2 represents the variance of the observation or the noise power. If the noise power is very large, it reduces our confidence that $\boldsymbol{z}$ accurately reflects $\boldsymbol{s}$, so the weight decreases.

Suppose that $\boldsymbol{s}$ and $\boldsymbol{z}$ are uncorrelated, so that $\rho_{s,z}=0$. The linear estimator relies only upon the second-order moments between $\boldsymbol{s}$ and $\boldsymbol{z}$, but uncorrelatedness means that $\boldsymbol{z}$ and $\boldsymbol{s}$ are not related via their second-order moments. Therefore, a linear estimate does not provide any new information about $\boldsymbol{s}$. The LMMSE estimate is just zero, which is the mean of $\boldsymbol{s}$. On the other hand, the greater the correlation between $\boldsymbol{s}$ and $\boldsymbol{z}$, the greater $|\rho_{s,z}|$ becomes, with a maximum magnitude of one. As a result, more weight is given to $\boldsymbol{z}$ if $\boldsymbol{s}$ and $\boldsymbol{z}$ are highly correlated.

Now let us examine the performance of this estimator. The average error is

$$\mathrm{E}[s - \hat{s}_{\mathrm{MMSE}}] = \mathrm{E}[s - \lambda z] = 0.$$

Hence, the LMMSE estimate of s given z is unbiased.

The associated MSE is

$$\mathrm{MSE} = \mathrm{E}\left[(s - \hat{s}_{\mathrm{LMMSE}})^2\right].$$

Substituting Equation (3.55) for $\hat{s}_{\mathrm{LMMSE}}$ and expanding the square, we obtain

$$\begin{aligned}\mathrm{MSE} &= \mathrm{E}\left[s^2\right] + \frac{\sigma_s^2}{\sigma_z^2}\rho_{s,z}^2 \mathrm{E}\left[z^2\right] - 2\frac{\sigma_s}{\sigma_z}\rho_{s,z}\mathrm{E}[sz] \\ &= \sigma_s^2 + \sigma_s^2\rho_{s,z}^2 - 2\frac{\sigma_s}{\sigma_z}\rho_{s,z}\mathrm{Cov}[s,z] \\ &= \sigma_s^2 + \sigma_s^2\rho_{s,z}^2 - 2\frac{\sigma_s}{\sigma_z}\rho_{s,z}\left(\rho_{s,z}\sigma_s\sigma_z\right) \\ &= \sigma_s^2\left(1 - \rho_{s,z}^2\right).\end{aligned}$$

We see that the MSE is greatest when $\rho_{s,z} = 0$, meaning s and z have no linear correlation. This result is sensible, since we cannot expect an affine estimator to perform very well if s and z are not themselves linearly related.

When $\rho_{s,z} = \pm 1$, the MSE is a minimum; in fact, it is zero. A value of $\rho_{s,z} = \pm 1$ implies that $z = \mu s$, for some constant μ; i.e., s and z are completely linearly correlated. We should expect our estimator to perform best in this situation since the form of the estimator exactly models the relationship between s and z.

Overall Optimality

Recall that the optimal MMSE estimate over all estimators is the conditional expectation, $\mathrm{E}[s|z]$. We perform *linear* MMSE (LMMSE) estimation because often the conditional expectation is difficult or impossible to compute. This compromise means that the LMMSE estimate is usually suboptimal to the optimal MMSE estimate (over all possible estimators). However, when s and z are jointly Gaussian, the LMMSE estimate is also the optimal MMSE estimate [1]. The next example demonstrates such a case.

Example 3.9 An Overall MMSE Estimate

Suppose that s and z have a zero-mean bivariate Gaussian distribution with covariance matrix P given by

$$P = \begin{bmatrix} \sigma_s^2 & \mathrm{Cov}[s,z] \\ \mathrm{Cov}[z,s] & \sigma_z^2 \end{bmatrix}.$$

From Section 3.3, the overall MMSE estimate of $\boldsymbol{s}$ given $\boldsymbol{z}$ is $\hat{s}_{\mathrm{MMSE}} = \mathrm{E}[\boldsymbol{s}|\boldsymbol{z}]$. We now compute $\mathrm{E}[\boldsymbol{s}|\boldsymbol{z}]$.

Let $\rho = \rho_{s,z}$. Then the joint PDF of $\boldsymbol{s}$ and $\boldsymbol{z}$ is

$$f_{\boldsymbol{s},\boldsymbol{z}}(s,z) = \frac{1}{2\pi\sigma_s\sigma_z\sqrt{(1-\rho)^2}} \exp\left\{-\frac{1}{2(1-\rho)^2}\left[\left(\frac{s}{\sigma_s}\right)^2 - \frac{2\rho s z}{\sigma_s\sigma_z} + \left(\frac{z}{\sigma_z}\right)^2\right]\right\}.$$

We find the conditional PDF of $\boldsymbol{s}$ given $\boldsymbol{z} = z$ to be

$$f_{\boldsymbol{s}}(s|\boldsymbol{z}=z) = \frac{1}{\sqrt{2\pi}\sigma_s\sqrt{(1-\rho)^2}} \exp\left\{-\frac{1}{2\sigma_s^2(1-\rho)^2}\left(s - \frac{\mathrm{E}[\boldsymbol{s}\boldsymbol{z}]}{\mathrm{E}[\boldsymbol{z}^2]}z\right)^2\right\}. \quad (3.56)$$

Observe that (3.56) is the PDF of $\mathcal{N}\left((\mathrm{E}[\boldsymbol{s}\boldsymbol{z}]/\mathrm{E}[\boldsymbol{z}^2])z, \sigma_s^2(1-\rho)^2\right)$. Hence,

$$\hat{s}_{\mathrm{MMSE}} = \mathrm{E}[\boldsymbol{s}|\boldsymbol{z}=z] = \frac{\mathrm{E}[\boldsymbol{s}\boldsymbol{z}]}{\mathrm{E}[\boldsymbol{z}^2]}z. \quad (3.57)$$

Now recall (3.43), the *linear* MMSE estimate,

$$\hat{s}_{\mathrm{LMMSE}} = \frac{\mathrm{E}[\boldsymbol{s}\boldsymbol{z}]}{\mathrm{E}[\boldsymbol{z}^2]}z,$$

Equations (3.57) and (3.43) are identical, so the conditional expectation is a linear function of $\boldsymbol{z}$ in this case. Thus, *for the zero-mean, jointly Gaussian case, the* ***linear*** *MMSE estimate is also the* ***best overall*** *MMSE estimate.*

3.5 Comparison of Estimation Methods

We have just presented several common optimality criteria and now pause to compare them. Table 3.2 presents a brief summary of the estimation criteria.

At first glance, ML and MAP estimation may appear identical, but they differ in several respects. ML asks, "Given z, what value of $\boldsymbol{s}$ is most likely to have generated z?", while MAP asks, "Given z, what value of $\boldsymbol{s}$ is most likely to have occurred?". There is a subtle but important difference here. In ML, we seek the value of s that maximizes the likelihood function $f_{\boldsymbol{z}}(z|\boldsymbol{s}=s)$, which is the *a priori* density $f_{\boldsymbol{z}}(z|\boldsymbol{s}=s)$ viewed as a function of s (rather than z). In MAP, we seek the value of s that maximizes the *a posteriori* density $f_{\boldsymbol{s}}(s|\boldsymbol{z}=z)$. Also note that ML is a non-Bayesian estimation method, while MAP is a Bayesian estimation technique.

Both ML and MAP estimation seek a most-likely value of s for a given measurement z. In this way, ML and MAP estimation find a best estimate for the current realization z. MMSE estimation differs from them in that it reduces the MSE. In this sense, MMSE estimation produces an estimate that is optimal over an ensemble. For a given realization z, the ML or MAP estimates may have a smaller square-error than the MMSE estimate. However, over a large number of realizations, the latter will have a smaller MSE.

LMMSE estimation represents a trade-off between overall optimality and tractability. Unlike the other methods, it does not require knowledge of likelihood functions or conditional densities. Instead, it relies only upon the second-order moments of $\boldsymbol{s}$ and $\boldsymbol{z}$, which are often readily available in many situations. As a result, LMMSE estimation is not directly dependent upon probability densities, unlike the other criteria. These properties make the estimator relatively simple to determine and implement, at the price of possibly lower performance compared to the MMSE estimator.

Example 3.10 MMSE and ML Estimates

Consider the RVs of Example 3.2. Let us now find the MMSE estimate. First, we need $f_{\boldsymbol{z}}(z)$ in order to find $f_{\boldsymbol{s}}(s|\boldsymbol{z}=z)$.

$$\begin{aligned} f_{\boldsymbol{z}}(z) &= \int_0^4 f_{\boldsymbol{s},\boldsymbol{z}}(s,z)\,ds = \int_0^4 \tfrac{1}{12}(s+z)e^{-z}\,ds \\ &= \tfrac{1}{12}e^{-z}\left[\tfrac{1}{2}s^2 + sz\right]_{s=0}^4 = \frac{z+2}{3}e^{-z}. \end{aligned}$$

Thus,

$$f_{\boldsymbol{s}}(s|\boldsymbol{z}=z) = \left(\frac{1}{4}\right)\frac{s+z}{z+2}, \quad 0 \le s \le 4,\ 0 \le z < \infty.$$

Then the MMSE estimate is

$$\begin{aligned} \hat{s}_{\mathrm{MMSE}} &= \mathrm{E}\left[\boldsymbol{s}|\boldsymbol{z}=z\right] = \int_0^4 s f_{\boldsymbol{s}}(s|\boldsymbol{z}=z)\,ds = \int_0^4 s\left(\frac{1}{4}\right)\frac{s+z}{z+2}\,ds \\ &= \frac{1}{4(z+2)}\left[\tfrac{1}{3}s^3 + \tfrac{1}{2}s^2 z\right]_{s=0}^4 = \frac{6z+16}{3z+6}. \end{aligned}$$

Let us compare the MSE of the maximum-likelihood estimate of Example 3.2 and the MMSE estimate. The MSE associated with $\hat{\boldsymbol{s}}_{\mathrm{ML}}$ is

$$\begin{aligned} \mathrm{MSE}\left(\hat{\boldsymbol{s}}_{\mathrm{ML}}\right) &= \int_0^\infty \int_0^4 \left(s - \hat{s}_{\mathrm{ML}}\right)^2 f_{\boldsymbol{s}\boldsymbol{z}}(s,z)\,ds\,dz \\ &= \int_0^1 \int_0^4 (s-4)^2\, \tfrac{1}{12}(s+z)e^{-z}\,ds\,dz \\ &\quad + \int_1^\infty \int_0^4 (s-0)^2\, \tfrac{1}{12}(s+z)e^{-z}\,ds\,dz \\ &\approx 4.8636, \end{aligned}$$

where we have evaluated the integrals numerically.

The MSE of the MMSE estimate is

$$\begin{aligned} \operatorname{MSE}(\hat{s}_{\text{MMSE}}) &= \int_0^\infty \int_0^4 \left(s - \frac{6z+16}{3z+6}\right)^2 \tfrac{1}{12}(s+z)e^{-z}\, ds\, dz \\ &\approx 1.1192. \end{aligned}$$

As expected, the MMSE estimate has a lower MSE than the ML estimate. Also, the MMSE estimate $\hat{s}_{\text{MMSE}}$ differs from the ML estimate $\hat{s}_{\text{ML}}$. This result is not surprising, since the two criteria define optimality in different ways.

Problems

3.1. Prove Theorem 4.1. Hint: Use equation (4.9), the orthogonality principle for LTI estimators.

3.2. Prove Theorem 4.2. Hint: Follow a procedure similar to the proof of Theorem 3.3, where α_1, α_2, and γ are restricted to the class of LTI functions of $\boldsymbol{Z}$.

3.3. Given RVs $\boldsymbol{s}$ and $\boldsymbol{z}$, assume an affine estimator of the form $\hat{\boldsymbol{s}} = a\boldsymbol{z} + b$.

(a) Find the values of a and b that produce the affine MMSE estimate. Show that this estimate reduces to (3.55) when $\boldsymbol{s}$ and $\boldsymbol{z}$ are both zero-mean.

(b) Show that the estimate may be written in the form

$$\hat{\boldsymbol{s}} = \mathrm{E}[\boldsymbol{s}] + \rho_{s,z}\frac{\sigma_s}{\sigma_z}(\boldsymbol{z} - \mathrm{E}[\boldsymbol{z}]).$$

(c) Show that the affine MMSE estimate is unbiased.

(d) Show that the MSE can be expressed as

$$\operatorname{MSE} = \sigma_s^2\left(1 - \rho_{s,z}^2\right).$$

3.4. Suppose that $\boldsymbol{s}$ and $\boldsymbol{z}$ have a bivariate Gaussian distribution with mean vector $\underline{\eta}$ and covariance matrix P,

$$\underline{\eta} = \begin{bmatrix} \eta_s \\ \eta_z \end{bmatrix} \quad \text{and} \quad P = \begin{bmatrix} \sigma_s^2 & \operatorname{Cov}[\boldsymbol{s},\boldsymbol{z}] \\ \operatorname{Cov}[\boldsymbol{z},\boldsymbol{s}] & \sigma_z^2 \end{bmatrix}.$$

With $\rho = \rho_{s,z}$, the joint PDF of $\boldsymbol{s}$ and $\boldsymbol{z}$ is

$$\begin{aligned} f_{\boldsymbol{s},\boldsymbol{z}}(s,z) = \frac{1}{2\pi\sigma_s\sigma_z\sqrt{(1-\rho)^2}} \exp\Bigg\{-\frac{1}{2(1-\rho)^2}\Bigg[\left(\frac{s-\eta_s}{\sigma_s}\right)^2 \\ - \frac{2\rho(s-\eta_s)(z-\eta_z)}{\sigma_s\sigma_z} + \left(\frac{z-\eta_z}{\sigma_z}\right)^2\Bigg]\Bigg\}. \end{aligned}$$

	Maximum likelihood (ML)	Maximum *a posteriori* (MAP)
Motivation	Given z, what value of $\boldsymbol{s}$ is most likely to have produced z?	Given z, what value of $\boldsymbol{s}$ is most likely to have occurred?
Objective	Maximize the likelihood function $f_{\boldsymbol{z}}(z \mid \boldsymbol{s}=s)$.	Maximize the conditional density $f_{\boldsymbol{s}}(s \mid \boldsymbol{z}=z)$. Via Bayes' rule, equivalently maximize $f_{\boldsymbol{z}}(z \mid \boldsymbol{s}=s) f_{\boldsymbol{s}}(s)$.
Estimate	$\hat{s}_{\text{ML}}$ $=$ value of s that maximizes $f_{\boldsymbol{z}}(z \mid \boldsymbol{s}=s)$ $=$ value of s such that $\partial f_{\boldsymbol{z}}(z \mid \boldsymbol{s}=s)/\partial s = 0$.	$\hat{s}_{\text{MAP}}$ $=$ value of s that maximizes $f_{\boldsymbol{s}}(s \mid \boldsymbol{z} = z)$ $=$ value of s such that $\partial f_{\boldsymbol{s}}(s \mid \boldsymbol{z} = z)/\partial s = 0$.
Required Knowledge	Likelihood function $f_{\boldsymbol{z}}(z \mid \boldsymbol{s}=s)$.	Density $f_{\boldsymbol{s}}(s \mid \boldsymbol{z}=z)$. Bayes' formula implies that knowledge about $\boldsymbol{s}$ is needed through $f_{\boldsymbol{s}}(s)$.

	Minimum mean-square error (MMSE)	Linear MMSE (LMMSE)
Motivation	Given the RV $\boldsymbol{z}$, what estimate of $\boldsymbol{s}$ gives the smallest MSE?	Given the RV $\boldsymbol{z}$, what linear function $\hat{\boldsymbol{s}} = \lambda \boldsymbol{z}$ gives the smallest MSE?
Objective	Minimize the MSE $\mathrm{E}\left[(\boldsymbol{s}-\hat{\boldsymbol{s}})^2\right]$.	Find λ to minimize $\mathrm{E}\left[(\boldsymbol{s}-\lambda \boldsymbol{z})^2\right]$.
Estimate	$\hat{\boldsymbol{s}}_{\text{MMSE}}$ $= \mathrm{E}[\boldsymbol{s} \mid \boldsymbol{z}]$ $= \int_{-\infty}^{\infty} s f_{\boldsymbol{s}}(s \mid \boldsymbol{z})\, ds$.	$\hat{\boldsymbol{s}}_{\text{LMMSE}}$ $= \lambda \boldsymbol{z}$, where $\lambda = \mathrm{E}[\boldsymbol{s}\boldsymbol{z}] / \mathrm{E}\left[\boldsymbol{z}^2\right]$.
Required Knowledge	Density $f_{\boldsymbol{s}}(s \mid \boldsymbol{z})$. Bayes' formula implies that knowledge about $\boldsymbol{s}$ is needed through $f_{\boldsymbol{s}}(s)$.	Cross-correlation of $\boldsymbol{s}$ and $\boldsymbol{z}$, $\mathrm{E}[\boldsymbol{s}\boldsymbol{z}]$; second moment of $\boldsymbol{z}$, $\mathrm{E}\left[\boldsymbol{z}^2\right]$.

Table 3.2. Comparison of optimality criteria

(a) Show that the conditional PDF of s given $z = z$ is

$$f_s(s|z=z) = \frac{1}{\sqrt{2\pi}\sigma_s\sqrt{(1-\rho)^2}} \exp\left\{-\frac{1}{2\sigma_s^2(1-\rho)^2} \times \left[s - \eta_s - \rho\frac{\sigma_s}{\sigma_z}(z-\eta_z)\right]^2\right\}.$$

(b) Find the estimate $\hat{s}_{\text{MMSE}} = \mathrm{E}[s|z]$, which is the overall MMSE estimate of s given z.

(c) Compare the expression for $\hat{s}_{\text{MMSE}}$ from part (b) with the result of Problem 3.3(b), the *affine* MMSE estimate of s given z. This result shows that for the arbitrary-mean Gaussian case, the affine MMSE estimate is also the best overall MMSE estimate.

3.5. Two measurements z_1 and z_2 of the random variable s are given by

$$z_1 = s + v_1, \quad \text{and} \quad z_2 = s + v_2,$$

where the noise terms v_1 and v_2 are zero-mean Gaussian random variables with

$$\mathrm{E}\left[v_1^2\right] = \mathrm{E}\left[v_2^2\right] = 2 \quad \text{and} \quad \mathrm{E}\left[v_1 v_2\right] = 1.$$

Compute the ML estimate $\hat{s}_{\text{ML}}$ of s based on the measurements z_1 and z_2. Express your solution as a function of z_1 and z_2 with all coefficients evaluated.

3.6. Consider the signal plus noise $z = s + v$, where s and v are independent RVs with density functions

$$f_s(s) = A\exp\left(-5(s-5)^2\right) \quad \text{and} \quad f_v(v) = B\exp\left(-5(v-5)^2\right),$$

where A and B are real numbers.

(a) Determine the ML estimate $\hat{s}_{\text{ML}}$.

(b) Determine the MAP estimate $\hat{s}_{\text{MAP}}$.

(c) Now suppose that $z = \alpha s + v$, where s and v are the RVs defined above, and α is a RV that takes on the values $\alpha = 1$ and $\alpha = 2$ with $P\{\alpha = 2\} = 0.75$. Determine the ML estimate $\hat{s}_{\text{ML}}$.

3.7. Consider the measurement $z = s + v$, where s and v are independent RVs, with both s and v uniformly distributed from 0 to 1. For each of the following parts, express the desired estimate as a function of the sample measurement $z = z$.

(a) Determine the LS estimate $\hat{s}_{\text{LS}}$ given $z = z$.

(b) Determine the ML estimate $\hat{s}_{\text{ML}}$ given $z = z$.

(c) Determine the MAP estimate $\hat{s}_{\text{MAP}}$ given $z = z$.

(d) Determine the LMMSE estimate $\hat{s}_{\text{LMMSE}}$ given $z = z$.

3.8. Measurements of two unknown constants c_1 and c_2 are given by

$$z_1(n) = c_1 + c_2 + v_1(n), \qquad z_2(n) = c_1 - c_2 + v_2(n),$$

where $v_1(n)$ and $v_2(n)$ are zero-mean white Gaussian noise with

$$\mathrm{E}\left[v_1^2\right] = \mathrm{E}\left[v_2^2\right] = 2 \quad \text{and} \quad \mathrm{E}\left[v_1 v_2\right] = 1.$$

Let $c = \begin{bmatrix} c_1 & c_2 \end{bmatrix}^{\mathrm{T}}$. Determine the ML estimate $\hat{c}_{\mathrm{ML}}$ of c at time n based on the measurements $z(1)$, $z(2)$, ... , $z(n)$. All coefficients in your expression should be evaluated.

3.9. A random signal $s(n)$ is modeled by $s(n+1) = as(n)$, where a is a real number, and measurements of $s(n)$ are given by $z(n) = s(n) + v(n)$. For all values of i and j, $s(i)$ and $v(j)$ are independent, $s(n)$ is Gaussian with mean b and variance σ_s^2, and $v(n)$ is zero-mean Gaussian white noise with variance σ_v^2 .

(a) Determine the LS estimate $\hat{s}_{\mathrm{LS}}(n)$ at time n given the measurements $z(i) = z(i)$ for $i = 1, 2, \ldots, n$.

(b) Determine the ML estimate $\hat{s}_{\mathrm{ML}}(n)$ at time n given the measurements $z(i) = z(i)$ for $i = 1, 2, \ldots, n$.

(c) Determine the MAP estimate $\hat{s}_{\mathrm{MAP}}(n)$ at time n given the measurements $z(i) = z(i)$ for $i = 1, 2, \ldots, n$.

(d) Determine the LMMSE estimate $\hat{s}_{\mathrm{LMMSE}}(n)$ at time n given the measurements $z(i) = z(i)$ for $i = 1, 2, \ldots, n$.

3.10. Measurements are given by $z(n) = (-1)^n a + v(n)$, where a is a Gaussian RV with mean b and variance σ_s^2, $v(n)$ is zero-mean Gaussian white noise with variance σ_v^2, and a and $v(n)$ are independent.

(a) Determine the MAP estimate $\hat{a}_{\mathrm{MAP}}(n)$ at time n given the measurements $z(i) = z(i)$ for $i = 1, 2, \ldots, n$.

(b) Given $z(i) = z(i)$ for $i = 1, 2, \ldots, n$, determine the estimate $\hat{a}(n)$ of a at time n that minimizes $\mathrm{E}\left[(a - \hat{a}(n))^2\right]$.

(c) Given $z(i) = z(i)$ for $i = 1, 2, \ldots, n$, determine the coefficient $h(n)$ in the estimate $\hat{a}(n) = h(n)z(n)$ so that $\mathrm{E}\left[(a - \hat{a}(n))^2\right]$ is minimized. Express $h(n)$ in the simplest possible form.

3.11. A signal process is given by $s(n) = a\cos(\pi n)$, where a is a RV with $\mathrm{E}\left[a^2\right] = 10$. Measurements of $s(n)$ are given by $z(n) = s(n) + v(n)$, where $v(n)$ is zero-mean white noise with variance 2 and $v(n)$ is independent of a for all n.

(a) Determine the orthogonality conditions for computing the LMMSE estimate $\hat{a}_{\mathrm{LMMSE}}$ at time n given the measurements $z(i) = z(i)$ for $i = 1, 2, \ldots, n$.

(b) Determine the LMMSE estimate $\hat{a}_{\mathrm{LMMSE}}$ at time n given the measurements $z(i) = z(i)$ for $i = 1, 2, \ldots, n$.

3.12. A random signal $s(n)$ has the two-dimensional state model $x(n+1) = \Phi x(n) + w(n)$, $s(n) = cx(n)$ where

$$\Phi = \begin{bmatrix} 2 & 1 \\ 1 & 1 \end{bmatrix}, \quad c = \begin{bmatrix} 1 & 1 \end{bmatrix}, \quad \mathrm{E}\left[x(0)\right] = \begin{bmatrix} 2 \\ 2 \end{bmatrix}, \quad \mathrm{Cov}\left[x(0)\right] = \begin{bmatrix} 1 & 0 \\ 0 & 1 \end{bmatrix}.$$

$w(n)$ is zero-mean Gaussian white noise with covariance $3I$, where I is the 2×2 identity matrix, and $x(0)$ is independent of $w(n)$ for all n. Noisy measurements $z(n)$ of $s(n)$ are given by $z(n) = s(n) + v(n)$, where $v(n)$ is zero-mean Gaussian white noise with variance 1, $v(n)$ is independent of $x(0)$ for all n, and $v(n)$ is independent of $w(n)$. Determine the MAP estimate of $x(n)$ at time n based on the two measurements $z(n)$ and $z(n-1)$.

Chapter 4

The Wiener Filter

This chapter introduces the Wiener filter, which is used in many control and signal-processing applications. The Wiener filter is a LTI filter, and it may have different forms, depending upon the constraints imposed on the filter (e.g., finite or infinite impulse response, and causality). For the given constraints, the Wiener filter produces the LMMSE estimate of a signal $s(n)$.

We first present the finite impulse response Wiener filter, which introduces the Wiener-Hopf equation. Then the Wiener filter is allowed to be noncausal and have an infinite impulse response. Next, to find the causal version of this filter, we introduce the ideas of spectral factorization and causal-part extraction. At the end of the chapter, the three forms of the Wiener filter are compared.

4.1 Linear Time-Invariant MMSE Filters

In this chapter we consider the case when the estimator is a *linear time-invariant* (LTI) system so that the estimate $\hat{s}(n)$ of a signal $s(n)$ is given by the convolution representation

$$\hat{s}(n) = h(n) * z(n) = \sum_{i=-\infty}^{\infty} h(n-i)z(i), \tag{4.1}$$

where the $z(i)$ are the measurements and $h(n)$ is the impulse response of the estimator, which is also called a *filter*. By limiting our estimator to be a LTI system, we may utilize frequency domain tools, namely the Fourier and z-transforms.

Since the convolution operation $*$ is commutative, the estimate $\hat{s}(n)$ given by (4.1) can be written in the form

$$\hat{s}(n) = \sum_{i=-\infty}^{\infty} h(i)z(n-i). \tag{4.2}$$

Let $\mathcal{H}$ denote the region of support (or simply the *support*) of $h(n)$, defined by

$$\mathcal{H} = \{n : h(n) \neq 0\}.$$

Then (4.2) can be rewritten as

$$\hat{s}(n) = \sum_{i\in\mathcal{H}} h(i)z(n-i). \tag{4.3}$$

Finally, we model the signal $s(n)$ and the measurement signal $z(n)$ as jointly-distributed random signals $\boldsymbol{s}(n)$ and $\boldsymbol{z}(n)$, respectively, defined on a probability space S. Then the estimate can be viewed as a random signal $\hat{\boldsymbol{s}}(n)$ given by

$$\hat{\boldsymbol{s}}(n) = \sum_{i\in\mathcal{H}} h(i)\boldsymbol{z}(n-i). \tag{4.4}$$

In this chapter it is also assumed that $\boldsymbol{s}(n)$ and $\boldsymbol{z}(n)$ are jointly wide-sense stationary (JWSS).

The goal is to determine the impulse response $h(n)$ of the estimator so that the mean-square estimation error is minimized; that is,

$$\text{MSE} = \text{E}\left[(\boldsymbol{s}(n) - \hat{\boldsymbol{s}}(n))^2\right] \tag{4.5}$$

should be minimized. Inserting (4.4) into (4.5) gives

$$\begin{aligned}\text{MSE} &= \text{E}\left[\left(\boldsymbol{s}(n) - \sum_{i\in\mathcal{H}} h(i)\boldsymbol{z}(n-i)\right)\left(\boldsymbol{s}(n) - \sum_{j\in\mathcal{H}} h(j)\boldsymbol{z}(n-j)\right)\right] \\ &= \text{E}\left[\boldsymbol{s}^2(n)\right] - 2\sum_{i\in\mathcal{H}} h(i)\text{E}\left[\boldsymbol{s}(n)\boldsymbol{z}(n-i)\right] \\ &\quad + \sum_{i\in\mathcal{H}}\sum_{j\in\mathcal{H}} h(i)h(j)\text{E}\left[\boldsymbol{z}(n-i)\boldsymbol{z}(n-j)\right]. \end{aligned} \tag{4.6}$$

To minimize the MSE we take the partial derivatives of the above expression with respect to $h(i)$, for each $h(i) \neq 0$. Then we set the result equal to zero, which yields

$$\begin{aligned}\frac{\partial \text{MSE}}{\partial h(i)} &= -2\text{E}\left[\boldsymbol{s}(n)\boldsymbol{z}(n-i)\right] + 2\sum_{j\in\mathcal{H}} h(j)\text{E}\left[\boldsymbol{z}(n-i)\boldsymbol{z}(n-j)\right] \\ &= 0.\end{aligned}$$

Solving this expression, we find

$$\sum_{j\in\mathcal{H}} h(j)\mathrm{E}\left[\boldsymbol{z}(n-i)\boldsymbol{z}(n-j)\right] = \mathrm{E}\left[\boldsymbol{s}(n)\boldsymbol{z}(n-i)\right], \quad i \in \mathcal{H} \tag{4.7}$$

which we may express in the form of

$$\sum_{j\in\mathcal{H}} h(j)R_z(j-i) = R_{sz}(i), \quad i \in \mathcal{H}, \tag{4.8}$$

where $R_z(k)$ is the autocorrelation function of $\boldsymbol{z}(n)$ and $R_{sz}(k)$ is the cross-correlation function of $\boldsymbol{s}(n)$ and $\boldsymbol{z}(n)$.

Equation (4.8) is the discrete-time *Wiener-Hopf equation*. It is the basis for the derivation of the Wiener filter, which will be presented in Sections 4.2–4.4. It describes a system of linear equations, one equation for each $i \in \mathcal{H}$. We solve this system to find the values of $h(n)$ and thereby determine the filter.

Some comments regarding the above procedure are in order and are analogous to Section 3.4. We have chosen to design a *LTI* estimator and adopted the MSE criterion for optimality. Several significant consequences arise from these two choices:

- Only the second-order moments (autocorrelation and cross-correlation) are necessary to specify the optimum LTI filter.

- The expression for the MSE, Equation (4.6), is quadratic in the values of $h(n)$. As a result,

  - the MSE over all filters of the form (4.1) has a unique minimum;

  - this minimum occurs where

  $$\frac{\partial \mathrm{MSE}}{\partial h(i)} = 0, \quad i \in \mathcal{H};$$

  - the impulse response $h(n)$ of the optimal LTI filter is unique.

- The solution for the impulse response $h(n)$ of the optimal LTI MMSE filter is described by the Wiener-Hopf equation (4.8). An important property of the Wiener-Hopf equation is that it specifies a system of *linear* equations, which can be solved in a straightforward manner.

The Orthogonality Principle for LTI Filters

Equation (4.7) may be rewritten as

$$\begin{aligned} \mathrm{E}\left[z(n-i) \left(s(n) - \sum_{j\in\mathcal{H}} h(j) z(n-j) \right) \right] &= 0 \\ \mathrm{E}\left[z(n-i)\left(s(n) - \hat{s}(n)\right)\right] &= 0 \\ \mathrm{E}\left[z(n-i)\tilde{s}(n)\right] &= 0, \quad i \in \mathcal{H}. \end{aligned} \tag{4.9}$$

Equation (4.9) is the *orthogonality principle* for LTI estimators. It shows that, given the observations $\boldsymbol{Z} = \{z(n-i) : i \in \mathcal{H}\}$, the error $\tilde{s}(n) = s(n) - \hat{s}(n)$ of the optimum LTI estimate $\hat{s}(n) = h(n) * z(n)$ is orthogonal to each of the observations in $\boldsymbol{Z}$. The next theorem follows immediately; compare it with Theorem 3.2.

Theorem 4.1 *Let $\hat{s}(n) = h(n) * z(n)$ be the optimum LTI estimate, and denote the support of $h(n)$ by $\mathcal{H}$. Suppose $g(n)$ is the impulse response of any LTI filter such that the support of $g(n)$ is a subset of $\mathcal{H}$. Then the error $s(n) - \hat{s}(n)$ is orthogonal to the output $g(n) * z(n)$; that is,*

$$E[(s(n) - \hat{s}(n))\,(g(n) * z(n))] = 0. \tag{4.10}$$

A theorem analogous to Theorem 3.3 may also be derived:

Theorem 4.2 *Let $h(n)$ be the impulse response of a LTI filter with support $\mathcal{H}$. Let $\mathbb{G}$ be the set of all LTI filters that have an impulse response whose support is a subset of $\mathcal{H}$. Then, among all filters in $\mathbb{G}$, $\hat{s}(n) = h(n) * z(n)$ is the MMSE estimate of $s(n)$ if and only if $s(n) - \hat{s}(n)$ is orthogonal to $g(n) * z(n)$, i.e.,*

$$E[(s(n) - h(n) * z(n))\,(g(n) * z(n))] = 0, \tag{4.11}$$

where $g(n)$ is the impulse response of any filter in $\mathbb{G}$.

Consider a filter with impulse response $g(n) = \delta(n-i)$, where i is one of the integers in $\mathcal{H}$. That is, $g(n)$ consists of a single unit impulse delayed by i. Certainly this filter belongs to the set $\mathbb{G}$ from Theorem 4.2, and the output of is

$$g(n) * z(n) = \delta(n-i) * z(n) = z(n-i).$$

As a result, we have a corollary to Theorem 4.2; this corollary is the orthogonality principle for LTI filters.

Corollary 4.1 *Let $h(n)$ be the impulse response of a LTI filter with support $\mathcal{H}$. Let $\mathbb{G}$ be the set of all LTI filters that have an impulse response whose support is a subset of $\mathcal{H}$. Then, among all filters in $\mathbb{G}$, $\hat{s}(n) = h(n) * z(n)$ is the MMSE estimate of $s(n)$ if and only if $s(n) - \hat{s}(n)$ is orthogonal to $z(n-i)$ for $i \in \mathcal{H}$, i.e.,*

$$E[(s(n) - h(b) * z(n))\, z(n-i)] = 0, \quad i \in \mathcal{H}. \tag{4.12}$$

Example 4.1 The Wiener-Hopf Equation via Orthogonality

We can derive the Wiener-Hopf equation (4.8) directly from the orthogonality principle (Corollary 4.1). From (4.2) and (4.12), we know that $\hat{s}(n) = h(n) * z(n)$ is the LTI MMSE estimate if and only if it satisfies

$$\mathrm{E}\left[\left(s(n) - \sum_{j\in\mathcal{H}} h(j) z(n-j)\right) z(n-i)\right] = 0, \quad i \in \mathcal{H}.$$

Working with this equation, we have

$$\mathrm{E}\left[s(n)z(n-i)\right] = \sum_{j\in\mathcal{H}} h(j)\mathrm{E}\left[z(n-j)z(n-i)\right], \quad i \in \mathcal{H},$$

$$R_{sz}(i) = \sum_{j\in\mathcal{H}} h(j) R_z(j-i), \quad i \in \mathcal{H},$$

which is the desired result.

4.2 The FIR Wiener Filter

We are now prepared to derive the various forms of the Wiener filter, which are the optimum LTI estimators for JWSS random processes. *For all the Wiener filters, we assume that $s(n)$, $v(n)$, and $z(n)$ are zero-mean JWSS random processes.* Then we have

$$\mathrm{E}\left[z(i)z(j)\right] = R_z(i-j) = R_z(k),$$

and

$$\mathrm{E}\left[s(i)z(j)\right] = R_{sz}(i-j) = R_{sz}(k).$$

Derivation

Suppose the filter has an impulse response $h(n)$ with support $\mathcal{H}$, where $\mathcal{H}$ is a finite set. For example, $\mathcal{H} = \{0, 1, 2, \ldots N-1\}$. Then the impulse response $h(n)$ has a finite duration, and this type of filter is called a *finite impulse response* (FIR) filter.

Here we derive the FIR Wiener filter. Let us assume a causal filter with an impulse response duration of N time indices. N is called the *filter length* and $N-1$ the *filter order*. This assumption is equivalent to setting $\mathcal{H} = \{0, 1, \ldots, N-1\}$ in Equation (4.2).

Derivation of the FIR Wiener filter is equivalent to the derivation of the Wiener-Hopf equation (4.8), so we do not repeat the details here. We have the estimate

$$\hat{s}(n) = \sum_{i=0}^{N-1} h(i) z(n-i), \tag{4.13}$$

and the MSE is

$$\text{MSE} = \text{E}\left[(s(n) - \hat{s}(n))^2\right]. \tag{4.14}$$

When we minimize the MSE or apply the orthogonality principle, we arrive at the following system of equations:

$$\sum_{j=0}^{N-1} h(j) R_z(j-i) = R_{sz}(i), \quad 0 \le i \le N-1. \tag{4.15}$$

Equation (4.15) describes a system of N linear equations in N unknowns. We may write it in matrix form as

$$\begin{bmatrix} R_z(0) & R_z(1) & \cdots & R_z(N-1) \\ R_z(1) & R_z(0) & \cdots & R_z(N-2) \\ \vdots & \vdots & \ddots & \vdots \\ R_z(N-1) & R_z(N-2) & \cdots & R_z(0) \end{bmatrix} \begin{bmatrix} h(0) \\ h(1) \\ \vdots \\ h(N-1) \end{bmatrix} = \begin{bmatrix} R_{sz}(0) \\ R_{sz}(1) \\ \vdots \\ R_{sz}(N-1) \end{bmatrix}, \tag{4.16}$$

where we have used the fact that $R_z(k) = R_z(-k)$.

Let R_z denote the matrix of $R_z(k)$ values, h the vector of $h(n)$ values, and r_{sz} the vector of $R_{sz}(k)$ values. Then we may write (4.16) as

$$R_z h = r_{sz}. \tag{4.17}$$

Since $R_z(k)$ is an autocorrelation function, the matrix R_z is a symmetric, positive-definite matrix and is therefore guaranteed to be invertible. The filter coefficients are given by

$$\boxed{h = R_z^{-1} r_{sz}} \tag{4.18}$$

or

$$\boxed{\begin{bmatrix} h(0) \\ h(1) \\ \vdots \\ h(N-1) \end{bmatrix} = \begin{bmatrix} R_z(0) & R_z(1) & \cdots & R_z(N-1) \\ R_z(1) & R_z(0) & \cdots & R_z(N-2) \\ \vdots & \vdots & \ddots & \vdots \\ R_z(N-1) & R_z(N-2) & \cdots & R_z(0) \end{bmatrix}^{-1} \times \begin{bmatrix} R_{sz}(0) \\ R_{sz}(1) \\ \vdots \\ R_{sz}(N-1) \end{bmatrix}} \tag{4.19}$$

Equations (4.18) and (4.19) give the FIR Wiener filter of order $N-1$.

In addition, the matrix R_z has a special structure: along each diagonal, all values of R_z are the same. This type of matrix is known as a Toeplitz matrix. The Toeplitz structure of R_z allows (4.19) to be solved using a computationally-efficient method known as the Levinson-Durbin recursion [2, 3, 4].

Computing the Mean-Square Error

Once we have the FIR Wiener filter, we may derive a simple expression for computing its associated MSE. We may write the MSE as

$$\begin{aligned} \text{MSE} &= \mathrm{E}\left[\tilde{\boldsymbol{s}}(n)\left(\boldsymbol{s}(n) - \sum_{i=0}^{N-1} h(i)\boldsymbol{z}(n-i)\right)\right] \\ &= \mathrm{E}\left[\tilde{\boldsymbol{s}}(n)\boldsymbol{s}(n)\right] - \sum_{i=0}^{N-1} h(i)\mathrm{E}\left[\tilde{\boldsymbol{s}}(n)\boldsymbol{z}(n-i)\right]. \end{aligned} \tag{4.20}$$

Since we have the MMSE estimate, we apply the orthogonality principle [Theorem 4.1]. Then the second term in (4.20) is zero, and (4.20) becomes

$$\begin{aligned}
\mathrm{MSE} &= \mathrm{E}\left[\tilde{s}(n)s(n)\right] \\
&= \mathrm{E}\left[s^2(n)\right] - \sum_{i=0}^{N-1} h(i)\mathrm{E}\left[s(n)z(n-i)\right] \\
&= R_s(0) - \sum_{i=0}^{N-1} h(i)R_{sz}(i) \qquad (4.21)\\
&= R_s(0) - h^{\mathrm{T}} r_{sz}. \qquad (4.22)
\end{aligned}$$

Equations (4.21–4.22) provide a convenient means for computing the MSE of the FIR Wiener filter.[1] Section 4.3 demonstrates a means for computing the MMSE among *all* LTI estimators.

Observe that if no filtering is performed and we simply use $\hat{s}(n) = z(n)$, the MSE is

$$\mathrm{MSE}_{\text{no filter}} = \mathrm{E}\left[(s(n) - z(n))^2\right] = R_s(0) - 2R_{sz}(0) + R_z(0). \qquad (4.23)$$

We can use (4.23) to measure the effectiveness of the Wiener filter. The reduction in MSE due to Wiener filtering is given by

$$\text{reduction in MSE} = 10\log_{10}\left(\frac{\mathrm{MSE}_{\text{no filter}}}{\mathrm{MSE}_{\text{filter}}}\right) \text{ dB}. \qquad (4.24)$$

We can inspect the MSE to determine whether the filter performance is acceptable. If the MSE is too high, we may wish to use a larger value of N. In practice we may experience diminishing returns, in which a FIR Wiener filter of high-order performs only slightly better than a low-order one. For example, see Problem 4.13.

Applying the Additive-Noise Model

When the noise corruption process is assumed to fit the additive-noise model, we can reduce the equations for $R_z(k)$ and $R_{sz}(k)$. Let us assume

$$z(n) = s(n) + v(n),$$

where $s(n)$ and $v(n)$ are uncorrelated, zero-mean, JWSS random processes.

[1] An alternative method (4.47) for computing the MSE is derived in Section 4.3. This method requires integration around a closed contour in the complex plane, so typically equations (4.21) and (4.22) are preferred for their simplicity. Note that equations (4.21) and (4.22) apply only to the FIR Wiener filter.

We may now easily compute $R_z(k)$ and $R_{sz}(k)$ as follows:

$$\begin{aligned} R_z(k) &= \mathrm{E}\left[\boldsymbol{z}(n)\boldsymbol{z}(n-k)\right] \\ &= \mathrm{E}\left[(\boldsymbol{s}(n)+\boldsymbol{v}(n))(\boldsymbol{s}(n-k)+\boldsymbol{v}(n-k))\right] \\ &= \mathrm{E}\left[\boldsymbol{s}(n)\boldsymbol{s}(n-k)\right] + \mathrm{E}\left[\boldsymbol{v}(n)\boldsymbol{v}(n-k)\right] \\ &= R_s(k) + R_v(k), \end{aligned} \tag{4.25}$$

and

$$\begin{aligned} R_{sz}(k) &= \mathrm{E}\left[\boldsymbol{s}(n)(\boldsymbol{s}(n-k)+\boldsymbol{v}(n-k))\right] \\ &= \mathrm{E}\left[\boldsymbol{s}(n)\boldsymbol{s}(n-k)\right] + \mathrm{E}\left[\boldsymbol{s}(n)\boldsymbol{v}(n-k)\right] \\ &= R_s(k). \end{aligned} \tag{4.26}$$

Then the filter coefficients are given by

$$h = (R_s + R_v)^{-1} r_s, \tag{4.27}$$

where

$$h = \begin{bmatrix} h(0) & h(1) & \cdots & h(N-1) \end{bmatrix}^{\mathrm{T}},$$

$$R_s = \begin{bmatrix} R_s(0) & R_s(1) & \cdots & R_s(N-1) \\ R_s(1) & R_s(0) & \cdots & R_s(N-2) \\ \vdots & \vdots & \ddots & \vdots \\ R_s(N-1) & R_s(N-2) & \cdots & R_s(0) \end{bmatrix},$$

$$R_v = \begin{bmatrix} R_v(0) & R_v(1) & \cdots & R_v(N-1) \\ R_v(1) & R_v(0) & \cdots & R_v(N-2) \\ \vdots & \vdots & \ddots & \vdots \\ R_v(N-1) & R_v(N-2) & \cdots & R_v(0) \end{bmatrix},$$

and

$$r_s = \begin{bmatrix} R_s(0) & R_s(1) & \cdots & R_s(N-1) \end{bmatrix}^{\mathrm{T}}.$$

Example 4.2 Signal Filtering

Suppose we have a signal $\boldsymbol{s}(n)$ with autocorrelation function $R_s(k) = 0.95^{|k|}$. The signal is observed in the presence of additive white noise with variance $\sigma_v^2 = 2$. Hence $R_v(k) = 2\delta(k)$. $\boldsymbol{s}(n)$ and $\boldsymbol{v}(n)$ are uncorrelated, zero-mean, JWSS random processes. We would like to design the optimum LTI second-order filter h.

Since the filter is second-order, we set $N = 3$. The matrix equation (4.16) to be solved is

$$\begin{bmatrix} R_z(0) & R_z(1) & R_z(2) \\ R_z(1) & R_z(0) & R_z(1) \\ R_z(2) & R_z(1) & R_z(0) \end{bmatrix} \begin{bmatrix} h(0) \\ h(1) \\ h(2) \end{bmatrix} = \begin{bmatrix} R_{sz}(0) \\ R_{sz}(1) \\ R_{sz}(2) \end{bmatrix}. \tag{4.28}$$

According to (4.25) we have

$$R_z(k) = 0.95^{|k|} + 2\delta(k).$$

Similary, equation (4.26) yields

$$R_{sz}(k) = 0.95^{|k|}.$$

Then (4.28) becomes

$$\begin{bmatrix} 3 & 0.95 & 0.9025 \\ 0.95 & 3 & 0.95 \\ 0.9025 & 0.95 & 3 \end{bmatrix} \begin{bmatrix} h(0) \\ h(1) \\ h(2) \end{bmatrix} = \begin{bmatrix} 1 \\ 0.95 \\ 0.9025 \end{bmatrix}, \tag{4.29}$$

which has solution

$$h = \begin{bmatrix} 0.2203 & 0.1919 & 0.1738 \end{bmatrix}^{\mathrm{T}}. \tag{4.30}$$

The MSE of this estimator can be computed via Equation (4.22),

$$\mathrm{MSE} = R_s(0) - h^{\mathrm{T}} r_{sz} = 0.4405.$$

For comparison, we remark that without any filtering, $\mathrm{MSE}_{\text{no filter}} = 2$. The FIR Wiener filter has reduced the MSE by about 6.5 dB.

In Examples 4.5 and 4.10 we compare the performance of this filter with the noncausal and causal Wiener filters, respectively.

Signal Prediction

We have derived the FIR Wiener filter for the filtering problem, where the estimate is $\hat{s}(n)$. As discussed in Section 3.1, we may also attempt to estimate the value of the signal process $s(\cdot)$ at some other time. Let us consider the case of *one-step prediction,* which makes our estimate $\hat{s}(n+1)$. We will see that the resulting equations are very similar to those associated with the previous filter.

The estimate is generated by

$$\hat{s}(n+1) = \sum_{i=0}^{N-1} h(i) z(n-i), \tag{4.31}$$

and the MSE is now

$$\mathrm{MSE} = \mathrm{E}\left[\left(s(n+1) - \hat{s}(n+1) \right)^2 \right]. \tag{4.32}$$

We compare Equations (4.31–4.32) with those of the filter, Equations (4.13–4.14). The sets of equations differ only by the time index of $s(\cdot)$ and $\hat{s}(\cdot)$.

To find the coefficients of h, we follow the familiar procedure of taking partial derivatives of (4.32) and setting them equal to zero. This procedure ultimately yields the equations

$$\sum_{j=0}^{N-1} h(j)R_z(j-i) = R_{sz}(i+1), \quad 0 \le i \le N-1. \tag{4.33}$$

The matrix form of (4.33) is

$$\begin{bmatrix} R_z(0) & R_z(1) & \cdots & R_z(N-1) \\ R_z(1) & R_z(0) & \cdots & R_z(N-2) \\ \vdots & \vdots & \ddots & \vdots \\ R_z(N-1) & R_z(N-2) & \cdots & R_z(0) \end{bmatrix} \begin{bmatrix} h(0) \\ h(1) \\ \vdots \\ h(N-1) \end{bmatrix} = \begin{bmatrix} R_{sz}(1) \\ R_{sz}(2) \\ \vdots \\ R_{sz}(N) \end{bmatrix}. \tag{4.34}$$

Compare Equations (4.33) and (4.34) with Equations (4.15) and (4.16), respectively. The only difference between these pairs of equations is the index of $R_{sz}(\cdot)$ on the right-hand side.

The MSE associated with this estimator can be computed via

$$\text{MSE} = R_s(0) - \sum_{i=0}^{N-1} h(i)R_{sz}(i+1). \tag{4.35}$$

The derivation of this result parallels that of (4.21).

As a result of the similarity between the filter and the one-step predictor, the reader may suspect that changing the type of estimate (filtering, smoothing, or prediction) simply changes the indexing of $R_{sz}(\cdot)$. This conclusion is correct and is addressed by Problem 4.3.

Example 4.3 Prediction with Noisy Resonance

We want to design a third-order one-step predictor for the following situation. Suppose we measure an acoustic signal $s(\cdot)$ inside a cavity. The signal is generated via

$$s(n) = 0.4s(n-1) + w(n),$$

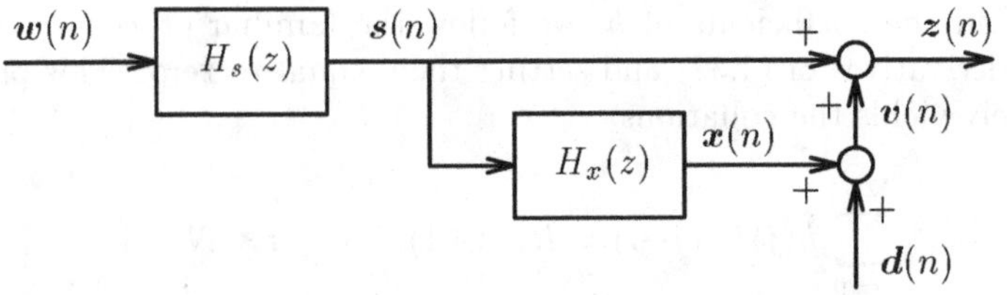

Figure 4.1. A model of the system that generates the signal $s(n)$ and observations $z(n)$.

where $w(n)$ is zero-mean white noise with unit variance. Due to resonances within the cavity, we model the measurement noise $v(n)$ by

$$x(n) = 0.5x(n-1) + 0.3s(n-1)$$
$$v(n) = x(n) + d(n).$$

$x(n)$ models the resonance due to the cavity, and $d(n)$ is zero-mean white-noise with unit variance that models the noise of our microphone. $d(n)$ and $w(n)$ are uncorrelated; all processes of interest are JWSS and zero-mean. Figure 4.3 diagrams this model.

In this example $s(n)$ and $v(n)$ are correlated due to the resonance term $x(n)$. We now set about determining $R_z(k)$ and $R_{sz}(k)$ by using power spectral densities. We first find $R_z(k)$.

$$\begin{aligned} R_z(k) &= \mathrm{E}\left[z(n)z(n-k)\right] \\ &= \mathrm{E}\left[(s(n) + 0.3s(n-1) + 0.5x(n-1) + d(n))\right. \\ &\qquad \left.\times\ (s(n-k) + 0.3s(n-k-1) + 0.5x(n-k-1) + d(n-k))\right] \\ &= 0.3R_s(k+1) + 1.09R_s(k) + 0.3R_s(k-1) \\ &\qquad + 0.25R_x(k) + R_y(k) + 0.5R_{sx}(k+1) + 0.5R_{xs}(k-1) \\ &\qquad + 0.15R_{sx}(k) + 0.15R_{xs}(k). \end{aligned} \tag{4.36}$$

Now we must compute expressions for $R_s(k)$, $R_x(k)$, and $R_{sx}(k)$. In order to find $R_s(k)$, we observe that

$$H_s(z) = \frac{S(z)}{W(z)} = \frac{1}{1-0.4z^{-1}},$$

so the power density spectrum of $s(n)$, $S_s(z)$, is given by

$$S_s(z) = H_s(z)H_s(z^{-1})S_w(z) = \frac{1}{(1-0.4z^{-1})(1-0.4z)}.$$

Taking the inverse z-transform, we find

$$R_s(k) = \tfrac{25}{21}(0.4)^{|k|}. \tag{4.37}$$

We can find $R_x(k)$ by relating its power density spectrum to $S_s(z)$:

$$\begin{aligned}
S_x(z) &= H_x(z)H_x(z^{-1})S_s(z) \\
&= \frac{0.3z^{-1}}{(1-0.5z^{-1})}\frac{0.3z}{(1-0.5z)}\frac{1}{(1-0.4z^{-1})(1-0.4z)} \\
&= \frac{0.09}{0.63}\frac{0.84}{(1-0.4z^{-1})(1-0.4z)}\frac{0.75}{(1-0.5z^{-1})(1-0.5z)} \\
&= \frac{1}{7}\left[\frac{(-\frac{125}{21})(0.84)}{(1-0.4z^{-1})(1-0.4z)} + \frac{(\frac{25}{3})(0.75)}{(1-0.5z^{-1})(1-0.5z)}\right],
\end{aligned}$$

and hence

$$R_x(k) = \tfrac{1}{7}\left[\tfrac{25}{3}(0.5)^{|k|} - \tfrac{125}{21}(0.4)^{|k|}\right]. \tag{4.38}$$

Finally, we need $R_{sx}(k)$, which we find from its power density spectrum.

$$\begin{aligned}
S_{sx}(z) = H_x(z^{-1})S_s(z) &= \frac{0.3z}{(1-0.5z)}\frac{1}{(1-0.4z^{-1})(1-0.4z)} \\
&= \frac{-0.6z}{(1-2z^{-1})}\left(\frac{1}{0.84}\right)\frac{0.84}{(1-0.4z^{-1})(1-0.4z)} \\
&= \frac{-0.6}{0.84}\left[\left(\tfrac{125}{42}z - \tfrac{125}{84}\right)\frac{0.84}{(1-0.4z^{-1})(1-0.4z)} + \frac{\frac{25}{4}}{(1-2z^{-1})}\right].
\end{aligned}$$

Then the inverse z-transform produces $R_{sx}(k)$,

$$R_{sx}(k) = \tfrac{5}{7}\left\{\tfrac{25}{4}(2)^k 1(-k-1) - \tfrac{125}{84}\left[2(0.4)^{|k+1|} - (0.4)^{|k|}\right]\right\}, \tag{4.39}$$

and we recall that $R_{xs}(k) = R_{sx}(-k)$.

Next we need to determine $R_{sz}(k)$. Fortunately, we have computed all the correlation functions we will need.

$$\begin{aligned}
R_{sz}(k) &= \mathrm{E}\left[s(n)\left(s(n-k) + 0.5x(n-k-1) + 0.3s(n-k-1) + d(n-k)\right)\right] \\
&= R_s(k) + 0.5R_{sx}(k+1) + 0.3R_s(k+1).
\end{aligned} \tag{4.40}$$

We can compute the values of $R_z(k)$ and $R_{sz}(k)$ from (4.36–4.40). h is a third-order predictor, so $N = 4$. Substituting into Equation (4.34), we have

$$\begin{bmatrix} 2.8172 & 1.2129 & 0.7567 & 0.4385 \\ 1.2129 & 2.8172 & 1.2129 & 0.7567 \\ 0.7567 & 1.2129 & 2.8172 & 1.2129 \\ 0.4385 & 0.7567 & 1.2129 & 2.8172 \end{bmatrix} \begin{bmatrix} h(0) \\ h(1) \\ h(2) \\ h(3) \end{bmatrix} = \begin{bmatrix} 0.5503 \\ 0.2201 \\ 0.0881 \\ 0.0352 \end{bmatrix},$$

which gives the filter coefficients

$$\begin{aligned}
&\begin{bmatrix} h(0) & h(1) & h(2) & h(3) \end{bmatrix} \\
&\qquad = \begin{bmatrix} 0.2010 & 0.0029 & -0.0191 & -0.0113 \end{bmatrix}^{\mathrm{T}}.
\end{aligned}$$

The MSE is 1.0813.

Example 4.4 Pure Prediction

Let us consider the case of signal prediction without any noise, which is called *pure prediction.* We must still guess the future value of the signal, which is a random process.

Because we have no noise, the observations $z(n)$ are simply the signal $s(n)$. Then the correlation functions of interest become

$$R_z(k) = R_{sz}(k) = R_s(k).$$

Suppose the signal $s(n)$ has autocorrelation function

$$R_s(k) = 0.5\delta(k) + (0.9)^{|k|}\cos(\pi k/8),$$

and let us use a second-order predictor.

We compute the filter coefficients via equation (4.34), which becomes

$$\begin{bmatrix} 1.5000 & 0.8315 & 0.5728 \\ 0.8315 & 1.5000 & 0.8315 \\ 0.5728 & 0.8315 & 1.5000 \end{bmatrix} \begin{bmatrix} h(0) \\ h(1) \\ h(2) \end{bmatrix} = \begin{bmatrix} 0.8315 \\ 0.5728 \\ 0.2790 \end{bmatrix}.$$

The filter coefficients are

$$\begin{bmatrix} h(0) & h(1) & h(2) \end{bmatrix} = \begin{bmatrix} 0.5045 & 0.1528 & -0.0914 \end{bmatrix}^{\mathrm{T}},$$

and the MSE is 1.0185. Without filtering, the MSE would be 1.337, so the FIR Wiener filter has reduced the MSE by 1.2 dB.

4.3 The Noncausal Wiener Filter

Suppose now that the filter impulse response $h(n)$ has support $\mathcal{H}$, where $\mathcal{H}$ is an infinite set. The corresponding filter is an *infinite impulse response* (IIR) filter. There are many possible infinite sets for $\mathcal{H}$, but let us specifically consider the case $\mathcal{H} = \mathbb{Z}$, where $\mathbb{Z}$ is the set of integers: $\mathbb{Z} = \{ \ldots, -2, -1, 0, 1, 2, \ldots \}$. This filter is noncausal and known as the noncausal IIR Wiener filter; it is usually called the *noncausal Wiener filter*[2]. It corresponds to the ideal case in which all the observations $\{z(n) : n \in \mathbb{Z}\}$ are available.

For real-time applications, of course, the noncausal Wiener filter is unrealizable. However, in some situations we may perform filtering off-line on a large collection of observations so that noncausal filtering can be applied. In either case, the noncausal Wiener filter tells us the best estimate $\hat{s}(n)$ we can expect from all LTI filters.

[2] Any FIR filter can be made causal by sufficiently delaying its impulse response, so the FIR Wiener filter can always be made causal. A noncausal IIR filter, however, can never be made causal. As a result, we usually drop the "IIR" in "noncausal IIR Wiener filter."

Derivation

The noncausal Wiener filter generates an estimate via

$$\hat{s}(n) = \sum_{i=-\infty}^{\infty} h(i)z(n-i).$$

The Wiener-Hopf equation (4.8) becomes

$$\sum_{j=-\infty}^{\infty} h(j)R_z(j-i) = R_{sz}(i), \quad \text{for all } i \in \mathbb{Z}. \tag{4.41}$$

Equation (4.41) represents an infinite-dimensional system of linear equations, so we cannot solve it by matrix methods. However, because $s(n)$ and $z(n)$ are zero-mean and JWSS, we can take the z-transform of (4.41). Doing so produces

$$H(z)S_z(z) = S_{sz}(z)$$

so that

$$\boxed{H(z) = \frac{S_{sz}(z)}{S_z(z)}.} \tag{4.42}$$

Equation (4.42) is the solution for the noncaual Wiener filter, and the solution is given in terms of the system function for h. In general, $S_z(z)$ will have zeros outside the unit circle, so $H(z)$ will have poles outside the unit circle. Hence $H(z)$ will usually be noncausal.

However, $S_z(z)$ may have zeros on the unit circle, which means the noncausal Wiener filter *may be unstable*. Often we eliminate this potential difficulty by adopting the additive-noise model with white measurement noise.

Computing the Mean-Square Error

As with the FIR Wiener filter of Section 4.2, we can derive an expression for the MSE. The MSE can be written as[3]

$$\text{MSE} = R_s(0) - \sum_{i=-\infty}^{\infty} h(i)R_{sz}(i). \tag{4.43}$$

If we can find $h(n)$ from $H(z)$, we may use (4.43) to compute the MSE. In some cases, however, it will be very difficult to find the inverse z-transform of $H(z)$.

[3] The derivation parallels the derivation of equation (4.21).

Using some properties of the z-transform, we can derive an alternative method for computing the MSE. We can compute $R_s(0)$ via

$$\begin{aligned} R_s(0) &= \mathcal{Z}^{-1}\{S_s(z)\}\big|_{n=0} \\ &= \frac{1}{j2\pi}\oint_C S_s(z)z^{n-1}\,dz\bigg|_{n=0} \\ &= \frac{1}{j2\pi}\oint_C S_s(z)z^{-1}\,dz, \end{aligned} \tag{4.44}$$

where C is a closed contour that lies entirely within the region of convergence (ROC) of $S_s(z)$, and the complex integral is taken on C in the counterclockwise direction. Because the ROC of $S_s(z)$ must include the unit circle in order for $S_s(e^{j\omega})$ to exist, usually we take C to be the unit circle

Parseval's relation allows us to write the summation in (4.43) as

$$\sum_{i=-\infty}^{\infty} h(i)R_{sz}(i) = \frac{1}{j2\pi}\oint_C H(z)S_{sz}^*(1/z^*)z^{-1}\,dz. \tag{4.45}$$

If we take C to be the unit circle, $z^* = z^{-1}$ so $S_{sz}(1/z^*) = S_{sz}(z)$. If, in addition, $S_{sz}(z)$ is rational in z, then $S_{sz}^*(z) = S_{sz}(z^{-1})$, and

$$\sum_{i=-\infty}^{\infty} h(i)R_{sz}(i) = \frac{1}{j2\pi}\oint_C H(z)S_{sz}(z^{-1})z^{-1}\,dz. \tag{4.46}$$

By Equations (4.44), (4.45), and (4.46), the MSE can be calculated via

$$\text{MSE} = \begin{cases} \dfrac{1}{j2\pi}\displaystyle\oint_C \left[S_s(z) - H(z)S_{sz}(z^{-1})\right] z^{-1}\,dz, & S_{sz}(z) \text{ rational;} \\ \dfrac{1}{j2\pi}\displaystyle\oint_C \left[S_s(z) - H(z)S_{sz}^*(z)\right] z^{-1}\,dz, & \text{otherwise.} \end{cases} \tag{4.47}$$

In fact, this equation can be used to compute the MSE for any of the Wiener filters (FIR, noncausal, or causal).

Fourier Transform Form

The noncausal Wiener filter may also be derived by using the discrete-time Fourier transform (DTFT) of (4.41). In this case,

$$H(e^{j\omega}) = \frac{S_{sz}(e^{j\omega})}{S_z(e^{j\omega})},$$

and the MSE can be computed by

$$\text{MSE} = \begin{cases} \dfrac{1}{2\pi}\displaystyle\int_{2\pi} \left[S_s(e^{j\omega}) - H(e^{j\omega})S_{sz}(e^{-j\omega})\right] d\omega, & S_{sz}(e^{j\omega}) \text{ rational;} \\ \dfrac{1}{2\pi}\displaystyle\int_{2\pi} \left[S_s(e^{j\omega}) - H(e^{j\omega})S_{sz}^*(e^{j\omega})\right] d\omega, & \text{otherwise.} \end{cases} \tag{4.48}$$

Applying the Additive-Noise Model

Let us assume the additive-noise model and assume that the signal and noise are uncorrelated. As a result of Equations (4.25) and (4.26), we find

$$S_z(z) = S_s(z) + S_v(z),$$

and

$$S_{sz}(z) = S_s(z).$$

Then the noncausal Wiener filter is

$$H(z) = \frac{S_s(z)}{S_s(z) + S_v(z)}.$$

Next, we simplify the equations for the MSE. Since $S_{sz}(e^{j\omega}) = S_s(e^{j\omega}) = S_s^*(e^{j\omega})$, we may write (4.48) as

$$\begin{aligned} \text{MSE} &= \frac{1}{2\pi}\int_{2\pi} S_s(e^{j\omega})\left[1 - H(e^{j\omega})\right] d\omega \\ &= \frac{1}{2\pi}\int_{2\pi} S_s(e^{j\omega})\frac{S_v(e^{j\omega})}{S_s(e^{j\omega}) + S_v(e^{j\omega})}\, d\omega \\ &= \frac{1}{2\pi}\int_{2\pi} S_v(e^{j\omega})H(e^{j\omega})\, d\omega. \end{aligned}$$

In the z-transform domain, we write this result as

$$\text{MSE} = \frac{1}{j2\pi}\oint_C S_v(z)H(z)z^{-1}\, dz. \tag{4.49}$$

Example 4.5 Comparison with an FIR Wiener Filter

Let us apply a noncausal Wiener filter to the filtering problem of Example 4.2. In that example we found that

$$R_z(k) = 0.95^{|k|} + 2\delta(k), \quad \text{and} \quad R_{sz}(k) = 0.95^{|k|}.$$

We take z-transforms of these correlation functions and obtain

$$\begin{aligned} S_z(z) &= \frac{1-(0.95)^2}{(1-0.95z^{-1})(1-0.95z)} + 2 \\ &= 2.3955 \frac{(1-0.7931z^{-1})(1-0.7931z)}{(1-0.95z^{-1})(1-0.95z)}, \end{aligned}$$

and

$$S_{sz}(z) = \frac{1-(0.95)^2}{(1-0.95z^{-1})(1-0.95z)}.$$

The noncausal Wiener filter (4.42) is given by

$$\begin{aligned} H(z) = \frac{S_{sz}(z)}{S_z(z)} &= \frac{0.0975}{2.3955(1-0.7931z^{-1})(1-0.7931z)} \\ &= 0.1097 \frac{1-(0.7931)^2}{(1-0.7931z^{-1})(1-0.7931z)}. \end{aligned}$$

The impulse response of this filter is

$$h(n) = 0.1097\,(0.7931)^{|n|},$$

which indicates that the filter is noncausal.

Now we compare the mean-square errors associated with the noncausal Wiener filter and the FIR Wiener filter of Example 4.2. From Example 4.2, $\text{MSE}_{\text{FIR}} = 0.4405$. For the noncausal Wiener filter we use (4.43):

$$\begin{aligned} \text{MSE}_{\text{NC}} &= (0.95)^0 - \sum_{n=-\infty}^{\infty} 0.1097\,(0.7931)^{|n|}\,(0.95)^{|n|} \\ &= 1 - 0.1097 \left[\sum_{n=0}^{\infty} (0.7931)^n (0.95)^n \right. \\ &\qquad \left. + \sum_{n=-1}^{-\infty} (0.7931)^{-n} (0.95)^{-n} \right] \\ &= 1 - 0.1097 \left[2 \sum_{n=0}^{\infty} (0.7931)^n (0.95)^n - 1 \right] \\ &= 1 - 0.1097 \left[\frac{2}{1-(0.7931)(0.95)} - 1 \right] \\ &= 0.2195. \end{aligned}$$

The improvement of the noncausal filter over no filtering is 9.6 dB. Also, MSE_{NC} is slightly less than half of MSE_{FIR}. Compared to the second-order filter, the noncausal Wiener filter reduces the estimation error by an additional 3 dB. We should expect the noncausal Wiener filter to perform best among all LTI filters because it is the least-constrained LTI estimator. Also see Example 4.10.

Example 4.6 Computing the MSE by Contour Integration

In this example we use (4.49) to compute the MSE of the noncausal Wiener filter in Example 4.5. The contour of integration C is chosen to be the unit circle, taken in the counterclockwise direction.

$$\begin{aligned}\mathrm{MSE}_{\mathrm{NC}} &= \frac{1}{j2\pi}\oint_C 2 \times \frac{0.0407}{(1-0.7931z^{-1})(1-0.7931z)} z^{-1}\, dz \\ &= 0.0814\, \frac{1}{j2\pi}\oint_C \frac{1}{(z-0.7931)(1-0.7931z)}\, dz.\end{aligned}$$

We now apply the Cauchy residue theorem for complex contour integration (B.15) [5] and find

$$\begin{aligned}\mathrm{MSE}_{\mathrm{NC}} &= 0.0814 \left[\sum \text{residues of integrand inside unit circle}\right] \\ &= 0.0814\, \frac{1}{(1-0.7931z)}\bigg|_{z=0.7931} \\ &= 0.2195,\end{aligned}$$

which is what we obtained via (4.43).

4.4 Toward the Causal Wiener Filter

For many applications we do not have the luxury of noncausal filtering, and we cannot collect an infinite number of measurements. We now confront the problem of designing the optimum causal LTI filter. The estimate $\hat{s}(n)$ is produced via

$$\hat{s}(n) = \sum_{i=0}^{\infty} h(i) z(n-i). \tag{4.50}$$

The Wiener-Hopf equation (4.8) now has the form

$$\sum_{j=0}^{\infty} h(j) R_z(j-i) = R_{sz}(i), \quad i \geq 0. \tag{4.51}$$

Now i is restricted to the set of natural numbers $\mathbb{N}$ ($\mathbb{N} = \{1, 2, 3, \dots\}$) instead of $\mathbb{Z}$. Hence, Equation (4.51) cannot be treated as simple convolution, and we cannot apply the methods used to derive the noncausal Wiener filter in Section 4.3, and we must adopt another technique.

Before we can continue the derivation, we must introduce two important concepts: (1) *the spectral factorization theorem*, and (2) *causal-part extraction*, also known as the plus operation.

The Spectral Factorization Theorem

The first tool we require is spectral factorization. The spectral factorization theorem states that any rational power spectrum can be factored into the product of two rational terms that are conveniently related to one another. Knowledge of one term is sufficient to specify the entire power density spectrum.

All poles of the first term lie strictly within the unit circle, and all of its zeros lie on or inside the unit circle. As a result, this term is stable and causal. For the second term, its poles lie strictly outside the unit circle and its zeros lie on or outside the unit circle.

Theorem 4.3 (Spectral Factorization) *Let $\boldsymbol{x}(n)$ be a real-valued, zero-mean, WSS random process with power density spectrum $S_x(z)$, where $S_x(z)$ is rational in z and has no poles on the unit circle. Then $S_x(z)$ can be factored into the product*

$$S_x(z) = S_x^+(z) S_x^-(z), \tag{4.52}$$

where

- *$S_x^+(z)$ and $S_x^-(z)$ are rational in z,*
- *if z_i is a pole of $S_x^+(z)$, then $|z_i| < 1$,*
- *if z_i is a zero of $S_x^+(z)$, then $|z_i| \leq 1$,*
- *if z_i is a pole of $S_x^-(z)$, then $|z_i| > 1$,*
- *if z_i is a zero of $S_x^-(z)$, then $|z_i| \geq 1$, and*
- *$S_x^+(z) = S_x^-(z^{-1})$.*

Proof. The first part of the proof shows that the poles of $S_x(z)$ occur in reciprocal pairs. $S_x(z) = S_x(z^{-1})$, and by hypothesis $S_x(z)$ has no poles on the unit circle. Let a be a pole of $S_x(z)$ such that $|a| < 1$; then a^{-1} is also a pole of $S_x(z)$ and $|a^{-1}| > 1$. Similarly, let a be a pole such that $|a| > 1$; then a^{-1} is also a pole and $|a^{-1}| < 1$. Hence, *the poles of $S_x(z)$ occur in reciprocal pairs.*

The following four paragraphs establish that all zeros of $S_x(z)$ occur in reciprocal pairs. First, consider the zeros of $S_x(z)$ strictly inside or outside of the unit circle. The preceding argument regarding the poles applies. Hence the zeros of $S_x(z)$ not on the unit circle occur in reciprocal pairs.

Second, let a be a zero of $S_x(z)$ on the unit circle but not at ± 1; that is, $|a| = 1$ and $a \neq \pm 1$. Since $S_x(z) = S_x(z^{-1})$, it follows that a^{-1} must also be a zero. Therefore zeros of $S_x(z)$ on the unit circle occur in reciprocal pairs, except perhaps for zeros at $z = \pm 1$.

Third, assume $S_x(1) = 0$ and $S_x(-1) = 0$, and neither 1 nor -1 forms part of a reciprocal pair. That is, the zero at $z = 1$ has odd degree (it may be a single zero, a triple zero, etc.), and the zero at $z = -1$ also has odd degree. Since $S_x(z)$ is rational, it can be written in the form

$$S_x(z) = (z-1)(z+1)F(z) = (z^2-1)F(z),$$

where $F(z)$ is rational in z, and all poles and zeros of $F(z)$ occur in reciprocal pairs. Then $F(z) = F(z^{-1})$ and $F(e^{j\omega}) = F(e^{-j\omega})$, and

$$S_x(e^{j\omega}) = (e^{j2\omega}-1)F(e^{j\omega}) = (\cos 2\omega + j\sin 2\omega - 1)F(e^{j\omega}),$$

and

$$S_x(e^{-j\omega}) = (e^{-j2\omega}-1)F(e^{-j\omega}) = (\cos 2\omega - j\sin 2\omega - 1)F(e^{j\omega}).$$

It follows that $S_x(e^{j\omega}) \neq S_x(e^{-j\omega})$ for some ω between $-\pi$ and π. Since $S_x(e^{j\omega})$ is not even, this result contradicts the hypothesis that $S_x(z)$ is a power density spectrum. As a result, the zeros at $z = 1$ must have even degree (a double zero, a quadruple zero, etc.), and likewise for the zero at $z = -1$.

Fourth, assume that $z = 1$ is a zero of odd degree and that $z = -1$ either is not a zero or is a zero with even degree. Then $S_x(z)$ can be expressed as

$$S_x(z) = (z-1)G(z),$$

where $G(z)$ is rational in z and $G(z) = G(z^{-1})$. Then

$$S_x(e^{j\omega}) = (e^{j\omega}-1)G(e^{j\omega}), \quad \text{and} \quad S_x(e^{-j\omega}) = (e^{-j\omega}-1)G(e^{-j\omega}).$$

Now $G(e^{j\omega}) = G(e^{-j\omega})$, but $(e^{j\omega}-1)$ is not even in ω. Hence, $S_x(e^{j\omega}) \neq S_x(e^{-j\omega})$ for some ω between $-\pi$ and π. This result contradicts the hypothesis that $S_x(z)$ is a power density spectrum. Thus the zero $z = 1$ must have even degree, making it part of a reciprocal pair. By the same argument a zero at $z = -1$ must have even degree. Hence, all zeros of $S_x(z)$ on the unit circle have even degree.

Therefore *all poles and zeros of $S_x(z)$ occur in reciprocal pairs*, and knowledge of the poles and zeros on and within the unit circle is sufficient to specify all poles and zeros of $S_x(z)$. Call this property the "reciprocal property."

The next four paragraphs construct the spectral factorization. Let $2k$ equal the number of zeros of $S_x(z)$ at $z = 1$ and 2ℓ equal the number of zeros at $z = -1$. There are N distinct poles on $S_x(z)$ that are strictly inside the unit circle. Denote each pole and its corresponding degree by a_i and n_i, respectively, $i = 1, \ldots, N$.

Next consider the zeros of $S_x(z)$. There are M_1 distinct zeros of $S_x(z)$ strictly inside the unit circle. Denote these zeros and their corresponding degrees by b_i and m_i, respectively, $i = 1, \ldots, M_1$. The zeros on the unit circle must still be considered. There are $2M_2$ distinct zeros on the unit circle but not at 1 or -1. Denote them by

$$b_{M_1+1}, b_{M_1+2}, \ldots, b_{M_1+M_2}, b_{M_1+M_2+1}, \ldots, b_{M_1+2M_2},$$

where $b_{M_1+i} = b^{-1}_{M_1+M_2+i}$. This requirement can always be satisfied because of the reciprocal property. Also let m_i denote the corresponding degree of b_i, $i = M_1 + 1$, $\ldots$, $M_1 + 2M_2$. Finally, let $M = M_1 + M_2$ (so M is the minimum number of zeros needed to specify all zeros of $S_x(z)$).

Then from this construction, $S_x(z)$ can be directly placed in the form

$$S_x(z) = \sigma^2(1-z^{-1})^{2k}(1+z^{-1})^{2\ell}\frac{\prod_{j=1}^{M}(1-b_jz^{-1})^{m_j}(1-b_jz)^{m_j}}{\prod_{i=1}^{N}(1-a_iz^{-1})^{n_i}(1-a_iz)^{n_i}}$$

Define

$$S_x^+(z) = \sigma(1-z^{-1})^{k}(1+z^{-1})^{\ell}\frac{\prod_{j=1}^{M}(1-b_jz^{-1})^{m_j}}{\prod_{i=1}^{N}(1-a_iz^{-1})^{n_i}},$$

and

$$S_x^-(z) = \sigma(1-z^{-1})^{k}(1+z^{-1})^{\ell}\frac{\prod_{j=1}^{M}(1-b_jz)^{m_j}}{\prod_{i=1}^{N}(1-a_iz)^{n_i}}.$$

It immediately follows that

$$S_x(z) = S_x^+(z)S_x^-(z).$$

Certainly $S_x^+(z)$ and $S_x^-(z)$ are rational in z. By construction, all poles of $S_x^+(z)$ lie strictly inside the unit circle, and all zeros of $S_x^+(z)$ lie on or inside the unit circle. Similarly, all poles of $S_x^-(z)$ lie strictly outside the unit circle, and all zeros of $S_x^-(z)$ lie on or outside the unit circle. Finally, it is clear that $S_x^+(z) = S_x^-(z^{-1})$. **Q.E.D.** ■

The latter part of the proof is constructive, so it provides a means for finding the spectral factorization of a power spectrum. An example demonstrates the procedure.

Example 4.7 Spectral Factorization

Let $s(n)$ be a random process described by the difference equation

$$s(n) = 1.1s(n-1) - 0.24s(n-2) + 2w(n) + 3w(n-1),$$

where $w(n)$ is a zero-mean, WSS random process with autocorrelation function $R_w(n) = 5(0.6)^{|n|}$. We want to find the spectral factorization of $S_s(z)$.

We begin by finding the system function relating $s(n)$ to $w(n)$. The z-transform of the above difference equation is

$$S(z) = 1.1z^{-1}S(z) - 0.24z^{-2}S(z) + 2W(z) + 3z^{-1}W(z),$$

so the system function is

$$H_s(z) = \frac{S(z)}{W(z)} = \frac{2 + 3z^{-1}}{1 - 1.1z^{-1} + 0.24z^{-2}} = \frac{2(1 + 1.5z^{-1})}{(1 - 0.3z^{-1})(1 - 0.8z^{-1})}.$$

Then the power spectrum of $s(n)$ is given by

$$S_s(z) = H_s(z)H_s(z^{-1})S_w(z),$$

and from $R_w(n)$ we have

$$S_w(z) = 5 \times \frac{0.64}{(1 - 0.6z^{-1})(1 - 0.6z)}.$$

Hence

$$S_s(z) = \frac{2(1 + 1.5z^{-1})}{(1 - 0.3z^{-1})(1 - 0.8z^{-1})} \frac{2(1 + 1.5z)}{(1 - 0.3z)(1 - 0.8z)} \times \frac{3.2}{(1 - 0.6z^{-1})(1 - 0.6z)}.$$

We collect the terms of $S_s(z)$ that have poles or zeros inside the unit circle to form $S_s^+(z)$. $S_s(z)$ has a zero at $z = -2/3$ and poles at $z = 0.3$, $z = 0.6$, and $z = 0.8$. As a result we have

$$S_s^+(z) = \frac{2\sqrt{3.2}(1 + 1.5z)}{(1 - 0.3z^{-1})(1 - 0.8z^{-1})(1 - 0.6z^{-1})}.$$

Since $S_s^-(z) = S_s^+(z^{-1})$, we have effectively specified $S_s^-(z)$.

Causal-Part Extraction: The Plus Operation

The second tool we require is causal-part extraction. Consider the impulse response $h(n)$ of a real-valued LTI system with rational z-transform $H(z)$. In the time domain we can split $h(n)$ into its causal and anticausal parts trivially:

$$h(n) = \text{causal part of } h(n) + \text{anticausal part of } h(n), \tag{4.53}$$

where

$$\text{causal part of } h(n) \triangleq [h(n)]_+ = h(n)1(n), \tag{4.54}$$

$$\text{anticausal part of } h(n) \triangleq [h(n)]_- = h(n)1(-n-1). \tag{4.55}$$

We would like to perform this decomposition in the z-domain. However, the operation becomes more complicated. We are particularly interested in obtaining the causal part of $h(n)$ because it is the realizable part of the system. By the linearity of the z-transform, Equation (4.53) becomes

$$H(z) = [H(z)]_+ + [H(z)]_- , \tag{4.56}$$

where

$$[H(z)]_+ = \mathcal{Z}\{h(n)1(n)\}, \tag{4.57}$$
$$[H(z)]_- = \mathcal{Z}\{h(n)1(-n-1)\}. \tag{4.58}$$

The operation represented by Equation (4.57) is called the *plus operator* or *causal-part extractor.* Similarly, Equation (4.58) is the minus operator or anticausal-part extractor.

Extraction for Rational $H(z)$

We now develop a method for determining $[H(z)]_+$. We begin with a general form for $H(z)$. Then we perform some basic algebra to convert $H(z)$ into a form that reflects its causal and anticausal parts.

We assume $H(z)$ has the form[4]

$$H(z) = \frac{\sum_{n=1}^{L} \beta_{-n} z^n + \sum_{n=0}^{M} \beta_n z^{-n}}{\sum_{n=0}^{N} a_n z^{-n}}. \tag{4.59}$$

By long division in z, z^{-1}, or both, we can always convert $H(z)$ into the form

$$H(z) = \underbrace{\sum_{n=1}^{L} c_{-n} z^n}_{P_A(z)} + \underbrace{\sum_{n=0}^{M-N} c_n z^{-n}}_{P_C(z)} + \underbrace{\left(\frac{\sum_{n=0}^{N-1} b_n z^{-n}}{\sum_{n=0}^{N} a_n z^{-n}} \right)}_{Q(z)}. \tag{4.60}$$

At this point, $H(z)$ consists of two polynomial terms, $P_A(z)$ and $P_C(z)$, and a *proper* fraction in z^{-1}, $Q(z)$. Clearly, $P_A(z)$ corresponds to a purely anticausal sequence, and $P_C(z)$ to a purely causal sequence. L is the number of poles of $H(z)$ at infinity, and $M-N-1$ the number of poles at zero.

Now we examine $Q(z)$. Equation (4.60) indicates that $Q(z)$ has at most N distinct poles and at most $N-1$ distinct zeros. Let $K \leq N$ be the number of distinct poles of $Q(z)$, and denote these poles and their associated degrees

[4] Any rational z-transform can be put in this form by multiplying it by z^n/z^n.

by p_k and m_k, respectively, $k = 1, \ldots, K$. Then $Q(z)$ may be written as

$$Q(z) = \left(\frac{1}{a_0}\right) \frac{\sum_{n=0}^{N-1} b_n z^{-n}}{\prod_{k=1}^{K} (1 - p_k z^{-1})^{m_k}}.$$

Via partial fractions, we can write this expression as [6]

$$Q(z) = \sum_{k=1}^{K} \sum_{m=1}^{m_k} \frac{q_{k,m}}{(1 - p_k z^{-1})^m} = \sum_{k=1}^{K} Q_k(z),$$

where

$$Q_k(z) = \sum_{m=1}^{m_k} \frac{q_{k,m}}{(1 - p_k z^{-1})^m}, \tag{4.61}$$

for some scalars $q_{k,m}$. Equation (4.61) completely describes the effect of the kth pole p_k in the system function $H(z)$. We may rewrite this expression as follows:

$$Q_k(z) = \underbrace{\left(\frac{q_{k,1}}{1 - p_k z^{-1}}\right)}_{Q'_k(z)} + \sum_{m=2}^{m_k} \frac{q_{k,m}}{(1 - p_k z^{-1})^m}. \tag{4.62}$$

Compare $Q'_k(z)$ with the z-transform pairs

$$\mu^n 1(n) \longleftrightarrow \frac{1}{1 - \mu z^{-1}}, \quad \text{ROC} = \{z : |z| > |\mu|\}, \tag{4.63}$$

and

$$-\mu^n 1(-n-1) \longleftrightarrow \frac{1}{1 - \mu z^{-1}}, \quad \text{ROC} = \{z : |z| < |\mu|\}. \tag{4.64}$$

These transform pairs differ only in their ROCs. For $Q'_k(z)$ we should choose the pair whose ROC intersects the ROC of $H(z)$. In this way we can determine the causality or anticausality of $Q'_k(z)$.

The product of two z-transforms corresponds to convolution of their associated sequences. Also, the convolution of two causal sequences yields a causal sequence, while the convolution of two anticausal sequences produces an anticausal sequence. Thus, if $Q'_k(z)$ is causal, then so is $Q_k(z)$; if $Q'_k(z)$ is anticausal, so is $Q_k(z)$.

Extraction for Stable Rational $H(z)$

We now introduce the requirement that $H(z)$ be *stable*, which will simplify the extraction procedure. If $H(z)$ is stable, its ROC includes the unit circle, which means that $H(z)$ has no poles on the unit circle. In addition, the ROC of each $Q_k(z)$ must include the unit circle. We use this fact to find the casual and anticausal parts of $Q(z)$.

Suppose the pole p_k of $Q_k(z)$ lies inside the unit circle. Then $Q_k'(z)$ must correspond to (4.63), which means that $Q_k(z)$ represents a causal portion of $H(z)$. Conversely, if p_k lies outside the unit circle, $Q_k(z)$ is anticausal.

Let N_A be the total order of the poles of $H(z)$ outside the unit circle, and N_C be the total order of the poles of $H(z)$ inside the unit circle with the exception of the origin. Then $N_A + N_C = N$ and we may write

$$Q_A(z) = \sum_{\substack{k=1 \\ |p_k|>1}}^{K} Q_k(z) = \frac{\sum_{n=0}^{N_A-1} \lambda_n z^{-n}}{\prod_{\substack{k=1 \\ |p_k|>1}}^{K} (1 - p_k z^{-1})^{m_k}},$$

and

$$Q_C(z) \triangleq \sum_{\substack{k=1 \\ 0<|p_k|<1}}^{K} Q_k(z) = \frac{\sum_{n=0}^{N_C-1} \mu_n z^{-n}}{\prod_{\substack{k=1 \\ 0<|p_k|<1}}^{K} (1 - p_k z^{-1})^{m_k}}.$$

In other words, $Q_A(z)$ is the anticausal, proper, rational portion of $H(z)$, and $Q_C(z)$ is the causal, proper, rational portion of $H(z)$. Then for a stable rational $H(z)$, we have

$$[H(z)]_+ = P_C(z) + Q_C(z),$$

which is the sum of a constant term and all terms in the partial fraction expansion of $H(z)$ that have poles inside the unit circle.

Extraction Methods

Now we can define two procedures[5] for finding $[H(z)]_+$ or $[H(z)]_-$ for any stable $H(z)$ in the form of (4.59). The first procedure is:

[5] A third method, based on the modulation property of the Fourier transform, exists. However, it is usually more difficult than the methods presented here, so the latter are preferred. An example appears in Problem 4.10.

1. If necessary, perform long division in z, z^{-1}, or both to convert $H(z)$ into the form

$$H(z) = P_A(z) + P_C(z) + Q(z).$$

2. Find the poles of $Q(z)$ and use partial fractions to decompose $Q(z)$ into the sum of two proper fractions in z^{-1},

$$Q(z) = Q_A(z) + Q_C(z).$$

3. Combine $Q_A(z)$ and $Q_C(z)$ to form a single rational expression in z^{-1}. Collect the coefficients of each z^{-n} in the numerator to form a system of linear equations.

4. Solve this system of equations for λ_n and μ_n.

5. Then

$$[Q(z)]_+ = P_C(z) + Q_C(z) \quad \text{and} \quad [Q(z)]_- = P_A(z) + Q_A(z).$$

An equivalent procedure is the following:

1. Assume the form

$$H(z) = P_A(z) + P_C(z) + Q_A(z) + Q_C(z).$$

2. Combine these terms to form a single rational expression in z and z^{-1}. Collect the coefficients of each z^{-n} in the numerator to form a system of linear equations.

3. Solve this system of equations for c_n, λ_n, and μ_n.

4. Then

$$[Q(z)]_+ = P_C(z) + Q_C(z) \quad \text{and} \quad [Q(z)]_- = P_A(z) + Q_A(z).$$

The only difference between these methods lies in the process of finding c_{-n} and c_n. In the first method, they are determined by long division; λ_n and μ_n are found by solving a system of linear equations. In the second method c_n, λ_n, and μ_n are found by solving a different (and larger) system of linear equations.

Two examples demonstrate these procedures.

Example 4.8 Causal-Part Extraction 1

Let us find the causal part of

$$H(z) = \frac{5z - \frac{1}{2} + 5z^{-1} - \frac{5}{4}z^{-2}}{1 - \frac{1}{4}z^{-1}},$$

with ROC $= \left\{z : \frac{1}{4} < |z| < \infty\right\}$. We will show both procedures in this example.

First procedure. Since the numerator of $H(z)$ contains a polynomial in z, we do long division in z:

$$\begin{array}{r|rrrrrrr}
 & 5z & + & \frac{3}{4} & & & & \\ \hline
1 - \frac{1}{4}z^{-1} & 5z & - & \frac{1}{2} & + & 5z^{-1} & - & \frac{5}{4}z^{-2} \\
 & 5z & - & \frac{5}{4} & & & & \\ \cline{2-4}
 & & & \frac{3}{4} & + & 5z^{-1} & - & \frac{5}{4}z^{-2} \\
 & & & \frac{3}{4} & - & \frac{3}{16}z^{-1} & & \\ \cline{4-6}
 & & & & & \frac{83}{16}z^{-1} & - & \frac{5}{4}z^{-2}
\end{array}$$

Then

$$H(z) = 5z + \tfrac{3}{4} + \frac{\frac{83}{16}z^{-1} - \frac{5}{4}z^{-2}}{1 - \frac{1}{4}z^{-1}}.$$

The rational term here is not proper, so we perform long division in z^{-1}:

$$\begin{array}{r|rrrrr}
 & & 5z^{-1} & & - & \frac{3}{4} \\ \hline
-\frac{1}{4}z^{-1} + 1 & - & \frac{5}{4}z^{-2} & + & \frac{83}{16}z^{-1} & \\
 & - & \frac{5}{4}z^{-2} & + & 5z^{-1} & \\ \cline{2-5}
 & & & & \frac{3}{16}z^{-1} & \\
 & & & & \frac{3}{16}z^{-1} & - \frac{3}{4} \\ \cline{5-6}
 & & & & & \frac{3}{4}
\end{array}$$

Now we have

$$H(z) = 5z + \tfrac{3}{4} + \left(-\tfrac{3}{4}\right) + 5z^{-1} + \frac{\frac{3}{4}}{1 - \frac{1}{4}z^{-1}},$$

so

$$[H(z)]_+ = 5z^{-1} + \frac{\frac{3}{4}}{1 - \frac{1}{4}z^{-1}}.$$

Note that $[H(z)]_+$ corresponds to the sequence $5\delta(n{-}1) + \frac{3}{4}(\frac{1}{4})^n 1(n)$, which is indeed causal.

Second procedure. Comparing $H(z)$ with (4.59), we see that $L = 1$, $M = 2$, and $N = 1$, and the only pole not at zero or infinity is a first-order pole at $\frac{1}{4}$ (inside the unit circle). Hence, $N_A = 0$ and $N_C = 1$. We therefore assume the form

$$H(z) = c_{-1}z + c_0 + c_1 z^{-1} + \frac{\mu_0}{1 - \frac{1}{4}z^{-1}}.$$

The right-hand side of this equation becomes

$$H(z) = \frac{c_{-1}z + (-\frac{1}{4}c_{-1} + c_0 + \mu_0) + (-\frac{1}{4}c_0 + c_1)z^{-1} + (-\frac{1}{4}c_1)z^{-2}}{1 - \frac{1}{4}z^{-1}},$$

which gives the system of linear equations

$$\begin{matrix} z^1 \\ z^0 \\ z^{-1} \\ z^{-2} \end{matrix} : \begin{bmatrix} 1 & 0 & 0 & 0 \\ -\frac{1}{4} & 1 & 0 & 1 \\ 0 & -\frac{1}{4} & 1 & 0 \\ 0 & 0 & -\frac{1}{4} & 0 \end{bmatrix} \begin{bmatrix} c_{-1} \\ c_0 \\ c_1 \\ \mu_0 \end{bmatrix} = \begin{bmatrix} 5 \\ -\frac{1}{2} \\ 5 \\ -\frac{5}{4} \end{bmatrix}.$$

The solution of this system is

$$\begin{bmatrix} c_{-1} & c_0 & c_1 & \mu_0 \end{bmatrix}^{\mathrm{T}} = \begin{bmatrix} 5 & 0 & 5 & \frac{3}{4} \end{bmatrix}^{\mathrm{T}}.$$

These values give the same expansion for $H(z)$ as with the first method.

Example 4.9 Causal Part Extraction 2

We want the causal part of

$$H(z) = \frac{6z^2 - 51z + 128 - 109z^{-1} + 197z^{-2} - 232z^{-3} + 80z^{-4}}{2 - 17z^{-1} + 40z^{-2} - 16z^{-3}},$$

with ROC $= \left\{z : \frac{1}{2} < |z| < 4\right\}$. We use only the first procedure in this example.

Long division in z gives

$$H(z) = 3z^2 + 4 + \frac{7z^{-1} + 37z^{-2} - 168z^{-3} + 80z^{-4}}{2 - 17z^{-1} + 40z^{-2} - 16z^{-3}}.$$

Then long division in z^{-1} produces

$$H(z) = 3z^2 + 4 + (-2) - 5z^{-1} + \frac{4 - 17z^{-1} + 32z^{-2}}{2 - 17z^{-1} + 40z^{-2} - 16z^{-3}}.$$

We let $Q(z)$ be the rational term and find

$$Q(z) = \frac{4 - 17z^{-1} + 32z^{-2}}{2 - 17z^{-1} + 40z^{-2} - 16z^{-3}} = \frac{2 - \frac{17}{2}z^{-1} + 16z^{-2}}{(1 - \frac{1}{2}z^{-1})(1 - 4z^{-1})^2}.$$

$Q(z)$ has a double pole at 4 and a single pole at 1/2. Hence, $N_A = 2$ and $N_C = 1$. So we may assume

$$Q_A(z) = \frac{\lambda_0 + \lambda_1 z^{-1}}{(1 - 4z^{-1})^2}, \quad \text{and} \quad Q_C(z) = \frac{\mu_0}{1 - \frac{1}{2}z^{-1}}.$$

Solving for the unknowns, we find $\lambda_0 = 1$, $\lambda_1 = 0$, and $\mu_0 = 1$. Hence

$$H(z) = 3z^2 + 2 - 5z^{-1} + \frac{1}{(1 - 4z^{-1})^2} + \frac{1}{1 - \frac{1}{2}z^{-1}}.$$

So the causal part of $H(z)$ is

$$[H(z)]_+ = 2 - 5z^{-1} + \frac{1}{1 - \frac{1}{2}z^{-1}},$$

which corresponds to the sequence $2\delta(n) - 5\delta(n-1) + (\frac{1}{2})^n 1(n)$. Also, the anticausal part of $H(z)$ is

$$[H(z)]_- = 3z^2 + \frac{1}{(1 - 4z^{-1})^2},$$

which corresponds to $3\delta(n+2) + [(-4)^n 1(-n-1)] * [(-4)^n 1(-n-1)]$.

4.5 Derivation of the Causal Wiener Filter

Derivation by Causality

The previous section has presented the tools that we will need to derive the causal Wiener filter. Two common derivations of this filter exist, and we present both of them. Because we will require the ideas of spectral factorization and causal-part extraction, *we assume that $S_z(z)$ is rational in z and has no poles or zeros on the unit circle.*[6]

We restate (4.51), the equation that we must solve to obtain the causal Wiener filter:

$$\sum_{j=0}^{\infty} h(j) R_z(j-i) = R_{sz}(i), \quad i \geq 0.$$

As stated earlier, we cannot take the z-transform of (4.51) because it holds only for $i \geq 0$, not for all $i \in \mathbb{Z}$. Suppose, however, that we define

$$h'(i) = R_{sz}(i) - \sum_{j=0}^{\infty} h(j) R_z(j-i), \quad \text{for all } i \in \mathbb{Z}. \tag{4.65}$$

By (4.51), if h is the causal Wiener filter, then

$$h'(i) = \begin{cases} 0, & i \geq 0; \\ R_{sz}(i) - \sum_{j=0}^{\infty} h(j) R_z(j-i), & i < 0. \end{cases} \tag{4.66}$$

[6] The spectral factorization theorem, Theorem 4.3, permits zeros on the unit circle. Once we have derived the causal Wiener filter, however, we will see that prohibiting them ensures stability of the filter.

Observe that $h'(i)$ is defined for all $i \in \mathbb{Z}$. As a result, we can take the z-transform of both sides of (4.65). We obtain

$$\begin{aligned} H'(z) &= S_{sz}(z) - H(z)S_z(z) \\ &= S_{sz}(z) - H(z)S_z^+(z)S_z^-(z). \end{aligned}$$

Dividing by $S_z^-(z)$, we have

$$\frac{H'(z)}{S_z^-(z)} = \frac{S_{sz}(z)}{S_z^-(z)} - H(z)S_z^+(z).$$

We extract the causal part of both sides of this equation and find

$$\left[\frac{H'(z)}{S_z^-(z)}\right]_+ = \left[\frac{S_{sz}(z)}{S_z^-(z)}\right]_+ - \left[H(z)S_z^+(z)\right]_+ . \tag{4.67}$$

Now we examine the terms in (4.67). First, from (4.66), $h'(i)$ is purely anticausal. Therefore $H'(z)$ contains only poles outside the unit circle. Since the zeros of $S_z^-(z)$ lie outside the unit circle, the poles of $1/S_z^-(z)$ also lie outside the unit circle. We conclude that all the poles of $H'(z)/S_z^-(z)$ are outside the unit circle, and thus

$$\left[\frac{H'(z)}{S_z^-(z)}\right]_+ = 0.$$

Second, $H(z)$ is causal by definition, so its poles lie within the unit circle. The poles of $S_z^+(z)$ are also within the unit circle. Hence all the poles of $H(z)S_z^+(z)$ are inside the unit circle, and

$$\left[H(z)S_z^+(z)\right]_+ = H(z)S_z^+(z).$$

Finally, we cannot conclude anything about $S_{sz}(z)/S_z^-(z)$. Then (4.67) becomes

$$0 = \left[\frac{S_{sz}(z)}{S_z^-(z)}\right]_+ - H(z)S_z^+(z),$$

which produces the causal Wiener filter, described by its system function $H(z)$:

$$\boxed{H(z) = \frac{1}{S_z^+(z)}\left[\frac{S_{sz}(z)}{S_z^-(z)}\right]_+ .} \tag{4.68}$$

Derivation by Applying a Whitening Filter

This section presents an alternative derivation of the causal Wiener filter. The derivation by causality is purely mathematical, while this derivation introduces the application of a whitening filter.

Again, we begin with Equation (4.51):

$$\sum_{j=0}^{\infty} h(j)R_z(j-i) = R_{sz}(i), \quad i \geq 0.$$

Suppose we replace $\boldsymbol{z}(n)$ with $\boldsymbol{d}(n)$, where $\boldsymbol{d}(n)$ is zero-mean *white noise* with unit variance: $R_d(k) = \delta(k)$. It is easy to find the causal Wiener filter in this case, diagrammed in Figure 4.5. We use $h'(n)$ to denote the impulse response of the causal Wiener filter for $\boldsymbol{d}(n)$. Then (4.51) becomes

$$\sum_{j=0}^{\infty} h'(j)\delta(j-i) = R_{sd}(i), \quad i \geq 0,$$

or equivalently,

$$h'(i) = R_{sd}(i), \quad i \geq 0.$$

Because $h'(i)$ is causal,

$$h'(i) = \begin{cases} R_{sd}(i), & i \geq 0, \\ 0, & i < 0, \end{cases} = R_{sd}(i)1(i),$$

and therefore the causal Wiener filter *for unit-variance white noise* can be described by its system function $H'(z)$:

$$H'(z) = [S_{sd}(z)]_+ \,. \tag{4.69}$$

We will refer to this filter as the *white-noise causal Wiener filter*.

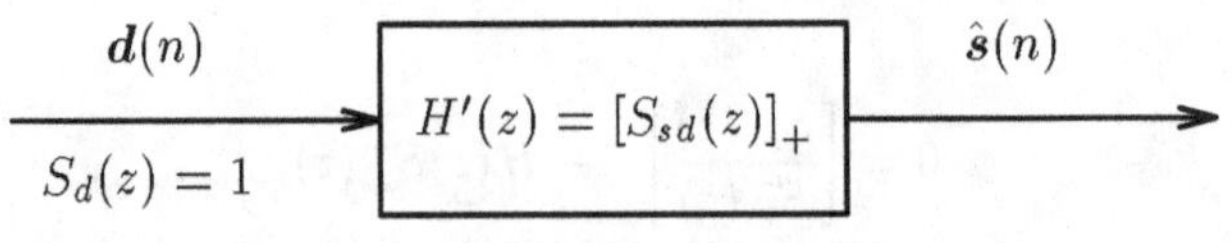

Figure 4.2. The causal Wiener filter for a white noise input. White noise, which is correlated with the signal process $\boldsymbol{s}(n)$, is input to a filter with system function $H'(z)$. The output of the filter is the causal LTI MMSE estimate $\hat{\boldsymbol{s}}(n)$.

In general $\boldsymbol{z}(n)$ is not white noise, so we must find a stable filter with the property that when $\boldsymbol{z}(n)$ is the filter input, the filter output is the white noise process $\boldsymbol{d}(n)$. For this reason we call this filter a *whitening filter*. Let $H_d(z)$ denote the system function of the whitening filter.

Once we have the whitening filter, we cascade it with the white-noise causal Wiener filter to produce the causal Wiener filter, which has system function $H(z)$:

$$H(z) = H_d(z)H'(z). \tag{4.70}$$

Figure 4.5 demonstrates this strategy.

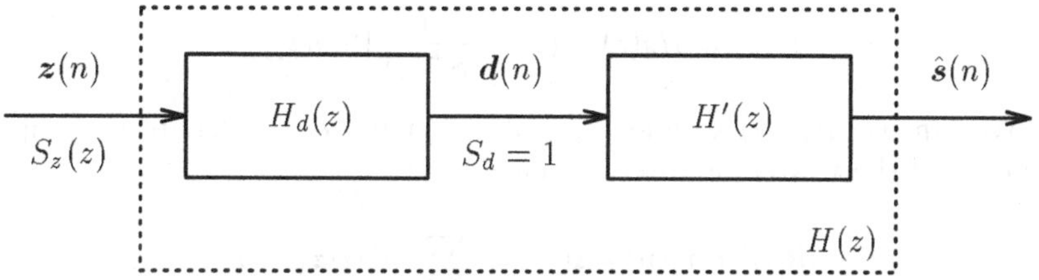

Figure 4.3. Cascading a whitening filter with the white-noise causal Wiener filter to form the general causal Wiener filter with system function $H(z)$. The observation process enters a whitening filter, which has system function $H_d(z)$. The output is $\boldsymbol{d}(n)$, white noise correlated with the signal $\boldsymbol{s}(n)$. Next, $\boldsymbol{d}(n)$ is filtered by the white-noise causal Wiener filter with system function $H'(z)$ to generate the estimate $\hat{\boldsymbol{s}}(n)$.

The system function $H_d(z)$ must satisfy

$$S_d(z) = 1 = H_d(z)H_d(z^{-1})S_z(z).$$

Applying the spectral factorization theorem to $S_z(z)$, this condition becomes

$$H_d(z)H_d(z^{-1})S_z^+(z)S_z^-(z) = 1. \tag{4.71}$$

Observe that $H_d(z)H_d(z^{-1})$ matches the form of a spectral factorization. So, let us choose

$$H_d(z) = \frac{1}{S_z^+(z)}. \tag{4.72}$$

Then (4.71) becomes

$$\frac{1}{S_z^+(z)}\frac{1}{S_z^+(z^{-1})}S_z^+(z)S_z^-(z) = \frac{1}{S_z^+(z)}\frac{1}{S_z^-(z)}S_z^+(z)S_z^-(z) = 1,$$

which satisfies our condition. Hence (4.72) is our whitening filter, shown in Figure 4.5.

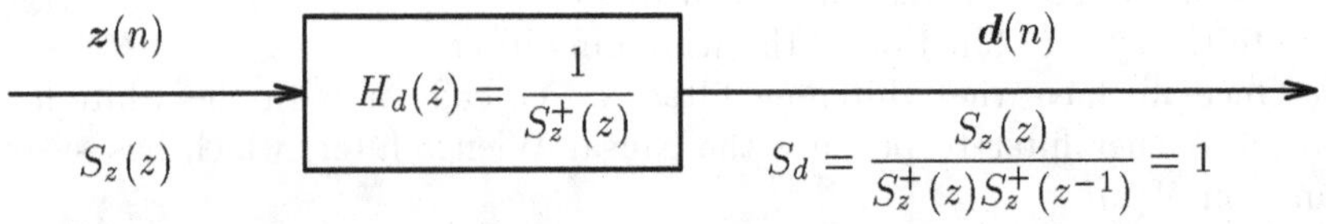

Figure 4.4. A whitening filter. Colored noise is input to a filter with system function $H_d(z)$. The output of the filter is white noise with unit variance.

Now by (4.69), (4.70), and (4.72) we have

$$H(z) = H_d(z)H'(z) = \frac{1}{S_z^+(z)}[S_{sd}(z)]_+ . \tag{4.73}$$

We can use $R_{sd}(n)$ to find $[S_{sd}(z)]_+$. Letting $h_d(n)$ denote the impulse response of the whitening filter, we have

$$\boldsymbol{d}(n) = h_d(n) * \boldsymbol{z}(n) = \sum_{i=-\infty}^{\infty} h_d(i)\boldsymbol{z}(n-i).$$

Then

$$\begin{aligned} R_{sd}(n) &= \mathrm{E}\left[\boldsymbol{s}(n)\boldsymbol{d}(0)\right] = \mathrm{E}\left[\boldsymbol{s}(n)\sum_{i=-\infty}^{\infty} h_d(i)\boldsymbol{z}(-i)\right] \\ &= \sum_{i=-\infty}^{\infty} h_d(i)\mathrm{E}\left[\boldsymbol{s}(n)\boldsymbol{z}(-i)\right] = \sum_{i=-\infty}^{\infty} h_d(i)R_{sz}(n+i). \end{aligned}$$

Letting $j = -i$,

$$R_{sd}(n) = \sum_{j=-\infty}^{\infty} h_d(-j)R_{sz}(n-j) = h_d(-n) * R_{sz}(n),$$

so that

$$S_{sd}(z) = H_d(z^{-1})S_{sz}(z) = \frac{1}{S_z^+(z^{-1})}S_{sz}(z) = \frac{S_{sz}(z)}{S_z^-(z)},$$

and the causal part of this expression is

$$[S_{sd}(z)]_+ = \left[\frac{S_{sz}(z)}{S_z^-(z)}\right]_+ .$$

Finally, (4.73) gives us the system function $H(z)$ for the causal Wiener filter,

$$\boxed{H(z) = \frac{1}{S_z^+(z)}\left[\frac{S_{sz}(z)}{S_z^-(z)}\right]_+},$$

which is the same result as that derived by causality (4.68).

Filter Properties

Equation (4.68) is the general solution for the causal Wiener filter. Because we have assumed that $S_z(z)$ has no zeros on the unit circle, the zeros of $S_z^+(z)$ lie strictly within the unit circle. By causal-part extraction the poles of $[S_{sz}(z)/S_z^-(z)]_+$ are strictly inside the unit circle as well. Therefore $H(z)$ is guaranteed to be stable.

Computing the Mean-Square Error

The equations for computing the MSE may be derived exactly the same way as for the noncausal Wiener filter of Section 4.3. When the impulse response $h(n)$ is available, we may compute the MSE via

$$\text{MSE} = R_s(0) - \sum_{i=0}^{\infty} h(i) R_{sz}(i), \tag{4.74}$$

provided that we can compute the infinite sum.

Using $H(z)$ and restating (4.47), we have

$$\text{MSE} = \begin{cases} \dfrac{1}{j2\pi} \displaystyle\oint_C \left[S_s(z) - H(z) S_{sz}(z^{-1})\right] z^{-1}\, dz, & S_{sz}(z) \text{ rational}, \\ \dfrac{1}{j2\pi} \displaystyle\oint_C \left[S_s(z) - H(z) S_{sz}^*(z)\right] z^{-1}\, dz, & \text{otherwise}. \end{cases} \tag{4.75}$$

Example 4.10 Causal Wiener Filtering

Let us again consider the situation of Examples 4.2 and 4.5. $s(n)$ and $v(n)$ are zero-mean JWSS random processes with

$$S_s(z) = \frac{0.0975}{(1 - 0.95z^{-1})(1 - 0.95z)},$$

and

$$S_v(z) = 2.$$

We observe $z(n) = s(n) + v(n)$ and estimate $s(n)$.

We have already demonstrated that $S_{sz}(z) = S_s(z)$, and from Example 4.5 we have

$$S_z(z) = 2.3955\, \frac{(1 - 0.7931z^{-1})(1 - 0.7931z)}{(1 - 0.95z^{-1})(1 - 0.95z)}.$$

We now perform a spectral factorization on $S_z(z)$ and obtain

$$S_z^+(z) = 1.5477\,\frac{1-0.7931z^{-1}}{1-0.95z^{-1}},$$

and

$$S_z^-(z) = 1.5477\,\frac{1-0.7931z}{1-0.95z}.$$

Next we need to find $\left[S_{sz}(z)/S_z^-(z)\right]_+$. We have

$$\begin{aligned}\frac{S_{sz}(z)}{S_z^-(z)} &= \frac{0.0630}{(1-0.95z^{-1})(1-0.7931z)}\\ &= \frac{-0.0794z^{-1}}{(1-0.95z^{-1})(1-1.2608z^{-1})}\\ &= \frac{0.2555}{1-0.95z^{-1}} - \frac{0.2555}{1-1.2608z^{-1}},\end{aligned}$$

so $\left[S_{sz}(z)/S_z^-(z)\right]_+$ is

$$\left[\frac{S_{sz}(z)}{S_z^-(z)}\right]_+ = \frac{0.2555}{1-0.95z^{-1}}.$$

The causal Wiener filter is then

$$\begin{aligned}H(z) &= \frac{1}{S_z^+(z)}\left[\frac{S_{sz}(z)}{S_z^-(z)}\right]_+\\ &= \frac{1-0.95z^{-1}}{1.5477(1-0.7931z^{-1})}\times\frac{0.2555}{1-0.95z^{-1}}\\ &= \frac{0.1651}{1-0.7931z^{-1}},\end{aligned}$$

so

$$h(n) = 0.1651\,(0.7931)^n\,1(n).$$

The MSE associated with this filter can be computed using (4.74).

$$\begin{aligned}\mathrm{MSE_C} &= 1 - 0.1651\sum_{i=0}^{\infty}(0.7931)^i(0.95)^i\\ &= 1 - \frac{0.1651}{1-(0.7931)(0.95)}\\ &= 0.3302.\end{aligned}$$

Compared to no filtering (MSE = 2), the causal Wiener filter reduces the MSE by 7.8 dB. We compare this error with that of Examples 4.2 and 4.5, where we found $\mathrm{MSE_{FIR}} = 0.4405$ and $\mathrm{MSE_{NC}} = 0.2195$. Since the causal Wiener filter uses an infinite number of observations to determine $\hat{s}(n)$, we expect it to perform better than the FIR Wiener filter. The noncausal Wiener filter uses all observations (both causal and anticausal), so it performs better than its causal counterpart.

Signal Prediction

Let us estimate $\boldsymbol{s}(n+m)$, where m is some positive integer. This estimation problem is the basic m-step prediction problem, which we have seen in Sections 3.1 and 4.2. The m-step causal Wiener predictor is derived in a manner analogous to the causal Wiener filter.

Now the estimate is

$$\hat{\boldsymbol{s}}(n+m) = \sum_{i=0}^{\infty} h(i)\boldsymbol{z}(n-i).$$

Via the same procedure as in Section 4.1, the Wiener-Hopf equation (4.8) for the causal Wiener predictor becomes

$$\sum_{j=0}^{\infty} h(j)R_z(j-i) = R_{sz}(m+i), \quad i \geq 0.$$

We define

$$h'(i) = R_{sz}(m+i) - \sum_{j=0}^{\infty} h(j)R_z(j-i), \quad \text{for all } i \in \mathbb{Z}, \tag{4.76}$$

and we take the z-transform of both sides of (4.76) to obtain

$$\begin{aligned} H'(z) &= z^m S_{sz}(z) - H(z)S_z(z) \\ &= z^m S_{sz}(z) - H(z)S_z^+(z)S_z^-(z). \end{aligned}$$

Hence, we have

$$\frac{H'(z)}{S_z^-(z)} = \frac{z^m S_{sz}(z)}{S_z^-(z)} - H(z)S_z^+(z),$$

and we apply the causal-part extractor to this equation:

$$\left[\frac{H'(z)}{S_z^-(z)}\right]_+ = \left[\frac{z^m S_{sz}(z)}{S_z^-(z)}\right]_+ - \left[H(z)S_z^+(z)\right]_+. \tag{4.77}$$

Compare (4.77) with (4.67), the corresponding equation for the causal Wiener filter. They differ only by a time advance of z^m in one term. As a result, we may use the same arguments as in the derivation of the causal Wiener filter. Doing so, we obtain the m-step causal Wiener predictor,

$$H(z) = \frac{1}{S_z^+(z)}\left[\frac{z^m S_{sz}(z)}{S_z^-(z)}\right]_+. \tag{4.78}$$

The equations for computing the MSE are identical to those for the causal Wiener filter [Equations (4.74–4.75)].

Example 4.11 Two-Step Causal Wiener Prediction

Suppose we have uncorrelated JWSS random processes $s(n)$ and $v(n)$ and use the additive-noise model. We are given

$$S_s(z) = \frac{0.36}{(1-0.8z^{-1})(1-0.8z)},$$

and

$$S_v(z) = -0.2z + 0.5 - 0.2z^{-1} = 0.4(1-0.5z^{-1})(1-0.5z).$$

Let us find the two-step causal Wiener predictor.

Because we have used the additive-noise model and $s(n)$ and $v(n)$ are uncorrelated, $S_{sz}(z) = S_s(z)$ and $S_z(z) = S_s(z) + S_v(z)$. Then

$$\begin{aligned} S_z(z) &= \frac{0.36}{(1-0.8z^{-1})(1-0.8z)} + 0.4(1-0.5z^{-1})(1-0.5z) \\ &= 0.16\,\frac{z^2 - 4.55z + 9.375 - 4.55z^{-1} + z^{-2}}{(1-0.8z^{-1})(1-0.8z)} \\ &= \frac{0.16}{r^2}\,\frac{(1-re^{j\theta}z^{-1})(1-re^{-j\theta}z^{-1})(1-re^{j\theta}z)(1-re^{-j\theta}z)}{(1-0.8z^{-1})(1-0.8z)}, \end{aligned}$$

where $r = 0.3792$ and $\theta = 0.2280\pi = 0.7164$. We find the spectral factorization to be

$$S_z^+(z) = \frac{0.4}{r}\,\frac{(1-re^{j\theta}z^{-1})(1-re^{-j\theta}z^{-1})}{(1-0.8z^{-1})},$$

and $0.4/r = 1.0550$.

It follows that

$$\begin{aligned} \frac{z^2 S_{sz}(z)}{S_z^-(z)} &= \frac{z^2(0.36)}{(1-0.8z^{-1})(1-0.8z)}\,\frac{(1-0.8z)}{(1.0550)(1-re^{j\theta}z)(1-re^{-j\theta}z)} \\ &= 0.3412\,\frac{z^2}{(1-0.8z^{-1})(1-re^{j\theta}z)(1-re^{-j\theta}z)} \\ &= \frac{0.3412}{r^2}\,\frac{1}{(1-0.8z^{-1})(1-r^{-1}e^{j\theta}z^{-1})(1-r^{-1}e^{-j\theta}z^{-1})} \\ &= \frac{0.3442}{1-0.8z^{-1}} + \frac{2.2689\,e^{-j0.3524\pi}}{1-r^{-1}e^{j\theta}z^{-1}} + \frac{2.2689\,e^{j0.3524\pi}}{1-r^{-1}e^{-j\theta}z^{-1}}. \end{aligned}$$

Since $r^{-1} = 2.6375$, the last two terms have poles outside the unit circle. Then $\left[z^2 S_{sz}(z)/S_z^-(z)\right]_+$ is simply

$$\left[\frac{z^2 S_{sz}(z)}{S_z^-(z)}\right]_+ = \frac{0.3442}{1-0.8z^{-1}}.$$

The system function for the causal Wiener predictor is given by

$$\begin{aligned}
H(z) &= \frac{1}{S_z^+(z)} \left[\frac{z^2 S_{sz}(z)}{S_z^-(z)} \right]_+ \\
&= \frac{(1 - 0.8z^{-1})}{(1.0550)(1 - re^{j\theta}z^{-1})(1 - re^{-j\theta}z^{-1})} \times \frac{0.3442}{1 - 0.8z^{-1}} \\
&= \frac{0.3263}{(1 - re^{j\theta}z^{-1})(1 - re^{-j\theta}z^{-1})} \\
&= \frac{0.2484e^{-j0.2720\pi}}{1 - re^{j\theta}z^{-1}} + \frac{0.2484e^{j0.2720\pi}}{1 - re^{-j\theta}z^{-1}}.
\end{aligned}$$

The impulse response for this filter is

$$h(n) = \left[(0.2484\, e^{-j0.2720\pi})(re^{j\theta})^n + (0.2484\, e^{j0.2720\pi})(re^{-j\theta})^n \right] 1(n).$$

We can compute the MSE associated with this filter. Combining Equations (4.43) and (4.46), we have

$$\begin{aligned}
\text{MSE} &= R_s(0) - \frac{1}{j2\pi} \oint_C H(z) S_{sz}(z^{-1}) z^{-1}\, dz \\
&= (0.8)^0 - \frac{1}{j2\pi} \oint_C \frac{0.3263}{(1 - re^{j\theta}z^{-1})(1 - re^{-j\theta}z^{-1})} \\
&\qquad \times \frac{0.36}{(1 - 0.8z)(1 - 0.8z^{-1})} z^{-1}\, dz \\
&= 1 - \frac{1}{j2\pi} \oint_C \frac{0.1175z^2}{(z - re^{j\theta})(z - re^{-j\theta})(1 - 0.8z)(z - 0.8)}\, dz \\
&= 1 - 0.1175 \left[\sum \text{residues of integrand inside unit circle} \right] \\
&= 1 - 0.1175 \left[\left. \frac{z^2}{(z - re^{-j\theta})(1 - 0.8z)(z - 0.8)} \right|_{z = re^{j\theta}} \right. \\
&\qquad + \left. \frac{z^2}{(z - re^{j\theta})(1 - 0.8z)(z - 0.8)} \right|_{z = re^{-j\theta}} \\
&\qquad \left. + \left. \frac{z^2}{(z - re^{j\theta})(z - e^{-j\theta})(1 - 0.8z)} \right|_{z = 0.8} \right] \\
&= 0.4858.
\end{aligned}$$

By contrast, without filtering, the MSE is 1.22. The Wiener predictor reduces the MSE by 4 dB.

4.6 Summary of Wiener Filters

Overall Optimality

The derivation of the Wiener filter makes it the optimal LTI MMSE estimator. In some cases the Wiener filter is also the overall MMSE estimator (including

	FIR	**Noncausal (IIR)**	**Causal (IIR)**
Filter	$h = R_z^{-1} r_{sz}$ from (4.18)	$H(z) = \dfrac{S_{sz}(z)}{S_z(z)}$	$H(z) = \dfrac{1}{S_z^+(z)} \left[\dfrac{S_{sz}(z)}{S_z^-(z)} \right]_+$
Causality	Causal	Noncausal	Causal
Stability	Always stable.	Stable if $S_z(z)$ has no zeros on the unit circle. Unstable if $S_z(z)$ has a zero on the unit circle.	Stable if $S_z(z)$ has no zeros on the unit circle. May be unstable if $S_z(z)$ has a zero on the unit circle.
MSE	$R_s(0) - h^{\mathrm{T}} r_{sz}$	$R_s(0) - \sum_{i=\infty}^{\infty} h(i) R_{sz}(i)$ or (4.47)	$R_s(0) - \sum_{i=0}^{\infty} h(i) R_{sz}(i)$ or (4.75)

Table 4.1. Characteristics of Wiener filters

nonlinear, time-varying filters). Example 3.9 showed that when $\boldsymbol{s}$ and $\boldsymbol{z}$ are zero-mean, jointly-distributed Gaussian RVs, then the LMMSE estimate $\hat{\boldsymbol{s}}_{\text{LMMSE}}$ is also the overall MMSE estimate, i.e., $\hat{\boldsymbol{s}}_{\text{LMMSE}} = \hat{\boldsymbol{s}}_{\text{MMSE}}$. Similarly, when $\boldsymbol{s}(n)$ and $\boldsymbol{z}(n)$ are zero-mean, JWSS, jointly Gaussian random signals, then the Wiener filter is not only the LMMSE filter, but also the overall MMSE filter. A common example in which the Wiener filter achieves overall optimality is when both the signal process $\boldsymbol{s}(n)$ and the noise process $\boldsymbol{v}(n)$ are WSS Gaussian random signals and combine via the additive-noise model.

Remarks

The three Wiener filters are summarized in Table 4.1. We can interpret these filters in the following way. First, the noncausal Wiener filter is the ideal case where all observations $\{\boldsymbol{z}(n) : n \in \mathbb{Z}\}$ are available for estimating $\boldsymbol{s}(n)$, and the filter is noncausal, capable of using observations in the infinite future. In some applications, we have a very large set of observations, so we can reasonably assume that we have an "infinite" number of observations, and we filter off-line, so noncausal filtering becomes possible. Then the noncausal Wiener filter can be applied. Audio and image processing include many applications (e.g., noise removal) where off-line filtering is permissible. For applications where causality is required, we cannot use the noncausal Wiener filter.

Second, the causal Wiener filter is a useful estimator for real-time applications. Although the filter no longer uses future observations, ideally it requires all observations from the infinite past, $\{\boldsymbol{z}(n) : n \leq 0\}$. For a large set of observations, this requirement is often reasonable enough to make the causal Wiener filter a useful, practical estimator. In practice the filter displays a transient response, but usually the transient is acceptably brief and bounded.

Finally, the FIR Wiener filter is a realizable filter within a delay. As discussed in Section 4.3, any FIR filter can be made causal by delaying its impulse response. Often the FIR filter provides an acceptable compromise between computational complexity, which is lower than an IIR Wiener filter, and performance, which may be only slightly poorer than an IIR Wiener filter.

Comparison of IIR Wiener Filters

We may make an interesting interpretation regarding the noncausal and causal Wiener filters. For the causal filter, Equation (4.68) states that the system function is

$$H_{\mathrm{C}}(z) = \frac{1}{S_z^+(z)} \left[\frac{S_{sz}(z)}{S_z^-(z)} \right]_+ .$$

In Section 4.4 we viewed this filter as the cascade of two filters. We first have a whitening filter with system function $1/S_z^+(z)$, and this filter is cascaded with a *causal* LTI MMSE estimator, which is designed for for unit-variance white noise $\boldsymbol{d}(n)$ and which has system function $[S_{sz}(z)/S_z^-(z)]_+$. Note that the whitening filter described by $1/S_z^+(z)$ is causal because the zeros of $S_z^+(z)$ lie strictly within the unit circle.

We can write the system function of the noncausal filter (4.42) as

$$H_{\mathrm{NC}}(z) = \frac{S_{sz}(z)}{S_z(z)} = \frac{1}{S_z^+(z)} \frac{S_{sz}(z)}{S_z^-(z)} .$$

We may interpret the noncausal Wiener filter as the cascade of a causal whitening filter with system function $1/S_z^+(z)$ and a *noncausal* LTI MMSE estimator (with system function $S_{sz}(z)/S_z^-(z)$) for $\boldsymbol{d}(n)$.

Notice that if we take the noncausal Wiener filter and replace its estimator with its causal part, we have exactly the causal Wiener filter. This idea is illustrated in Figure 4.6. In other words, both IIR Wiener filters have the same causal whitening filter; only the estimator differs. The estimator following the whitening filter determines the causality properties of the Wiener filter.

Problems

4.1. Section 4.1 derived the Wiener-Hopf equation (4.8) for the smoothing problem, estimating $\boldsymbol{s}(n)$. The Wiener-Hopf equation can be generalized to estimate $\boldsymbol{s}(\ell)$ at any time ℓ, where ℓ may selected from Table 3.1. Derive the general form of the Wiener-Hopf equation.

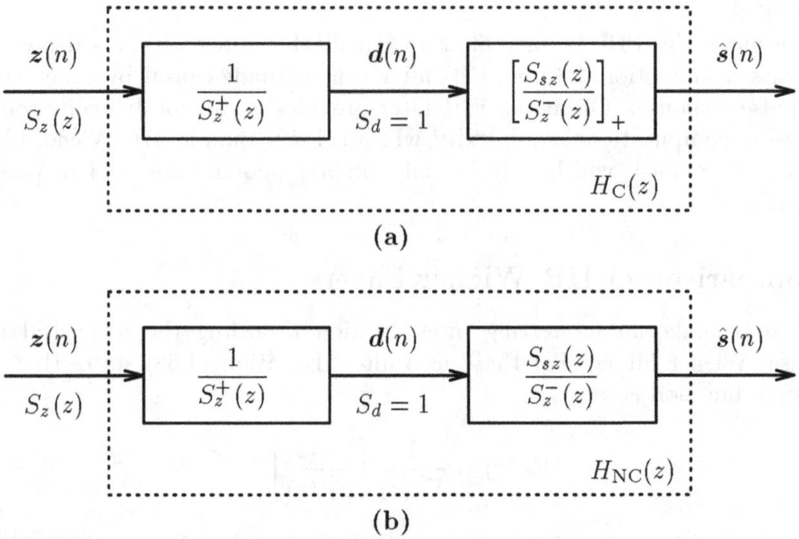

Figure 4.5. Interpretation of the IIR Wiener filters. **(a)** The causal Wiener filter with system function $H_C(z)$ may be viewed as a causal whitening filter followed by a causal estimator for white noise. **(b)** The noncausal Wiener filter with system function $H_{NC}(z)$ may similarly be viewed as the same whitening filter followed by a noncausal estimator for white noise.

4.2. Equation (4.8) gives the Wiener-Hopf equation for LTI systems. For linear time-varying (LTV) systems, a similar system of equations can be obtained. Denote the filter by its *time-varying* impulse response $h_m(n)$:

$$h_m(n) = \sum_{i=-\infty}^{\infty} h_m(i)\delta(n-i),$$

where $h_m(n)$ is read as "the value of h at time m in response to a unit impulse at time n." Denote the filter's region of support at time m by $\mathcal{H}_m$. Then

$$h_m(n) = \sum_{i \in \mathcal{H}_m} h_m(i)\delta(n-i).$$

Likewise, the estimate $\hat{s}(n)$ will vary with time. We can express the estimate $\hat{s}_m(n)$ in terms of a *time-dependent* convolution sum,

$$\hat{s}_m(n) = \sum_{i \in \mathcal{H}_m} h_m(i)z(n-i),$$

which is "the estimate of $s(n)$ made at time m."

For each m we must find the filter coefficients $h_m(n)$.

(a) Holding m fixed, write the expression for the MSE between $\hat{s}(n)$ and $\hat{s}_m(n)$.

(b) Minimize the MSE in part (a) to derive the *time-varying discrete-time Wiener-Hopf* equation:

$$\sum_{j \in \mathcal{H}_m} h_m(j) R_z(k-i, k-j) = R_{sz}(k, k-i), i \in \mathcal{H}_m.$$

4.3. The FIR Wiener filter can be generalized to estimate $s(\ell)$ at any time ℓ. Show that the Wiener-Hopf equation for the FIR Wiener filter becomes

$$\begin{bmatrix} R_z(0) & R_z(1) & \cdots & R_z(N-1) \\ R_z(1) & R_z(0) & \cdots & R_z(N-2) \\ \vdots & \vdots & \ddots & \vdots \\ R_z(N-1) & R_z(N-2) & \cdots & R_z(0) \end{bmatrix} \begin{bmatrix} h(0) \\ h(1) \\ \vdots \\ h(N-1) \end{bmatrix} = \begin{bmatrix} R_{sz}(\ell-n) \\ R_{sz}(\ell-n+1) \\ \vdots \\ R_{sz}(\ell-n+N-1) \end{bmatrix}.$$

4.4. Differential pulse-code modulation, or DPCM, is a technique used in speech processing, image and video compression, and some digital communication systems.

Suppose the signal $s(n)$ changes slowly relative to the time index or sampling rate n. Then subsequent samples of $s(n)$ will display some correlation, and $s(n+1)$ may be approximately predicted. Only the error signal

$$\tilde{s}(n+1) = \hat{s}(n+1) - s(n+1)$$

must be transmitted.

If $\hat{s}(n+1)$ is a reasonably accurate estimate of $s(n+1)$, then the range of $\tilde{s}(n+1)$ will be considerably smaller than the range of $s(n+1)$ itself. This property is advantageous because digital systems must quantize information to a finite number of values. With the number of quantization levels held constant, $\tilde{s}(n+1)$ can be transmitted with greater resolution than $s(n+1)$.

(a) Consider a signal modeled by

$$s(n) = \lambda s(n-1) + w(n),$$

where $w(n)$ is white noise $\sim (0, \sigma^2)$, and $|\lambda| < 1$. Find $R_s(n)$.

(b) Often the DPCM system must run in real-time, so a 1- or 2-tap predictor is employed. That is,

$$h(n) = h(0)s(n),$$

or

$$h(n) = h(0)s(n) + h(1)s(n-1).$$

Find the optimum 1- and 2-tap LTI predictors for the signal in part (a).

4.5. Consider the following familiar scenario: a signal $s(n)$ is corrupted by additive noise $v(n)$ to form the observations $z(n)$,

$$z(n) = s(n) + v(n).$$

However, suppose that $R_{sz}(k)$ is *not known.* Instead, a sensor located away from the signal source measures $v'(n)$, which is correlated with the measurement noise $v(n)$ but not with the signal. Then it is possible to estimate $v(n)$ from $v'(n)$ and produce an estimate of the signal via

$$\hat{s}(n) = z(n) - \hat{v}(n).$$

This approach forms the basis for *noise cancellation* and is diagrammed in Figure 4.5. Rather than estimate the signal, we attempt to estimate the noise and remove it. Such a system might be employed by military aircraft radio systems or manufacturing plants near residential areas.

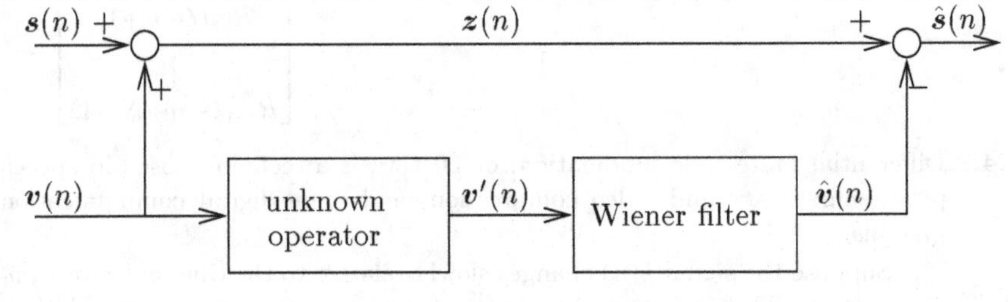

Figure 4.6. A simple noise cancellation system. Signal $s(n)$ is corrupted by additive white noise $v(n)$ to produce measurements $z(n)$. Another sensor measures $v'(n)$, which is related to $v(n)$ in some unknown manner. A Wiener filter estimates $v(n)$ from $v'(n)$ and the estimate of the noise $\hat{v}(n)$ is removed from $z(n)$.

This problem uses an FIR Wiener filter to estimate the noise. The Wiener-Hopf equation (4.17) becomes

$$R_{v'}h = r_{vv'}.$$

(a) Assume $v'(n)$ is zero-mean and uncorrelated with $s(n)$. Show that

$$R_{vv'}(k) = R_{zv'}(k),$$

so that it is possible to set up the Wiener-Hopf equation strictly from $z(n)$ and $v'(n)$.

The remainder of this problem develops a MATLAB simulation. Let the signal be given by

$$s(n) = \cos(0.04\pi n).$$

Let the noise processes $v(n)$ and $v'(n)$ be produced via

$$v(n) = 0.7v(n-1) + y(n),$$

and

$$v'(n) = -0.5v'(n-1) + 0.36v'(n-2) + y(n),$$

where $y(n)$ is zero-mean white Gaussian noise with unit variance.

(b) Use MATLAB to generate 100-sample realizations of $s(n)$ and $y(n)$; the results should be 100-element row vectors `s` and `y`.

(c) Use `filter` to produce sample realizations `v` and `vprime` from `y`. Then add `s` and `v` to form `z`. [Problem 2.5 introduces `filter`.] Plot `s` and `z` on the same graph. On a separate graph, plot `vprime`.

(d) Estimate $R_{v'}(k)$ and $R_{vv'}(k)$ from `vprime` and `z`. [See Problem 2.4.] Store the estimates in vectors `Rvprime` and `Rvvprime`, respectively.

(e) From `Rvprime` and `Rvvprime`, set up the Wiener-Hopf equation for a 4-tap ($N = 4$) FIR Wiener smoothing filter. Solve the Wiener-Hopf equation to obtain the filter coefficients `h`. [The function `inv` will compute a matrix inverse.]

(f) Use `filter` to apply the Wiener filter to `vprime` and call the output of the filter `vhat`. Then estimate the signal `shat` from `z` and `vhat`. Plot `s` and `shat` on the same graph to see the effect of the noise cancellation system.

(g) Use ensemble averaging to estimate $\sigma_v^2 = R_v(0)$. Call the estimate `varv`. [See Problem 2.2 for computing ensemble averages.] Then compute the mean square error,

$$\mathrm{E}\left[(v(n) - \hat{v}(n))^2\right] = \sigma_v^2 - h^{\mathrm{T}} r_{vv'}.$$

Note that the MSE may be negative since σ_v^2, $R_{v'}(k)$, and $R_{vv'}(k)$ are all estimated quantities. Also compute the experimental MSE for the sample realization

experimental MSE = `(v-vhat)*(v-vhat)'/100`.

Depending upon the values in the sample realization, the experimental MSE may be larger or smaller than the theoretical MSE. However, the two quantities should be somewhat similar.

(h) Repeat parts (e)–(g) for $N = 6$ and $N = 8$.

4.6. Derive the noncausal IIR Wiener filter for the general case where the estimate is $\hat{s}(\ell)$ for any ℓ from Table 3.1. The resulting filter is a shifted version of the noncausal Wiener filter. Is this result surprising? Explain.

4.7. Let a signal process $s(n)$ have the following power spectrum:

$$S_s(e^{j\omega}) = \begin{cases} 0, & \frac{3\pi}{4} < |\omega| < \pi, \\ 8, & |\omega| \le \frac{3\pi}{4}, \end{cases}$$

and suppose measurements of the signal are modeled by additive noise with power spectrum

$$S_v(e^{j\omega}) = \begin{cases} 5, & \left||\omega| - \frac{3\pi}{8}\right| \le \frac{\pi}{8}, \\ 0, & \text{otherwise.} \end{cases}$$

Find the frequency response of the optimum noncausal LTI filter and sketch it.

4.8. Find the causal part of the following z-transforms:

(a) $H(z) = \dfrac{2z + \frac{1}{2}}{1 - \frac{1}{4}z^{-1}}$, ROC $= \{\frac{1}{4} < |z| < \infty\}$,

(b) $Y(z) = \dfrac{4z^{-3} + 22z^{-4} - 2z^{-5}}{1 - 4z^{-1} - 5z^{-2}}$, ROC $= \{1 < |z| < 5\}$.

4.9. Find the causal part of $H(z) = N(z)/D(z)$ with ROC $= \{\frac{1}{2} < |z| < 2\}$. $N(z)$ and $D(z)$ are given by

$$N(z) = 256z - 64 - 3920z^{-1} - 6372z^{-2} - 2400z^{-3} - 671z^{-4} + 4093z^{-5} - 2031z^{-6} + 411z^{-7} - 30z^{-8},$$

and

$$D(z) = 128 - 544z^{-1} - 728z^{-2} + 1370z^{-3} - 677z^{-4} + 137z^{-5} - 10z^{-6}.$$

4.10. Section 4.4 presented two methods for finding $[H(z)]_+$, the transform-domain form of the causal part of an impulse response $h(n)$. A third method for computing $[H(z)]_+$ is based on the modulation property of the Fourier transform.

From (4.54), we have

$$[h(n)]_+ = h(n)1(n).$$

Hence, $[h(n)]_+$ is just $h(n)$ modulated by the unit step $1(n)$. By the modulation property of the Fourier transform, the frequency-domain version of this equation is

$$\left[X(e^{j\omega})\right]_+ = X(e^{j\omega}) * \mathrm{I}(e^{j\omega}) = \frac{1}{2\pi}\int_{-\pi}^{\pi} X(e^{j\theta})\mathrm{I}(e^{j(\omega-\theta)})d\theta,$$

where $\mathrm{I}(e^{j\omega})$ denotes the Fourier transform of $1(n)$:

$$\mathrm{I}(e^{j\omega}) = \frac{1}{1 - e^{-j\omega}} + \sum_{k=-\infty}^{\infty} \pi\delta(\omega - 2\pi k).$$

(a) Use the modulation property to find the causal part of $h(n) = \lambda^{n+1}1(n+1)$, $|\lambda| < 1$. Note that

$$X(e^{j\omega}) = \frac{e^{j\omega}}{1 - \lambda e^{-j\omega}}.$$

(Answer: $[h(n)]_+ = \lambda^{n+1}1(n)$.)

(b) Use either z-transform method in Section 4.4 to determine $[H(z)]_+$. Note that the z-transform of $h(n)$ is

$$H(z) = \frac{z}{1 - \lambda z^{-1}}.$$

4.11. Suppose a signal $s(n)$ is generated by the stochastic difference equation

$$s(n) = 0.75s(n-1) + w(n),$$

where $w(n)$ is white noise $\sim \mathcal{N}(0, 4.9)$. The signal is measured in the presence of additive white noise $\sim \mathcal{N}(0, 8)$ that is uncorrelated with the signal.

Find the causal Wiener filter for estimating $\hat{s}(n)$ from the noisy measurements. Give the filter's system function and impulse response, and compute the MSE associated with $\hat{s}(n)$. Also compute the MSE if no filtering is performed and compare it to the Wiener filter MSE.

4.12. Consider the signal $s(n)$ with power density spectrum

$$S_s(z) = \frac{0.925}{(1 - 0.9z^{-1})(1 - 0.9z)}.$$

The signal is corrupted by additive, zero-mean, white noise $v(n)$ with autocorrelation function

$$R_v(k) = \tfrac{45}{14}(0.4)^{|k|}.$$

Determine the causal Wiener filter for estimating $s(n)$ from the noisy measurements. Give the system function of the filter and compute the associated MSE.

4.13. Consider a signal process described by the difference equation

$$s(n) = aw(n) + bw(n-1) + cw(n-2) + w(n-3),$$

where $w(n)$ is white noise $\sim (0, \sigma_w^2)$, and a, b, and c are nonzero real numbers.

(a) Find the optimum causal LTI estimators for predicting $s(n+3)$ and $s(n+4)$. Each filter should be expressed as a system function.

(b) Determine the autocorrelation function $R_s(n)$ and sketch it.

(c) What is the correlation between $s(n)$ and $s(n+4)$? What is the LMMSE estimate $\hat{s}(n+4)$?

(d) Based on the results of part (c), how much information regarding $s(n+k)$, $k \geq 4$, does $s(n)$ contain? What is $\hat{s}(n+k)$ for $k \geq 4$? How does $\hat{s}(n+k)$ relate to $\mathrm{E}[\hat{s}(n+k)]$ for $k \geq 4$?

Chapter 5

Recursive Estimation and the Kalman Filter

In this chapter we study estimation based on the causal, available past. Chapter 4 discussed the three Wiener filters: noncausal, causal, and FIR. Certainly, the noncausal Wiener filter cannot be employed for causal estimation. The causal Wiener filter requires all observations from the entire past: from time $n = -\infty$ to the present. Finally, the FIR filter uses only the N most-recent observations. At time N, the observation $\boldsymbol{z}(0)$ is discarded, at time $N+1$, $\boldsymbol{z}(1)$ is discarded, and so on. In other words, past observations that might contain information about $\boldsymbol{s}(n)$ are abandoned. However, if we could store the entire past, we would expect our estimator to perform better.

We begin by considering re-computing our estimate $\hat{\boldsymbol{s}}(n)$ each time we obtain a new measurement $\boldsymbol{z}(n)$. We begin by examining LMMSE, ML, and MAP estimation and discover that these estimators require larger and larger amounts of memory to store the past measurements. We then consider a model problem, estimating a constant signal, which leads to the notion of recursive estimation without the need to store an ever-increasing number of measurements.

The notion of recursive LMMSE estimation leads naturally to the Kalman filter. After deriving the Kalman filter, we discuss its main properties. In particular, it is a time-varying system. We then consider the case of a time-invariant signal/measurement model, which leads to the steady-state Kalman filter. Finally, we show that for this case, the steady-state Kalman filter and the causal Wiener filter are equivalent.

5.1 Estimation with Growing Memory

LMMSE Estimation

As discussed in the previous chapter, the causal time-invariant LMMSE estimator based on the measurements $z(1)$, $z(2)$, ..., $z(n)$ has the form

$$\hat{s}(n) = \sum_{i=0}^{n-1} h(i)z(n-i), \tag{5.1}$$

where $h(n)$ is the impulse response of the estimator. However, the LMMSE estimator is not generally time invariant, and thus the filter given by (5.1) is constrained because it is time invariant. The general (time-varying) form of the causal LMMSE filter is

$$\hat{s}(n) = \sum_{i=0}^{n-1} h_n(i)z(n-i), \tag{5.2}$$

where now the coefficients $h_n(i)$ in (5.2) depend on the time index n. To determine the $h_n(i)$, (5.2) can be inserted into the MSE criterion, and then the appropriate partial derivatives can be taken as was done in Chapter 4. The result is a time-varying version of the Wiener-Hopf equation (4.8) (see Problem 4.2) given by

$$\sum_{j=0}^{n} h_n(j)R_z(n-i, n-j) = R_{sz}(n, n-i), \quad 0 \leq i \leq n. \tag{5.3}$$

At each time n, equation (5.3) says that we must solve n linear equations to find the n filter coefficients $\{h_n(i) : 0 \leq i \leq n-1\}$. We then apply (5.2) to the set of observations $\{z(i) : 1 \leq i \leq n\}$ and obtain $\hat{s}(n)$. Because we must store every observation, the filter described by Equations (5.2) and (5.3) is called the *growing-memory LMMSE filter*.

Our filter must have sufficient memory to store the observations and filter coefficients. In addition, it must be able to solve the n equations implied by (5.3). As n becomes very large, we will undoubtedly exceed the memory capacity and computational capabilities of the filter. As a result, we cannot realize this filter in its current form.

Estimation Based on a State Model

Recall also that the ML and MAP estimators require knowledge of $f_z(z|s = s)$ and $f_s(s|z = z)$, respectively. If we have sufficient knowledge about how $s(n)$ is generated, we can determine $f_z(z|s = s)$ and $f_s(s|z = z)$ and then find the ML and MAP estimates.

Specifically, let us assume that the signal $\boldsymbol{s}(n)$ can be expressed in the form of a state model:

$$\boldsymbol{x}(n+1) = \Phi \boldsymbol{x}(n) + \boldsymbol{w}(n)$$
$$\boldsymbol{s}(n) = C\boldsymbol{x}(n),$$

where the state $\boldsymbol{x}(n)$ is an N-vector, $\boldsymbol{w}(n)$ is an N-element noise vector, Φ is an $N \times N$ real matrix, and C is an N-element row vector of real numbers. Finally, assume that the measurements are given by

$$\boldsymbol{z}(n) = \boldsymbol{s}(n) + \boldsymbol{v}(n).$$

Now the problem of interest is to estimate the state $\boldsymbol{x}(n)$ based on the measurements $\boldsymbol{z}(1), \boldsymbol{z}(2), \ldots, \boldsymbol{z}(n)$. (If $\boldsymbol{x}(n)$ is known, we can immediately obtain $\boldsymbol{s}(n)$ via $\boldsymbol{s}(n) = C\boldsymbol{x}(n)$.) The general form of the estimate $\hat{\boldsymbol{x}}(n)$ is

$$\hat{\boldsymbol{x}}(n) = \alpha_n(\hat{x}(0), \hat{P}(0), \boldsymbol{Z}_n), \tag{5.4}$$

where

$$\hat{x}(0) = \text{a guess of } \mathrm{E}\left[\boldsymbol{x}(0)\right],$$
$$\hat{P}(0) = \text{a guess of } \mathrm{Cov}\left[\boldsymbol{x}(0)\right],$$

and

$$\boldsymbol{Z}_n = \begin{bmatrix} \boldsymbol{z}(1) \\ \boldsymbol{z}(2) \\ \vdots \\ \boldsymbol{z}(n) \end{bmatrix}.$$

In general, the function α_n in (5.4) is nonlinear and time-varying (dependent on n). Of course, once we have $\hat{\boldsymbol{x}}(n)$, we can estimate $\boldsymbol{s}(n)$ via $\hat{\boldsymbol{s}}(n) = C\hat{\boldsymbol{x}}(n)$.

If Z_n is a sample realization of $\boldsymbol{Z}_n$, the value $\hat{x}(n)$ of the estimate at time n is given by

$$\hat{x}(n) = \alpha_n(\hat{x}(0), P(0), Z_n).$$

The objective is to determine $\hat{x}(n)$ so that some optimality criterion is satisfied.

ML Estimation

For simplicity, let us assume that $\boldsymbol{w}(n) = 0$; that is, there is no input driving the system. (The Kalman filter will incorporate the driving term $\boldsymbol{w}(n)$.) Then the signal model becomes

$$\boldsymbol{x}(n+1) = \Phi \boldsymbol{x}(n) \tag{5.5}$$
$$\boldsymbol{s}(n) = C\boldsymbol{x}(n). \tag{5.6}$$

In addition, assume that Φ is invertible so that

$$\boldsymbol{x}(n) = \Phi^{-1}\boldsymbol{x}(n+1) \quad \text{and} \quad \boldsymbol{x}(n-1) = \Phi^{-1}\boldsymbol{x}(n), \tag{5.7}$$

with $\boldsymbol{s}(n)$ and $\boldsymbol{v}(n)$ independent for all n. Then, given the measurements $\boldsymbol{z}(n) = \boldsymbol{s}(n) + \boldsymbol{v}(n)$, we have that

$$\boldsymbol{z}(i) = \boldsymbol{s}(i) + \boldsymbol{v}(i) = C\Phi^{-i+1}\boldsymbol{x}(n) + \boldsymbol{v}(i), \quad i = 1, 2, \ldots, n.$$

We can gather the past observations together into a batch to create the batch form. Then, corresponding to the constructions in Section 1.3, we have

$$\boldsymbol{Z}_n = \begin{bmatrix} \boldsymbol{z}(1) \\ \boldsymbol{z}(2) \\ \vdots \\ \boldsymbol{z}(n) \end{bmatrix} = \begin{bmatrix} C\Phi^{-n+1} \\ C\Phi^{-n+2} \\ \vdots \\ C \end{bmatrix} \boldsymbol{x}(n) + \begin{bmatrix} \boldsymbol{v}(1) \\ \boldsymbol{v}(2) \\ \vdots \\ \boldsymbol{v}(n) \end{bmatrix},$$

or equivalently,

$$\boldsymbol{Z}_n = U_n\boldsymbol{x}(n) + \boldsymbol{V}_n,$$

where U_n is the $n \times N$ matrix

$$U_n = \begin{bmatrix} C\Phi^{-n+1} \\ C\Phi^{-n+2} \\ \vdots \\ C \end{bmatrix},$$

and

$$\boldsymbol{V}_n = \begin{bmatrix} \boldsymbol{v}(1) \\ \boldsymbol{v}(2) \\ \vdots \\ \boldsymbol{v}(n) \end{bmatrix}.$$

We assume that $\boldsymbol{v}(n)$ is zero-mean Gaussian noise for all n, and thus $\boldsymbol{V}_n$ is zero-mean n-variate Gaussian with density function

$$f_{\boldsymbol{V}_n}(V_n) = \frac{1}{(2\pi)^{n/2}\,|R_n|^{1/2}} \exp\left(-\tfrac{1}{2}V_n^{\mathrm{T}}R_n^{-1}V_n\right), \tag{5.8}$$

where $R_n = \mathrm{Cov}\left[\boldsymbol{V}_n\right]$. Then the conditional density function of $\boldsymbol{Z}_n$ given $\boldsymbol{x}(n) = x$ becomes

$$\begin{aligned} f_{\boldsymbol{Z}_n}(Z_n|\boldsymbol{x}(n) = x) &= f_{\boldsymbol{V}_n}(V_n)|_{V_n = Z_n - U_n x} \\ &= \frac{1}{(2\pi)^{n/2}\,|R_n|^{1/2}} \\ &\quad \times \exp\left[-\tfrac{1}{2}\left(Z_n - U_n x\right)^{\mathrm{T}} R_n^{-1}\left(Z_n - U_n x\right)\right]. \end{aligned} \tag{5.9}$$

Recall that the ML estimate $\hat{x}_{\mathrm{ML}}(n)$ of $\boldsymbol{x}(n)$, given realization $\boldsymbol{Z}_n = Z_n$, is the value of x for which the likelihood function $f_{\boldsymbol{Z}_n}(Z_n|\boldsymbol{x}(n) = x)$ is maximum. Clearly, the maximum occurs when

$$(Z_n - U_n x)^{\mathrm{T}} R_n^{-1} (Z_n - U_n x)$$

is minimized, and the solution is that of a weighted least-squares problem (see Section 1.3). The solution is

$$\hat{x}_{\mathrm{ML}}(n) = \left[U_n^{\mathrm{T}} R_n^{-1} U_n\right]^{-1} U_n^{\mathrm{T}} R_n^{-1} Z_n. \tag{5.10}$$

Note that information about $\boldsymbol{v}(n)$, in the form of the covariance R_n, is incorporated into the estimate. This behavior is in contrast with the least-squares estimate derived in Section 1.3, which does not take $\boldsymbol{v}(n)$ into account.

From (5.10) we have an ML estimate given the available past. Again, the problem is that we must store every past observation $z(n)$ that becomes available.

MAP Estimation

Now consider the MAP estimate of $\boldsymbol{x}(n)$, again using the unforced state model of (5.5) and (5.6) and assuming additive noise with density (5.8). Since MAP estimation requires some *a priori* knowledge about the signal, let us assume that $\boldsymbol{x}(n)$ is n-variate Gaussian with mean $\bar{x}(n)$ and covariance $P(n)$. Then

$$f_{\boldsymbol{x}(n)}(x) = \frac{1}{(2\pi)^{n/2}|P(n)|^{1/2}} \exp\left[-\tfrac{1}{2}\left(x - \bar{x}(n)\right)^{\mathrm{T}} P^{-1}(n)\left(x - \bar{x}(n)\right)\right], \tag{5.11}$$

and since

$$f_{\boldsymbol{x}(n)}(x|\boldsymbol{Z}_n = Z_n) = \frac{f_{\boldsymbol{Z}_n}(Z_n|\boldsymbol{x}(n) = x) f_{\boldsymbol{x}(n)}(x)}{f_{\boldsymbol{Z}_n}(Z_n)},$$

given the realization $\boldsymbol{Z}_n = Z_n$, the MAP estimate $\hat{x}_{\mathrm{MAP}}$ is the value of x for which $f_{\boldsymbol{Z}_n}(Z_n|\boldsymbol{x}(n) = x) f_{\boldsymbol{x}(n)}(x)$ is maximum. Using (5.9) and (5.11), we have

$$\begin{aligned} f_{\boldsymbol{Z}_n}(Z_n|\boldsymbol{x}(n) = x) f_{\boldsymbol{x}(n)}(x) &= \\ \frac{1}{(2\pi)^n |R_n|^{1/2}|P(n)|^{1/2}} &\exp\left[-\tfrac{1}{2}(Z_n - U_n x)^{\mathrm{T}} R_n^{-1}(Z_n - U_n x)\right. \\ &\left. -\tfrac{1}{2}\left(x - \bar{x}(n)\right)^{\mathrm{T}} P^{-1}(n)\left(x - \bar{x}(n)\right)\right]. \end{aligned}$$

Hence, we must find the value of x that minimizes

$$(Z_n - U_n x)^{\mathrm{T}} R_n^{-1} (Z_n - U_n x) + \left(x - \bar{x}(n)\right)^{\mathrm{T}} P^{-1}(n)\left(x - \bar{x}(n)\right).$$

Taking the derivative with respect to x and setting it to zero gives

$$-2U_n^{\mathrm{T}}R_n^{-1}Z_n + 2U_n^{\mathrm{T}}R_n^{-1}U_n x - 2P^{-1}(n)\bar{x}(n) + 2P^{-1}(n)x = 0.$$

Then solving for x gives the MAP estimate,

$$\hat{x}_{\mathrm{MAP}}(n) = \left[U_n^{\mathrm{T}}R_n^{-1}U_n + P^{-1}(n)\right]^{-1}\left[P^{-1}(n)\bar{x}(n) + U_n^{\mathrm{T}}R_n^{-1}Z_n\right]. \quad (5.12)$$

Note that the matrix $\left[U_n^{\mathrm{T}}R_n^{-1}U_n + P^{-1}(n)\right]$ in (5.12) is always invertible since $P^{-1}(n) > 0$.

This estimate can be expressed in terms of an initial estimate $\hat{x}(n)$ of $\bar{x}(0)$ and an initial guess of $P(0)$ as follows. First, by propagation of the mean and covariance (see Section 2.4), we have

$$\bar{x}(n) = \Phi^n \bar{x}(0) \quad (5.13)$$

$$P(n) = \Phi^n P(0)\left(\Phi^{\mathrm{T}}\right)^n. \quad (5.14)$$

Then setting $\bar{x}(0) = \hat{x}(0)$ and $P(0) = \hat{P}(0)$ in (5.13) and (5.14) and inserting the result into (5.12) yields

$$\hat{x}_{\mathrm{MAP}}(n) = \left[U_n^{\mathrm{T}}R_n^{-1}U_n + \Phi^n\hat{P}(0)(\Phi^{\mathrm{T}})^n\right]^{-1} \times \left[\left(\Phi^{\mathrm{T}}\right)^{-n}\hat{P}^{-1}(0)\hat{x}(0) + U_n^{\mathrm{T}}R_n^{-1}Z_n\right]. \quad (5.15)$$

This is the batch form of the MAP estimate. Note that this estimate is in the form of (5.4), namely,

$$\hat{x}_{\mathrm{MAP}}(n) = \alpha_n(\hat{x}(0), \hat{P}(0), Z_n).$$

Again, however, we must accumulate all the past observations.

5.2 Estimation of a Constant Signal

The methods presented in the preceding section require ever-increasing storage as we accumulate measurement $\boldsymbol{z}(n)$. Of course, we would like to avoid this problem of growing memory. In this section, we consider the problem of estimating a constant signal, which leads to a recursive estimator with fixed memory requirements. We consider only the LMMSE case and set $\boldsymbol{s}(n) = \boldsymbol{s}$, where $\boldsymbol{s}$ is a random variable.

The LMMSE estimate at time n from observations $\boldsymbol{z}(1), \boldsymbol{z}(2), \ldots, \boldsymbol{z}(n)$ is

$$\hat{\boldsymbol{s}}(n) = \sum_{i=0}^{n-1} h_n(i)\boldsymbol{z}(n-i)$$

with associated MSE

$$\text{MSE} = \mathrm{E}\left[\left(\boldsymbol{s}(n) - \sum_{i=0}^{n-1} h_n(i)\boldsymbol{z}(n-i)\right)^2\right].$$

We assume

$$\mathrm{E}\left[\boldsymbol{s}^2\right] = P > 0,$$

where the actual value of P may not be known. It follows that

$$R_s(i-j) = \mathrm{E}\left[\boldsymbol{s}(i)\boldsymbol{s}(j)\right] = \mathrm{E}\left[\boldsymbol{s}^2\right] = P, \quad \forall\, i,j \in \mathbb{Z}.$$

We adopt the additive-noise model $\boldsymbol{z}(n) = \boldsymbol{s}(n) + \boldsymbol{v}(n) = \boldsymbol{s} + \boldsymbol{v}(n)$, where $\boldsymbol{s}(n)$ and $\boldsymbol{v}(n)$ are independent and JWSS, but $\boldsymbol{s}(n)$ is *not necessarily zero-mean,* and $\boldsymbol{v}(n)$ is white noise with known variance σ_v^2. Then

$$R_v(i-j) = \mathrm{E}\left[\boldsymbol{v}(i)\boldsymbol{v}(j)\right] = \sigma_v^2\delta(i-j),$$

and

$$R_z(i-j) = R_s(i-j) + R_v(i-j) = P + \sigma_v^2\delta(i-j).$$

The filter coefficients are found by taking

$$\begin{aligned}\frac{\partial \text{MSE}}{\partial h_n(i)} &= -2\mathrm{E}\left[\boldsymbol{s}(n)\boldsymbol{z}(n-i)\right] \\ &\quad + 2\sum_{j=0}^{n-1} h_n(j)\mathrm{E}\left[\boldsymbol{z}(n-i)\boldsymbol{z}(n-j)\right],\end{aligned}$$

and setting this expression equal to zero for each $i = 0, 1, \ldots, n-1$. Doing so gives

$$\sum_{j=0}^{n-1} h_n(j)R_z(j-i) = R_{sz}(i) = R_s(i) = P, \quad i = 0, 1, \ldots, n-1.$$

We write this result in matrix form as

$$\begin{bmatrix} P+\sigma_v^2 & P & \cdots & P \\ P & P+\sigma_v^2 & \cdots & P \\ \vdots & \vdots & \ddots & \vdots \\ P & P & \cdots & P+\sigma_v^2 \end{bmatrix} \begin{bmatrix} h_n(0) \\ h_n(1) \\ \vdots \\ h_n(n-1) \end{bmatrix} = \begin{bmatrix} P \\ P \\ \vdots \\ P \end{bmatrix}$$

From linear algebra and the obvious symmetry of the problem, the solution of this equation is

$$h_n(0) = h_n(1) = \cdots = h_n(n-1) \tag{5.16}$$

$$= \frac{P}{nP + \sigma_v^2} = \frac{1}{n + \sigma_v^2/P}. \tag{5.17}$$

Since by (5.16) the $h_n(i)$ are identical, we denote them by h_n. Then the estimate is given by

$$\hat{s}(n) = \sum_{i=0}^{n-1} h_n z(n-i) = h_n \sum_{i=0}^{n-1} z(n-i). \tag{5.18}$$

Equation (5.18) still requires an infinite amount of storage as $n \to \infty$. However, let us consider the subsequent estimate $\hat{s}(n+1)$ when the observation $z(n+1)$ becomes available. From (5.18), we find

$$\hat{s}(n+1) = h_{n+1} \sum_{i=0}^{n} z(n+1-i), \tag{5.19}$$

and

$$h_{n+1} = \frac{P}{(n+1)P + \sigma_v^2} = \frac{1}{n+1+\sigma_v^2/P}. \tag{5.20}$$

Next, we let $j = i - 1$ in (5.19) to obtain

$$\hat{s}(n+1) = h_{n+1} \sum_{j=-1}^{n-1} z(n-j)$$

$$= h_{n+1} z(n+1) + h_n \sum_{j=0}^{n-1} z(n-j). \tag{5.21}$$

By (5.18),

$$\sum_{j=0}^{n-1} z(n-j) = \frac{\hat{s}(n)}{h_n},$$

so (5.21) becomes

$$\hat{s}(n+1) = \frac{h_{n+1}}{h_n} \hat{s}(n) + h_{n+1} z(n+1). \tag{5.22}$$

Observe that (5.22) no longer requires that we store every observation. Instead, we simply combine our most-recent estimate $\hat{s}(n)$ with the new observation $z(n+1)$ with appropriate weights. *We only need to store the most-recent estimate.* Hence, (5.22) forms a *recursive estimate.*

We now attempt to find a recursive form for the filter coefficients h_n. From (5.17) and (5.20) we write

$$\begin{aligned} \frac{h_{n+1}}{h_n} &= \frac{n+\sigma_v^2/P}{n+1+\sigma_v^2/P} = 1 - \frac{1}{n+1+\sigma_v^2/P} \\ &= 1 - h_{n+1}. \end{aligned} \tag{5.23}$$

We solve (5.23) for h_{n+1} and find

$$h_{n+1} = \frac{h_n}{1+h_n} = h_n\left(1+h_n\right)^{-1}. \tag{5.24}$$

Equation (5.24) allows us to compute the filter coefficients recursively. Notice that the most-recent filter coefficients h_{n-1} can be used to determine the new filter coefficients h_n. From (5.24), our recursive estimator (5.22) can now be written as

$$\hat{s}(n+1) = \hat{s}(n) + h_{n+1}\left[z(n+1) - \hat{s}(n)\right]. \tag{5.25}$$

The coefficient h_{n+1} is known as the *filter gain.*

We now investigate the MSE. Since our estimate $\hat{s}(n)$ changes over time, the MSE varies over time and we denote it by MSE(n),

$$\begin{aligned} \mathrm{MSE}(n) &= \mathrm{E}\left[(s-\hat{s}(n))^2\right] \\ &= \mathrm{E}\left[s^2\right] - 2\mathrm{E}\left[s\hat{s}(n)\right] + \mathrm{E}\left[\hat{s}^2(n)\right] \\ &= P - 2h_n\sum_{i=0}^{n-1}\mathrm{E}\left[sz(n-i)\right] \\ &\quad + h_n^2\mathrm{E}\left[\left(\sum_{i=0}^{n-1} z(n-i)\right)\left(\sum_{j=0}^{n-1} z(n-j)\right)\right] \end{aligned} \tag{5.26}$$

Now the middle term requires $\mathrm{E}\left[sz(n-i)\right]$, which is given by

$$\begin{aligned} \mathrm{E}\left[sz(n-i)\right] &= \mathrm{E}\left[s(s+v(n-i))\right] = \mathrm{E}\left[s^2\right] + \mathrm{E}\left[sv(n-i)\right] \\ &= P + \mathrm{E}\left[s\right]\mathrm{E}\left[v(n-i)\right] = P. \end{aligned}$$

To find the third term, we note that

$$\mathrm{E}\left[z(n-i)z(n-j)\right] = \mathrm{E}\left[(s+v(n-i))(s+v(n-j))\right] = P + \sigma_v^2\delta(i-j),$$

so that

$$h_n^2\mathrm{E}\left[\left(\sum_{i=0}^{n-1} z(n-i)\right)\left(\sum_{j=0}^{n-1} z(n-j)\right)\right] = h_n^2(n^2P + n\sigma_v^2).$$

Thus, (5.26) becomes

$$\mathrm{MSE}(n) = P - 2nh_nP + h_n^2(n^2P + n\sigma_v^2) \tag{5.27}$$

This expression for the MSE holds whether or not P is known.

Now, *assuming that P is known*, we substitute (5.17) into (5.27). After some simplification, we find

$$\mathrm{MSE}(n) = \frac{P\sigma_v^2}{nP + \sigma_v^2} = h_n\sigma_v^2. \tag{5.28}$$

It follows from (5.23) that

$$\mathrm{MSE}(n+1) = h_{n+1}\sigma_v^2 = \frac{h_{n+1}}{h_n}\mathrm{MSE}(n) = (1 - h_{n+1})\,\mathrm{MSE}(n). \tag{5.29}$$

which means that we can compute the MSE recursively as well.

Equations (5.24), (5.25), and (5.29) give the recursion for the optimum recursive LMMSE filter for estimation of a constant signal. The filter is diagrammed in Figure 5.2, and we summarize the recursion below.

1. Initialization. Set $n = 0$. Set $\hat{s}(0)$ to some initial value; often $\mathrm{E}[\boldsymbol{s}]$ or 0 is used. Since P is not known, choose an initial value V that is a guess of P. Then set $h_0 = V/\sigma_v^2$, and $\mathrm{MSE}(0) = V$.

2. Acquire observation $z(n+1)$.

3. Compute

$$\begin{aligned} h_{n+1} &= h_n\left(1 + h_n\right)^{-1}, \\ \hat{s}(n+1) &= \hat{s}(n) + h_{n+1}\left[z(n+1) - \hat{s}(n)\right], \\ \mathrm{MSE}(n+1) &= (1 - h_{n+1})\,\mathrm{MSE}(n). \end{aligned}$$

4. Increment n and return to step 2.

It is important to realize that, in practice, *P is not known*, so we must use some V that approximates P, and usually $V \neq P$. Then the actual filter gain is

$$h_n = \frac{1}{v + \sigma_v^2/V} = \frac{V}{nV + \sigma_v^2}. \tag{5.30}$$

Let us analyze the estimator to determine how well it performs. We would like an unbiased estimate, and we would like to know how large the MSE is.

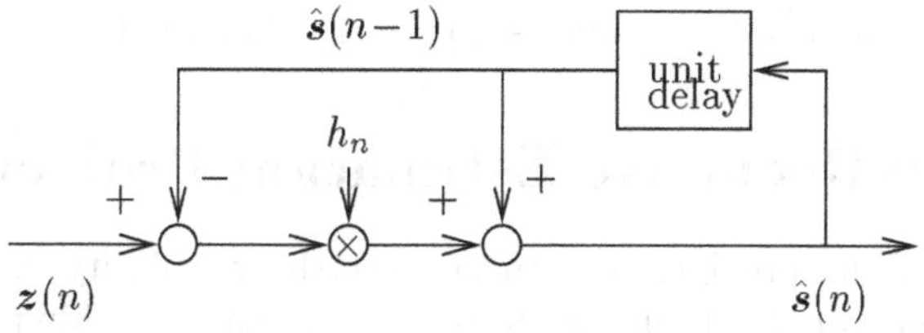

Figure 5.1. Recursive estimation of a constant.

The expected value of the estimate $\hat{s}(n)$ is

$$\begin{aligned} \mathrm{E}[\hat{s}(n)] &= \mathrm{E}\left[h_n \sum_{i=0}^{n-1} z(n-i)\right] = h_n \sum_{i=0}^{n-1} \mathrm{E}[z(n-i)] \\ &= h_n \sum_{i=0}^{n-1} (\mathrm{E}[s] + \mathrm{E}[v(n-i)]) = h_n \sum_{i=0}^{n-1} \mathrm{E}[s] \\ &= nh_n \mathrm{E}[s] = \frac{n}{n+\sigma_v^2/V} \mathrm{E}[s]. \end{aligned} \tag{5.31}$$

Equation (5.31) indicates that the estimate is not unbiased. However,

$$\lim_{n\to\infty} \mathrm{E}[\hat{s}(n)] = \lim_{n\to\infty} \frac{n}{n+\sigma_v^2/V} \mathrm{E}[s] = \mathrm{E}[s]. \tag{5.32}$$

From (5.32), even though P is not known, the estimator produces an *asymptotically unbiased* estimate of s. Although initial estimates may be inaccurate, gradually the estimate will converge to the expected value of s.

As for the MSE, from (5.27) it follows that

$$\begin{aligned} \mathrm{MSE}(n) &= P - (2nP)h_n + (n^2P + n\sigma_v^2)h_n^2 \\ &= P - (2nP)\frac{V}{nV+\sigma_v^2} + (n^2P + n\sigma_v^2)\left(\frac{V}{nV+\sigma_v^2}\right)^2 \\ &= \frac{(nV+\sigma_v^2)^2P - 2nPV(nV+\sigma_v^2) + (n^2P+n\sigma_v^2)V^2}{(nV+\sigma_v^2)^2} \\ &= \frac{nV^2\sigma_v^2 + P\sigma_v^2}{(nV+\sigma_v^2)^2} \end{aligned}$$

Therefore,

$$\begin{aligned} \lim_{n\to\infty} \mathrm{MSE}(n) &= \lim_{n\to\infty} \frac{nV^2\sigma_v^2 + P\sigma_v^4}{n^2V^2 + 2nV\sigma_v^2 + \sigma_v^4} = \lim_{n\to\infty} \frac{nV^2\sigma_v^2}{n^2V^2} \\ &= \lim_{n\to\infty} \frac{\sigma_v^2}{n} = 0. \end{aligned} \tag{5.33}$$

Equations (5.32) and (5.33) show that, as $n \to \infty$, the recursive estimator given by (5.24) and (5.25) produces a perfect estimate of $\boldsymbol{s}$.

5.3 The Recursive Estimation Problem

The results of the previous section suggest the possibility of developing a *recursive estimator*. Such an estimator uses the updated filter gain h_{n+1} to combine the most-recent estimate $\hat{\boldsymbol{s}}(n)$ and the new observation $\boldsymbol{z}(n+1)$ and create $\hat{\boldsymbol{s}}(n+1)$. As a result, the storage and computational requirements remain fixed as n increases.

Formally, we have the general *recursive estimation problem*, stated below.

Recursive Estimation Problem

> Given the optimum estimate $\hat{\boldsymbol{s}}(n)$ at time n of a random process $\boldsymbol{s}(n)$, the observation $\boldsymbol{z}(n+1)$, and the operator f, design a causal (time-varying) estimator that produces the optimum estimate $\hat{\boldsymbol{s}}(n+1)$ at time $n+1$ such that
>
> $$\hat{\boldsymbol{s}}(n+1) = \alpha_{n+1}(\hat{\boldsymbol{s}}(n), \boldsymbol{z}(n+1)). \tag{5.34}$$

Note that the estimator may be nonlinear and is generally time-varying. As in the optimal filtering problem, the definition of optimality remains open to the designer. However, we will concern ourselves only with linear filters and optimality in the MMSE sense. Hence, we seek the optimum recursive LMMSE estimator.

5.4 The Signal/Measurement Model

In this section we introduce some assumptions about the signal $\boldsymbol{s}(n)$ and the measurements $\boldsymbol{z}(n)$. The primary assumption is that the signal and measurements can be reasonably modeled as the state and output, respectively, of a state model.

We allow $\boldsymbol{s}(n)$ and $\boldsymbol{z}(n)$ to be *vector* random processes. Letting p be the length of $\boldsymbol{s}(n)$ and $\boldsymbol{z}(n)$, we have

$$\boldsymbol{s}(n) = \begin{bmatrix} s_1(n) \\ s_2(n) \\ \vdots \\ s_p(n) \end{bmatrix}, \quad \text{and} \quad \boldsymbol{z}(n) = \begin{bmatrix} z_1(n) \\ z_2(n) \\ \vdots \\ z_p(n) \end{bmatrix}. \tag{5.35}$$

Next, we introduce a state model for the signal process. In general, $\boldsymbol{s}(n)$ is a colored (i.e., non-white) random process. We have seen that white noise injected into a linear system produces an output that is a colored random process. Hence, it is reasonable to model $\boldsymbol{s}(n)$ as the output of a linear system excited by white noise.

We assume that the signal $\boldsymbol{s}(n)$ can be described by a linear time-invariant state model and adopt the additive-noise model for the measurements. Then we have the *signal/measurement model* (SMM):

$$\boldsymbol{x}(n+1) = \Phi \boldsymbol{x}(n) + \Gamma \boldsymbol{w}(n), \quad \boldsymbol{x}(0) = \boldsymbol{x}_0, \tag{5.36}$$

$$\boldsymbol{s}(n) = C\boldsymbol{x}(n), \tag{5.37}$$

and

$$\boldsymbol{z}(n) = \boldsymbol{s}(n) + \boldsymbol{v}(n), \tag{5.38}$$

where

- Φ, Γ, and C are known deterministic matrices with respective dimensions $N \times N$, $N \times m$, and $p \times N$;
- $\boldsymbol{x}[n]$ is an N-vector random variable for the current state;
- $\boldsymbol{x}_0$ is an N-vector random variable representing the initial state;
- $\boldsymbol{w}(n)$ is stationary *process noise*, an m-vector zero-mean white-noise random process with $m \times m$ covariance matrix Q, i.e.,

$$\mathrm{E}\left[\boldsymbol{w}(i)\boldsymbol{w}^{\mathrm{T}}(j)\right] = Q\delta(i-j); \tag{5.39}$$

- $\boldsymbol{v}(n)$ is stationary *measurement noise*, a p-vector zero-mean white-noise random process with $p \times p$ covariance matrix R, i.e.,

$$\mathrm{E}\left[\boldsymbol{v}(i)\boldsymbol{v}^{\mathrm{T}}(j)\right] = R\delta(i-j); \tag{5.40}$$

- $\boldsymbol{x}_0$, $\boldsymbol{w}(n)$, and $\boldsymbol{v}(n)$ are mutually independent for all n.

Q and R are covariance matrices, so they are positive-semidefinite and symmetric for all n.

Since Φ, Γ, and C are known, we know the state dynamics completely. However, the initial state $\boldsymbol{x}_0$, process noise $\boldsymbol{w}(n)$, and measurement noise $\boldsymbol{v}(n)$ are not known. Also, because C is known, we can always obtain $\boldsymbol{s}(n)$ via $\boldsymbol{s}(n) = C\boldsymbol{x}(n)$. Hence, *the problem becomes that of estimating* $\boldsymbol{x}(n)$, not $\boldsymbol{s}(n)$. Rather than estimating the signal itself, we attempt to estimate the state. This reformulation, however slight, is crucial to the Kalman filter.

The SMM of (5.36), (5.37), and (5.38), along with the covariance matrices of (5.39) and (5.40), are the basic assumptions from which we will derive the Kalman filter.

The freedom provided by these properties allow the Kalman filter to operate in cases where the Wiener filters of Chapter 4 cannot. Recall that the Wiener filters require stationary, zero-mean signal and noise processes. These constraints require that the systems that generate the processes be stable. For the Kalman filter, however, the SMM can be unstable, and $\boldsymbol{x}(n)$ is not required to have zero mean. In addition, the Kalman filter allows for a time-varying SMM with nonstationary noises (discussed below).

On the other hand, the Wiener filters are derived purely based upon the statistics of the signal and noise processes. Only the appropriate correlation functions and/or power spectra are required. No knowledge of the system that generates the signal is necessary. In contrast, the Kalman filter requires a state model, which is a complete description of the dynamics that generate $\boldsymbol{x}(n)$ and $\boldsymbol{z}(n)$.

Finally, recall that the Wiener filters permit colored measurement noise and correlated signal and noise processes. We remark that the Kalman filter can be modified to handle these cases. Sections 7.2, 7.3, and 7.4 address such situations.

a priori and *a posteriori* Estimates

Finally, we present some definitions that we will use throughout the discussion of the Kalman filter. Recall the recursive estimator of a constant signal from Section 5.2. This estimator consisted of three parts: a filter gain update (5.24), an estimate update (5.25), and a MSE update (5.29). We will find that the Kalman filter consists of similar parts. However, it employs estimates of $\boldsymbol{x}(n)$ made before and after the measurement $\boldsymbol{z}(n)$ is available.

One simple but important idea is the notion of *a priori* and *a posteriori* estimates. Basically, an *a priori* estimate of $\boldsymbol{x}(n)$ is an estimate made *before* the measurement $\boldsymbol{z}(n)$ becomes available. An *a posteriori* estimate is made *after* $\boldsymbol{z}(n)$ has become available. Figure 5.4 illustrates this idea.

Suppose we have the set of measurements $\boldsymbol{Z}^- = \{\boldsymbol{z}(1), \ldots, \boldsymbol{z}(n-1)\}$. (Recall that actual realizations will be denoted Z^-.) From $\boldsymbol{Z}^-$ we predict $\boldsymbol{x}(n)$ to obtain an *a priori* estimate, which we denote by a superscript minus ($^-$). We define the *a priori* estimate by

$$\hat{\boldsymbol{x}}^-(n) = \text{the LMMSE estimate of } \boldsymbol{x}(n) \text{ given } \boldsymbol{Z}^-.$$

When $\boldsymbol{z}(n)$ becomes available, we have the set of measurements $\boldsymbol{Z} = \{\boldsymbol{z}(1), \ldots, \boldsymbol{z}(n)\}$. (Again, realizations of these measurements are indicated by Z.) Now we may make an *a posteriori* estimate, defined by

$$\hat{\boldsymbol{x}}(n) = \text{the LMMSE estimate of } \boldsymbol{x}(n) \text{ given } \boldsymbol{Z}.$$

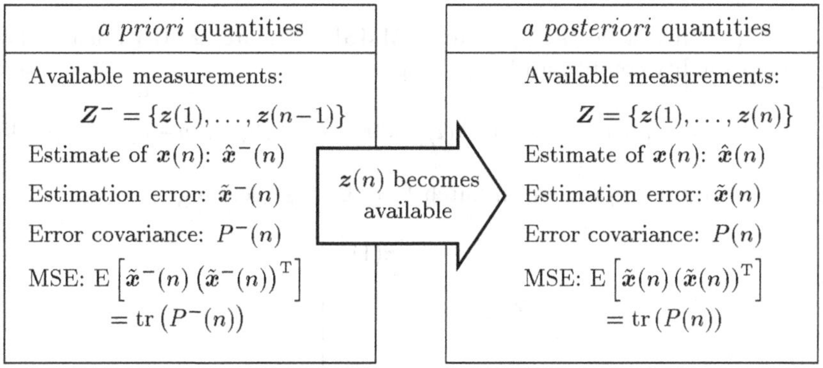

Figure 5.2. Conceptual view of *a priori* and *a posteriori* estimates.

Both of these estimates have an associated error. We denote the estimation errors by a tilde (˜) and define the appropriate errors by

$$\begin{aligned} \tilde{x}^-(n) &= x(n) - \hat{x}^-(n) &&= \text{the } a\ priori \text{ error}, \\ \tilde{x}(n) &= x(n) - \hat{x}(n) &&= \text{the } a\ posteriori \text{ error}. \end{aligned}$$

We also define the estimation error covariance matrices associated with $\hat{x}^-(n)$ and $\hat{x}(n)$ by

$$P^-(n) = \mathrm{Cov}\left[\tilde{x}^-(n)\right] = \mathrm{E}\left[\tilde{x}^-(n)(\tilde{x}^-(n))^{\mathrm{T}}\right], \tag{5.41}$$

and

$$P(n) = \mathrm{Cov}\left[\tilde{x}(n)\right] = \mathrm{E}\left[\tilde{x}(n)\tilde{x}^{\mathrm{T}}(n)\right]. \tag{5.42}$$

From the definition of the trace of a matrix (3.45), we can write the MSE of the estimates as

$$\mathrm{MSE}(\hat{x}^-(n)) = \mathrm{E}\left[(\tilde{x}^-(n))^{\mathrm{T}}\tilde{x}^-(n)\right] = \mathrm{tr}\left(P^-(n)\right), \tag{5.43}$$

and

$$\mathrm{MSE}(x(n)) = \mathrm{E}\left[\tilde{x}^{\mathrm{T}}(n)\tilde{x}(n)\right] = \mathrm{tr}\left(P(n)\right). \tag{5.44}$$

5.5 Derivation of the Kalman Filter

Time Update

Let us assume that we have the LMMSE estimate $\hat{x}(n-1)$ and consider the problem of predicting $\hat{x}(n)$ from Z^-. That is, we want to find the *a priori* estimate $\hat{x}^-(n)$ from the *a posteriori* estimate $\hat{x}(n-1)$. We also want to find the error covariance matrix $P^-(n)$.

The *a priori* Estimate

Since $\hat{\boldsymbol{x}}^-(n)$ should be the optimum LMMSE estimate, $\hat{\boldsymbol{x}}^-(n)$ must satisfy the orthogonality condition (3.52). That is,

$$\mathrm{E}\left[\left(\boldsymbol{x}(n) - \hat{\boldsymbol{x}}^-(n)\right)\boldsymbol{z}^{\mathrm{T}}(i)\right] = \mathbf{0}, \quad i = 1, 2, \ldots, n-1. \tag{5.45}$$

We can write this expression in batch form by first defining

$$\boldsymbol{Z}_{n-1} = \begin{bmatrix} \boldsymbol{z}(1) \\ \boldsymbol{z}(2) \\ \vdots \\ \boldsymbol{z}(n-1) \end{bmatrix}.$$

Then (5.45) can be written compactly as

$$\mathrm{E}\left[\left(\boldsymbol{x}(n) - \hat{\boldsymbol{x}}^-(n)\right)\boldsymbol{Z}_{n-1}^{\mathrm{T}}\right] = \mathbf{0}, \tag{5.46}$$

where $\mathbf{0}$ is the $N \times (n-1)p$ zero matrix.

From (5.36) and since $\boldsymbol{x}(n-1) = \hat{\boldsymbol{x}}(n-1) + \tilde{\boldsymbol{x}}(n-1)$, we have

$$\boldsymbol{x}(n) = \Phi\hat{\boldsymbol{x}}(n-1) + \Phi\tilde{\boldsymbol{x}}(n-1) + \Gamma\boldsymbol{w}(n-1).$$

Then (5.46) becomes

$$\mathrm{E}\left[\left(\Phi\hat{\boldsymbol{x}}(n-1) - \hat{\boldsymbol{x}}^-(n)\right)\boldsymbol{Z}_{n-1}^{\mathrm{T}}\right] + \Phi\mathrm{E}\left[\tilde{\boldsymbol{x}}(n-1)\boldsymbol{Z}_{n-1}^{\mathrm{T}}\right] + \Gamma\mathrm{E}\left[\boldsymbol{w}(n-1)\boldsymbol{Z}_{n-1}^{\mathrm{T}}\right] = \mathbf{0}.$$

Since $\hat{\boldsymbol{x}}(n-1)$ is the LMMSE estimate of $\boldsymbol{x}(n-1)$ given $\boldsymbol{Z}_{n-1}$,

$$\mathrm{E}\left[\tilde{\boldsymbol{x}}(n-1)\boldsymbol{Z}_{n-1}^{\mathrm{T}}\right] = \mathbf{0}.$$

Similarly, $\boldsymbol{w}(n-1)$ is independent of all measurements in $\boldsymbol{Z}_{n-1}$, so

$$\mathrm{E}\left[\boldsymbol{w}(n-1)\boldsymbol{Z}_{n-1}^{\mathrm{T}}\right] = \mathbf{0}.$$

Therefore,

$$\mathrm{E}\left[\left(\Phi\hat{\boldsymbol{x}}(n-1) - \hat{\boldsymbol{x}}^-(n)\right)\boldsymbol{Z}_{n-1}^{\mathrm{T}}\right] = \mathbf{0},$$

so that

$$\boxed{\hat{\boldsymbol{x}}^-(n) = \Phi\hat{\boldsymbol{x}}(n-1).} \tag{5.47}$$

The *a priori* Error Covariance

We also study the relationship between the error covariance matrices, $P(n-1)$ and $P^-(n)$. From Definition (5.41),

$$\begin{aligned} P^-(n) &= \mathrm{Cov}\left[\boldsymbol{x}(n) - \hat{\boldsymbol{x}}^-(n)\right] \\ &= \mathrm{Cov}\left[\Phi\boldsymbol{x}(n-1) + \Gamma\boldsymbol{w}(n-1) - \Phi\hat{\boldsymbol{x}}(n-1)\right] \\ &= \mathrm{Cov}\left[\Phi\tilde{\boldsymbol{x}}(n-1) + \Gamma\boldsymbol{w}(n-1)\right]. \end{aligned} \tag{5.48}$$

Now $\tilde{\boldsymbol{x}}(n-1)$ and $\boldsymbol{w}(n-1)$ are independent since $\tilde{\boldsymbol{x}}(n-1)$ depends only on $\boldsymbol{x}(i)$ and $\boldsymbol{w}(i)$ for $i = 1, 2, \ldots, n-2$. Then

$$\mathrm{E}\left[\tilde{\boldsymbol{x}}(n-1)\boldsymbol{w}^{\mathrm{T}}(n-1)\right] = \boldsymbol{0},$$

and (5.48) becomes

$$\boxed{P^-(n) = \Phi P(n-1)\Phi^{\mathrm{T}} + \Gamma Q\Gamma^{\mathrm{T}}.} \tag{5.49}$$

The recursions (5.47) and (5.49) together are called the *time update*. They describe how the estimate and error covariance evolve from time $n-1$ to time n before $\boldsymbol{z}(n)$ becomes available.

Measurement Update

When the measurement $\boldsymbol{z}(n)$ becomes available, we want to update the estimate to produce the LMMSE estimate of $\boldsymbol{x}(n)$ given the set of observations $\boldsymbol{Z} = \{\boldsymbol{z}(1), \ldots, \boldsymbol{z}(n)\}$. We also need the corresponding *a posteriori* error covariance matrix $P(n)$. This update is called the *measurement update* because it describes how the new measurement $\boldsymbol{z}(n)$ is used to adjust the estimate of $\boldsymbol{x}(n)$ and how the error covariance matrix evolves.

The *a posteriori* Estimate

We assume that we already have $\hat{\boldsymbol{x}}(n-1)$, the LMMSE estimate of $\boldsymbol{x}(n-1)$ given $\boldsymbol{z}(1), \ldots, \boldsymbol{z}(n-1)$. We then obtain the new measurement $\boldsymbol{z}(n)$ and wish to find $\hat{\boldsymbol{x}}(n)$.

We note that $\hat{\boldsymbol{x}}(n-1)$ satisfies the orthogonality condition (3.52) in batch form:

$$\mathrm{E}\left[\left(\boldsymbol{x}(n-1) - \hat{\boldsymbol{x}}(n-1)\right)\boldsymbol{Z}_{n-1}^{\mathrm{T}}\right] = \boldsymbol{0}. \tag{5.50}$$

We write $\hat{\boldsymbol{x}}(n-1)$ as

$$\hat{\boldsymbol{x}}(n-1) = \sum_{j=1}^{n-1} H_{n-1}(j)\boldsymbol{z}(j) = J(n-1)\boldsymbol{Z}_{n-1}, \tag{5.51}$$

for some $N \times p$ matrices $H_{n-1}(j)$, and with

$$J(n-1) = \left[H_{n-1}(1) \quad H_{n-1}(2) \quad \cdots \quad H_{n-1}(n-1)\right].$$

Now to find $\hat{\boldsymbol{x}}(n)$, we have the form

$$\hat{\boldsymbol{x}}(n) = \sum_{j=1}^{n} H_n(j)\boldsymbol{z}(j). \tag{5.52}$$

Since $\hat{\boldsymbol{x}}(n)$ should be the LMMSE estimate of $\boldsymbol{x}(n)$, $\hat{\boldsymbol{x}}(n)$ must satisfy the orthogonality condition (3.52):

$$\mathrm{E}\left[(\boldsymbol{x}(n) - \hat{\boldsymbol{x}}(n))\, \boldsymbol{z}^{\mathrm{T}}(i)\right] = \mathbf{0}, \quad i = 1, 2, \ldots, n,$$

which we write in batch form as

$$\mathrm{E}\left[(\boldsymbol{x}(n) - \hat{\boldsymbol{x}}(n))\, \boldsymbol{Z}_n^{\mathrm{T}}\right] = \mathbf{0}, \tag{5.53}$$

where $\mathbf{0}$ is the $N \times np$ zero matrix and

$$\boldsymbol{Z}_n = \begin{bmatrix} \boldsymbol{z}(1) \\ \boldsymbol{z}(2) \\ \vdots \\ \boldsymbol{z}(n) \end{bmatrix} = \begin{bmatrix} \boldsymbol{Z}_{n-1} \\ \boldsymbol{z}(n) \end{bmatrix}.$$

We write (5.52) as

$$\hat{\boldsymbol{x}}(n) = K(n)\boldsymbol{z}(n) + G(n)\boldsymbol{Z}_{n-1}, \tag{5.54}$$

where

$$K(n) = H_n(n), \qquad \text{and} \qquad G(n) = \left[H_n(1) \quad H_n(2) \quad \cdots \quad H_n(n-1)\right].$$

We need to determine $K(n)$ and $G(n)$.

Let us consider $\hat{\boldsymbol{x}}(n)$ and $\boldsymbol{x}(n)$ in (5.53). Using (5.37) and (5.38) for $\boldsymbol{z}(n)$, we write (5.54) as

$$\begin{aligned} \hat{\boldsymbol{x}}(n) &= K(n)\left[C\boldsymbol{x}(n) + \boldsymbol{v}(n)\right] + G(n)\boldsymbol{Z}_{n-1} \\ &= K(n)C\Phi\boldsymbol{x}(n-1) + K(n)C\Gamma\boldsymbol{w}(n-1) \\ &\quad + K(n)\boldsymbol{v}(n) + G(n)\boldsymbol{Z}_{n-1} \end{aligned} \tag{5.55}$$

We also substitute $\boldsymbol{x}(n-1) = \hat{\boldsymbol{x}}(n-1) + \tilde{\boldsymbol{x}}(n-1)$ into (5.36) to obtain

$$\boldsymbol{x}(n) = \Phi\hat{\boldsymbol{x}}(n-1) + \Phi\tilde{\boldsymbol{x}}(n-1) + \Gamma\boldsymbol{w}(n-1). \tag{5.56}$$

Now from (5.55) and (5.56) we have

$$\begin{aligned}\tilde{\boldsymbol{x}}(n) &= \boldsymbol{x}(n) - \hat{\boldsymbol{x}}(n) \\ &= [I - K(n)C]\,\Phi\hat{\boldsymbol{x}}(n-1) + [I - K(n)C]\,\Phi\tilde{\boldsymbol{x}}(n-1) \\ &\qquad + [I - K(n)C]\,\Gamma\boldsymbol{w}(n-1) - K(n)\boldsymbol{v}(n) - G(n)\boldsymbol{Z}_{n-1}. \end{aligned} \tag{5.57}$$

By (5.51), $\hat{\boldsymbol{x}}(n-1) = J(n-1)\boldsymbol{Z}_{n-1}$, so (5.57) becomes

$$\begin{aligned}\tilde{\boldsymbol{x}}(n) = [I - K(n)C]\,\Phi J(n-1)\boldsymbol{Z}_{n-1} + [I - K(n)C]\,\Phi\tilde{\boldsymbol{x}}(n-1) \\ + [I - K(n)C]\,\Gamma\boldsymbol{w}(n-1) - K(n)\boldsymbol{v}(n) - G(n)\boldsymbol{Z}_{n-1}. \end{aligned} \tag{5.58}$$

Now we consider Condition (5.53). Let us write it as

$$\mathrm{E}\left[\tilde{\boldsymbol{x}}(n)\boldsymbol{Z}_n^{\mathrm{T}}\right] = \mathrm{E}\left[\tilde{\boldsymbol{x}}(n)\begin{bmatrix}\boldsymbol{Z}_{n-1} \\ \boldsymbol{z}(n)\end{bmatrix}^{\mathrm{T}}\right] = \mathbf{0}.$$

This expression is equivalent to the two simultaneous conditions

$$\mathrm{E}\left[\tilde{\boldsymbol{x}}(n)\boldsymbol{Z}_{n-1}^{\mathrm{T}}\right] = \mathbf{0}, \tag{5.59}$$

and

$$\mathrm{E}\left[\tilde{\boldsymbol{x}}(n)\boldsymbol{z}^{\mathrm{T}}(n)\right] = \mathbf{0}. \tag{5.60}$$

From (5.58) we have

$$\begin{aligned}&\mathrm{E}\left[\tilde{\boldsymbol{x}}(n)\boldsymbol{Z}_{n-1}^{\mathrm{T}}\right] \\ &\qquad = [I - K(n)C]\,\Phi J(n-1)\mathrm{E}\left[\boldsymbol{Z}_{n-1}\boldsymbol{Z}_{n-1}^{\mathrm{T}}\right] - G(n)\mathrm{E}\left[\boldsymbol{Z}_{n-1}\boldsymbol{Z}_{n-1}^{\mathrm{T}}\right],\end{aligned}$$

where we have exploited the orthogonality assumption (5.50), and the fact that both $\boldsymbol{w}(n-1)$ and $\boldsymbol{v}(n)$ are independent of $\boldsymbol{Z}_{n-1}$. Then Condition (5.59) becomes

$$[I - K(n)C]\,\Phi J(n-1)\mathrm{E}\left[\boldsymbol{Z}_{n-1}\boldsymbol{Z}_{n-1}^{\mathrm{T}}\right] - G(n)\mathrm{E}\left[\boldsymbol{Z}_{n-1}\boldsymbol{Z}_{n-1}^{\mathrm{T}}\right] = \mathbf{0}.$$

Therefore,

$$G(n)\mathrm{E}\left[\boldsymbol{Z}_{n-1}\boldsymbol{Z}_{n-1}^{\mathrm{T}}\right] = [I - K(n)C]\,\Phi J(n-1)\mathrm{E}\left[\boldsymbol{Z}_{n-1}\boldsymbol{Z}_{n-1}^{\mathrm{T}}\right] \tag{5.61}$$

Since $R > 0$, $\mathrm{E}\left[\boldsymbol{Z}_{n-1}\boldsymbol{Z}_{n-1}^{\mathrm{T}}\right]$ is positive-definite and thus always invertible. Then we multiply both sides of (5.61) by $\left(\mathrm{E}\left[\boldsymbol{Z}_{n-1}\boldsymbol{Z}_{n-1}^{\mathrm{T}}\right]\right)^{-1}$ to obtain

$$G(n) = [I - K(n)C]\,\Phi J(n-1). \tag{5.62}$$

Substituting this result into (5.54) and using (5.51), we have

$$\hat{\boldsymbol{x}}(n) = \Phi\hat{\boldsymbol{x}}(n-1) + K(n)\left[\boldsymbol{z}(n) - C\Phi\hat{\boldsymbol{x}}(n-1)\right]. \tag{5.63}$$

By (5.47), we can also relate $\hat{\boldsymbol{x}}(n)$ to $\hat{\boldsymbol{x}}^-(n)$ via

$$\boxed{\hat{\boldsymbol{x}}(n) = \hat{\boldsymbol{x}}^-(n) + K(n)\left[\boldsymbol{z}(n) - C\hat{\boldsymbol{x}}^-(n)\right].} \tag{5.64}$$

This expression and (5.63) show that it is possible to compute $\hat{\boldsymbol{x}}(n)$ *recursively* from $\boldsymbol{z}(n)$ and $\hat{\boldsymbol{x}}^-(n)$. Note that $K(n)$ still must be determined.

The Kalman Gain

The next step is to satisfy the remaining condition (5.60) and determine $K(n)$. From (5.36) and (5.63) we have

$$\begin{aligned}\tilde{\boldsymbol{x}}(n) &= \Phi\boldsymbol{x}(n-1) + \Gamma\boldsymbol{w}(n-1)\\ &\quad - \Phi\hat{\boldsymbol{x}}(n-1) - K(n)\boldsymbol{z}(n) + K(n)C\Phi\hat{\boldsymbol{x}}(n-1).\end{aligned} \tag{5.65}$$

From (5.37) and (5.38), we write $\boldsymbol{z}(n)$ as

$$\boldsymbol{z}(n) = C\Phi\boldsymbol{x}(n-1) + C\Gamma\boldsymbol{w}(n-1) + \boldsymbol{v}(n) \tag{5.66}$$

$$= C\Phi\tilde{\boldsymbol{x}}(n-1) + C\Phi\hat{\boldsymbol{x}}(n-1) + C\Gamma\boldsymbol{w}(n-1) + \boldsymbol{v}(n). \tag{5.67}$$

Now substituting (5.66) into (5.65) gives

$$\tilde{\boldsymbol{x}}(n) = \Phi\tilde{\boldsymbol{x}}(n-1) + \left[I - K(n)C\right]\Gamma\boldsymbol{w}(n-1) - K(n)\boldsymbol{v}(n) \tag{5.68}$$

Next we substitute (5.68) and (5.67) into Condition (5.60). In the process, we note that $\tilde{\boldsymbol{x}}(n-1)$ is orthogonal to $\hat{\boldsymbol{x}}(n-1)$, and that $\hat{\boldsymbol{x}}(n-1)$, $\tilde{\boldsymbol{x}}(n-1)$, $\boldsymbol{w}(n-1)$, and $\boldsymbol{v}(n)$ are all mutually independent. We obtain

$$\begin{aligned}\mathrm{E}\left[\tilde{\boldsymbol{x}}(n)\boldsymbol{z}^{\mathrm{T}}(n)\right] &= \Phi P(n-1)\Phi^{\mathrm{T}}C^{\mathrm{T}} + \Gamma Q\Gamma^{\mathrm{T}}C^{\mathrm{T}}\\ &\quad - K(n)C\Phi P(n-1)\Phi^{\mathrm{T}}C^{\mathrm{T}} - K(n)C\Gamma Q\Gamma^{\mathrm{T}}C^{\mathrm{T}} - K(n)R = \mathbf{0},\end{aligned} \tag{5.69}$$

where Q, R, and $P(n-1)$ are given by (5.39), (5.40), and (5.42), respectively.

Using (5.49), (5.69) reduces to

$$P^-(n)C^{\mathrm{T}} = K(n)\left[CP^-(n)C^{\mathrm{T}} + R\right],$$

Because $R > 0$, the term $\left[CP^-(n)C^{\mathrm{T}} + R\right]$ is positive-definite and hence invertible. Solving this expression for $K(n)$, we have

$$\boxed{K(n) = P^-(n)C^{\mathrm{T}}\left[CP^-(n)C^{\mathrm{T}} + R\right]^{-1}.} \tag{5.70}$$

The matrix $K(n)$ is known as the *Kalman gain*. Note that (5.70) must be computed before $\hat{\boldsymbol{x}}(n)$ can be computed in (5.64).

The *a posteriori* Error Covariance

Finally, we study the relationship between $P(n)$ and $P^-(n)$. Equation (5.49) provides a convenient method for computing $P^-(n)$ from $P(n-1)$, but we would also like to relate $P(n)$ to $P^-(n)$.

From (5.42),

$$P(n) = \mathrm{Cov}\left[\boldsymbol{x}(n) - \hat{\boldsymbol{x}}(n)\right].$$

Applying (5.37), (5.38), and (5.70), we have

$$\begin{aligned} P(n) &= \mathrm{Cov}\left[\boldsymbol{x}(n) - \hat{\boldsymbol{x}}^-(n) - K(n)\left[C\boldsymbol{x}(n) + \boldsymbol{v}(n) - C\hat{\boldsymbol{x}}^-(n)\right]\right] \\ &= \mathrm{Cov}\left[\left[I - K(n)C\tilde{\boldsymbol{x}}^-(n)\right] - K(n)\boldsymbol{v}(n)\right] \\ &= P^-(n) - K(n)CP^-(n) - P^-(n)C^{\mathrm{T}}K^{\mathrm{T}}(n) \\ &\quad + K(n)\left[CP^-(n)C^{\mathrm{T}} + R\right]K^{\mathrm{T}}(n). \end{aligned} \tag{5.71}$$

We substitute (5.70) into the last term in (5.71). This term becomes

$$\begin{aligned} P^-(n)C^{\mathrm{T}}\left[CP^-(n)C^{\mathrm{T}} + R\right]^{-1}\left[CP^-(n)C^{\mathrm{T}} + R\right]K^{\mathrm{T}}(n) \\ = P^-(n)C^{\mathrm{T}}K^{\mathrm{T}}(n). \end{aligned}$$

Then (5.71) becomes

$$\boxed{P(n) = P^-(n) - K(n)CP^-(n),} \tag{5.72}$$

which describes the relationship between $P^-(n)$ and $P(n)$.

5.6 Summary of Kalman Filter Equations

Here we organize the filter equations to form the Kalman filter recursion. We start the recursion at time $n = 0$. Ideally, we should initialize the filter with

$$\hat{x}^-(0) = \mathrm{E}\left[\boldsymbol{x}(0)\right], \tag{5.73}$$

and

$$P^-(0) = \mathrm{E}\left[\left(\boldsymbol{x}(0) - \hat{x}^-(0)\right)\left(\boldsymbol{x}(0) - \hat{x}^-(0)\right)^{\mathrm{T}}\right], \tag{5.74}$$

where the non-bold $\hat{x}^-(0)$ indicates that it is a value, not a RV.

In actual practice, we do not know the expected values in (5.73) and (5.74). Thus, we initialize $\hat{x}^-(0)$ with

$$\hat{x}^-(0) = \text{guess of the value of } \hat{\boldsymbol{x}}^-(0) \text{ or of } \mathrm{E}\left[\boldsymbol{x}(0)\right]. \tag{5.75}$$

For the error covariance, we use

$$P^-(0) = \text{guess of E}\left[(\boldsymbol{x}(0) - \hat{x}^-(0))\,(\boldsymbol{x}(0) - \hat{x}^-(0))^{\mathrm{T}}\right], \tag{5.76}$$

where $P^-(0)$ is a positive-definite, symmetric matrix. Often it is sufficient to use

$$P^-(0) = \lambda I, \quad \lambda > 0, \tag{5.77}$$

or

$$P^-(0) = \text{diag}\left(\begin{bmatrix}\lambda_1 & \lambda_2 & \cdots & \lambda_m\end{bmatrix}\right), \quad \lambda_i > 0 \text{ for } i = 1, 2, \ldots, m, \tag{5.78}$$

where

$$\text{diag}\left(\begin{bmatrix}a_1 & a_2 & \cdots & a_m\end{bmatrix}\right) = \begin{bmatrix} a_1 & 0 & \cdots & 0 \\ 0 & a_2 & \cdots & 0 \\ \vdots & \vdots & \ddots & \vdots \\ 0 & 0 & \cdots & a_m \end{bmatrix}. \tag{5.79}$$

Further justification for these initial conditions is given in Section 5.7.

The filter recursion is

- *Measurement update.* Acquire $z(n)$ and compute *a posteriori* quantities:

$$K(n) = P^-(n)C^{\mathrm{T}}\left[CP^-(n)C^{\mathrm{T}} + R\right]^{-1} \tag{5.80}$$

$$\hat{x}(n) = \hat{x}^-(n) + K(n)\left[z(n) - C\hat{x}^-(n)\right] \tag{5.81}$$

$$P(n) = P^-(n) - K(n)CP^-(n) \tag{5.82}$$

- *Time update.* Compute *a priori* quantities for time $n+1$:

$$\hat{x}^-(n+1) = \Phi\hat{x}(n) \tag{5.83}$$

$$P^-(n+1) = \Phi P(n)\Phi^{\mathrm{T}} + \Gamma Q \Gamma^{\mathrm{T}} \tag{5.84}$$

- *Time increment.* Increment n and repeat.

Of course, we can always obtain the signal estimates via

$$\hat{s}(n) = C\hat{x}(n), \tag{5.85}$$

$$\hat{s}^-(n+1) = C\hat{x}^-(n+1). \tag{5.86}$$

Thus we have the Kalman filter for filtering and one-step prediction, although it is generally referred to simply as *the Kalman filter.* Each time we obtain a new measurement $z(n)$, we employ this sequence of computations to estimate $\boldsymbol{x}(n)$ and $\boldsymbol{s}(n)$. For filtering, $\hat{x}(n)$ is of interest; for prediction, $\hat{x}^-(n)$. We study the m-step Kalman predictor in Section 6.5 and the Kalman smoother in Section 6.6.

Alternate Initialization

Alternatively, we may start the recursion at Equation (5.83). In this case, we initialize the filter by setting

$$\hat{x}(0) = \text{guess of } \hat{\boldsymbol{x}}(0), \tag{5.87}$$

and

$$P(0) = \text{guess of E}\left[(\boldsymbol{x}(0) - \hat{x}(0))\,(\boldsymbol{x}(0) - \hat{x}(0))^{\mathrm{T}}\right], \tag{5.88}$$

We set $n = 1$, and the Kalman filter recursion becomes:

- *Time update.* Compute:

$$\begin{aligned} \hat{x}^{-}(n) &= \Phi\hat{x}(n-1) \\ P^{-}(n) &= \Phi P(n-1)\Phi^{\mathrm{T}} + \Gamma Q \Gamma^{\mathrm{T}} \end{aligned}$$

- *Measurement update.* Acquire $z(n)$ and compute:

$$\begin{aligned} K(n) &= P^{-}(n)C^{\mathrm{T}}\left[CP^{-}(n)C^{\mathrm{T}} + R\right]^{-1} \\ \hat{x}(n) &= \hat{x}^{-}(n) + K(n)\left[z(n) - C\hat{x}^{-}(n)\right] \\ P(n) &= P^{-}(n) - K(n)CP^{-}(n) \end{aligned}$$

- *Time increment.* Increment n and repeat.

5.7 Kalman Filter Properties

This section discusses several important properties of the Kalman filter of Section 5.6. Topics include the overall optimality of the Kalman filter in certain cases, the Riccati equation, initialization of the filter, and an equivalent form for the filter.

General Properties

The characteristics presented here arise directly from inspection of the filter equations.

- The Kalman filter is a time-varying system. Even if the SMM matrices Φ, Γ, C, Q, and R are constant, the expressions for $P(n)$, $P^{-}(n)$, and $K(n)$ depend upon n.

- By regrouping the Kalman filter equations, we find two independent recursions. The Kalman gain $K(n)$ appears in both of them.

  - The first, which consists of Equations (5.80), (5.81), and (5.83), computes the actual estimate. We call this recursion the *estimate recursion.*
  - The second, Equations (5.80), (5.82), and (5.84), calculates the error covariance matrices. It forms the *covariance recursion,* and allows us to compute the MSE of the estimates.

- The covariance recursion has no dependence upon the observation process $\boldsymbol{z}(n)$. Because of this independence, we can pre-compute the error covariances and Kalman gains without actually making any observations.

Overall Optimality

Section 4.6 discussed the overall optimality of the Wiener filter. Recall that when the signal $\boldsymbol{s}(n)$ and measurements $\boldsymbol{z}(n)$ are zero-mean, stationary, jointly Gaussian random processes, the Wiener filter is the overall optimum MMSE estimator. Similarly, under certain conditions, the Kalman filter is not only the MMSE estimator among causal, linear time-varying systems, but also the overall MMSE estimator. This result occurs because, for a jointly Gaussian signal $\boldsymbol{s}(n)$ and measurement $\boldsymbol{z}(n)$, the overall MMSE estimate is a linear estimate (see Example 3.9). It follows that, *when $\boldsymbol{w}(n)$ and $\boldsymbol{v}(n)$ are zero-mean (possibly nonstationary) Gaussian random processes, the Kalman filter is the overall optimal MMSE estimator.*

The Riccati Equation

We now introduce an important property regarding the matrices $P^-(n)$ and $P^-(n-1)$. From (5.84) and (5.82) we have

$$P^-(n+1) = \Phi P^-(n)\Phi^{\mathrm{T}} - \Phi K(n) C P^-(n)\Phi^{\mathrm{T}} + \Gamma Q(n)\Gamma^{\mathrm{T}}.$$

Substituting for $K(n)$ we get

$$P^-(n+1) = \Phi\left\{P^-(n) - P^-(n)C^{\mathrm{T}}\left[CP^-(n)C^{\mathrm{T}} + R\right]^{-1} CP^-(n)\right\}\Phi^{\mathrm{T}} + \Gamma Q(n)\Gamma^{\mathrm{T}}. \tag{5.89}$$

Equation (5.89) is called the *Riccati equation.* It is a matrix quadratic equation that appears in a class of optimal control problems known as the linear quadratic regulator (LQR) problem. Of course, the Riccati equation also appears in optimal estimation problems.

Error Systems

In this section we consider the systems that describe the estimation errors $\tilde{\boldsymbol{x}}(n)$ and $\tilde{\boldsymbol{x}}^-(n)$. These systems relate to the stability properties of the Kalman filter and suggest its initial conditions.

The mean of the *a posteriori* estimation error $\tilde{\boldsymbol{x}}(n)$ is given by

$$\begin{aligned}
\mathrm{E}\left[\tilde{\boldsymbol{x}}(n)\right] &= \mathrm{E}\left[\boldsymbol{x}(n) - \hat{\boldsymbol{x}}(n)\right] \\
&= \mathrm{E}\left[\Phi\boldsymbol{x}(n-1) + \Gamma(n-1)\boldsymbol{w}(n-1) - \Phi\hat{\boldsymbol{x}}(n-1)\right. \\
&\qquad \left. -K(n)\left[\boldsymbol{z}(n) - C\Phi\hat{\boldsymbol{x}}(n-1)\right]\right] \\
&= \mathrm{E}\left[\Phi\boldsymbol{x}(n-1) + \Gamma(n-1)\boldsymbol{w}(n-1) - \Phi\hat{\boldsymbol{x}}(n-1)\right. \\
&\qquad - K(n)\left\{C\left[\Phi\boldsymbol{x}(n-1) + \Gamma(n-1)\boldsymbol{w}(n-1)\right] + \boldsymbol{v}(n)\right. \\
&\qquad \left.\left. - C\Phi\hat{\boldsymbol{x}}(n-1)\right\}\right] \\
&= \left[I - K(n)C\right]\Phi\mathrm{E}\left[\tilde{\boldsymbol{x}}(n-1)\right] \qquad (5.90)
\end{aligned}$$

Equation (5.90) describes the behavior of $\mathrm{E}\left[\tilde{\boldsymbol{x}}(n]\right)$.

From (5.83), the expected value of the *a priori* error $\tilde{\boldsymbol{x}}^-(n)$ is described by

$$\mathrm{E}\left[\tilde{\boldsymbol{x}}^-(n)\right] = \Phi\left[I - K(n-1)C(n-1)\right]\mathrm{E}\left[\tilde{\boldsymbol{x}}^-(n-1)\right], \qquad (5.91)$$

and the two errors are related by

$$\mathrm{E}\left[\tilde{\boldsymbol{x}}(n)\right] = \left[I - K(n)C\right]\mathrm{E}\left[\tilde{\boldsymbol{x}}^-(n)\right]. \qquad (5.92)$$

In the remainder of this section we will refer to these results, which will be useful in studying the behavior of the estimation errors and the stability of the Kalman filter.

Initialization

Section 5.6 presented the general initialization values of the Kalman filter without justification. Here we explain the reasoning behind them.

First, we consider the choice of $\tilde{\boldsymbol{x}}^-(0)$. From (5.91) and (5.92), we observe that if $\mathrm{E}\left[\tilde{\boldsymbol{x}}^-(n_0)\right] = 0$, then

$$\mathrm{E}\left[\tilde{\boldsymbol{x}}^-(n)\right] = \mathrm{E}\left[\tilde{\boldsymbol{x}}(n)\right] = 0, \quad n \geq n_0.$$

In other words, if we make an unbiased estimate at time n_0, then all subsequent estimates—*a priori* and *a posteriori* —will be unbiased. Therefore we should (ideally) initialize the Kalman filter with (5.75),

$$\hat{\boldsymbol{x}}^-(0) = \mathrm{E}\left[\boldsymbol{x}(0)\right],$$

so that all our estimates are unbiased.

Second, we explain the choice of $P^-(0)$. By definition (5.41) $P^-(n)$ is a covariance matrix, so it must be positive-semidefinite and symmetric. Hence (5.76) follows, namely, that $P^-(0) \geq 0$.

Equivalent Form

This section presents an equivalent form of the Kalman filter. This form is useful in the derivation of the Kalman smoother (Section 6.6) and in the proof of the stability of Kalman filter (Appendix C). It provides alternative expressions for computing $\hat{\boldsymbol{x}}(n)$ (5.81) and $P(n)$ (5.82).

The derivation requires the matrix inversion lemma, stated below.

Lemma 5.1 (Matrix Inversion Lemma) *Let A_{11} be a nonsingular $p \times p$ matrix, A_{12} and A_{21} be $p \times q$ and $q \times p$ matrices, respectively, and A_{22} be a nonsingular $q \times q$ matrix. Then*

$$\begin{aligned}&\left(A_{11}^{-1} + A_{12}A_{22}A_{21}\right)^{-1} \\ &\qquad = A_{11} - A_{11}A_{12}\left(A_{21}A_{11}A_{12} + A_{22}^{-1}\right)^{-1} A_{21}A_{11}. \quad (5.93)\end{aligned}$$

First we find a new expression for $\hat{\boldsymbol{x}}(n)$. In (5.81) we multiply by $P^{-}(n)$ and $\left(P^{-}(n)\right)^{-1}$ to obtain

$$\begin{aligned}\hat{\boldsymbol{x}}(n) &= [I - K(n)C]\, P^{-}(n)(P^{-}(n))^{-1}\hat{\boldsymbol{x}}^{-}(n) + K(n)\boldsymbol{z}(n) \\ &= P(n)(P^{-}(n))^{-1}\Phi\hat{\boldsymbol{x}}(n-1) + K(n)\boldsymbol{z}(n) \\ &= P(n)\left[(P^{-}(n))^{-1}\Phi\hat{\boldsymbol{x}}(n-1) + P^{-1}(n)K(n)\boldsymbol{z}(n)\right]. \quad (5.94)\end{aligned}$$

Now $P(n)$ can be written as

$$P(n) = P^{-}(n) - P^{-}(n)C^{\mathrm{T}}\left[CP^{-}(n)C^{\mathrm{T}} + R\right]^{-1} CP^{-}(n).$$

We apply the matrix inversion lemma to this equation, which yields

$$\begin{aligned}P^{-1}(n) = (P^{-}(n))^{-1} &+ (P^{-}(n))^{-1}P^{-}(n)C^{\mathrm{T}} \\ &\times \left[-CP^{-}(n)(P^{-}(n))^{-1}P^{-}(n)C^{\mathrm{T}}\right. \\ &\left. + (CP^{-}(n)C^{\mathrm{T}} + R)\right]^{-1} CP^{-}(n)(P^{-}(n))^{-1}.\end{aligned}$$

Next we multiply by $K(n)$ and substitute for $K(n)$ to obtain

$$\begin{aligned}P^{-1}(n)K(n) &= \left[(P^{-}(n))^{-1} + C^{\mathrm{T}}R^{-1}C\right]P^{-}(n)C^{\mathrm{T}}\left[CP^{-}(n)C^{\mathrm{T}} + R\right]^{-1} \\ &= C^{\mathrm{T}}\left[I + R^{-1}CP^{-}(n)C^{\mathrm{T}}\right]\left[CP^{-}(n)C^{\mathrm{T}} + R\right]^{-1} \\ &= C^{\mathrm{T}}R^{-1}\left[R + CP^{-}(n)C^{\mathrm{T}}\right]\left[CP^{-}(n)C^{\mathrm{T}} + R\right]^{-1} \\ &= C^{\mathrm{T}}R^{-1}.\end{aligned}$$

Therefore (5.94) can be written as

$$\boxed{\hat{\boldsymbol{x}}(n) = P(n)\left[(P^{-}(n))^{-1}\Phi\hat{\boldsymbol{x}}(n-1) + C^{\mathrm{T}}R^{-1}\boldsymbol{z}(n)\right].} \quad (5.95)$$

Next we find an equivalent form for $P(n)$. By (5.80) and (5.82) we have

$$P(n) = P^-(n) - P^-(n)C^{\mathrm{T}}\left[CP^-(n)C^{\mathrm{T}} + R\right]^{-1} CP^-(n).$$

The right-hand side of this equation fits the form of the right-hand side of (5.93). Then we conclude

$$\boxed{P(n) = \left[(P^-(n))^{-1} + C^{\mathrm{T}}R^{-1}C\right]^{-1}.} \tag{5.96}$$

We may now write the Kalman filter equations in the alternate form below.

$$\hat{\boldsymbol{x}}^-(n) = \Phi\hat{\boldsymbol{x}}(n-1) \tag{5.97}$$

$$P(n) = \left[(P^-(n))^{-1} + C^{\mathrm{T}}R^{-1}C\right]^{-1} \tag{5.98}$$

$$\hat{\boldsymbol{x}}(n) = P(n)\left[(P^-(n))^{-1}\Phi\hat{\boldsymbol{x}}(n-1) + C^{\mathrm{T}}R^{-1}\boldsymbol{z}(n)\right] \tag{5.99}$$

$$P^-(n+1) = \Phi P(n)\Phi^{\mathrm{T}} + \Gamma Q(n)\Gamma^{\mathrm{T}}. \tag{5.100}$$

This formulation of the Kalman filter is sometimes referred to as the *inverse covariance form* since the matrix inverse may be taken to the left-hand side of Equation (5.98), thus relating $P^{-1}(n)$ and $(P^-(n))^{-1}$.

5.8 The Steady-state Kalman Filter

The Kalman filter is inherently a time-varying system because the covariance matrices $P^-(n)$ and $P(n)$ and the Kalman gain $K(n)$ are updated at each time increment. However, in the case of a time-invariant SMM with stationary noises, these quantities can reach steady-state and no longer change with time. At this point the Kalman filter becomes time-invariant, too, and is known as the *steady-state Kalman filter* (SSKF).

If $P^-(n)$ reaches a constant, steady-state value, then both $P(n)$ and $K(n)$ also become constant. The Riccati equation (5.89) relates $P^-(n)$ and $P(n)$ and is useful is understanding the SSKF. We have

$$P^-(n+1) = \Phi\left\{P^-(n) - P^-(n)C^{\mathrm{T}}\left[CP^-(n)C^{\mathrm{T}} + R\right]^{-1} CP^-(n)\right\}\Phi^{\mathrm{T}} + \Gamma Q\Gamma^{\mathrm{T}}. \tag{5.101}$$

Suppose that $P^-(n)$ reaches steady-state, so it approaches a constant matrix as $n \to \infty$. Let

$$P_\infty^- = \lim_{n\to\infty} P^-(n). \tag{5.102}$$

Now for large n, $P_\infty^- = P^-(n) = P^-(n+1)$, and (5.101) becomes

$$P_\infty^- = \Phi\left[P_\infty^- - P_\infty^-C^{\mathrm{T}}\left(CP_\infty^-C^{\mathrm{T}}+R\right)^{-1} CP_\infty^-\right]\Phi^{\mathrm{T}} + \Gamma Q\Gamma^{\mathrm{T}}. \tag{5.103}$$

Equation (5.103) does not depend on n, and hence it is an algebraic matrix equation, called the *algebraic Riccati equation* (ARE). Given P_∞^-, the Kalman gain and *a posteriori* estimation error covariance matrix become constant matrices, defined by

$$K_\infty = P_\infty^- C^{\mathrm{T}} \left(C P_\infty^- C^{\mathrm{T}} + R\right)^{-1}, \tag{5.104}$$

and

$$P_\infty = P_\infty^- - K_\infty C P_\infty^- \tag{5.105}$$

$$= P_\infty^- - P_\infty^- C^{\mathrm{T}} \left(C P_\infty^- C^{\mathrm{T}} + R\right)^{-1} C P_\infty^-. \tag{5.106}$$

Since both K_∞ and P_∞ depend only upon P_∞^-, *we can find the SSKF by solving the ARE (5.103).*

Example 5.1 Finding a State Model 1

Suppose we have a WSS signal process $s(n)$ with autocorrelation function

$$R_s(n) = 0.5\delta(n+1) + \delta(n) + 0.5\delta(n-1), \tag{5.107}$$

and power spectral density

$$S_s(z) = 0.5z + 1 + 0.5z^{-1}. \tag{5.108}$$

We measure the signal in the presence of additive white noise $w(n)$ with variance $\sigma_w^2 = 2$. The processes $s(n)$ and $w(n)$ are uncorrelated.

Before we can determine the optimum causal linear estimator, we must find a state model for generating $z(n)$. This example and the next example demonstrate two methods for accomplishing this task. In subsequent examples, we compare the predictors and filters.

The first method requires performing a spectral factorization of $S_s(z)$. Recall that $S_s(z) = S_s^+(z) S_s^+(z^{-1})$. Once we have $S_s^+(z)$, we can treat it as an LTI system excited by white noise with unit variance, which is just $w(n)$. Then we can use canonical forms to find a state model for generating $s(n)$.

Let us factor $S_s(z)$. We want

$$S_s(z) = 0.5z + 1 + 0.5z^{-1} = a(1 - bz)a(1 - bz^{-1})$$
$$= a^2 \left[-bz + (1 + b^2) - bz^{-1}\right].$$

Then we must solve the system of nonlinear equations

$$a^2(1 + b^2) = 1, \quad \text{and} \quad -a^2 b = 1/2.$$

It follows that $b = -1/2a^2$, so $a^2(1 + 1/4a^4) = 1$, and thus

$$4a^4 - 4a^2 + 1 = 0$$

This equation has solution $a^2 = 1/2$, so $a = 1/\sqrt{2}$ and $b = -1$.

We choose

$$S_s^+(z) = \frac{1+z^{-1}}{\sqrt{2}}.$$

From Appendix A, the observable canonical form for this system function is

$$x'(n+1) = 0x'(n) + 1/\sqrt{2}w'(n)$$
$$s(n) = x'(n) + 1/\sqrt{2}w'(n),$$

where $w'(n)$ is zero-mean white noise with unit variance.

However, the SMM does not include a direct-feed term for the process noise $w'(n)$. Let us define a new state vector

$$x(n) \triangleq \begin{bmatrix} s(n) \\ x'(n+1) \end{bmatrix},$$

so

$$x(n+1) = \begin{bmatrix} 0 & 1 \\ 0 & 0 \end{bmatrix} x(n) + \begin{bmatrix} 1/\sqrt{2} \\ 1/\sqrt{2} \end{bmatrix} w'(n+1).$$

Let $w(n) = w'(n+1)$ and the model becomes

$$x(n+1) = \begin{bmatrix} 0 & 1 \\ 0 & 0 \end{bmatrix} x(n) + \begin{bmatrix} 1/\sqrt{2} \\ 1/\sqrt{2} \end{bmatrix} w(n).$$
$$s(n) = \begin{bmatrix} 1 & 0 \end{bmatrix} x(n).$$

Finally, we include the additive noise and have

$$z(n) = \begin{bmatrix} 1 & 0 \end{bmatrix} x(n) + v(n),$$

where $R_v(n) = 2\delta(n)$, and $w(n)$ and $v(n)$ are uncorrelated.

Example 5.2 Finding a State Model 2

The second method for constructing a state model results from our knowledge about the autocorrelation function $R_s(n)$ and linear systems. Suppose we have a LTI system with impulse response $h(n)$, and white noise $w'(n)$ with unit variance is applied to the system input. Designating the output by $s(n)$, we know that

$$R_s(n) = h(n) * h(-n).$$

In this example $R_s(n)$ has a finite region of support. Recall that if $h(n)$ has duration N, $R_s(n)$ has duration $2N-1$. The region of support for $R_s(n)$ is three samples, so $h(n)$ has a duration of two samples.

Based on this reasoning, we assume that the signal process is generated via

$$s(n) = aw'(n) + bw'(n-1).$$

Then we plug into $R_s(n) = \mathrm{E}[s(0)s(n)]$ and find

$$R_s(n) = \begin{cases} a^2 + b^2 = 1, & n = 0, \\ ab = 1/2, & n = \pm 1, \\ 0, & \text{otherwise.} \end{cases}$$

This system of nonlinear equations has solution $a = b = 1/\sqrt{2}$.

Next we define the state vector

$$x(n) \triangleq \begin{bmatrix} s(n) \\ w'(n) \end{bmatrix},$$

so that

$$x(n+1) = \begin{bmatrix} 0 & 1/\sqrt{2} \\ 0 & 0 \end{bmatrix} x(n) + \begin{bmatrix} 1/\sqrt{2} \\ 1 \end{bmatrix} w'(n+1)$$
$$s(n) = \begin{bmatrix} 1 & 0 \end{bmatrix} x(n).$$

Letting $w(n) = w'(n+1)$, we have

$$x(n+1) = \begin{bmatrix} 0 & 1/\sqrt{2} \\ 0 & 0 \end{bmatrix} x(n) + \begin{bmatrix} 1/\sqrt{2} \\ 1 \end{bmatrix} w(n) \tag{5.109}$$
$$s(n) = \begin{bmatrix} 1 & 0 \end{bmatrix} x(n) \tag{5.110}$$
$$z(n) = \begin{bmatrix} 1 & 0 \end{bmatrix} x(n) + v(n), \tag{5.111}$$

where once again $R_v(n) = 2\delta(n)$, and $w(n)$ and $v(n)$ are uncorrelated. In the following examples we use the SMM of Equations (5.109–5.111).

Example 5.3 Determining the SSKF

Let us find P_∞^- and K_∞ for the model of Example 5.2. In practice we usually set up the Kalman filter and then apply Equations (5.80), (5.82) and (5.84) until $P^-(n)$ converges to its steady-state value, which corresponds to P_∞^-. In this example the mathematics are simple enough that we can find P_∞^- analytically.

P_∞^- is symmetric and positive-definite, so we assume

$$P_\infty^- = \begin{bmatrix} a & c \\ c & b \end{bmatrix}.$$

When we substitute this expression into (5.103) we find that the ARE reduces to

$$\begin{bmatrix} a & c \\ c & b \end{bmatrix} = \begin{bmatrix} \dfrac{(a+2)b - c^2}{2(a+2)} & 0 \\ 0 & 0 \end{bmatrix} + \begin{bmatrix} 1/2 & 1/\sqrt{2} \\ 1/\sqrt{2} & 1 \end{bmatrix}.$$

Hence, $b = 1$, and $c = 1/\sqrt{2}$.

Equating the upper left-hand elements, we have

$$a = \frac{(a+2)(1) - (1/\sqrt{2})^2}{2a+4},$$

which gives

$$2a^2 + 2a - 7/2 = 0.$$

Then $a = -1/2 \pm \sqrt{2}$. Since $P_\infty^- > 0$, we choose the positive value for a.

Then

$$P_\infty^- = \begin{bmatrix} \sqrt{2} - 1/2 & 1/\sqrt{2} \\ 1/\sqrt{2} & 1 \end{bmatrix} \approx \begin{bmatrix} 0.9142 & 0.7071 \\ 0.7071 & 1 \end{bmatrix}, \qquad (5.112)$$

and from (5.104), we find

$$K_\infty = \begin{bmatrix} \frac{\sqrt{2}-1/2}{\sqrt{2}+3/2} \\ \frac{1/\sqrt{2}}{\sqrt{2}+3/2} \end{bmatrix} \approx \begin{bmatrix} 0.3137 \\ 0.2426 \end{bmatrix}. \qquad (5.113)$$

If we employ Equations (5.82) and (5.84), we find that $P^-(n)$ and $K(n)$ converge to the above values in less than twenty iterations.

Input-Output Form

Since the SSKF is an LTI system, we can apply frequency-domain techniques to find an input-output representation for it, namely, a system function. From Equations (5.81), (5.83), and (5.86), the SSKF may be written in the form:

$$\hat{\boldsymbol{x}}^-(n+1) = \Phi\,[I - K_\infty C]\,\hat{\boldsymbol{x}}^-(n) + \Phi K_\infty \boldsymbol{z}(n) \qquad (5.114)$$

$$\hat{\boldsymbol{s}}^-(n) = C\hat{\boldsymbol{x}}^-(n). \qquad (5.115)$$

Let $H_{\mathrm{SSKF}}(z)$ denote the system function of the SSKF. From Appendix A we then have

$$H_{\mathrm{SSKF}}(z) = C\left\{zI - \Phi\,[I - K_\infty C]\right\}^{-1} \Phi K_\infty. \qquad (5.116)$$

The filter described by (5.116) may be realized using any of a variety of digital filter implementations.

Example 5.4 A Transfer Function for the SSKF

We continue working with the models in Examples 5.2 and 5.3. We plug into (5.116) to find the system function for the SSKF.

$$\begin{aligned} H_{\mathrm{SSKF}}(z) &= \begin{bmatrix} 1 & 0 \end{bmatrix} \left\{ zI - \begin{bmatrix} -0.1716 & 0.7071 \\ 0 & 0 \end{bmatrix} \right\}^{-1} \begin{bmatrix} 0.1716 \\ 0 \end{bmatrix} \\ &= \frac{0.1716}{z + 0.1716}. \end{aligned}$$

Relation to the Causal Wiener Filter

In the course of presenting the SSKF, we made the SMM LTI. Then the SMM may be characterized by a system function $H(z)$ instead of a state model. As a result, the signal and observations may be described by power spectral densities $S_s(z)$ and $S_z(z)$. The causal Wiener filter of Chapter 3 already handles this problem formulation.

Recall that the optimum causal linear estimator is unique. Does it follow that the SSKF and causal Wiener filter are identical? Provided we develop the question appropriately, the answer is yes.

We may convert $H_{\mathrm{SSKF}}(z)$ from a system function into a convolution sum. The output equation (5.115) does not contain a direct-feed term. That is,

$$\hat{\boldsymbol{s}}^-(n) = C\hat{\boldsymbol{x}}^-(n) + 0\boldsymbol{z}(n),$$

which means that the observation $\boldsymbol{z}(n)$ does not directly contribute to $\hat{\boldsymbol{s}}^-(n)$. Then the convolution sum for the SSKF has the form

$$\hat{\boldsymbol{s}}^-(n) = \sum_{i=1}^{n} b(i)\boldsymbol{z}(n-i). \tag{5.117}$$

However, the convolution sum associated with the causal Wiener filter is given by equation (4.50),

$$\hat{\boldsymbol{s}}(n) = \sum_{i=0}^{n} a(i)\boldsymbol{z}(n-i). \tag{5.118}$$

Equations (5.117) and (5.118) possess two slight but important differences. First, the SSKF has $b(0) = 0$, but the Wiener filter has $a(0) \neq 0$ in general. Second, the SSKF produces a one-step *predicted* estimate $\hat{\boldsymbol{s}}^-(n)$ while the Wiener filter yields a *filtered* estimate $\hat{\boldsymbol{s}}(n)$.

In order to make a meaningful comparison between the SSKF and the causal Wiener filter, we need to add a direct-feed term to the SSKF. Additionally, we should compare corresponding estimators, i.e., compare two predictors or two filters. If we do so, we find that the SSKF and causal Wiener filter have identical system functions.

A complete proof of this result involves MIMO system function matrices, but we have only considered the Wiener filter for SISO system functions. For a more detailed discussion, consult [7]. We present a pair of examples that demonstrate the equivalance of the SSKF and causal Wiener filter for the SISO case.

Example 5.5 Predicting Estimator Comparison

Let us compare the one-step predictor form of the SSKF and the causal Wiener filter for Examples 5.2 and 5.3. The Wiener filter generates $s(n+1)$ given $z(0)$, ..., $z(n)$, so we need to find the output equation for $\hat{s}^-(n+1)$. Equation (5.114) gives

$$\hat{s}^-(n+1) = C\hat{x}^-(n+1) = C\Phi\left[I - K_\infty C\right]\hat{x}^-(n) + C\Phi K_\infty z(n).$$

Therefore the system function of the SSKF predictor with a direct feed is

$$\begin{aligned} H_{\mathrm{SSKF,pred}}(z) = {} & C\Phi\left(I - K_\infty C\right)\left[zI - \Phi\left(I - K_\infty C\right)\right]^{-1}\Phi K_\infty \\ & + C\Phi K_\infty. \end{aligned}$$

When we plug (5.112) and (5.113) into this expression, we obtain

$$H_{\mathrm{SSKF,pred}}(z) = \frac{0.1716}{1 + 0.1716z^{-1}},$$

where "pred" stands for "predicted estimate."

The Wiener predictor is given by

$$H_{\text{Wiener, pred}}(z) = \frac{1}{S_z^+(z)}\left[\frac{zS_{sz}(z)}{S_z^-(z)}\right]_+$$

We need the spectral factorization for $S_z(z)$. In the same manner as Example 5.2, we find that

$$S_z^+(z) = 1.7071(1 + 0.1716z^{-1}).$$

For the additive noise model with uncorrelated processes, $S_{sz}(z) = S_s(z)$, so

$$\begin{aligned} \left[\frac{zS_{sz}(z)}{S_z^-(z)}\right]_+ &= \left[\frac{0.5z + 1 + 0.5z^{-1}}{1.7071(1 + 0.1716z)}\right]_+ \\ &= \left[0.2929 + 1.7029z - \frac{1.1714}{z^{-1} + 0.1716}\right]_+ \\ &= 0.2929. \end{aligned}$$

Then

$$\begin{aligned} H_{\text{Wiener, pred}}(z) &= \frac{1}{1.7071(1 + 0.1716z^{-1})} \times 0.2929 \\ &= \frac{0.1716}{1 + 0.1716z^{-1}}. \end{aligned}$$

A quick comparison of $H_{\text{Wiener, pred}}(z)$ with $H_{\mathrm{SSKF,pred}}(z)$ verifies that the estimators are identical.

Example 5.6 Filtering Estimator Comparison

Now we compare the filtered estimate forms of the filters for Examples 5.2 and 5.3. The Wiener filter generates $\hat{s}(n)$. We have (5.81), which states

$$\hat{\boldsymbol{x}}(n) = (I - K_\infty C)\,\hat{\boldsymbol{x}}^-(n) + K_\infty \boldsymbol{z}(n).$$

From the above expression and (5.114), the output equation for $\hat{s}(n)$ is

$$\hat{s}(n) = C\hat{\boldsymbol{x}}(n) = C\,(I - K_\infty C)\,\hat{\boldsymbol{x}}^-(n) + CK_\infty \boldsymbol{z}(n),$$

and we have the system function of interest:

$$H_{\text{SSKF,filt}}(z) = C\,(I - K_\infty C)\,[zI - \Phi\,(I - K_\infty C)]^{-1}\,\Phi K_\infty + CK_\infty,$$

and "filt" abbreviates "filtered estimate."

Equations (5.112) and (5.113) give

$$H_{\text{SSKF,filt}}(z) = 1 - \frac{0.6863}{1 + 0.1716z^{-1}}.$$

Now consider the causal Wiener filter. Its system function is

$$\begin{aligned}
H_{\text{Wiener, filt}}(z) &= \frac{1}{S_z^+(z)}\left[\frac{S_{sz}(z)}{S_z^-(z)}\right]_+ \\
&= \frac{1}{S_z^+(z)}\left[\frac{0.5z + 1 + 0.5z^{-1}}{1.7071(1 + 0.1716z)}\right]_+ \\
&= \frac{1}{S_z^+(z)}\left[0.2929z^{-1} + 0.5355 + \frac{0.2010}{z^{-1} + 0.1716}\right]_+ \\
&= \frac{1}{1.7071(1 + 0.1716z^{-1})} \times \left(0.2929z^{-1} + 0.5355\right) \\
&= 1 - \frac{0.6863}{1 + 0.1716z^{-1}}.
\end{aligned}$$

Hence, the filtered-estimate forms of the SSKF and the causal Wiener filter are the same.

5.9 The SSKF as an Unbiased Estimator

The preceding section has presented the SSKF and demonstrated its equivalence with the causal Wiener filter. However, we are also interested in the conditions which guarantee the existence of the SSKF and which ensure that it is an asymptotically unbiased estimator.

The existence of the SSKF depends only upon the existence of P_∞^-. It does *not* depend upon $\hat{\boldsymbol{x}}^-(n)$, $\hat{\boldsymbol{x}}(n)$, or $\boldsymbol{z}(n)$. We should be able to determine whether or not the Kalman filter reaches steady-state directly from the time-invariant SMM with stationary noises (5.36–5.40).

When the Kalman filter is asymptotically unbiased, the expected values of the estimation errors $\tilde{\boldsymbol{x}}(n)$ and $\tilde{\boldsymbol{x}}^-(n)$ approach zero as time progresses. By its very construction, the Kalman filter is the optimum causal LMMSE estimator. However, it is possible that the optimum filter may not be unbiased. In some cases, $\tilde{\boldsymbol{x}}(n)$ and $\tilde{\boldsymbol{x}}^-(n)$ may remain bounded, but in other cases, $\tilde{\boldsymbol{x}}^-(n)$ or $\tilde{\boldsymbol{x}}(n)$ may grow without bound. These situations are known as *filter divergence* (see Section 6.4).

We have several important questions to address:

Existence *When does the SSKF exist?*

Asymptotically Unbiased *When is the SSKF asymptotically unbiased?* In other words, when does

$$\lim_{n\to\infty} \mathrm{E}\left[\hat{\boldsymbol{x}}(n)\right] = \mathrm{E}\left[\boldsymbol{x}(n)\right]?$$

Bounded Error *When is P_∞^- bounded for all initial guesses $P^-(0)$?* We want $P(n)$ to remain bounded for all n, which implies that the MSE is bounded.

Independence of P_∞^- and $P^-(0)$ *When is P_∞^- independent of the initial guess for $P^-(0)$?* In practice we do not know the exact value of $P^-(0)$. We would like to know the conditions that ensure that, for large n, an inaccurate guess for $P^-(0)$ will not severely affect the performance of the SSKF.

We have the following theorem that answers these questions. The proof appears in Appendix D.

Theorem 5.1 *Let $\sqrt{Q}$ be any matrix such that $\sqrt{Q}\sqrt{Q}^T = Q$, where Q is positive definite, and let $(\Phi, \Gamma\sqrt{Q})$ be stabilizable. Then (Φ, C) is detectable if and only if*

- *the steady-state Kalman filter exists;*
- *the steady-state Kalman filter is asymptotically unbiased;*
- *P_∞^- is the* unique *finite positive-semidefinite solution to the algebraic Riccati equation (5.103);*
- *P_∞^- is independent of $P^-(0)$, provided $P^-(0) \geq 0$.*

Briefly, stabilizability [8] means that if a system has any unstable modes, they are controllable by the process noise $\boldsymbol{w}(n)$, and any uncontrollable modes are inherently stable (they are bounded or approach zero). Detectability [8] means that no unobservable mode can tend toward infinity without manifesting its growth in the measurements $\boldsymbol{z}(n)$. Stabilizability and detectability are reviewed in Appendix A.

Theorem 5.1 is particularly appealing because it provides *necessary and sufficient* conditions rather than sufficient conditions only. Since we are attempting to estimate $\boldsymbol{x}(n)$, the condition of detectability is not unexpected. However, we also have the surprising condition of stabilizability: the process noise $\boldsymbol{w}(n)$ must excite the controllable modes of the SMM. We actually require process noise to drive the system.

Let us assume the hypotheses of stabilizability and detectability are satisfied. The first and second conclusions of Theorem 5.1 show that the SSKF exists and is asymptotically unbiased. The third conclusion indicates that P_∞^- will be bounded, and thus the MSE will remain bounded as well. The fourth conclusion is useful because states that it does not matter what initial guess $P^-(0)$ we use, as long as $P^-(0)$ is positive-semidefinite.

Also, note that for an LTI state model, observability implies detectability and controllability implies stabilizability (see Appendix A). Then we have the following corollary to Theorem 5.1:

Corollary 5.1 *If Q is positive definite, $(\Phi, \Gamma\sqrt{Q})$ is controllable, and (Φ, C) is observable, then*

- *the SSKF filter exists;*
- *the SSKF is asymptotically unbiased;*
- *P_∞^- is a solution of the ARE (5.103), and it is unique, finite, and positive-semidefinite;*
- *P_∞^- is independent of $P^-(0)$, provided $P^-(0) \geq 0$.*

This corollary is useful because controllability and observability of a state model can often be determined more easily than stabilizability and detectability. Note, however, that Corollary 5.1 only provides sufficient conditions for the existence of the SSKF as an unbiased estimator.

The results for the time-varying SMM with nonstationary noises (6.29–6.33) are considerably more complicated. Conditions for the Kalman filter to be an unbiased estimator for this case appear in Section 6.3.

5.10 Summary

This chapter presented the Kalman filter, which is motivated by our desire to employ the information in every observation rather than discarding past observations. Doing so directly creates a growing-memory filter that eventually exceeds all practical storage and computational facilities. This shortcoming suggests that we search for a recursive filter.

We assume a linear state model for generating the observations. By exploiting the orthogonality principle, we derive the optimum causal LMMSE estimator, the Kalman filter. The Kalman filter provides a recursive algorithm and generates both filtered and predicted estimates.

Besides being optimal, the Kalman filter enables the Kalman gains and error covariances to be computed off-line. Hence the filter performance can be evaluated quickly and inexpensively. Optimality does not imply stability, but a set of moderate sufficient conditions guarantee stability of the Kalman filter.

Remarks

The utility of the Kalman filter lies in its algorithmic form. The causal Wiener filter requires an understanding of spectral factorization and causal-part extraction. With the Kalman filter, we merely construct an SMM, add process and measurement noise, and start the algorithm.

This simplicity offers a mixed blessing. On the one hand, the Kalman filter may be applied with a minimal understanding of the underlying theory and still produce acceptable estimates. On the other hand, the Kalman filter has acquired the reputation as a panacea for estimation. In many cases less computationally-intensive estimators perform comparably. A great deal of effort may be expended on constructing the SMM when a Wiener filter suffices. Most important, an understanding of the theory behind the estimation problem is essential to the development of new solutions.

Problems

5.1. This problem shows how we can simulate the Kalman filter with MATLAB. The state model is

$$\boldsymbol{x}(n+1) = \begin{bmatrix} 0.15 & 0.35 & 0.35 \\ 0.15 & 0.25 & -0.15 \\ 0.20 & -0.10 & 0.30 \end{bmatrix} \boldsymbol{x}(n) + \begin{bmatrix} 0.4 & 0 \\ 0 & 0.5 \\ 0 & 0 \end{bmatrix} \boldsymbol{w}(n)$$

$$\boldsymbol{s}(n) = \boldsymbol{x}(n)$$

$$\boldsymbol{z}(n) = \boldsymbol{s}(n) + \boldsymbol{v}(n),$$

where $\boldsymbol{w}(n)$ is zero-mean white noise with covariance matrix $Q = 15I_2$, $\boldsymbol{v}(n)$ is zero-mean white noise with covariance $R = 4I_3$, and $\boldsymbol{w}(n)$ and $\boldsymbol{v}(n)$ are uncorrelated.

(a) Use **randn** to generate a 50-element sample realization of the signal $\boldsymbol{x}(n)$ (since $\boldsymbol{s}(n) = \boldsymbol{x}(n)$). Call the realization **x**; it should be a 3×50 matrix.

(b) Use **randn** again to create a sample realization **z** of the observations.

(c) Assume initial conditions $\hat{x}^-(0) = 0$ and $P^-(0) = 200I_3$. Following the algorithm given in Section 5.6, write a MATLAB routine (script or function) that implements the Kalman filter and generates $\hat{x}(n)$ for $0 \le n \le 49$. The routine should produce a 3×50 matrix called `xhat`.

(d) The actual state $x_1(n)$ is the first row of `x`, `x(1,n)`. The estimate $\hat{x}_1(n)$ is `xhat(1,n)`. Plot `x(1,n)` and `xhat(1,n)` versus the time index n, $0 \le n \le 49$. Does the estimate converge to the true state?

(e) Repeat (d) for the second and third states. What is the final error covariance matrix $P(49)$?

(f) Repeat (d) and (e) with $P^-(0) = 10I_3$ and $1500I_3$. How quickly do the estimates converge?

5.2. One practical consideration is the computer generation of the correlated process noise vector $\boldsymbol{w}(n)$ with covariance matrix Q. Suppose that there exists a matrix Λ such that $\boldsymbol{w}(n) = \Lambda \boldsymbol{d}(n)$, where $\boldsymbol{d}(n)$ is a white noise vector process with $\mathrm{E}\left[\boldsymbol{d}(n)\boldsymbol{d}^{\mathrm{T}}(n)\right] = I$. Then

$$Q = \mathrm{E}\left[\boldsymbol{w}(n)\boldsymbol{w}^{\mathrm{T}}(n)\right] = \Lambda \mathrm{E}\left[\boldsymbol{d}(n)\boldsymbol{d}^{\mathrm{T}}(n)\right]\Lambda^{\mathrm{T}} = \Lambda\Lambda^{\mathrm{T}}.$$

Typical random number generators produce uncorrelated random numbers like $\boldsymbol{d}(n)$. Hence, if Λ can be found, $\boldsymbol{w}(n)$ can be generated easily by $\boldsymbol{w}(n) = \Lambda \boldsymbol{d}(n)$.

Since Q is a covariance matrix, it is positive definite. Therefore there exists a matrix $\sqrt{Q}$ such that $Q = \sqrt{Q}\left(\sqrt{Q}\right)^{\mathrm{T}}$, so Λ exists and is equal to $\sqrt{Q}$. Note that $\sqrt{Q}$ is the square root of the matrix Q, which is *not* equivalent to the square root of each element of Q. In MATLAB, the square root of a matrix is provided by the `sqrtm` function.

(a) Using MATLAB, generate 500 samples of a 3-vector white Gaussian noise process `d`. Compute the sample correlation matrix `Isam` for `d` and compare it with the 3×3 identity matrix. `Isam` can be computed via

$$I_{\mathrm{sam}}(i,j) = \frac{1}{500}\sum_{n=1}^{500} d(i,n)d(j,n).$$

(b) Suppose the desired correlation matrix Q is

$$Q = \begin{bmatrix} 50 & 10 & 6 \\ 10 & 22 & 5 \\ 6 & 5 & 40 \end{bmatrix}.$$

Use `sqrtm` to find a suitable matrix $\sqrt{Q}$.

(c) Generate correlated noise `w` using the relation $w(n) = \sqrt{Q}d(n)$. Then compute the sample correlation matrix `Qsam` for `w`. Compare `Qsam` with Q.

5.3. Consider the measurements $z(n) = s(n)+v(n)$ where $v(n)$ is zero-mean white noise with variance 1 and $s(n)$ is a zero-mean WSS random signal with the autocorrelation

$$R_s(k) = 0.5\delta(k+1) + \delta(k) + 0.5\delta(k-1),$$

where $\delta(k)$ is the unit pulse. The signal $s(n)$ and the noise $v(n)$ are independent. Give the equations for the linear time-varying recursive filter that generates the LMMSE estimate $\hat{s}_{\text{LMMSE}}(n)$ of $s(n)$ at time n based on the measurements $z(i) = z(i)$ for $i = 1, 2, \ldots, n$. All coefficients in your equations must be evaluated.

5.4. Measurements $z(n)$ of the signal process $s(n) = a\cos(\pi n)$ are given by $z(n) = s(n) + v(n)$ where a is a RV, $v(n)$ is zero-mean white noise with variance 2, and $v(n)$ is independent of a. Give the equations for the linear time-varying recursive filter that generates the LMMSE estimate $\hat{s}_{\text{LMMSE}}(n)$ of $s(n)$ at time n based on the measurements $z(i) = z(i)$ for $i = 1, 2, \ldots, n$. All coefficients in your equations must be evaluated.

5.5. A signal process $s(n)$ is given by $s(n) = w(n-1)-w(n-2)$ where $w(n)$ is zero-mean white noise with variance 2. Measurements $z(n)$ of the signal process $s(n)$ are given by $z(n) = s(n) + v(n)$, where $v(n)$ is zero-mean white noise with variance 1, and $v(n)$ is independent of $w(n)$. Determine the equations for a Kalman filter that produces an estimate of $s(n)$ at time n based on the measurements $z(i) = z(i)$ for $i = 1, 2, \ldots, n$.

5.6. A signal process $s(n)$ is given by $s(n) = (0.5)^n c_1 + (-0.5)^n c_2$ where c_1 and c_2 are RVs. Measurements $z(n)$ of $s(n)$ are given by $z(n) = s(n) + v(n)$, where $v(n)$ is zero-mean white noise with variance 5, and $v(n)$ is independent of $s(n)$.

(a) Set up a linear time-invariant state model for estimating $s(n)$. Express your answer by giving the definition of the state $x(n)$ and the coefficient matrices of the state model and the covariances of the noise terms.

(b) Using your result in Part (a), determine the equations for the Kalman filter that produces an estimate of $s(n)$ at time n based on the measurements $z(i) = z(i)$ for $i = 1, 2, \ldots, n$.

5.7. Repeat Problem 5.6 for the signal given by $s(n) = [a+(-1)^n]b$, where a and b are random variables with $\mathrm{E}\left[b^2\right] = 1$, $\mathrm{E}\left[a(b^2)\right] = 1$, and $\mathrm{E}\left[a^2 b^2\right] = 2$. Measurements $z(n)$ of $s(n)$ are given by $z(n) = s(n) + v(n)$, where $v(n)$ is zero-mean white noise with variance 2 and $v(n)$ is independent of a and b.

5.8. Repeat Problem 5.6 for the continuous-time random signal given by $s(t) = c_0 + c_1 t + c_2 t^2$, where c_0, c_1, and c_2 are RVs with $\mathrm{E}\left[c_0^2\right] = 1$, $\mathrm{E}\left[c_1^2\right] = 2$, $\mathrm{E}\left[c_2^2\right] = 3$, and $\mathrm{E}[c_i c_j] = 1$ for all $i \neq j$. Noisy measurements $z(n) = s(n) + v(n)$ are taken of the signal $s(t)$ at the sample times $t = n$, $i = 1, 2, \ldots,$ where $v(n)$ is zero-mean white noise with variance 1 and $v(n)$ is independent of c_0, c_1, and c_2.

5.9. A signal process $s(n)$ is given by

$$s(n) = s(n-1) + 3s(n-2) + w(n) - 2w(n-1) + w(n-2),$$

where $w(n)$ is zero-mean white noise with variance 1 and $s(j)$ is independent of $w(n)$ for all $n > j$. Noisy measurements $z(n)$ are given by

$$z(n) = s(n-1) + w(n) + v(n),$$

where $v(n)$ is zero-mean white noise with variance 4, $s(n)$ and $v(n)$ are independent, and $v(n)$ and $w(n)$ are independent.

(a) Set up a linear time-invariant state model having the smallest possible dimension for estimating $s(n)$. Express your answer by giving the definition of the state $x(n)$ and the coefficient matrices of the state model and the covariances of the noise terms.

(b) Using your result in Part (a), determine the equations for a Kalman filter that produces an estimate of $s(n)$ at time n based on the measurements $z(i) = z(i)$ for $i = 1, 2, \ldots, n$.

5.10. Repeat Problem 5.9 with the signal $s(n)$ given by $s(n) = s(n-1) + w(n)$ and with the measurements $z(n) = s(n) + v(n)$ where

$$w(n+1) = 0.5w(n) + \mu(n)$$
$$v(n+1) = v(n) + \alpha(n),$$

and $\mu(n)$ and $\alpha(n)$ are independent zero-mean unit-variance white noise signals.

5.11. Consider the measurements $z(n) = s(n) + v(n)$ where $v(n)$ is zero-mean white noise with variance 5 and $s(n)$ is given by

$$x(n+1) = \Phi x(n) + \Gamma w(n)$$
$$s(n) = c x(n),$$

where

$$\Phi = \begin{bmatrix} 0 & 1 \\ 1 & 0 \end{bmatrix}, \quad \Gamma = \begin{bmatrix} 0 \\ 1 \end{bmatrix}, \quad c = \begin{bmatrix} 0 & 1 \end{bmatrix}.$$

Here $w(n)$ is zero-mean white noise with variance 1, the noise signals $w(n)$ and $v(n)$ are independent, and $w(n)$ and $v(n)$ are independent of $x(0)$. Determine the transfer function of the linear time-invariant MMSE filter for generating the estimate $\hat{s}^{-}(n)$ of $s(n)$ based on the measurements $z(i) = z(i)$ for $i = 1, 2, \ldots, n-1$.

5.12. A random signal $s(n)$ has the two-dimensional state model $x(n+1) = \Phi x(n) + w(n)$, $s(n) = c x(n)$ where

$$\Phi = \begin{bmatrix} 1 & 2 \\ 1 & 1 \end{bmatrix}, \quad c = \begin{bmatrix} 1 & 1 \end{bmatrix}.$$

Here $w(n)$ is zero-mean white noise, and $x(0)$ is independent of $w(n)$ for all $n \geq 1$. Noisy measurements $z(n)$ of $s(n)$ are given by $z(n) = s(n) + v(n)$, where $v(n)$ is zero-mean white noise, $v(n)$ is independent of $x(0)$ for all n, and $v(n)$ is independent of $w(n)$. The gain K_∞ of the steady-state Kalman filter for generating the estimate $\hat{x}^-(n)$ of the state $x(n)$ is $K_\infty = \begin{bmatrix} 2.5 & -1 \end{bmatrix}^{\mathrm{T}}$.

(a) Determine the input/output difference equation of the linear time- invariant MMSE filter for generating an estimate $\hat{s}^-(n)$ of $s(n)$ based on the measurements $z(i) = z(i)$ for $i = 1, 2, \ldots, n-1$.

(b) (b) Does $\hat{s}^-(n) \to s(n)$ as $n \to \infty$? Justify your answer.

5.13. A random signal $s(n)$ has the two-dimensional state model $x(n+1) = x(n) + w(n)$, $s(n) = cx(n)$, where $c = \begin{bmatrix} 1 & 1 \end{bmatrix}^{\mathrm{T}}$, $w(n)$ is zero-mean white noise with covariance Q, and $x(0)$ is independent of $w(n)$ for all $n \geq 1$. Noisy measurements $z(n)$ of $s(n)$ are given by $z(n) = s(n) + v(n)$ where $v(n)$ is zero-mean white noise with variance 1, $v(n)$ is independent of $x(0)$ for all n, and $v(n)$ is independent of $w(n)$. It is known that the solution P_∞^- of the algebraic Riccati equation is equal to $I_2 = 2 \times 2$ identity matrix.

(a) Determine Q.

(b) Determine the transfer function of the linear time-invariant MMSE filter for generating the estimate $\hat{s}^-(n)$ of $s(n)$ based on the measurements $z(i) = z(i)$ for $i = 1, 2, \ldots, n-1$.

Chapter 6

Further Development of the Kalman Filter

The previous chapter took us from the idea of recursive estimation to the Kalman filter. This chapter contains several extensions of the Kalman filter. It begins with a discussion of the innovations, which are the estimation error for predicting the measurements. The innovations can be viewed as the "new information" about $\boldsymbol{x}(n)$ that is conveyed by $\boldsymbol{z}(n)$ and have a number of interesting and useful properties. An alternate derivation, based on the innovations, of the Kalman filter follows.

This chapter then shows how the Kalman filter can be applied to a time-varying SMM. Next, we discuss the effects of inaccuracies in the SMM; these inaccuracies can lead to a condition known as divergence, in which the Kalman filter fails to reflect the true state of the system being estimated. The chapter concludes with extensions of the Kalman filter to multistep prediction and smoothing.

6.1 The Innovations

We have seen that the Kalman filter recursion operates by combining the *a priori* estimate $\hat{\boldsymbol{x}}^{-}(n)$ with the new measurement $\boldsymbol{z}(n)$ to form $\hat{\boldsymbol{x}}(n)$. In a sense, the measurement $\boldsymbol{z}(n)$ carries information about $\boldsymbol{x}(n)$ that was not already available from the set of measurements $\boldsymbol{Z}^{-} = \{\boldsymbol{z}(1), \ldots, \boldsymbol{z}(n-1)\}$. The new information in $\boldsymbol{z}(n)$ is called the *innovation*, and the new information in each measurement forms a random signal called the *innovations*. In this section we investigate the innovations, which have several important properties and provide additional insight into estimation. For example, Section 6.2 shows how the innovations can be used to derive the Kalman filter.

Definition and Properties

The innovations are defined by

$$\varepsilon(n) = z(n) - C\hat{x}^{-}(n) \tag{6.1}$$

$$= z(n) - \hat{z}^{-}(n), \tag{6.2}$$

where $\hat{z}^{-}(n)$ is the LMMSE estimate of $z(n)$ given Z^{-} and is given by

$$\hat{z}^{-}(n) = C\hat{x}^{-}(n). \tag{6.3}$$

To see that (6.3) holds, recall that $z(n) = Cx(n) + v(n)$, and $x(n)$ and $v(n)$ are independent, so $\hat{z}^{-}(n) = C\hat{x}^{-}(n) + \hat{v}^{-}(n)$, where $\hat{v}^{-}(n)$ is the LMMSE estimate of $v(n)$ given Z^{-}. Since $v(n)$ is independent of the measurements in Z^{-}, $\hat{v}^{-}(n) = \mathrm{E}[v(n)] = 0$.

The innovations $\varepsilon(n)$ have a number of important properties. We list them here and provide derivations later in this section:

New Information The innovation $\varepsilon(n)$ contains the new information in measurement $z(n)$ that is not linearly predictable from $Z^{-} = \{z(1), \ldots, z(n-1)\}$. This property results directly from (6.2), since

$$z(n) = \hat{z}^{-}(n) + \varepsilon(n).$$

Orthogonality The innovations $\varepsilon(n)$ are orthogonal to $z(i)$:

$$\mathrm{E}\left[\varepsilon(n) z^{\mathrm{T}}(i)\right] = 0, \quad i = 1, 2, \ldots, n-1. \tag{6.4}$$

Uncorrelatedness/White Noise The innovations are uncorrelated with each other and therefore are white noise, i.e,

$$\mathrm{E}\left[\varepsilon(i) \varepsilon^{\mathrm{T}}(j)\right] = 0, \quad i \neq j. \tag{6.5}$$

Equivalent Information For any $n > 0$, the measurements $z(1), \ldots, z(n)$ can be obtained from a linear combination of the innovations $\varepsilon(1), \ldots, \varepsilon(n)$.

LMMSE Estimation As a consequence of the previous property, $\hat{x}(n)$, the LMMSE estimate of $x(n)$ given $z(1), \ldots, z(n)$, can also be computed from $\varepsilon(1), \ldots, \varepsilon(n)$, via

$$\hat{x}(n) = \sum_{j=1}^{n} \mathrm{E}\left[x(n)\varepsilon^{\mathrm{T}}(j)\right] \left(\mathrm{E}\left[\varepsilon(j)\varepsilon^{\mathrm{T}}(j)\right]\right)^{-1} \varepsilon(j). \tag{6.6}$$

Recursive LMMSE Estimation Given $\hat{\boldsymbol{x}}^-(n)$ and the innovation $\boldsymbol{\varepsilon}(n)$, $\hat{\boldsymbol{x}}(n)$ is given by

$$\hat{\boldsymbol{x}}(n) = \hat{\boldsymbol{x}}^-(n) + \mathrm{E}\left[\boldsymbol{x}(n)\boldsymbol{\varepsilon}^{\mathrm{T}}(n)\right]\left(\mathrm{E}\left[\boldsymbol{\varepsilon}(n)\boldsymbol{\varepsilon}^{\mathrm{T}}(n)\right]\right)^{-1}\boldsymbol{\varepsilon}(n). \tag{6.7}$$

The properties of equivalent information and new information also make sense on an intuitive level. The observations $\{\boldsymbol{z}(0), \dots, \boldsymbol{z}(n)\}$ contain some redundancy, which manifests itself as statistical correlation. In the course of obtaining the innovations $\{\boldsymbol{\varepsilon}(0), \dots, \boldsymbol{\varepsilon}(n)\}$, the redundant information is eliminated; only the new information is preserved. Therefore, the overall information content is the same.

Orthogonality

We first show that the innovations $\boldsymbol{\varepsilon}(n)$ are orthogonal to the measurements $\boldsymbol{z}(1), \dots, \boldsymbol{z}(n-1)$. To prove that (6.4) holds, we must show that $\hat{\boldsymbol{z}}^-(n)$ satisfies the orthogonality condition (3.52). Now, for $i = 1, 2, \dots, n-1$,

$$\begin{aligned}
\mathrm{E}\left[\boldsymbol{\varepsilon}(n)\boldsymbol{z}^{\mathrm{T}}(i)\right] &= \mathrm{E}\left[\left(\boldsymbol{z}(n) - \hat{\boldsymbol{z}}^-(n)\right)\boldsymbol{z}^{\mathrm{T}}(i)\right] \\
&= \mathrm{E}\left[\left(C\boldsymbol{x}(n) - \boldsymbol{v}(n) - C\hat{\boldsymbol{x}}^-(n)\right)\boldsymbol{z}^{\mathrm{T}}(i)\right] \\
&= C\mathrm{E}\left[\left(\boldsymbol{x}(n) - \hat{\boldsymbol{x}}^-(n)\right)\boldsymbol{z}^{\mathrm{T}}(i)\right] - \mathrm{E}\left[\boldsymbol{v}(n)\boldsymbol{z}^{\mathrm{T}}(i)\right].
\end{aligned} \tag{6.8}$$

Now $(\boldsymbol{x}(n) - \hat{\boldsymbol{x}}^-(n))$ is orthogonal to $\boldsymbol{z}(i)$, and $\boldsymbol{v}(n)$ is independent of $\boldsymbol{z}(i)$. Hence, both terms on the right-hand side of (6.8) are zero matrices and (6.4) follows.

A Direct Form for $\hat{\boldsymbol{z}}^-(n)$

We next present a more rigorous approach, which will prove useful later in this section. Let us estimate $\boldsymbol{z}(n)$ directly from $\boldsymbol{Z}^-$, so we have

$$\hat{\boldsymbol{z}}_1^-(n) = \sum_{i=1}^{n-1} A_n(i)\boldsymbol{z}(i), \tag{6.9}$$

where each $A_n(i)$ is a matrix to be determined. Define

$$P_z^-(n) = \mathrm{E}\left[\left(\boldsymbol{z}(n) - \hat{\boldsymbol{z}}_1^-(n)\right)\left(\boldsymbol{z}(n) - \hat{\boldsymbol{z}}_1^-(n)\right)^{\mathrm{T}}\right]. \tag{6.10}$$

Then the MSE associated with $\hat{\boldsymbol{z}}_1^-(n)$ can be written as $\mathrm{MSE}(n) = \mathrm{tr}\left(P_z^-(n)\right)$, where $\mathrm{tr}\left(\cdot\right)$ is defined as in (3.45).

To find the $A_n(i)$, we substitute (6.9) into (6.10) and get

$$P_z^-(n) = \mathrm{E}\left[\boldsymbol{z}(n)\boldsymbol{z}^{\mathrm{T}}(n)\right] - \sum_{i=1}^{n-1} A_n(i)\mathrm{E}\left[\boldsymbol{z}(i)\boldsymbol{z}^{\mathrm{T}}(n)\right]$$
$$- \sum_{j=1}^{n-1} \mathrm{E}\left[\boldsymbol{z}(n)\boldsymbol{z}^{\mathrm{T}}(j)\right] A_n^{\mathrm{T}}(n) + \sum_{i=1}^{n-1}\sum_{j=1}^{n-1} A_n(i)\mathrm{E}\left[\boldsymbol{z}(i)\boldsymbol{z}^{\mathrm{T}}(j)\right] A_n^{\mathrm{T}}(j).$$

Next we set $\partial \mathrm{tr}\left(P_z^-(n)\right)/\partial A_n(m) = \mathbf{0}$. We apply the matrix calculus results (3.48) and (3.49) and obtain

$$-2\mathrm{E}\left[\boldsymbol{z}(n)\boldsymbol{z}^{\mathrm{T}}(m)\right] + 2\sum_{i=1}^{n-1} A_n(i)\mathrm{E}\left[\boldsymbol{z}(i)\boldsymbol{z}^{\mathrm{T}}(m)\right] = \mathbf{0}, \quad \text{for } m = 1, 2, \ldots, n-1. \tag{6.11}$$

In fact, (6.11) is just the Wiener-Hopf equation (4.8) for vector processes (and we are estimating $\boldsymbol{z}(n)$ rather than $\boldsymbol{s}(n)$).

We can write the system of $n-1$ equations represented by (6.11) in block form as

$$\begin{bmatrix} \mathrm{E}\left[\boldsymbol{z}(1)\boldsymbol{z}^{\mathrm{T}}(1)\right] & \mathrm{E}\left[\boldsymbol{z}(1)\boldsymbol{z}^{\mathrm{T}}(2)\right] & \cdots & \boldsymbol{z}(1)\boldsymbol{z}^{\mathrm{T}}(n-1) \\ \mathrm{E}\left[\boldsymbol{z}(2)\boldsymbol{z}^{\mathrm{T}}(1)\right] & \mathrm{E}\left[\boldsymbol{z}(2)\boldsymbol{z}^{\mathrm{T}}(2)\right] & \cdots & \boldsymbol{z}(2)\boldsymbol{z}^{\mathrm{T}}(n-1) \\ \vdots & \vdots & \ddots & \vdots \\ \mathrm{E}\left[\boldsymbol{z}(n-1)\boldsymbol{z}^{\mathrm{T}}(1)\right] & \mathrm{E}\left[\boldsymbol{z}(n-1)\boldsymbol{z}^{\mathrm{T}}(2)\right] & \cdots & \mathrm{E}\left[\boldsymbol{z}(n-1)\boldsymbol{z}^{\mathrm{T}}(n-1)\right] \end{bmatrix}$$
$$\times \begin{bmatrix} A_n^{\mathrm{T}}(1) \\ A_n^{\mathrm{T}}(2) \\ \vdots \\ A_n^{\mathrm{T}}(n-1) \end{bmatrix} = \begin{bmatrix} \mathrm{E}\left[\boldsymbol{z}(1)\boldsymbol{z}^{\mathrm{T}}(n)\right] \\ \mathrm{E}\left[\boldsymbol{z}(2)\boldsymbol{z}^{\mathrm{T}}(n)\right] \\ \vdots \\ \mathrm{E}\left[\boldsymbol{z}(n-1)\boldsymbol{z}^{\mathrm{T}}(n)\right] \end{bmatrix} \tag{6.12}$$

The block matrix on the left-hand side (6.12) is always invertible, so we can solve this equation to find the matrices $A_n(1), \ldots, A_n(n-1)$. Then $\hat{\boldsymbol{z}}_1^-(n)$ is given by (6.9).

It follows that $\hat{\boldsymbol{z}}_1^-(n)$ satisfies the orthogonality principle (3.52), i.e.,

$$\mathrm{E}\left[\left(\boldsymbol{z}(n) - \hat{\boldsymbol{z}}_1^-(n)\right)\boldsymbol{z}^{\mathrm{T}}(m)\right]$$
$$= \mathrm{E}\left[\left(\boldsymbol{z}(n) - \sum_{i=1}^{n-1} A_n(i)\boldsymbol{z}(i)\right)\boldsymbol{z}^{\mathrm{T}}(m)\right] = \mathbf{0}, \quad m = 1, 2, \ldots, n-1. \tag{6.13}$$

From (6.8), $\hat{\boldsymbol{z}}^-(n) = C\hat{\boldsymbol{x}}^-(n)$ satisfies the same orthogonality condition as $\hat{\boldsymbol{z}}_1^-(n)$. Since the LMMSE estimate is unique, it follows that $\hat{\boldsymbol{z}}_1^-(n) = \hat{\boldsymbol{z}}^-(n) = C\hat{\boldsymbol{x}}^-(n)$, which shows that (6.1) and (6.2) are indeed equivalent.

Uncorrelatedness/White Noise

We now show that the innovations $\boldsymbol{\varepsilon}(n)$ form a white-noise process, a result which will be used shortly. Let n and i be integers such that $n > i \geq 1$. Then

$$\begin{aligned}
\mathrm{E}\left[\boldsymbol{\varepsilon}(i)\boldsymbol{\varepsilon}^{\mathrm{T}}(n)\right] &= \mathrm{E}\left[(\boldsymbol{z}(i) - \hat{\boldsymbol{z}}^{-}(i))\boldsymbol{\varepsilon}^{\mathrm{T}}(n)\right] \\
&= \mathrm{E}\left[\boldsymbol{z}(i)\boldsymbol{\varepsilon}^{\mathrm{T}}(n)\right] - \mathrm{E}\left[\hat{\boldsymbol{z}}^{-}(i)\boldsymbol{\varepsilon}^{\mathrm{T}}(n)\right] \\
&= \mathrm{E}\left[\boldsymbol{z}(i)\boldsymbol{\varepsilon}^{\mathrm{T}}(n)\right] \\
&\quad - \sum_{j=1}^{i-1} A_i(j)\mathrm{E}\left[\boldsymbol{z}(j)\boldsymbol{\varepsilon}^{\mathrm{T}}(n)\right].
\end{aligned} \tag{6.14}$$

Because $1 \leq j < i \leq n-1$, by (6.4), all terms on the right-hand side of (6.14) are zero matrices. Then the desired result (6.5) follows.

Intuitively, the white noise property is sensible. A statistical correlation between two RVs means that they contain some information about one another. If $\boldsymbol{\varepsilon}(i)$ and $\boldsymbol{\varepsilon}(j)$, $i \neq j$, are correlated, then $\boldsymbol{\varepsilon}(i)$ contains some information about $\boldsymbol{\varepsilon}(j)$. Hence, some available information about $\boldsymbol{z}(j)$ is not being incorporated into the estimate $\hat{\boldsymbol{z}}^{-}(j)$, so $\hat{\boldsymbol{z}}^{-}(j)$ cannot be the LMMSE estimate.

The innovations suggest another important idea in optimum linear filtering. It is the notion that when we have found the LMMSE estimate, all that is left—the error—is white noise, which cannot be predicted by linear estimation.

Equivalent Information

Here we explain the notion that the innovation $\boldsymbol{\varepsilon}(1), \ldots, \boldsymbol{\varepsilon}(n)$ contain the same information as the measurements $\boldsymbol{z}(1), \ldots, \boldsymbol{z}(n)$. From (6.9), the single innovation $\boldsymbol{\varepsilon}(n)$ can be obtained from the a linear combination of the measurements $\boldsymbol{z}(1), \ldots, \boldsymbol{z}(n)$. Moreover, the linear combination is causal, meaning that none of the measurements $\boldsymbol{z}(n+1), \boldsymbol{z}(n+2), \ldots$, are required. It is not surprising, then, to find that the reverse is true: the measurement $\boldsymbol{z}(n)$ can be obtained from a causal linear combination of $\boldsymbol{\varepsilon}(1), \ldots, \boldsymbol{\varepsilon}(n)$. This characteristic is the property of *equivalent information*.

Let us consider the first few measurements and innovations. In the absence of any information about $\boldsymbol{x}(1)$, we let $\hat{\boldsymbol{x}}^{-}(1) = \mathbf{0}$, so that $\boldsymbol{\varepsilon}(1) = \boldsymbol{z}(1)$.

For $\boldsymbol{\varepsilon}(2)$ we have

$$\boldsymbol{\varepsilon}(2) = \boldsymbol{z}(2) - \hat{\boldsymbol{z}}^{-}(1) = \boldsymbol{z}(2) - A_2(1)\boldsymbol{z}(1) = \boldsymbol{z}(2) - A_2(1)\boldsymbol{\varepsilon}(1).$$

Clearly, we can find $\boldsymbol{z}(2)$ via

$$z(2) = \boldsymbol{\varepsilon}(2) + A_2(1)\boldsymbol{\varepsilon}(1),$$

and this expression is causal because $\boldsymbol{z}(2)$ depends only on $\boldsymbol{\varepsilon}(n)$ for $n \leq 2$.

Similary, for $\boldsymbol{\varepsilon}(3)$ we find

$$\begin{aligned}\boldsymbol{\varepsilon}(3) &= \boldsymbol{z}(3) - A_3(1)\boldsymbol{z}(1) - A_3(2)\boldsymbol{z}(2) \\ &= \boldsymbol{z}(3) - A_3(1)\boldsymbol{\varepsilon}(1) - A_3(2)\left[\boldsymbol{\varepsilon}(2) + A_2(1)\boldsymbol{\varepsilon}(1)\right].\end{aligned}$$

Then $\boldsymbol{z}(3)$ can be computed from the causal linear equation

$$\boldsymbol{z}(3) = \boldsymbol{\varepsilon}(3) + A_3(1)\boldsymbol{\varepsilon}(2) + \left[A_3(1) + A_3(2)A_2(1)\right]\boldsymbol{\varepsilon}(1).$$

We can continue this process indefinitely for any $\boldsymbol{z}(n)$.

LMMSE Estimation Using the Innovations

One consequence of the equivalent-information property is a theorem that demonstrates that we can estimate $\boldsymbol{x}(n)$ from the innovations $\boldsymbol{\varepsilon}(1), \ldots, \boldsymbol{\varepsilon}(n)$, rather than from the measurements $\boldsymbol{z}(1), \ldots, \boldsymbol{z}(n)$.

Theorem 6.1 (Equivalence of Orthogonality) *If the random vector $\boldsymbol{y}$ is orthogonal to the innovations $\boldsymbol{\varepsilon}(1), \ldots, \boldsymbol{\varepsilon}(n)$ and $\boldsymbol{\varepsilon}(1) = \boldsymbol{z}(1)$, then $\boldsymbol{y}$ is also orthogonal to the measurements $\boldsymbol{z}(1), \ldots, \boldsymbol{z}(n)$. That is,*

$$E\left[\boldsymbol{y}\boldsymbol{\varepsilon}^T(i)\right] = \mathbf{0}, \quad i = 1, 2, \ldots, n, \tag{6.15}$$

implies that

$$E\left[\boldsymbol{y}\boldsymbol{z}^T(i)\right] = \mathbf{0}, \quad i = 1, 2, \ldots, n. \tag{6.16}$$

Proof. Assume (6.15) holds and $\boldsymbol{\varepsilon}(1) = \boldsymbol{z}(1)$. Then

$$\mathbf{0} = \mathrm{E}\left[\boldsymbol{y}\boldsymbol{z}^{\mathrm{T}}(1)\right] = \mathrm{E}\left[\boldsymbol{y}\boldsymbol{\varepsilon}^{\mathrm{T}}(1)\right].$$

Now let n be any integer greater than 1 and assume

$$\mathrm{E}\left[\boldsymbol{y}\boldsymbol{z}^{\mathrm{T}}(i)\right] = \mathbf{0}, \quad i = 1, 2, \ldots, n-1.$$

Then

$$\begin{aligned}\mathbf{0} = \mathrm{E}\left[\boldsymbol{y}\boldsymbol{\varepsilon}^{\mathrm{T}}(n)\right] &= \mathrm{E}\left[\boldsymbol{y}\boldsymbol{z}^{\mathrm{T}}(n)\right] - \sum_{i=1}^{n-1} \mathrm{E}\left[\boldsymbol{y}\boldsymbol{z}^{\mathrm{T}}(i)\right] A_n^{\mathrm{T}}(i) \\ &= \mathrm{E}\left[\boldsymbol{y}\boldsymbol{z}^{\mathrm{T}}(n)\right].\end{aligned}$$

Q.E.D. ■

Applying Theorem 6.1, we now estimate $\boldsymbol{x}(n)$ from the innovations $\boldsymbol{\varepsilon}(1)$, ..., $\boldsymbol{\varepsilon}(n)$. We write $\hat{\boldsymbol{x}}(n)$ as

$$\hat{\boldsymbol{x}}(n) = \sum_{j=1}^{n} B_n(j)\boldsymbol{\varepsilon}(j), \tag{6.17}$$

where we must find the matrices $B_n(j)$. The $B_n(j)$ must satisfy the orthogonality condition (3.52)

$$\mathrm{E}\left[\left(\boldsymbol{x}(n) - \sum_{j=1}^{n} B_n(j)\boldsymbol{\varepsilon}(j)\right) z^{\mathrm{T}}(i)\right] = \mathbf{0}, \quad i = 1, \ldots, n,$$

but by Theorem 6.1, this condition is equivalent to

$$\mathrm{E}\left[\left(\boldsymbol{x}(n) - \sum_{j=1}^{n} B_n(j)\boldsymbol{\varepsilon}(j)\right) \boldsymbol{\varepsilon}^{\mathrm{T}}(i)\right] = \mathbf{0}, \quad i = 1, \ldots, n. \tag{6.18}$$

Then (6.18) gives

$$\sum_{j=1}^{n} B_n(j)\mathrm{E}\left[\boldsymbol{\varepsilon}(j)\boldsymbol{\varepsilon}^{\mathrm{T}}(i)\right] = \mathrm{E}\left[\boldsymbol{x}(n)\boldsymbol{\varepsilon}^{\mathrm{T}}(i)\right], \quad i = 1, \ldots, n.$$

From (6.5), $EV\boldsymbol{\varepsilon}(j)\boldsymbol{\varepsilon}^{\mathrm{T}}(i) = \mathbf{0}$ for $i \neq j$, so we have

$$B_n(i)\mathrm{E}\left[\boldsymbol{\varepsilon}(i)\boldsymbol{\varepsilon}^{\mathrm{T}}(i)\right] = \mathrm{E}\left[\boldsymbol{x}(n)\boldsymbol{\varepsilon}^{\mathrm{T}}(i)\right] \quad i = 1, \ldots, n.$$

Therefore, the desired matrices are

$$B_n(j) = \mathrm{E}\left[\boldsymbol{x}(n)\boldsymbol{\varepsilon}^{\mathrm{T}}(j)\right] \left(\mathrm{E}\left[\boldsymbol{\varepsilon}(j)\boldsymbol{\varepsilon}^{\mathrm{T}}(j)\right]\right)^{-1}, \quad j = 1, \ldots, n. \tag{6.19}$$

Substituting (6.19) into (6.17), we find that $\hat{\boldsymbol{x}}(n)$ is given by (6.6):

$$\hat{\boldsymbol{x}}(n) = \sum_{j=1}^{n} \mathrm{E}\left[\boldsymbol{x}(n)\boldsymbol{\varepsilon}^{\mathrm{T}}(j)\right] \left(\mathrm{E}\left[\boldsymbol{\varepsilon}(j)\boldsymbol{\varepsilon}^{\mathrm{T}}(j)\right]\right)^{-1} \boldsymbol{\varepsilon}(j). \tag{6.20}$$

Recursive LMMSE Estimation Using the Innovations

The final result we present will enable us to derive the Kalman filter in a simple way in the next section. Suppose we have $\hat{\boldsymbol{x}}^{-}(n)$, the LMMSE estimate of $\boldsymbol{x}(n)$ given $\boldsymbol{Z}^{-}$. By a derivation analogous to that just presented above, we find that

$$\hat{\boldsymbol{x}}^{-}(n) = \sum_{j=1}^{n-1} \mathrm{E}\left[\boldsymbol{x}(n)\boldsymbol{\varepsilon}^{\mathrm{T}}(j)\right] \left(\mathrm{E}\left[\boldsymbol{\varepsilon}(j)\boldsymbol{\varepsilon}^{\mathrm{T}}(j)\right]\right)^{-1} \boldsymbol{\varepsilon}(j), \tag{6.21}$$

so (6.20) can be written in a recursive form, namely (6.7):

$$\hat{\boldsymbol{x}}(n) = \hat{\boldsymbol{x}}^{-}(n) + \mathrm{E}\left[\boldsymbol{x}(n)\boldsymbol{\varepsilon}^{\mathrm{T}}(n)\right]\left(\mathrm{E}\left[\boldsymbol{\varepsilon}(n)\boldsymbol{\varepsilon}^{\mathrm{T}}(n)\right]\right)^{-1}\boldsymbol{\varepsilon}(n). \tag{6.22}$$

Equations (6.20) and (6.22) are the fundamental relationships in all recursive linear estimation problems. Equation (6.22) says that given $\hat{\boldsymbol{x}}^{-}(n)$, $\hat{\boldsymbol{x}}(n)$ is $\hat{\boldsymbol{x}}^{-}(n)$ plus the weighted innovation $\boldsymbol{\varepsilon}(n)$.

We briefly interpret the meaning of the weighting factors in (6.22). If $\boldsymbol{x}(n)$ and $\boldsymbol{\varepsilon}(n)$ are highly correlated, $\boldsymbol{\varepsilon}(n)$ contains a large amount of information regarding $\boldsymbol{x}(n)$. The greater the correlation between $\boldsymbol{x}(n)$ and the $\boldsymbol{\varepsilon}(n)$, the larger $\mathrm{E}\left[\boldsymbol{x}(n)\boldsymbol{\varepsilon}^{\mathrm{T}}(n)\right]$ becomes. On the other hand, if the variance of $\boldsymbol{\varepsilon}(n)$ is large, then $\boldsymbol{\varepsilon}(n)$ may suffer from large errors. This property reduces our confidence in $\boldsymbol{\varepsilon}(n)$. The greater the variance of the error, the smaller $\left(\mathrm{E}\left[\boldsymbol{\varepsilon}(n)\boldsymbol{\varepsilon}^{\mathrm{T}}(n)\right]\right)^{-1}$ becomes. These two terms balance each other in the MMSE sense.

6.2 Derivation of the Kalman Filter from the Innovations

We are now prepared to derive the Kalman filter from the innovations $\boldsymbol{\varepsilon}(n)$. Most of the work has been done in Section 6.1. We derive only the updates for $\hat{\boldsymbol{x}}^{-}(n)$ and $\boldsymbol{x}(n)$, since the error covariance updates for $P^{-}(n)$ and $P(n)$ have been derived in Sections 5.5 and 5.5. Also, the covariance updates do not depend upon the measurements $\boldsymbol{z}(n)$, and hence do not depend upon the innovations $\boldsymbol{\varepsilon}(n)$.

The *a priori* Estimate

First, let us assume we have $\hat{\boldsymbol{x}}(n)$ and now seek $\hat{\boldsymbol{x}}^{-}(n+1)$. From (6.21),

$$\hat{\boldsymbol{x}}^{-}(n+1) = \sum_{j=1}^{n} \mathrm{E}\left[\boldsymbol{x}(n+1)\boldsymbol{\varepsilon}^{\mathrm{T}}(j)\right]\left(\mathrm{E}\left[\boldsymbol{\varepsilon}(j)\boldsymbol{\varepsilon}^{\mathrm{T}}(j)\right]\right)^{-1}\boldsymbol{\varepsilon}(j). \tag{6.23}$$

From (5.36), we have

$$\begin{aligned}\mathrm{E}\left[\boldsymbol{x}(n+1)\boldsymbol{\varepsilon}^{\mathrm{T}}(i)\right] &= \mathrm{E}\left[\left(\Phi\boldsymbol{x}(n) + \Gamma\boldsymbol{w}(n)\right)\boldsymbol{\varepsilon}^{\mathrm{T}}(i)\right] \\ &= \Phi\mathrm{E}\left[\boldsymbol{x}(n)\boldsymbol{\varepsilon}^{\mathrm{T}}(i)\right] + \Gamma\mathrm{E}\left[\boldsymbol{w}(n)\boldsymbol{\varepsilon}^{\mathrm{T}}(i)\right]\end{aligned}$$

Now

$$\boldsymbol{\varepsilon}(i) = C\boldsymbol{x}(i) + \boldsymbol{v}(i) - C(i)\hat{\boldsymbol{x}}^{-}(i),$$

and for $i = 1, 2, \ldots, n$, $\boldsymbol{w}(n)$ is independent of $\boldsymbol{x}(i)$ and $\hat{\boldsymbol{x}}^-(i)$; $\boldsymbol{w}(n)$ is independent of $\boldsymbol{v}(i)$ for all i. Thus,

$$\mathrm{E}\left[\boldsymbol{x}(n+1)\boldsymbol{\varepsilon}^{\mathrm{T}}(i)\right] = \Phi\mathrm{E}\left[\boldsymbol{x}(n)\boldsymbol{\varepsilon}^{\mathrm{T}}(i)\right], \quad i = 1, 2, \ldots, n.$$

Then (6.23) becomes

$$\hat{\boldsymbol{x}}^-(n+1) = \Phi\left[\sum_{j=1}^{n} \mathrm{E}\left[\boldsymbol{x}(n)\boldsymbol{\varepsilon}^{\mathrm{T}}(j)\right]\left(\mathrm{E}\left[\boldsymbol{\varepsilon}(j)\boldsymbol{\varepsilon}^{\mathrm{T}}(j)\right]\right)^{-1}\boldsymbol{\varepsilon}(j)\right]$$

so that

$$\boxed{\hat{\boldsymbol{x}}^-(n+1) = \Phi\hat{\boldsymbol{x}}(n),}$$

which is the same result as (5.83) from Section 5.5.

The *a posteriori* Estimate

Next consider the problem of finding $\hat{\boldsymbol{x}}(n)$ from $\hat{\boldsymbol{x}}^-(n)$ and the innovation $\boldsymbol{\varepsilon}(n)$. Given Equation (6.7), we must find $\mathrm{E}\left[\boldsymbol{\varepsilon}(n)\boldsymbol{\varepsilon}^{\mathrm{T}}(n)\right]$ and $\mathrm{E}\left[\boldsymbol{x}(n)\boldsymbol{\varepsilon}^{\mathrm{T}}(n)\right]$. From (6.1) and (5.38), we have

$$\begin{aligned}\boldsymbol{\varepsilon}(n) &= C\boldsymbol{x}(n) + \boldsymbol{v}(n) - C\hat{\boldsymbol{x}}^-(n) \\ &= C\tilde{\boldsymbol{x}}^-(n) + \boldsymbol{v}(n). \end{aligned} \tag{6.24}$$

Then $\mathrm{E}\left[\boldsymbol{\varepsilon}(n)\boldsymbol{\varepsilon}^{\mathrm{T}}(n)\right]$ is calculated to be

$$\begin{aligned}\mathrm{E}\left[\boldsymbol{\varepsilon}(n)\boldsymbol{\varepsilon}^{\mathrm{T}}(n)\right] &= \mathrm{E}\left[\left(C\tilde{\boldsymbol{x}}^-(n) + \boldsymbol{v}(n)\right)\left(C\tilde{\boldsymbol{x}}^-(n) + \boldsymbol{v}(n)\right)^{\mathrm{T}}\right] \\ &= C\mathrm{E}\left[\tilde{\boldsymbol{x}}^-(n)(\tilde{\boldsymbol{x}}^-(n))^{\mathrm{T}}\right]C^{\mathrm{T}} + \mathrm{E}\left[\boldsymbol{v}(n)\boldsymbol{v}^{\mathrm{T}}(n)\right]. \\ &= CP^-(n)C^{\mathrm{T}} + R. \end{aligned} \tag{6.25}$$

To find $\mathrm{E}\left[\boldsymbol{x}(n)\boldsymbol{\varepsilon}^{\mathrm{T}}(n)\right]$ we use (6.24) again:

$$\begin{aligned}\mathrm{E}\left[\boldsymbol{x}(n)\boldsymbol{\varepsilon}^{\mathrm{T}}(n)\right] &= \mathrm{E}\left[\boldsymbol{x}(n)\left(C\tilde{\boldsymbol{x}}^-(n) + \boldsymbol{v}(n)\right)^{\mathrm{T}}\right] \\ &= \mathrm{E}\left[\boldsymbol{x}(n)(\tilde{\boldsymbol{x}}^-(n))^{\mathrm{T}}\right]C^{\mathrm{T}} + \mathrm{E}\left[\boldsymbol{x}(n)\boldsymbol{v}^{\mathrm{T}}(n)\right] \\ &= \mathrm{E}\left[\left(\hat{\boldsymbol{x}}^-(n) + \tilde{\boldsymbol{x}}^-(n)\right)(\tilde{\boldsymbol{x}}^-(n))^{\mathrm{T}}\right]C^{\mathrm{T}} \\ &= \mathrm{E}\left[\hat{\boldsymbol{x}}^-(n)(\tilde{\boldsymbol{x}}^-(n))^{\mathrm{T}}\right] + P^-(n)C^{\mathrm{T}} \\ &= P^-(n)C^{\mathrm{T}}. \end{aligned} \tag{6.26}$$

Next we substitute (6.1), (6.25), and (6.26) into (6.7). The *a posteriori* estimate is given by

$$\begin{aligned}\hat{\boldsymbol{x}}(n) = \hat{\boldsymbol{x}}^-(n) &+ P^-(n)C^{\mathrm{T}}\left[CP^-(n)C^{\mathrm{T}} + R\right]^{-1} \\ &\times \left[\boldsymbol{z}(n) - C\hat{\boldsymbol{x}}^-(n)\right]. \end{aligned} \tag{6.27}$$

Define the *Kalman gain* by (5.80):

$$\boxed{K(n) = P^{-}(n)C^{\mathrm{T}}\left[CP^{-}(n)C^{\mathrm{T}} + R\right]^{-1},}$$

so (6.27) yields the *a posteriori* estimate given by (5.81):

$$\boxed{\hat{\boldsymbol{x}}(n) = \hat{\boldsymbol{x}}^{-}(n) + K(n)\left[\boldsymbol{z}(n) - C\hat{\boldsymbol{x}}^{-}(n)\right].}$$

Except for the covariance updates (derived in Sections 5.5 and 5.5), the derivation of the Kalman filter is complete. Refer back to Section 5.6 for the complete filter recursion.

Error Behavior

One interesting consequence of the innovations interpretation is that we can observe $\boldsymbol{\varepsilon}(n)$ to see whether the Kalman filter is performing well. We have seen in Section 6.1 that the innovations $\boldsymbol{\varepsilon}(n) = \boldsymbol{z}(n) - \hat{\boldsymbol{z}}^{-}(n)$ behave like zero-mean white noise.

Suppose we have implemented a Kalman filter. We choose $\tilde{\boldsymbol{x}}^{-}(0) = \mathrm{E}\left[\boldsymbol{x}(0)\right]$ to guarantee that $\hat{\boldsymbol{x}}^{-}(n)$ is unbiased. Then at time n we may compute

$$\begin{aligned} \boldsymbol{\varepsilon}(n) &= \boldsymbol{z}(n) - C\hat{\boldsymbol{z}}^{-}(n) \\ &= \boldsymbol{z}(n) - C\hat{\boldsymbol{x}}^{-}(n). \end{aligned} \tag{6.28}$$

If the filter is operating correctly, the sequence generated by (6.28) should be white. We can use this property to check the performance of the filter. If the sequence is not white, then the filter design may need revision. Perhaps the SMM is not accurate enough, or the processes $\boldsymbol{w}(n)$ and $\boldsymbol{v}(n)$ may be colored or correlated.[1]

6.3 Time-varying State Model and Nonstationary Noises

Section 5.4 presented the SMM using a time-invariant state model (5.36–5.38) with stationary process noise (5.39) and stationary measurement noise (5.40). For many applications, these assumptions are adequate. The derivation of the Kalman filter and many of the examples presented in the text assume the time-invariant, stationary SMM. However, in some cases, the time-invariant state model with stationary noises is too restrictive. It is possible to generalize the SMM to incorporate a time-varying state model, nonstationary noises, or both.

[1] Chapter 7 covers the cases of colored or correlated processes.

Time-varying SMM with Nonstationary Noises

To obtain a time-varying state model, we let the matrices Φ, Γ, and C vary over time. Then we have

$$\boldsymbol{x}(n+1) = \Phi(n+1, n)\boldsymbol{x}(n) + \Gamma(n)\boldsymbol{w}(n), \quad \boldsymbol{x}(0) = \boldsymbol{x}_0, \tag{6.29}$$

$$\boldsymbol{s}(n) = C(n)\boldsymbol{x}(n), \tag{6.30}$$

$$\boldsymbol{z}(n) = \boldsymbol{s}(n) + \boldsymbol{v}(n), \tag{6.31}$$

where it is assumed that $\Phi(n+1, n)$, $\Gamma(n)$, and $C(n)$ are deterministic matrices that are known for all n.

The noises $\boldsymbol{w}(n)$ and $\boldsymbol{v}(n)$ can be made nonstationary by allowing Q and R to vary with time. Thus,

$$\mathrm{E}\left[\boldsymbol{w}(i)\boldsymbol{w}^{\mathrm{T}}(j)\right] = Q(i)\delta(i-j), \tag{6.32}$$

and

$$\mathrm{E}\left[\boldsymbol{v}(i)\boldsymbol{v}^{\mathrm{T}}(j)\right] = R(i)\delta(i-j). \tag{6.33}$$

The covariance matrices $Q(n)$ and $R(n)$ are assumed known for all n.

The derivation of the Kalman filter in Section 5.5 (and in Section 6.2 can be extended directly to the time-varying SMM of (6.29), (6.30), and (6.31) with the nonstationary noises of (6.32) and (6.33). It is only necessary to make Φ, Γ, C, Q, and R into functions of the time index n, but the derivation remains exactly the same.

Kalman Filter Recursion

The Kalman filter readily handles a time-varying SMM or nonstationary noises. The filter is initialized as in (5.75) and (5.76). The recursion incorporates the time-varying state model matrices $\Phi(n+1, n)$, $\Gamma(n)$, and $C(n)$ and time-varying noise covariance matrices $Q(n)$ and $R(n)$. Then we have the following recursion starting at time $n = 0$:

- *Measurement update.* Acquire $z(n)$ and compute *a posteriori* quantities:

$$K(n) = P^{-}(n)C^{\mathrm{T}}(n)\left[C(n)P^{-}(n)C^{\mathrm{T}}(n) + R(n)\right]^{-1} \tag{6.34}$$

$$\hat{x}(n) = \hat{x}^{-}(n) + K(n)\left[z(n) - C(n)\hat{x}^{-}(n)\right] \tag{6.35}$$

$$P(n) = P^{-}(n) - K(n)C(n)P^{-}(n) \tag{6.36}$$

- *Time update.* Compute *a priori* quantities for time $n+1$:

$$\hat{x}^{-}(n+1) = \Phi(n+1, n)\hat{x}(n) \tag{6.37}$$

$$P^{-}(n+1) = \Phi(n+1, n)P(n)\Phi^{\mathrm{T}}(n+1, n) + \Gamma(n)Q(n)\Gamma^{\mathrm{T}}(n) \tag{6.38}$$

- *Time increment.* Increment n and repeat.

Properties

This section considers the following questions about the Kalman filter for the case of a time-varying SMM with nonstationary noises (6.29–6.31). They are analogous to the questions in Section 5.9 for the SSKF.

Asymptotically Unbiased *When is the Kalman filter asymptotically unbiased?* In other words, when does

$$\lim_{n\to\infty} \mathrm{E}\left[\hat{\boldsymbol{x}}(n)\right] = \mathrm{E}\left[\boldsymbol{x}(n)\right]?$$

Bounded Error *When is $P(n)$ bounded for all initial guesses $P^-(0)$?*

Independence of $P^-(n)$ *When is $P^-(n)$ independent of the initial guess for $P^-(0)$?*

Let us assume

$$|\Phi(n+1,n)| \neq 0, \quad \forall\, n, \tag{6.39}$$

which implies that $\Phi^{-1}(n+1,n)$ exists for all n. If the SMM is a properly discretized model of a continuous-time physical system, this property holds.

We also assume that certain matrices are bounded below for all n, and that there exist constant positive real numbers a_Φ, a_Q, and a_R such that for all n,

$$\Phi^{\mathrm{T}}(n+1,n)\Phi(n+1,n) \geq a_\Phi I > 0, \tag{6.40}$$

$$Q(n) \geq a_Q I > 0, \tag{6.41}$$

and

$$R(n) \geq a_R I > 0. \tag{6.42}$$

Finally, we assume that all matrices in the SMM are bounded above in norm.

The theorems we present employ the concepts of uniform complete observability (UCO) and uniform complete controllability (UCC). Appendix C contains a discussion of UCO, UCC, and complete proofs of the theorems described in this section. For now, we offer a brief description of UCO and UCC.

A system is observable at n_0 if the initial state $\boldsymbol{x}(n_0)$ can be estimated from a finite number of observations $\boldsymbol{z}(n_0), \ldots, \boldsymbol{z}(n_0 + N(\boldsymbol{x}(n_0))$. If we can determine $\hat{\boldsymbol{x}}(n_0)$ for every possible initial state $\boldsymbol{x}(n_0)$, the system is completely observable at n_0.

We define the *information matrix*[2] by

$$\mathcal{I}(n_0, n_0+N) \triangleq \sum_{i=n_0}^{n_0+N} \Phi^{\mathrm{T}}(i, n_0+N) C^{\mathrm{T}}(i) R^{-1}(i) C(i) \Phi(i, n_0+N).$$

It can be shown[3] that complete observability holds if and only if there is an integer $N > 0$ (independent of $\boldsymbol{x}(n_0)$) such that $\mathcal{I}(n_0, n_0+N)$ is positive-definite.

Suppose the information matrix grows within fixed bounds, i.e., there are real numbers $0 < \beta_1 < \beta_2 < \infty$ such that

$$\beta_1 I \leq \mathcal{I}(n-N, n) \leq \beta_2 I, \quad \forall\, n \geq N.$$

Then we say the system is *uniformly completely observable* (UCO).

We say a system is controllable at n_0 if an input sequence $\boldsymbol{w}(n)$ exists that will drive $\boldsymbol{x}(n)$ from $\boldsymbol{x}(n_0)$ to zero in a finite amount of time. Complete controllability at n_0 means that for all $\boldsymbol{x}(n_0)$, $\boldsymbol{x}(n)$ can be forced to zero after N time steps .

We define the *controllability Gramian* by

$$\mathcal{C}(n_0, n_0+N) \triangleq \sum_{i=n_0}^{n_0+N-1} \Phi(n_0+N, i+1) \Gamma(i) Q(i) \times \Gamma^{\mathrm{T}}(i) \Phi^{\mathrm{T}}(n_0+N, i+1).$$

Positive-definiteness of $\mathcal{C}(n_0, n_0+N)$ for a fixed integer $N > 0$ provides a necessary and sufficient condition for complete controllability.

When there are an integer $N > 0$ and real numbers $0 < \alpha_1 < \alpha_2 < \infty$ such that

$$\alpha_1 I \leq \mathcal{C}(n-N, n) \leq \alpha_2 I, \quad \forall\, n \geq N,$$

we have *uniform complete controllability* (UCC), the dual of UCO.

Now we may answer, at least partially, the three questions of interest. In answer to the first question, we have the following theorem:

Theorem 6.2 *Let the state model (6.29–6.33), (6.39–6.42) be UCO and UCC, and let* $P^{-}(0) \geq 0$. *Then,* for any $\hat{x}(0)$, *the Kalman filter is globally uniformly asymptotically stable (and hence asymptotically unbiased), i.e.,*

$$E[\hat{\boldsymbol{x}}(n)] \to E[\boldsymbol{x}(n)] \quad \text{as} \quad n \to \infty.$$

---

[2] The information matrix is *not* identical to the observability Gramian, usually defined (cf. [8, 9, 7]) as $\mathcal{O}(m, n) \triangleq \sum_{i=m}^{n} \Phi^{\mathrm{T}}(i, m) C^{\mathrm{T}}(i) R^{-1}(i) C(i) \Phi(i, m)$.

[3] Cf. Section C.1, [9], or [10].

An important consequence of this theorem is that, provided the conditions of UCO and UCC are satisfied, then for *any* initial guess $\hat{x}(0)$, the Kalman filter is asymptotically unbiased. This result is extremely powerful, because in practice we do not know $\mathrm{E}[\hat{\boldsymbol{x}}(0)]$ or $\mathrm{E}[\hat{\boldsymbol{x}}^-(0)]$ (see (5.75) and (5.87)).

The question of bounded error is answered by the following theorem:

Theorem 6.3 *Let the state model (6.29-6.33), (6.39-6.42) be UCO and UCC, and let $P^-(0) \geq 0$. Then $P(n)$ is uniformly bounded,*

$$\frac{\alpha_1^2}{\alpha_1 + N\alpha_2^2\beta_2} I \leq P(n) \leq \frac{\beta_1 + N\alpha_2\beta_2^2}{\beta_1^2} I, \quad \forall\, n \geq N.$$

Note that the MSE will never be zero since $P(n)$ is also bounded below.

The next theorem answers our third question, regarding the independence of $P^-(n)$ upon the initial matrix $P^-(0)$.

Theorem 6.4 *Let the state model (6.29-6.33), (6.39-6.42) be UCO and UCC. Let $P_1^-(n)$ and $P_2^-(n)$ correspond to solutions of (5.84) with initial conditions $P_1^-(0) \geq 0$ and $P_2^-(0) \geq 0$, respectively. Let $\Delta P^-(n) = P_1^-(n) - P_2^-(n)$. Then there exist real numbers $a, b > 0$ such that*

$$\lim_{n\to\infty} \|\Delta P^-(n)\| \leq \lim_{n\to\infty} a^2 e^{-2bn} \|P_1^-(0) - P_2^-(0)\| = 0.$$

Theorem 6.4 states that the initial covariance matrix $P^-(0)$ is gradually forgotten under proper conditions. In other words, for large n, $P^-(n)$ is independent of $P^-(0)$. It follows that $K(n)$ and $P(n)$ are also independent of $P^-(0)$, so the Kalman filter behaves independently of $P^-(0)$. A poor initial estimate of $P^-(0)$ will cause the Kalman filter to perform poorly at first, but the effect will diminish as n increases. This property is important because we rarely know $P^-(0)$ in practice. Assuming its hypotheses are satisfied, Theorem 6.4 tells us that we only need to choose $P^-(0) \geq 0$.

All of these theorems (Theorems 6.2, 6.3, and 6.4) provide only *sufficient* conditions to guarantee an asymptotically unbiased estimate, bounded covariance matrix $P(n)$, and independence of $P^-(n)$ from the initial value of $P^-(0)$. The theorems are weaker than Theorem 5.1 for the SSKF, which provided necessary and sufficient conditions.

Since the Kalman filter estimates the state vector $\boldsymbol{x}(n)$, it is not surprising to see the condition of UCO. However, the condition that the system be UCC is not at all expected. This requirement means that the states must be controllable by the process noise $\boldsymbol{w}(n)$; in other words, $\boldsymbol{w}(n)$ must excite every mode of the system. We actually require process noise to drive the system. Even if the actual system does not have any input, *we must supply process noise* to the SMM to make the Kalman filter asymptotically unbiased.

6.4 Modeling Errors

In the presentation of the Kalman filter, we have assumed that the SMM *exactly* describes the behavior of the system of interest. Rarely is such a model available in practice. Even when available, the exact model may require so many states that it becomes computationally impractical. Construction of an acceptable SMM usually depends upon the particular application and often can only be achieved through simulation.

System modeling remains an active area of research, and a complete discussion goes far beyond the scope of this text. The behavior of modeling errors is highly complicated as well. Nevertheless, this section presents a general framework for analyzing modeling errors, and it describes a few cases and their effect on the performance of the Kalman filter.

The material presented in this section focuses on the general, time-varying SMM with nonstationary process and measurement noises of Section 6.3.

Plant, Model, and Filter

To analyze modeling errors, we must distinguish between the *plant* and *model* (SMM). The plant refers the actual system, the true mathematical description of the system of interest. Often, *the model is an approximation to the plant.* The SMM serves as the model, so we may think of the model as the system that we use to design the Kalman filter. Differences between the model and plant introduce the errors of interest in this section.

Let us denote the plant dynamics by

$$\boldsymbol{x}_p(n+1) = \Phi_p(n+1, n)\boldsymbol{x}_p(n) + \Gamma_p(n)\boldsymbol{w}_p(n) + \Theta_p(n), \tag{6.43}$$

$$\boldsymbol{z}_p(n) = C_p(n)\boldsymbol{x}_p(n) + \boldsymbol{v}_p(n), \tag{6.44}$$

where $\boldsymbol{w}_p(n) \sim (0, Q_p(n))$, $\boldsymbol{v}_p(n) \sim (0, R_p(n))$, and the initial conditions are $\hat{\boldsymbol{x}}_p^-(0)$ and $P_p^-(0)$. $\Theta_p(n)$ represents an additional deterministic term that describes nonlinearities or the effect of reduced system order.

The model is then given by

$$\boldsymbol{x}_m(n+1) = \Phi_m(n+1, n)\boldsymbol{x}_m(n) + \Gamma_m(n)\boldsymbol{w}_m(n) + \Theta_m(n), \tag{6.45}$$

$$\boldsymbol{z}_m(n) = C_m(n)\boldsymbol{x}_m(n) + \boldsymbol{v}_m(n), \tag{6.46}$$

with $\boldsymbol{w}_m(n) \sim (0, Q_m(n))$, $\boldsymbol{v}_m(n) \sim (0, R_m(n))$, and initial conditions $\hat{\boldsymbol{x}}_m^-(0)$ and $P_m^-(0)$. Because of modeling errors, $\Phi_m(n+1, n)$ may be unequal to $\Phi_p(n+1, n)$, and likewise for the other system dynamics.

The Kalman filter is the usual filter in Section 5.6 with a subscript m to indicate quantities associated with the model.

$$K(n) = P_m^-(n)C_m^{\mathrm{T}}(n)\left[C_m(n)P_m^-(n)C_m^{\mathrm{T}}(n) + R_m(n)\right]^{-1} \tag{6.47}$$

$$\hat{\boldsymbol{x}}_m(n) = \hat{\boldsymbol{x}}_m^-(n) + K(n)\left[\boldsymbol{z}_p(n) - C_m(n)\hat{\boldsymbol{x}}_m^-(n)\right] \tag{6.48}$$

$$P_m(n) = \left[I - K(n)C_m(n)\right]P_m^-(n)\left[I - K(n)C_m(n)\right]^{\mathrm{T}} + K(n)R_m(n)K^{\mathrm{T}}(n) \tag{6.49}$$

$$\hat{\boldsymbol{x}}_m^-(n+1) = \Phi_m(n+1,n)\hat{\boldsymbol{x}}_m(n) \tag{6.50}$$

$$P_m^-(n+1) = \Phi_m(n+1,n)P_m(n)\Phi_m^{\mathrm{T}}(n+1,n) + \Gamma_m(n)Q_m(n)\Gamma_m^{\mathrm{T}}(n). \tag{6.51}$$

Notice that $P_m^-(n)$ and $P_m(n)$ represent the error covariances that the Kalman filter returns, but *they are not the actual error covariances.* Additionally, the observation $\boldsymbol{z}_p(n)$ appears in Equation (6.48) because the filter obtains its observations from the plant, but applies them to the model.

True Errors

Since we are interested in the actual filter error, we define new errors by

$$\tilde{\boldsymbol{x}}_{\text{true}}^-(n) \triangleq \boldsymbol{x}_p(n) - \hat{\boldsymbol{x}}_m^-(n),$$

and

$$\tilde{\boldsymbol{x}}_{\text{true}}(n) \triangleq \boldsymbol{x}_p(n) - \hat{\boldsymbol{x}}_m(n).$$

These errors give the difference between the true states and the state estimates produced by the Kalman filter. The associated error covariances are $P_{\text{true}}^-(n)$ and $P_{\text{true}}(n)$, respectively.

The true *a priori* error can be written in the form

$$\tilde{\boldsymbol{x}}_{\text{true}}^-(n+1) = \Delta\Phi(n+1,n)\boldsymbol{x}_p(n) + \Phi_m(n+1,n)\tilde{\boldsymbol{x}}_{\text{true}}(n) + \Gamma_p(n)\boldsymbol{w}_p(n) + \Delta\Theta(n), \tag{6.52}$$

where $\Delta\Phi(n+1,n) = \Phi_p(n+1,n) - \Phi_m(n+1,n)$ and $\Delta\Theta(n) = \Theta_p(n) - \Theta_m(n)$. Similarly, the *a posteriori* error becomes

$$\tilde{\boldsymbol{x}}_{\text{true}}(n) = \left[I - K(n)C_m(n)\right]\tilde{\boldsymbol{x}}_{\text{true}}^-(n) - K(n)\Delta C(n)\boldsymbol{x}_p(n) - K(n)\boldsymbol{v}_p(n), \tag{6.53}$$

where $\Delta C(n) = C_p(n) - C_m(n)$. With this formulation we can examine some simple types of modeling errors.

Filter Divergence

The Kalman filter provides error covariances $P_m^-(n)$ and $P_m(n)$ for its estimates; these covariances provide measures of the confidence we may place in the estimates. Modeling errors are significant because they introduce errors that are not accounted for by the Kalman filter and that may not be reflected by $P_m^-(n)$ and $P_m(n)$.

In the presence of modeling errors, it is possible for the true error $P_{\text{true}}(n)$ to exceed the value of $P_m(n)$ provided by the Kalman filter. In this case, the Kalman filter appears to be performing well, when in reality the estimates are inaccurate. This type of behavior is called *filter divergence*.

Divergence falls into two categories. *Apparent divergence* results when $P_{\text{true}}(n)$ is larger than $P_m(n)$ but remains bounded. *True divergence* occurs when $P_{\text{true}}(n)$ approaches infinity while $P_m(n)$ appears to be bounded.

General Inaccurate Model

Here we examine the general case where every system matrix may differ between the model and plant. This case identifies the many quantities can cause divergence. Substituting (6.53) into (6.52), we find

$$\begin{aligned}\tilde{\boldsymbol{x}}_{\text{true}}^-(n+1) = {}& \Phi_m(n+1,n)\,[I - K(n)C_m(n)]\,\tilde{\boldsymbol{x}}_{\text{true}}^-(n)\\ &+ [\Delta\Phi(n+1,n) - \Phi_m(n+1,n)K(n)\Delta C(n)]\,\boldsymbol{x}_p(n)\\ &+ \Gamma_p(n)\boldsymbol{w}_p(n)\\ &- \Phi_m(n+1,n)K(n)\boldsymbol{v}_p(n) + \Delta\Theta(n)\end{aligned} \qquad (6.54)$$

Observe that the true error is driven by several sources beyond the scope of our model. They are

- $\boldsymbol{x}_p(n)$: the actual state of the plant. If the plant is stable but not bounded as tightly as the model, apparent divergence may result. If the plant is unstable, it is possible for true divergence to occur.

- $\boldsymbol{w}_p(n)$ and $\boldsymbol{v}_p(n)$: the plant noise processes. The true error system is *not* driven by the noise processes of the model. If the model noise covariances do not match the plant noise covariances, the Kalman filter may improperly weight the terms in the Kalman gain. Again filter divergence may occur.

- $\Delta\Theta(n)$: nonlinearities or neglected states. The plant may possess nonlinearities that are unbounded or that are not adequately handled by the model process noise. In either case, true or apparent divergence can result. The plant may also have states that are neglected by the model. If these states are unstable, true divergence is possible.

In light of these inaccuracies, even though the state-transition matrix of the Kalman filter $(\Phi_m(n+1,n)\,[I - K(n)C_m(n)])$ is stable, the filter may still diverge. Since these problems are outside our control, usually the plant itself must be stable and the plant's inputs must be bounded to guarantee that $\tilde{x}^-_{\text{true}}(n)$ remains bounded. Even then, the Kalman filter may incorrectly report the estimation error $\tilde{x}^-_m(n)$.

Inaccurate Initial Conditions

As a special case, suppose the model and plant are identical, and the only errors lie in our selection of $\hat{x}^-_m(0)$, $P^-_m(0)$, $Q_m(n)$, and $R_m(n)$ (cf. [11, 12, 13]). That is, our model matches the plant except for our initial conditions and noise covariances.

The true error covariances become

$$P^-_{\text{true}}(n+1) = \Phi(n+1,n)P_{\text{true}}\Phi^{\mathrm{T}}(n+1,n) + \Gamma(n)Q_p(n)\Gamma^{\mathrm{T}}(n), \tag{6.55}$$

and

$$P_{\text{true}}(n) = [I - K(n)C(n)]\,P^-_{\text{true}}(n)\,[I - K(n)C(n)]^{\mathrm{T}} + K(n)R_p(n)K^{\mathrm{T}}(n). \tag{6.56}$$

Following the definition of Jazwinski [14], we define the difference between the true and reported error covariance matrices by

$$E(n) \triangleq P_{\text{true}}(n) - P_m(n),$$

and

$$E^-(n) \triangleq P^-_{\text{true}}(n) - P^-_m(n).$$

Direct substitution of (6.55) and (6.56) gives

$$E(n) = [I - K(n)C(n)]\,E^-(n)\,[I - K(n)C(n)]^{\mathrm{T}} + K(n)\,[R_p(n) - R_m(n)]\,K^{\mathrm{T}}(n). \tag{6.57}$$

and

$$E^-(n+1) = \Phi(n+1,n)E(n)\Phi^{\mathrm{T}}(n+1,n) + \Gamma(n)\,[Q_p(n) - Q_m(n)]\,\Gamma^{\mathrm{T}}(n), \tag{6.58}$$

Now suppose that the actual error covariance $P^-_{\text{true}}(0)$ is smaller than the model error covariance $P^-_m(0)$ and that the actual noise covariances $Q_p(n)$

and $R_p(n)$ are always smaller than the model noise covariances. That is, $P^-_{\text{true}}(0) \leq P^-_m(0)$, $Q_p(n) \leq Q_m(n)$, and $R_p(n) \leq R_m(n)$ for all n.

It follows that $E^-(0) \leq 0$, and equation (6.57) indicates that $E(0) \leq 0$. Then equation (6.58) shows that $E^-(1) \leq 0$ as well. We can continue this process indefinitely, which produces a theorem due to Jazwinski [14]:

Theorem 6.5 *If $P^-_{\text{true}}(0) \leq P^-_m(0)$, and $Q_p(n) \leq Q_m(n)$ and $R_p(n) \leq R_m(n)$ for all n, then $P^-_{\text{true}}(n) \leq P^-_m(n)$ and $P_{\text{true}}(n) \leq P_m(n)$ for all n.*

This theorem can be interpreted in the following way: Generally we do not have exact knowledge of $P^-_{\text{true}}(0)$, $Q_p(n)$, and $R_p(n)$, but often we can approximately determine these quantities. We can increase these approximations slightly to produce conservative estimates of $P^-_m(0)$, $Q_m(n)$, and $R_m(n)$. Next we design our Kalman filter with these model covariances, and the actual error covariance $P_{\text{true}}(n)$ will always be bounded by the reported error covariance $P_m(n)$. Thus, if the Kalman filter gives acceptable values of $P_m(n)$, we can be confident that the filter is producing reasonably accurate estimates.

Theorem 6.5 does *not* imply that $P_{\text{true}}(n)$ is bounded. If $P_m(n)$ approaches infinity, we cannot tell whether or not $P_{\text{true}}(n)$ is actually bounded. We would like to know conditions under which $P_{\text{true}}(n)$ remains bounded even in the presence of modeling errors. Fortunately we have the following corollary[4] [14]:

Corollary 6.1 *Let $P^-_{\text{true}}(0) \leq P^-_m(0)$, $Q_p(n) \leq Q_m(n)$, and $R_p(n) \leq R_m(n)$ for all n. Additionally let the system model (6.45–6.46) be UCO and UCC. Then there exists an integer $N > 0$ and a real number $\alpha > 0$ such that*

$$P_{\text{true}}(n) \leq \alpha I, \quad n \geq N.$$

As a final note, observe that the condition of UCC requires that some process noise be present, and it must excite the *model* sufficiently in order to ensure the bound on $P_{\text{true}}(n)$.

Nonlinearities and Neglected States

Finally, we consider the case where some states or nonlinearities may be neglected, in addition to errors in initial conditions and noise covariances. That is, $\Delta\Phi(n+1,n) = \Delta C(n) = 0$, and $\Gamma_p(n) = \Gamma_m(n)$ for all n, but $\Delta\Theta(n)$ is not identically zero.

In this case, the true error covariances are

$$\begin{aligned} P^-_{\text{true}}(n+1) = {} & \Phi(n+1,n)P_{\text{true}}(n)\Phi^{\mathrm{T}}(n+1,n) + \Gamma(n)Q_p(n)\Gamma^{\mathrm{T}}(n) \\ & + \Delta\Theta(n)\Delta\Theta^{\mathrm{T}}(n) + \Phi(n+1,n)\mathrm{E}\left[\tilde{\boldsymbol{x}}_{\text{true}}(n)\right]\Delta\Theta^{\mathrm{T}}(n) \\ & + \Delta\Theta(n)\mathrm{E}\left[\tilde{\boldsymbol{x}}^{\mathrm{T}}_{\text{true}}(n)\right]\Phi^{\mathrm{T}}(n+1,n), \end{aligned} \tag{6.59}$$

[4] Proof appears in Appendix E.

and

$$P_{\text{true}}(n) = [I - K(n)C(n)]\, P_{\text{true}}^{-}(n)\, [I - K(n)C(n)]^{\mathrm{T}} \\ + K(n)R_p(n)K^{\mathrm{T}}(n). \tag{6.60}$$

The true error depends on $\Delta\Theta(n)$ and $\mathrm{E}\left[\tilde{\boldsymbol{x}}_{\text{true}}(n)\right]$. Jazwinski [14] provides the following corollary:[5]

Corollary 6.2 *Suppose the model (6.45–6.46) is UCO and UCC, $\Delta\Theta(n)$ is uniformly bounded, and $P_{\text{true}}^{-}(0)$ is bounded. Then $P_{\text{true}}(n)$ is uniformly bounded for $n \geq N$.*

Just as in Corollary 6.1, the hypothesis of UCC requires that we must have adequate process noise exciting the model.

If we know that $\Theta_p(n)$ is bounded, we can simply set $\Theta_m(n) \equiv 0$ to bound $\Delta\Theta(n)$. For example, we may have an exact model of high dimensionality and reduce it to a practical model by neglecting states. Alternatively, we may know the nonlinear description of the plant but elect to approximate it with a linear state model. In either case, we can often compute bounds on $\Theta_p(n)$.

Summary

We have seen that modeling errors can significantly affect the performance of the Kalman filter. If the plant is unstable and inaccurately modeled, the Kalman filter is likely to diverge. However, in certain circumstances we can bound the true estimation error without being able to compute it exactly. In other situations we can at least guarantee that the true estimation error does not go to infinity.

More thorough treatments of this topic appear in [15, 14, 7]. A variety of methods for preventing divergence have been suggested. Jazwinski [14] proposes a limited-memory filter that slows the growth of $P_{\text{true}}(n)$. Anderson and Moore [15] suggest exponential data weighting, which gives recent observations more importance than those in the distant past. Lewis [7] uses fictitious process noise and shows that this technique is equivalent to that of Anderson and Moore.

6.5 Multistep Kalman Prediction

Section 5.6 presented the Kalman filter for filtering, which also includes a one-step predictor (5.83). In this section we extend the filter to the case of multistep prediction.

[5] Proof appears in Appendix E.

We wish to predict $\boldsymbol{s}(n{+}m)$, or equivalently $\boldsymbol{x}(n{+}m)$, where m is a positive integer. By arguments like those in Section 5.5, we find that

$$\hat{\boldsymbol{x}}^-(n+m) = \Phi^{m+1}\hat{\boldsymbol{x}}(n-1), \tag{6.61}$$

and the predicted signal is

$$\hat{\boldsymbol{s}}^-(n+m) = C\hat{\boldsymbol{x}}^-(n+m). \tag{6.62}$$

We can implement the Kalman predictor very easily. We begin with the ordinary Kalman filter recursion in Section 5.6. Immediately after Equation (5.83), we insert Equations (6.61) and (6.62), and the predictor is complete.

In the case of a time-varying state model, we have

$$\hat{\boldsymbol{x}}^-(n+m) = \Phi(n+m, n-1)\hat{\boldsymbol{x}}(n-1).$$

and

$$\hat{\boldsymbol{s}}^-(n+m) = C(n+m)\hat{\boldsymbol{x}}^-(n+m).$$

6.6 Kalman Smoothing

In this section we consider the problem of making a smoothed estimate with a Kalman filter. We have a collection of observations $\boldsymbol{Z}_N = \{\boldsymbol{z}(), \ldots, \boldsymbol{z}(N)\}$, and we wish to estimate $\boldsymbol{x}(n)$ for some n between 1 and N. Although the filtering and predicting forms of the Kalman filter are quite similar, the Kalman smoother is more complicated because of correlation *between* $\boldsymbol{z}(n)$ and $\boldsymbol{z}(n+1)$, $\boldsymbol{z}(n+2)$, ..., and $\boldsymbol{z}(N)$.

The approach we adopt involves three filters and is diagrammed in Figure 6.6. First, a Kalman filter uses the observations $\boldsymbol{Z}_f = \{\boldsymbol{z}(1), \ldots, \boldsymbol{z}(n)\}$ to estimate $\boldsymbol{x}(n)$. This filter is called the *forward filter,* and it operates like an ordinary Kalman filter. We use a subscript f to indicate forward-filtered quantities, such as $\hat{\boldsymbol{x}}_f(n)$.

Next, a second Kalman filter uses the remaining observations $\boldsymbol{Z}_b = \{\boldsymbol{z}(n+1), \ldots, \boldsymbol{z}(N)\}$ to estimate $\boldsymbol{x}(n)$. This filter can be viewed as a one-step Kalman predictor running *backwards* in time, so it is named the *backward predictor.* That is, the filter starts at observation $\boldsymbol{z}(N)$ and recurses back down to $\boldsymbol{z}(n+1)$. The resulting estimate is $\hat{\boldsymbol{x}}_b^-(n)$. It is an *a priori predicted* estimate because $\boldsymbol{z}(n)$ is not used. A subscript b denotes backward-filtered quantities.

Finally, a *smoother* combines $\hat{\boldsymbol{x}}_f(n)$ and $\hat{\boldsymbol{x}}_b^-(n)$ to produce the optimum smoothed estimate $\hat{\boldsymbol{x}}_s(n)$. A subscript s indicates smoothed quantities.

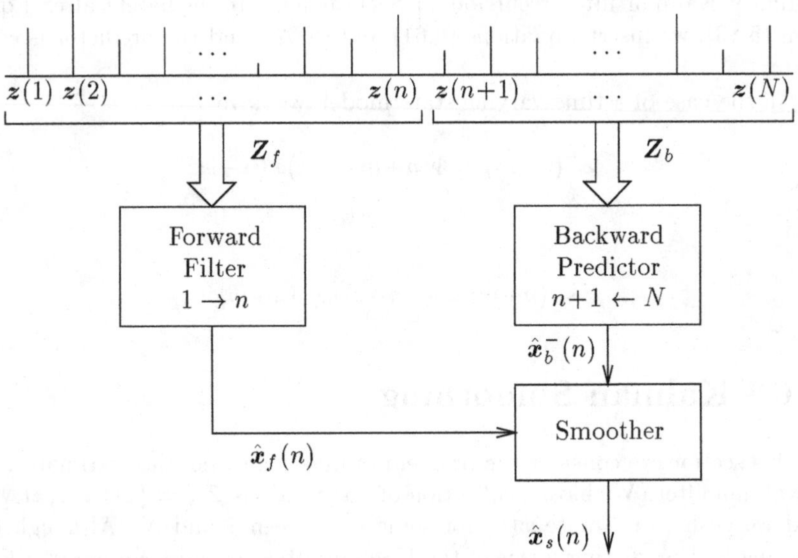

Figure 6.1. Kalman smoother diagram. A forward-filtering Kalman filter generates an estimate $\hat{\boldsymbol{x}}_f(n)$ from observations $\boldsymbol{z}(1), \boldsymbol{z}(2), \ldots, \boldsymbol{z}(n)$. A backward-predicting Kalman filter generates an backward-predicted estimate $\hat{\boldsymbol{x}}_b^-(n)$ from $\boldsymbol{z}(n+1), \ldots, \boldsymbol{z}(N)$. Finally, a smoothing filter combines $\hat{\boldsymbol{x}}_f(n)$ and $\hat{\boldsymbol{x}}_b^-(n)$ to form the estimate $\hat{\boldsymbol{x}}_s(n)$.

Backward Prediction

The forward filter is exactly the Kalman filter summarized in Section 5.6, so we do not discuss it again. In this section we present the backward predictor. It is simply a one-step Kalman predictor with the time direction reversed. We may use the results we have found for the Kalman filter as long as we remember that time runs backwards, starting at N and ending at $n+1$.

Let us define a few variables that facilitate the development of the backward predictor. Define

$$S(n) \triangleq P_b^{-1}(n) = (\mathrm{Cov}\left[\boldsymbol{x}(n) - \hat{\boldsymbol{x}}_b(n)\right])^{-1}, \tag{6.63}$$

and

$$S^-(n) \triangleq \left(P_b^-(n)\right)^{-1} = \left(\mathrm{Cov}\left[\boldsymbol{x}(n) - \hat{\boldsymbol{x}}_b^-(n)\right]\right)^{-1}. \tag{6.64}$$

It will also be convenient to define

$$\hat{\boldsymbol{y}}(n) \triangleq P_b^{-1}(n)\hat{\boldsymbol{x}}_b(n) = S(n)\hat{\boldsymbol{x}}_b(n), \tag{6.65}$$

and

$$\hat{\boldsymbol{y}}^-(n) \triangleq \left(P_b^-(n)\right)^{-1}\hat{\boldsymbol{x}}_b^-(n) = S^-(n)\hat{\boldsymbol{x}}_b^-(n). \tag{6.66}$$

Now we begin the construction of the predictor. We require that Φ be invertible. Then the signal model (5.36) gives

$$\boldsymbol{x}(n) = \Phi^{-1}\boldsymbol{x}(n+1) - \Phi^{-1}\Gamma\boldsymbol{w}(n).$$

Since $\hat{\boldsymbol{w}}(n) = 0$, the optimum *a priori* backward estimate is

$$\hat{\boldsymbol{x}}_b^-(n) = \Phi^{-1}\hat{\boldsymbol{x}}_b(n+1). \tag{6.67}$$

The equivalent form for the Kalman filter in Section 5.7 provides two useful results. First, Equation (5.98) gives

$$P_b^{-1}(n) = \left(P_b^-(n)\right)^{-1} + C^{\mathrm{T}}R^{-1}C.$$

As a result of Definitions (6.63) and (6.64), this equation becomes

$$\boxed{S(n) = S^-(n) + C^{\mathrm{T}}R^{-1}C.} \tag{6.68}$$

Second, from (5.99) we have

$$P_b^{-1}(n)\hat{\boldsymbol{x}}_b(n) = \left(P_b^-(n)\right)^{-1}\hat{\boldsymbol{x}}_b^-(n) + C^{\mathrm{T}}R^{-1}\boldsymbol{z}(n),$$

or equivalently,

$$\boxed{\hat{\boldsymbol{y}}(n) = \hat{\boldsymbol{y}}^-(n) + C^{\mathrm{T}} R^{-1} \boldsymbol{z}(n).} \tag{6.69}$$

At this point we have an expression for computing $S(n)$ and $\hat{\boldsymbol{y}}(n)$. We still need to find equations for $S^-(n)$ and $\hat{\boldsymbol{y}}^-(n)$.

To find $S^-(n)$, we write the covariance matrix $P_b^-(n)$ as

$$\begin{aligned} P_b^-(n) &= \operatorname{Cov}\left[\Phi^{-1}\boldsymbol{x}(n+1) - \Phi^{-1}\Gamma\boldsymbol{w}(n) - \Phi^{-1}\hat{\boldsymbol{x}}_b(n+1)\right] \\ &= \Phi^{-1}\left[P_b(n+1) + \Gamma Q(n)\Gamma^{\mathrm{T}}\right]\Phi^{-\mathrm{T}}. \end{aligned}$$

Then we apply Definition (6.63) to obtain

$$S^-(n) = \left\{\Phi^{-1}\left[S^{-1}(n+1) + \Gamma Q(n)\Gamma^{\mathrm{T}}\right]\Phi^{-\mathrm{T}}\right\}^{-1}.$$

By the matrix inversion lemma, this expression becomes

$$\begin{aligned} &S^-(n) \\ &= \Phi^{\mathrm{T}}\left\{S(n+1) - S(n+1)\Gamma\left[\Gamma^{\mathrm{T}}S(n+1)\Gamma + Q^{-1}(n)\right]^{-1}\Gamma^{\mathrm{T}}S(n+1)\right\}\Phi. \end{aligned}$$

Define a backward Kalman gain by

$$\boxed{K_b(n) = S(n+1)\Gamma\left[\Gamma^{\mathrm{T}}S(n+1)\Gamma + Q^{-1}(n)\right]^{-1},} \tag{6.70}$$

so that our expresion for $S^-(n)$ is

$$\boxed{S^-(n) = \Phi^{\mathrm{T}}\left[I - K_b(n)\Gamma^{\mathrm{T}}\right]S(n+1)\Phi.} \tag{6.71}$$

Finally, we need a means for computing $\hat{\boldsymbol{y}}^-(n)$ from $\hat{\boldsymbol{y}}(n+1)$. Using (6.66), (6.67), and (6.65) we find

$$\hat{\boldsymbol{y}}^-(n) = S^-(n)\Phi^{-1}S^{-1}(n+1)\hat{\boldsymbol{y}}(n+1)$$

When we substitute (6.71) for $S^-(n)$, we find

$$\boxed{\hat{\boldsymbol{y}}^-(n) = \Phi^{\mathrm{T}}\left[I - K_b(n)\Gamma^{\mathrm{T}}\right]\hat{\boldsymbol{y}}(n+1).} \tag{6.72}$$

Equations (6.68–6.72) form a recursion for the backward predictor. Later in this section we present the initialization for the predictor.

Combining the Estimates

Now we have recursive algorithms for computing $\hat{\boldsymbol{x}}_f(n)$ and $\hat{\boldsymbol{x}}_b^-(n)$. We still need to combine these estimates into a smoothed estimate $\hat{\boldsymbol{x}}_s(n)$. Since all three estimates are linear, $\hat{\boldsymbol{x}}_s(n)$ takes the form

$$\hat{\boldsymbol{x}}_s(n) = A_f(n)\hat{\boldsymbol{x}}_f(n) + A_b(n)\hat{\boldsymbol{x}}_b^-(n). \tag{6.73}$$

With proper initialization the estimates $\hat{\boldsymbol{x}}_f(n)$ and $\hat{\boldsymbol{x}}_b^-(n)$ will be unbiased. The processes $\boldsymbol{w}(n)$ and $\boldsymbol{v}(n)$ are white, so $\hat{\boldsymbol{x}}_f(n)$ and $\hat{\boldsymbol{x}}_b^-(n)$ are independent. We want $\hat{\boldsymbol{x}}_s(n)$ to be unbiased, so

$$\begin{aligned}
\mathrm{E}\left[\hat{\boldsymbol{x}}_s(n)\right] &= A_f(n)\mathrm{E}\left[\hat{\boldsymbol{x}}_f(n)\right] + A_b(n)\mathrm{E}\left[\hat{\boldsymbol{x}}_b^-(n)\right] \\
&= A_f(n)\mathrm{E}\left[\boldsymbol{x}(n)\right] + A_b(n)\mathrm{E}\left[\boldsymbol{x}(n)\right] \\
&= \mathrm{E}\left[\boldsymbol{x}(n)\right],
\end{aligned}$$

and thus,

$$A_f(n) + A_b(n) = I. \tag{6.74}$$

The estimation error for the smoothed estimate is

$$\begin{aligned}
\tilde{\boldsymbol{x}}_s(n) &= \boldsymbol{x}(n) - \hat{\boldsymbol{x}}_s(n) \\
&= \left[A_f(n) + A_b(n)\right]\boldsymbol{x}(n) - \left[A_f(n)\hat{\boldsymbol{x}}_f(n) + A_b(n)\hat{\boldsymbol{x}}_b^-(n)\right] \\
&= A_f(n)\tilde{\boldsymbol{x}}_f(n) + A_b(n)\tilde{\boldsymbol{x}}_b^-(n) \\
&= A_f(n)\tilde{\boldsymbol{x}}_f(n) + \left[I - A_f(n)\right]\tilde{\boldsymbol{x}}_b^-(n).
\end{aligned} \tag{6.75}$$

Then the estimation error covariance matrix is given by

$$\begin{aligned}
P_s(n) &= \mathrm{Cov}\left[A_f(n)\tilde{\boldsymbol{x}}_f(n) + \left[I - A_f(n)\right]\tilde{\boldsymbol{x}}_b^-(n)\right] \\
&= A_f(n)P_f(n)A_f^{\mathrm{T}}(n) + P_b^-(n) - A_f(n)P_b^-(n) \\
&\quad - P_b^-(n)A_f^{\mathrm{T}}(n) + A_f(n)P_b^-(n)A_f^{\mathrm{T}}(n).
\end{aligned} \tag{6.76}$$

To make $\hat{\boldsymbol{x}}_s(n)$ the MMSE estimate, we set

$$\frac{\partial \mathrm{tr}\left(P_s(n)\right)}{\partial A_f(n)} = 2A_f(n)P_f(n) - 2P_b^-(n) + 2A_f(n)P_b^-(n) = 0.$$

Therefore

$$A_f(n) = P_b^-(n)\left[P_f(n) + P_b^-(n)\right]^{-1}. \tag{6.77}$$

It follows from (6.74) that

$$\begin{aligned}
A_b(n) &= I - P_b^-(n)\left[P_f(n) + P_b^-(n)\right]^{-1} \\
&= \left[P_f(n) + P_b^-(n)\right]\left[P_f(n) + P_b^-(n)\right]^{-1} - P_b^-(n)\left[P_f(n) + P_b^-(n)\right]^{-1} \\
&= P_f(n)\left[P_f(n) + P_b^-(n)\right]^{-1}.
\end{aligned} \tag{6.78}$$

Substituting (6.77) and (6.78) into (6.76), we find

$$\begin{aligned} P_s(n) &= P_b^-(n)\left[P_f(n)+P_b^-(n)\right]^{-1}P_f(n)\left[P_f(n)+P_b^-(n)\right]^{-1}P_b^-(n) \\ &\quad + P_b^-(n) - 2P_b^-(n)\left[P_f(n)+P_b^-(n)\right]^{-1}P_b^-(n) \\ &\quad + P_b^-(n)\left[P_f(n)+P_b^-(n)\right]^{-1}P_b^-(n)\left[P_f(n)+P_b^-(n)\right]^{-1}P_b^-(n) \\ &= P_b^-(n)\left[P_f(n)+P_b^-(n)\right]^{-1}P_b^-(n) + P_b^-(n) \\ &\quad - 2P_b^-(n)\left[P_f(n)+P_b^-(n)\right]^{-1}P_b^-(n) \\ &= P_b^-(n) + P_b^-(n)\left\{-\left[P_f(n)+P_b^-(n)\right]\right\}^{-1}P_b^-(n). \end{aligned}$$

We apply the matrix inversion lemma to this expression and obtain

$$\begin{aligned} P_s^{-1}(n) &= \left(P_b^-(n)\right)^{-1} - \left(P_b^-(n)\right)^{-1}P_b^-(n)\left\{P_b^-(n)\left(P_b^-(n)\right)^{-1}P_b^-(n)\right. \\ &\quad \left. - \left[P_f(n)+P_b^-(n)\right]\right\}^{-1}\left(P_b^-(n)\right)^{-1}P_b^-(n) \\ &= \left(P_b^-(n)\right)^{-1} - \left[P_b^-(n) - P_f(n) - P_b^-(n)\right]^{-1} \\ &= P_f^{-1}(n) + \left(P_b^-(n)\right)^{-1} \end{aligned} \tag{6.79}$$

Next we multiply (6.79) by $\hat{\boldsymbol{x}}_s(n)$ and substitute (6.73), (6.77), and (6.78). We arrive at

$$\begin{aligned} P_s^{-1}\hat{\boldsymbol{x}}_s(n) &= \left[P_f^{-1}(n) + \left(P_b^-(n)\right)^{-1}\right]\hat{\boldsymbol{x}}_s(n) \\ &= \left[P_f^{-1}(n) + \left(P_b^-(n)\right)^{-1}\right]\left\{P_b^-(n)\left[P_f(n)+P_b^-(n)\right]^{-1}\hat{\boldsymbol{x}}_f(n)\right. \\ &\qquad \left. + P_f(n)\left[P_f(n)+P_b^-(n)\right]^{-1}\hat{\boldsymbol{x}}_b^-(n)\right\} \\ &= \left[P_f^{-1}(n)P_b^-(n) + P_f^{-1}(n)P_f(n)\right]\left[P_f(n)+P_b^-(n)\right]^{-1}\hat{\boldsymbol{x}}_f(n) \\ &\quad + \left[\left(P_b^-(n)\right)^{-1}P_b^-(n) + \left(P_b^-(n)\right)^{-1}P_b^-(n)\right] \\ &\quad \times \left[P_f(n)+P_b^-(n)\right]^{-1}\hat{\boldsymbol{x}}_b^-(n) \\ &= P_f^{-1}(n)\left[P_b^-(n)+P_f(n)\right]\left[P_f(n)+P_b^-(n)\right]^{-1}\hat{\boldsymbol{x}}_f(n) \\ &\quad + \left(P_b^-(n)\right)^{-1}\left[P_b^-(n)+P_f(n)\right]\left[P_f(n)+P_b^-(n)\right]^{-1}\hat{\boldsymbol{x}}_b^-(n) \\ &= P_f^{-1}(n)\hat{\boldsymbol{x}}_f(n) + \left(P_b^-(n)\right)^{-1}\hat{\boldsymbol{x}}_b^-(n). \end{aligned} \tag{6.80}$$

Then we can write

$$\begin{aligned}\hat{\boldsymbol{x}}_s(n) &= P_s(n)\left[P_f^{-1}(n)\hat{\boldsymbol{x}}_f(n) + \left(P_b^-(n)\right)^{-1}\hat{\boldsymbol{x}}_b^-(n)\right] \\ &= P_s(n)P_f^{-1}\hat{\boldsymbol{x}}_f(n) + P_s(n)\hat{\boldsymbol{y}}^-(n) \\ &= \left[P_f^{-1}(n) + S^-(n)\right]^{-1} P_f^{-1}(n)\hat{\boldsymbol{x}}_f(n) + P_s(n)\hat{\boldsymbol{y}}^-(n) \\ &= \left[I + P_f(n)S^-(n)\right]^{-1}\hat{\boldsymbol{x}}_f(n) + P_s(n)\hat{\boldsymbol{y}}^-(n)\end{aligned}$$

Note that $I + P_f(n)S^-(n) = I + [P_f(n)S^-(n)]\,II$. We apply the matrix inversion lemma and obtain

$$\hat{\boldsymbol{x}}_s(n) = \left\{I - P_f(n)S^-(n)\left[I + P_f(n)S^-(n)\right]^{-1}\right\}\hat{\boldsymbol{x}}_f(n) + P_s(n)\hat{\boldsymbol{y}}^-(n).$$

Let us define the smoothing gain by

$$\boxed{K_s(n) = P_f(n)S^-(n)\left[I + P_f(n)S^-(n)\right]^{-1},} \tag{6.81}$$

so the final form for the smoothed estimate is

$$\boxed{\hat{\boldsymbol{x}}_s(n) = \left[I - K_s(n)\right]\hat{\boldsymbol{x}}_f(n) + P_s(n)\hat{\boldsymbol{y}}^-(n).} \tag{6.82}$$

As an additional consequence of (6.81), we may write (6.79) as

$$\begin{aligned}P_s(n) &= \left[P_f^{-1}(n) + S^-(n)II\right]^{-1} \\ &= P_f(n) - P_f(n)S^-(n)\left[IP_f(n)S^-(n) + I\right]^{-1} IP_f(n),\end{aligned}$$

so that

$$\boxed{P_s(n) = \left[I - K_s(n)\right]P_f(n).} \tag{6.83}$$

Equations (6.81–6.83) complete the Kalman smoothing filter. Together they provide a recursion for combining $\hat{\boldsymbol{x}}_f(n)$ and $\hat{\boldsymbol{x}}_b^-(n)$ into $\hat{\boldsymbol{x}}_s(n)$.

Initialization

We now concern ourselves with the initial conditions required by the smoothing filter. The initial conditions for the forward predictor are the same as in Section 5.6: $\hat{x}_f^-(1)$ is a guess at $\mathrm{E}[\boldsymbol{x}(1)]$ and $P_f^-(1) \geq 0$.

For the backward predictor we observe that at time N, $P_s(N) = P_f(N)$. We write (6.79) as

$$P_s^{-1}(N) = P_f^{-1}(N) + S^-(N),$$

and it follows that $S^-(N) = 0$. Then definition (6.66) implies that $\hat{y}^-(N) = 0$. Notice that the initial condition $\hat{x}_b^-(N)$ is not necessary. This absence is desirable, because we do not know $\mathrm{E}\left[\hat{\boldsymbol{x}}_b^-(N)\right]$.

Summary

The Kalman smoothing filter equations are summarized below. Overall, it consists of an optimum forward filter, an optimum backward predictor, and a combining recursion.

Forward Filter

Initial conditions:

$$\hat{x}_f^-(1) = \text{guess of } \mathrm{E}[\boldsymbol{x}(1)]$$
$$P_f^-(1) \geq 0$$

Recursion:

$$K_f(n) = P_f^-(n)C^{\mathrm{T}}\left[CP_f^-(n)C^{\mathrm{T}} + R\right]^{-1}$$
$$\hat{x}_f(n) = \hat{x}_f^-(n) + K_f(n)[z(n) - C\hat{x}_f^-(n)]$$
$$P_f(n) = P_f^-(n) - K_f(n)CP_f^-(n)$$
$$\hat{x}_f^-(n+1) = \Phi\hat{x}_f(n)$$
$$P_f^-(n+1) = \Phi P_f(n)\Phi^{\mathrm{T}} + \Gamma Q(n)\Gamma^{\mathrm{T}}$$

Backward Predictor

Initial conditions:

$$\hat{y}^-(N) = 0$$
$$S^-(N) = 0$$

Recursion:

$$S(n) = S^-(n) + C^{\mathrm{T}}R^{-1}C$$
$$\hat{y}(n) = \hat{y}^-(n) + C^{\mathrm{T}}R^{-1}z(n)$$
$$K_b(n) = S(n+1)\Gamma\left[\Gamma^{\mathrm{T}}S(n+1)\Gamma + Q^{-1}(n)\right]^{-1}$$
$$S^-(n) = \Phi^{\mathrm{T}}\left[I - K_b(n)\Gamma^{\mathrm{T}}\right]S(n+1)\Phi$$
$$\hat{y}^-(n) = \Phi^{\mathrm{T}}\left[I - K_b(n)\Gamma^{\mathrm{T}}\right]\hat{y}(n+1)$$

Smoother

$$K_s(n) = P_f(n)S^-(n)\left[I + P_f(n)S^-(n)\right]^{-1}$$
$$P_s(n) = [I - K_s(n)]P_f(n)$$
$$\hat{x}_s(n) = [I - K_s(n)]\hat{x}_f(n) + P_s(n)\hat{y}^-(n).$$

Example 6.1 Kalman Smoothing

An example of Kalman smoothing using MATLAB appears here. We used the following state model

$$\Phi = \begin{bmatrix} -0.2 & 0 & 1 \\ 1.2 & 0.4 & -0.3 \\ 0 & 0.4 & 0.2 \end{bmatrix}, \qquad \Gamma = \begin{bmatrix} 0.4 \\ 0.5 \\ 0.3 \end{bmatrix}, \qquad C = \begin{bmatrix} 1 & 0.2 & 0.5 \\ 0.1 & 0.1 & 0.1 \\ 0 & 0 & 2 \end{bmatrix},$$

with $Q = 4$, and $R = \text{diag}\left(\begin{bmatrix} 80 & 66 & 25 \end{bmatrix}\right)$.

Simulation results appear in Figures 6.2 and 6.3. The first figure shows the actual states and the estimates produced by both the Kalman filter (no smoothing) and the Kalman smoother. The Kalman smoother produces more accurate estimates as well as a less jagged profile.

The second figure shows the square error, summed over all three states. It is clear that the Kalman smoother produces much more accurate estimates on average, particularly around $n = 20$, 50, and 80. The dashed curves show the estimated MSE for both filters. The curve for the Kalman filter drops from $n = 0$ and reaches an asymptote around 5. This behavior occurs because the estimate does not improve any, on average.

The curve for the Kalman smoother is large at both $n = 0$ and at $n = 100$ because the smoother incorporates information from both the past and the future relative to the current value of n. At $n = 100$, the smoother has no more information than the Kalman filter, and the curves are the same. However, the actual errors differ because the estimates are computed differently. The Kalman smoother MSE has an asymptote around 4, indicating that the use of additional information can reduce the MSE by about 20% versus regular Kalman filtering (which has an asymptote of about 5).

Problems

6.1. Consider the following state model

$$\Phi(n) = \begin{bmatrix} 1 & \phi_{12}(n) \\ 2 & 0.5 \end{bmatrix}, \quad \Gamma = \begin{bmatrix} 0.4 \\ 0.5 \end{bmatrix}, \quad C = \begin{bmatrix} 1 & 0 \\ 0 & 1 \end{bmatrix},$$

with $Q = 2$, $R = \text{diag}\left(\begin{bmatrix} 26 & 100 \end{bmatrix}\right)$, and $\phi_{12}(n) = -0.2\left(1 - e^{-0.11n}\right)$. This system is stable for all $n > 0$.

(a) Generate 200 samples $x(n)$ and noisy observations $z(n)$ for this model in MATLAB.

(b) Implement the Kalman filter from Section 6.3 to estimate the state of this system from the observations $z(n)$.

(c) Plot the true states $x_i(n)$ and estimated states $\hat{x}_i(n)$ for $1 \le n \le 200$.

6.2. Repeat Problem 6.1, but replace $\phi_{12}(n)$ by $\phi_{12}(n) = 0.35e^{-0.15|n-40|} - 0.15$. This system is stable when $n = 0$, becomes unstable for $36 \le n \le 46$, and again becomes stable for $n > 46$.

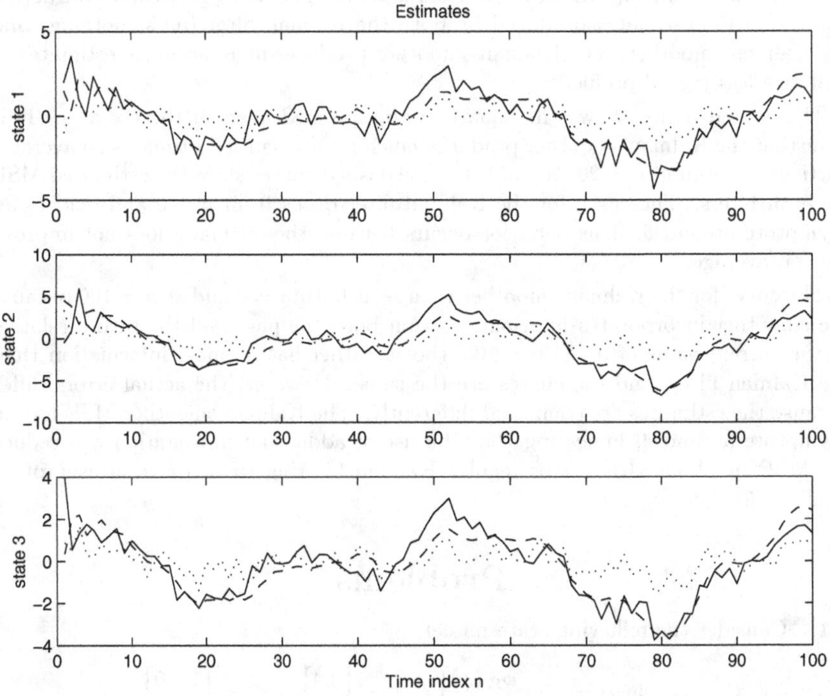

Figure 6.2. Kalman smoothing results for Example 6.1. Solid lines show actual states, dashed lines show the state estimates produced by the Kalman smoother, and dotted lines show the state estimates from the conventional Kalman filter.

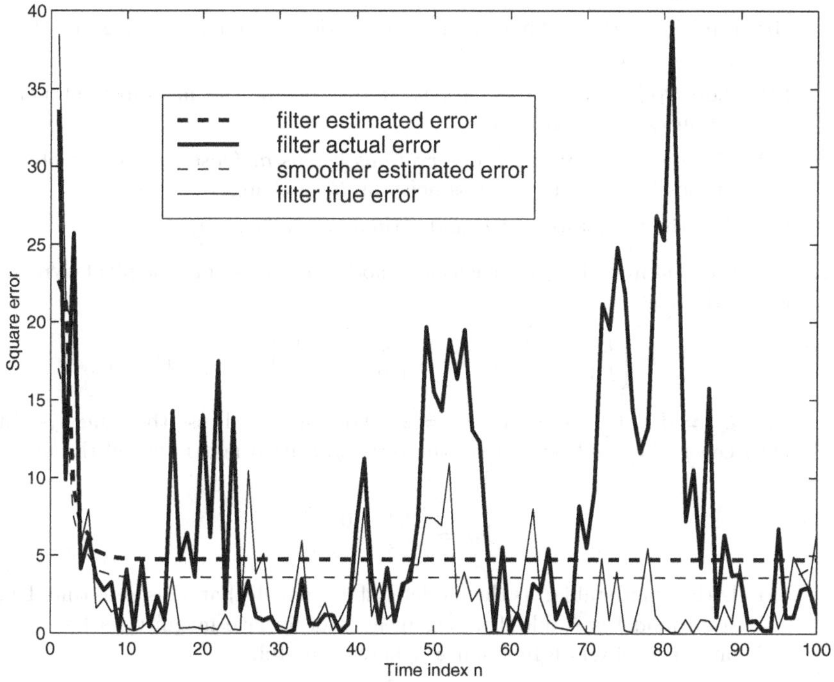

Figure 6.3. Total square error over all states for Example 6.1. Thick lines show the estimated and actual square error for the conventional Kalman filter. Thin lines show the square errors for the Kalman smoother.

6.3. Use the state model

$$\Phi = \begin{bmatrix} 0.3 & 0.1 \\ 3.2 & 0.5 \end{bmatrix}, \quad \Gamma = \begin{bmatrix} 0.5 & 0.2 \\ 0 & 0.1 \end{bmatrix}, \quad C = \begin{bmatrix} 1 & 0 \\ 0 & 1 \end{bmatrix},$$

with $Q = \text{diag}\left(\begin{bmatrix} 2 & 1.8 \end{bmatrix}\right)$, and

$$R(n) = \begin{bmatrix} 10\cos(0.025\pi n) + 12 & 0 \\ 0 & 40\cos(0.080\pi n) + 50 \end{bmatrix}.$$

(a) Using MATLAB, generate 150 samples and noisy observations for this system.

(b) Implement the Kalman filter from Section 6.3 for estimating the state of this system.

(c) Plot $R_{11}(n)$ and $R_{22}(n)$ versus n to see how the measurement noise variances change over time.

(d) Plot the true states and observations versus n. Observe how the observations become more or less noisy as $R(n)$ changes.

(e) Plot the true states $x_i(n)$ and estimated states $\hat{x}_i(n)$.

6.4. This problem investigates the effect of modeling errors. Let the plant dynamics be given by

$$\Phi_p = \begin{bmatrix} 0.4 & 0.2 \\ 2.3 & -0.5 \end{bmatrix}, \quad \Gamma_p = \begin{bmatrix} 0.2 & -0.15 \\ 0 & 0.4 \end{bmatrix}, \quad C_p = \begin{bmatrix} 1 & 0 \end{bmatrix},$$

with $Q_p = \text{diag}\left(\begin{bmatrix} 4 & 6 \end{bmatrix}\right)$, and $R_p = 20$. For the model, use the same system matrices as the plant, except set the state-transition matrix Φ_m of the *model* to

$$\Phi_m = \begin{bmatrix} 0.4 & 0.2 \\ 2.3 & -0.3 \end{bmatrix}.$$

(a) Is the plant stable? Is the model stable? Recall that a discrete-time LTI state model is stable if all eigenvalues of Φ have magnitudes less than one. The MATLAB function `eig` may be useful.

(b) Generate 120 samples $x_p(n)$ for the plant.

(c) Use the Kalman filter (with the incorrect model) to estimate the states $\hat{x}_m(n)$. Also compute $P_{11}(n)$ and $P_{22}(n)$ for $1 \leq n \leq 120$; these values give the Kalman filter's estimates of the MSE.

(d) Plot the true states $x_p(n)$ and estimated states $\hat{x}_m(n)$ versus n.

(e) The true MSE is $(x_{pi}(n) - \hat{x}_{mi}(n))^2$ for $i = 1, 2$. Plot the true MSE and $P_{ii}(n)$ against n.

(f) Does the MSE of the Kalman filter remain bounded?

6.5. Use the same state models in Problem 6.4 but replace the state-transition matrix Φ_p of the *plant* by

$$\Phi_p = \begin{bmatrix} 0.4 & 0.2 \\ 2.3 & -0.7 \end{bmatrix}.$$

(a) Is the plant stable?

(b) Generate 120 samples $x_p(n)$ for the plant.

(c) Use the Kalman filter (with the incorrect model) to estimate the states $\hat{x}_m(n)$. Also compute $P_{11}(n)$ and $P_{22}(n)$ for $1 \le n \le 120$; these values give the Kalman filter's estimates of the MSE.

(d) Plot the true states $x_p(n)$ and estimated states $\hat{x}_m(n)$ versus n.

(e) The true MSE is $(x_{pi}(n) - \hat{x}_{mi}(n))^2$ for $i = 1, 2$. Plot the true MSE and $P_{ii}(n)$ against n.

(f) Does the MSE of the Kalman filter remain bounded?

6.6. In some situations, the observations $z(n)$ may not always be available. For example, a sensor might briefly go off-line due to a momentary loss of power. Using the ideas of multistep Kalman prediction from Section 6.5, describe a Kalman filtering scheme that can continue running during occasional loss of measurements.

6.7. Define the m-step *a priori* error covariance matrix by

$$P_m^-(n) = \text{Cov}\left[\boldsymbol{x}(n+m) - \hat{\boldsymbol{x}}^-(n+m)\right].$$

(a) Show that $P_m^-(n)$ can be related to $P(n-1)$ by

$$P_m^-(n) = \Phi^{m+1} P(n-1) \left(\Phi^{\mathrm{T}}\right)^{m+1} + \sum_{k=0}^{m} \Phi^k \Gamma Q \Gamma^{\mathrm{T}} \left(\Phi^{\mathrm{T}}\right)^k.$$

(b) Just as the Kalman filter can reach steady-state under the proper conditions, so can the multistep predictor. Consult the ARE (5.103), (5.104) and (5.105) in Section 5.8 for the SSKF. Using the result from the previous part of this problem, give an expression for $\lim_{n\to\infty} P_m^-(n)$ if the multistep predictor reaches steady-state.

6.8. Using the state model in Example 6.1, implement the Kalman filter for multistep prediction as in Section 6.5. (Specify the initial state you used.)

(a) Implement the filter for steps of $m = 1, 2, 3$, and 4.

(b) Compare the estimates for these three predictors. How quickly does the error grow as m increases?

6.9. Implement the Kalman smoother of Example 6.1 in MATLAB. Specify the initial state you selected. Use $N = 100$ samples in your simulations.

6.10. For an LTI state model and stationary noises (Q and R constant), the Kalman filter converges to the SSKF (see Section 5.8). The SSKF can be determined by solving the algebraic Riccati equation (5.103). In a similar way, for $0 \ll n \ll N$, the Kalman smoother can reach steady-state.

Let S_{ss} denote $S(n)$ for the steady-state Kalman smoother. Show that, for the steady-state Kalman smoother,

$$S_{ss} = \Phi^{\mathrm{T}} \left[S_{ss} - S_{ss}\Gamma \left(\Gamma^{\mathrm{T}} S_{ss} \Gamma + Q^{-1}\right)^{-1} \Gamma^{\mathrm{T}} S_{ss}\right] \Phi + C^{\mathrm{T}} R^{-1} C.$$

Chapter 7

Kalman Filter Applications

At this point, we have derived the Kalman filter, presented some of its important properties, and demonstrated some simple examples. In this chapter, we examine some applications employing the Kalman filter. We first present the problem of tracking a single target based on noisy measurements. In this case, the SMM may be unstable, since the position of the target need not be zero-mean. We also consider three special cases of Kalman filtering: the case of colored (non-white) process noise, the case of correlated process and measurement noises, and the case of colored measurement noise. The target tracking problem is revisited for the case of measurements in polar, rather than Cartesian, form. Finally, we show how the Kalman filter can be used to estimate the parameters of a LTI system.

7.1 Target Tracking

The Kalman filter finds frequent application in target tracking problems. A target such as an aircraft is detected by a radar installation, such as an airport control tower or a surface-to-air missile system. The radar returns provide very noisy measurements of the position of the target. Given these measurements, we attempt to estimate the position and velocity of the target.

Let us present a simple example involving a single target. We assume that the target moves with approximately constant velocity, and measurements are provided in two-dimensional Cartesian coordinates, i.e, (x, y).

Constructing a Model

Let $x_c(t)$ and $y_c(t)$ denote the *continuous-time* horizontal and vertical Cartesian coordinates of the target, respectively. Then the target motion may be modelled via

$$\dot{x}_c(t) = v_x, \quad \text{and} \quad \dot{y}_c(t) = v_y,$$

where v_x and v_y are constants representing the target velocity in the x- and y-directions, respectively.

Certainly this model provides a mathematically valid description of the physical behavior of the target. However, in practice we do not know v_x or v_y, and this model is useless. Instead we take another derivative and find

$$\ddot{x}_c(t) = 0, \quad \text{and} \quad \ddot{y}_c(t) = 0.$$

This description also models a constant-velocity target, but it does not require knowledge of the actual velocity components.

Next we define a continuous-time state vector $\vec{x}_c(t)$ by

$$\vec{x}_c(t) = \begin{bmatrix} x_c(t) & \dot{x}_c(t) & y_c(t) & \dot{y}_c(t) \end{bmatrix}^{\mathrm{T}}.$$

(We adopt the arrow to distinguish the state vector $\vec{x}_c(t)$ from the scalar x-coordinate variable $x_c(t)$.) Note that $\vec{x}_c(t)$ is completely *deterministic* since it describes the *exact* position and velocity of the target.

Then the continuous-time state model for the position of the target is

$$\begin{aligned} \dot{\vec{\boldsymbol{x}}}_c(t) &= \begin{bmatrix} 0 & 1 & 0 & 0 \\ 0 & 0 & 0 & 0 \\ 0 & 0 & 0 & 1 \\ 0 & 0 & 0 & 0 \end{bmatrix} \vec{\boldsymbol{x}}_c(t) + \begin{bmatrix} 0 & 0 \\ 1 & 0 \\ 0 & 0 \\ 0 & 1 \end{bmatrix} \begin{bmatrix} \boldsymbol{w}_{c_x}(t) \\ \boldsymbol{w}_{c_y}(t) \end{bmatrix} \\ &= A\vec{\boldsymbol{x}}_c(t) + B_w \boldsymbol{w}_c(t). \end{aligned} \tag{7.1}$$

The process noise $\boldsymbol{w}_c(t)$ accounts for small maneuvers that do not otherwise fit the constant-velocity assumption. As a result, $\vec{\boldsymbol{x}}_c(t)$ and $\dot{\vec{\boldsymbol{x}}}_c(t)$ become random processes and are now boldfaced. We assume the covariance associated with $\boldsymbol{w}_c(t)$ is

$$Q_c = \begin{bmatrix} \sigma^2_{w_{cx}} & 0 \\ 0 & \sigma^2_{w_{cy}} \end{bmatrix}.$$

The measurement equation is

$$\boldsymbol{z}_c(t) = \begin{bmatrix} 1 & 0 & 0 & 0 \\ 0 & 0 & 1 & 0 \end{bmatrix} \vec{\boldsymbol{x}}_c(t) + \boldsymbol{v}_c(t), \tag{7.2}$$

with

$$\boldsymbol{v}_c(t) = \begin{bmatrix} \boldsymbol{v}_{c_x}(t) \\ \boldsymbol{v}_{c_y}(t) \end{bmatrix}, \quad \text{and} \quad R_c = \begin{bmatrix} \sigma^2_{v_{cx}} & 0 \\ 0 & \sigma^2_{v_{cy}} \end{bmatrix}.$$

Discretization

Since the Kalman filter is a discrete-time system, we must discretize the model (7.1–7.2). Our state vector is

$$\vec{\boldsymbol{x}}(n) = \begin{bmatrix}\boldsymbol{x}(n) & \dot{\boldsymbol{x}}(n) & \boldsymbol{y}(n) & \dot{\boldsymbol{y}}(n)\end{bmatrix}^{\mathrm{T}}.$$

Appendix A presents the details of discretizing a continuous-time system. The discretized state-transition matrix is

$$\Phi = \begin{bmatrix}1 & T & 0 & 0\\ 0 & 1 & 0 & 0\\ 0 & 0 & 1 & T\\ 0 & 0 & 0 & 1\end{bmatrix}. \tag{7.3}$$

To discretize the process noise we choose the second method presented in Appendix A It follows that $\boldsymbol{w}(n)$ is a zero-mean white noise process, and it is four-dimensional, i.e.,

$$\boldsymbol{w}(n) = \begin{bmatrix}\boldsymbol{w}_x(n) & \boldsymbol{w}_{\dot{x}}(n) & \boldsymbol{w}_y(n) & \boldsymbol{w}_{\dot{y}}(n)\end{bmatrix}^{\mathrm{T}}, \tag{7.4}$$

and $\Gamma = I_4$. We use (A.45) to find the covariance of $\boldsymbol{w}(n)$ and obtain

$$Q = \begin{bmatrix}\sigma^2_{w_{cx}}T^3/3 & \sigma^2_{w_{cx}}T^2/2 & 0 & 0\\ \sigma^2_{w_{cx}}T^2/2 & \sigma^2_{w_{cx}}T & 0 & 0\\ 0 & 0 & \sigma^2_{w_{cy}}T^3/3 & \sigma^2_{w_{cy}}T^2/2\\ 0 & 0 & \sigma^2_{w_{cy}}T^2/2 & \sigma^2_{w_{cy}}T\end{bmatrix}. \tag{7.5}$$

Hence the discretized signal state model takes the familiar form

$$\vec{\boldsymbol{x}}(n+1) = \Phi\vec{\boldsymbol{x}}(n) + \boldsymbol{w}(n), \tag{7.6}$$

where Φ is given by (7.3) and $\boldsymbol{w}(n)$ is zero-mean white noise with covariance given by (7.5).

Next we consider the measurement equation (7.2). From (A.12), we have

$$\boldsymbol{z}(n) = \begin{bmatrix}1 & 0 & 0 & 0\\ 0 & 0 & 1 & 0\end{bmatrix}\vec{\boldsymbol{x}}(n) + \boldsymbol{v}(n). \tag{7.7}$$

The measurement noise is two-dimensional ($\boldsymbol{v}(n) = \begin{bmatrix}\boldsymbol{v}_x(n) & \boldsymbol{v}_y(n)\end{bmatrix}^{\mathrm{T}}$), zero-mean white noise. We compute the covariance via (A.42) and obtain

$$R = \begin{bmatrix}\sigma^2_{v_{cx}}/T & 0\\ 0 & \sigma^2_{v_{cy}}/T\end{bmatrix}. \tag{7.8}$$

Filter Initialization

Having developed a model for the observations, we may now initialize the Kalman filter. We need to determine $\hat{\vec{x}}^-(0)$ and $P^-(0)$. From Equations (5.75) and (5.76), the ideal initial conditions are

$$\hat{\vec{x}}^-(0) = \mathrm{E}\left[\vec{x}(0)\right], \tag{7.9}$$

and

$$P^-(0) = \mathrm{Cov}\left[\vec{x}(0) - \hat{\vec{x}}^-(0)\right]. \tag{7.10}$$

In practice, we have little or no idea about these two quantities. In fact, they pose estimation problems of their own. The observation $\boldsymbol{z}(n)$ provides only positional information. We need at least two observations before we can make an initial estimate of the target's velocity. Several possible initial conditions follow.

First, we could simply take $\hat{\vec{x}}^-(0)$ to be the zero vector. Since we do not have any information regarding the true initial state $\vec{x}(0)$, this guess is just as likely as any other. This choice of $\hat{\vec{x}}^-(0)$ means that the Kalman filter may begin operating immediately.

Second, we could take the first measurement $\boldsymbol{z}(0)$ and assume the measured position is correct. This assumption is reasonable because $\boldsymbol{v}(n)$ has a mean of zero. As a result, the Kalman filter must wait for the measurement $\boldsymbol{z}(0)$ before it can be initialized. Then

$$\hat{\vec{x}}^-(0) = \begin{bmatrix} \boldsymbol{z}_1(0) & 0 & \boldsymbol{z}_2(0) & 0 \end{bmatrix}^{\mathrm{T}} \tag{7.11}$$

Third, we could take two measurements $\boldsymbol{z}(0)$ and $\boldsymbol{z}(1)$ and use them to approximate the target's position and velocity. In this case, we must wait for two observations before we can start the Kalman filter. We select the initial position as the midpoint of the two measured positions. The initial velocity is approximated by the difference between these positions, divided by the sampling period T. Then $\hat{\vec{x}}^-(0)$ is given by

$$\hat{\vec{x}}^-(0) = \begin{bmatrix} [\boldsymbol{z}_1(0) + \boldsymbol{z}_1(1)]/2 \\ [\boldsymbol{z}_1(1) - \boldsymbol{z}_1(0)]/T \\ [\boldsymbol{z}_2(0) + \boldsymbol{z}_2(1)]/2 \\ [\boldsymbol{z}_2(1) - \boldsymbol{z}_2(0)]/T \end{bmatrix}$$

Next we choose an appropriate matrix $P^-(0)$. Equation (7.10) becomes

$$P^-(0) = \mathrm{Cov}\left[\vec{x}(0)\right] - 2\mathrm{E}\left[\vec{x}(0)\left(\hat{\vec{x}}^-(0)\right)^{\mathrm{T}}\right] + \mathrm{Cov}\left[\hat{\vec{x}}^-(0)\right].$$

When implementing the filter, $\hat{\boldsymbol{x}}^-(0)$ is part of a sample realization of $\boldsymbol{z}(n)$, so $\hat{\boldsymbol{x}}^-(0)$ is actually a *deterministic* quantity. As a result, it can be removed from the expected values and is displayed as normal (non-bold) type. Then

$$P^-(0) = \operatorname{Cov}[\vec{\boldsymbol{x}}(0)] - 2\mathrm{E}[\vec{\boldsymbol{x}}(0)]\left(\hat{\vec{x}}^-(0)\right)^{\mathrm{T}} + \hat{\vec{x}}^-(0)\left(\hat{\vec{x}}^-(0)\right)^{\mathrm{T}},$$

but $\hat{\vec{x}}^-(0)$ is precisely $\mathrm{E}[\vec{\boldsymbol{x}}(0)]$, so

$$P^-(0) = \operatorname{Cov}[\vec{\boldsymbol{x}}(0)] - \hat{\vec{x}}^-(0)\left(\hat{\vec{x}}^-(0)\right)^{\mathrm{T}}.$$

Let us assume that the initial positions and velocities are independent, zero-mean random variables. Then we have

$$\begin{aligned} P^-(0) &= \mathrm{E}\left[\begin{bmatrix} \boldsymbol{x}^2(0) & 0 & 0 & 0 \\ 0 & \dot{\boldsymbol{x}}^2(0) & 0 & 0 \\ 0 & 0 & \boldsymbol{y}^2(0) & 0 \\ 0 & 0 & 0 & \dot{\boldsymbol{y}}^2(0) \end{bmatrix}\right] - \hat{\vec{x}}^-(0)\left(\hat{\vec{x}}^-(0)\right)^{\mathrm{T}} \\ &= \begin{bmatrix} \sigma_x^2 & 0 & 0 & 0 \\ 0 & \sigma_{\dot{x}}^2 & 0 & 0 \\ 0 & 0 & \sigma_y^2 & 0 \\ 0 & 0 & 0 & \sigma_{\dot{y}}^2 \end{bmatrix} - \hat{\vec{x}}^-(0)\left(\hat{\vec{x}}^-(0)\right)^{\mathrm{T}}. \end{aligned} \tag{7.12}$$

If we decide to adopt the first initialization for $\hat{\vec{x}}^-(0)$, the second term in (7.12) is zero. If we use the second or third initialization, then $P^-(0)$ may not be positive-definite due to the subtracted term. Of course, this result would make $P^-(0)$ an invalid covariance matrix. In order to avoid this problem, we choose

$$P^-(0) = \lambda I, \quad \lambda > 0,$$

which is simply equation (5.77).

Example 7.1 Single Target Tracking

Using MATLAB, we have simulated the tracking problem and implemented the Kalman filter. The sampling period was $T = 50$ ms. The process noise had $\sigma_{w_{cx}} = 50$ m/s and $\sigma_{w_{cy}} = 80$ m/s, and the measurement noise $\sigma_{v_{cx}} = 2$ km, $\sigma_{v_{cy}} = 2.25$ km. The initial position of the target was (30 km, 20 km) with initial velocity (100 km/s, 100 km/s), which is approximately 300 miles per hour. The filter is initialized with

$$\hat{\vec{x}}^-(0) = \begin{bmatrix} z_1(0) & 0 & z_2(0) & 0 \end{bmatrix}^{\mathrm{T}},$$

and

$$P^-(0) = \begin{bmatrix} 50 & 0 & 0 & 0 \\ 0 & 10 & 0 & 0 \\ 0 & 0 & 50 & 0 \\ 0 & 0 & 0 & 10 \end{bmatrix} = \text{diag}\left([50 \quad 10 \quad 50 \quad 10]\right).$$

Simulation results appear in Figures 7.1, 7.2, 7.3, and 7.4. Figure 7.1 shows the results for the target's horizontal position: noisy observations, true position, and the estimated position produced from the Kalman filter. Results for the vertical coordinate appear in Figure 7.2. In both cases, the Kalman filter tracked the target accurately.

Since the Kalman filter employs a state model, it also generates estimates of the target's velocity components (states $\vec{x}_2(n)$ and $\vec{x}_4(n)$). The actual and estimated velocities are shown in Figure 7.3. Again, the filter tracked the velocities well.

The Kalman filter also estimates the MSE, which is $\text{tr}\left(P(n)\right)$. It adjusts the gain factor $K(n)$ according to $P(n)$. Hence, it is worthwhile to compare the true MSE with the Kalman filter's estimated MSE. Results for a single simulation appear at the top of Figure 7.4. The estimated MSE was fairly close to the actual MSE.

Note, however, that MSE is an ensemble quantity, meaning it is the expected value of the results of many realizations. Hence, many simulations can be performed and the results averaged to form an experimental ensemble. This technique is known as the *Monte Carlo method*, with each realization called a *Monte Carlo simulation*. The resulting MSE and Kalman-estimated MSE from 100 Monte Carlo simulations appear at the bottom of Figure 7.4. The graph shows that Kalman-estimated ensemble MSE was very close to the true ensemble MSE.

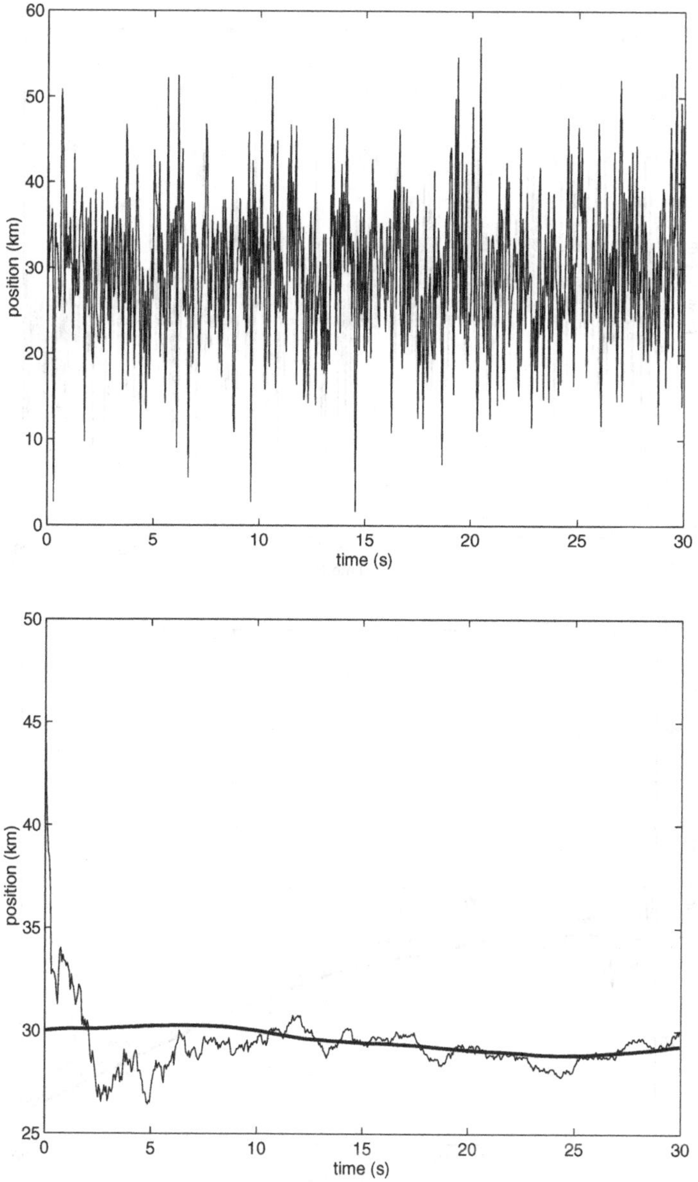

Figure 7.1. Target tracking results for Example 7.1. **Top:** Noisy observations $z_1(n)$ of the target's horizontal position. **Bottom:** Actual horizontal position $\vec{x}_1(n)$ or $x(n)$ (thick curve) and estimated position $\hat{\vec{x}}_1(n)$ or $\hat{x}(n)$ from the Kalman filter (thin curve).

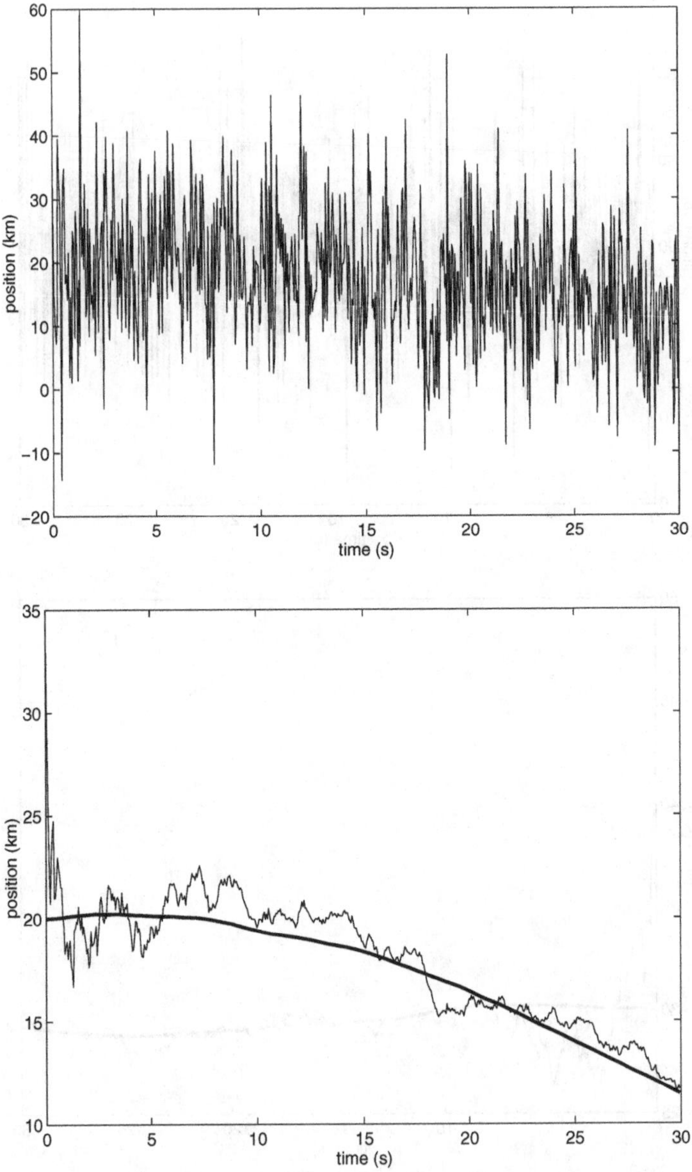

Figure 7.2. Target tracking results for Example 7.1, cont. **Top:** Noisy observations $z_2(n)$ of the target's vertical position. **Bottom:** Actual vertical position (thick curve) and estimate from the Kalman filter (thin curve).

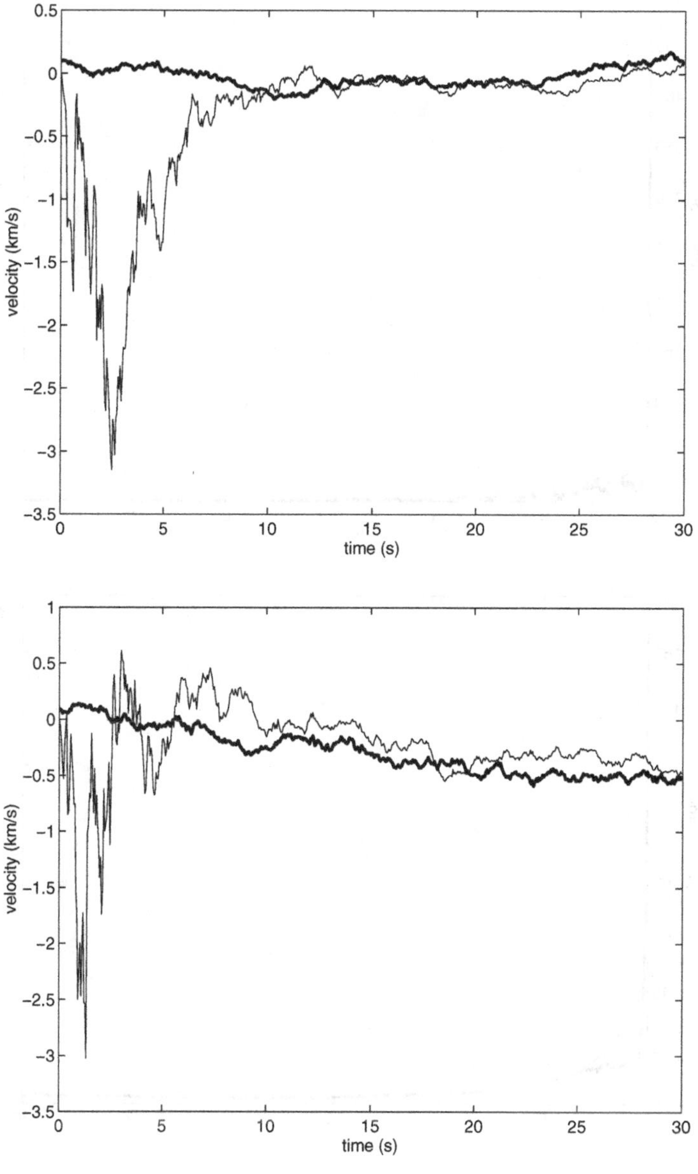

Figure 7.3. Target tracking results for Example 7.1, cont. **Top:** Actual horizontal velocity $\vec{x}_2(n)$ or $\dot{x}(n)$ (thick curve) and Kalman estimate (thin curve). **Bottom:** Vertical velocity and Kalman estimate.

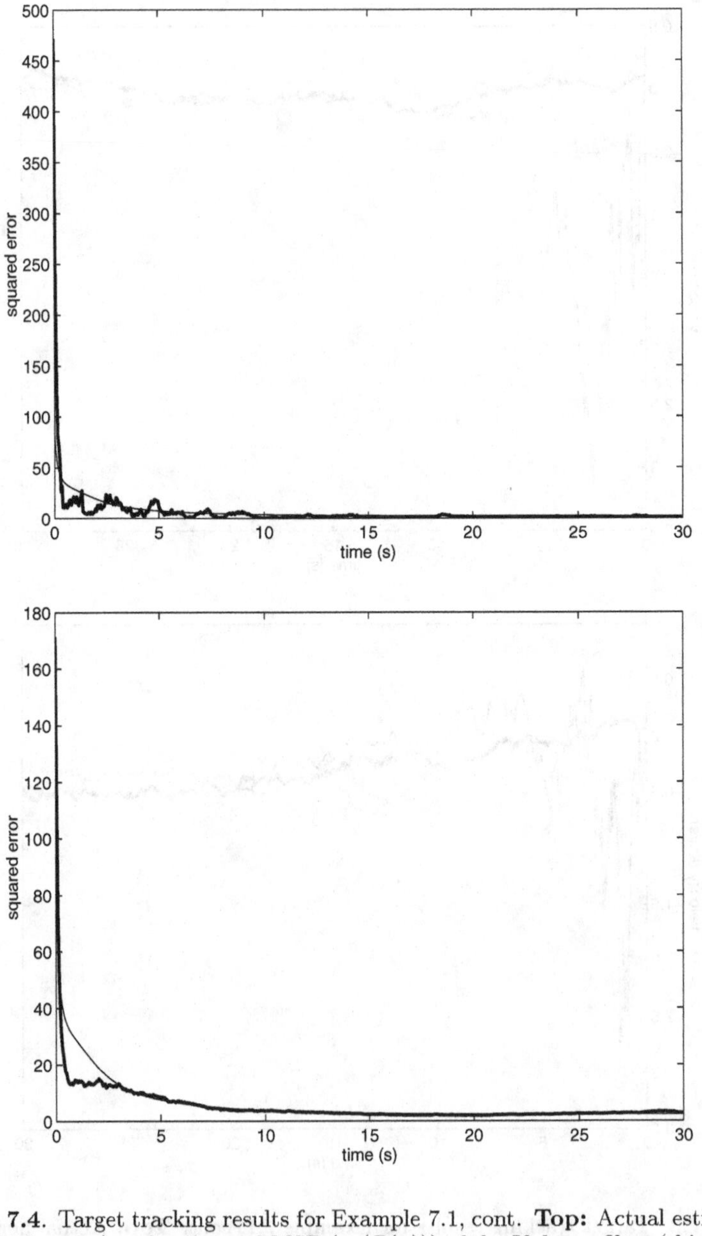

Figure 7.4. Target tracking results for Example 7.1, cont. **Top:** Actual estimation MSE (thick curve) and estimated MSE ($\text{tr}(P(n))$) of the Kalman filter (thin curve) for a single simulation. **Bottom:** Actual and Kalman-estimated MSEs averaged over 100 simulations.

Example 7.2 Tracking with Alternate Discretization

This example considers the tracking problem with an alternate discretization. From (A.9) and (7.1) we have

$$\Gamma = \int_0^T \begin{bmatrix} 1 & \tau & 0 & 0 \\ 0 & 1 & 0 & 0 \\ 0 & 0 & 1 & \tau \\ 0 & 0 & 0 & 1 \end{bmatrix} d\tau \begin{bmatrix} 0 & 0 \\ 1 & 0 \\ 0 & 0 \\ 0 & 1 \end{bmatrix} = \begin{bmatrix} T^2/2 & 0 \\ T & 0 \\ 0 & T^2/2 \\ 0 & T \end{bmatrix},$$

and (A.43) gives

$$Q = Q_c/T = \begin{bmatrix} \sigma^2_{w_{cx}}/T & 0 \\ 0 & \sigma^2_{w_{cy}}/T \end{bmatrix}.$$

The process noise vector is

$$\boldsymbol{w}(n) = \begin{bmatrix} w_x(n) & w_y(n) \end{bmatrix}^{\mathrm{T}}.$$

In this case, Q is diagonal, so the elements of $\boldsymbol{w}(n)$ are uncorrelated and are easily generated.

The other parameters of the simulation were the same as in Example 7.1, and the filter was initialized in the same way.

Simulation results with the same parameters as in Example 7.2 appear in Figures 7.5, 7.6, and 7.7. Position estimates appear in Figure 7.5 and velocity estimates in Figure 7.6. Figure 7.7 presents the actual and estimated MSE for this single simulation and for 100 Monte Carlo simulations. The estimated MSE was very close to the true MSE over the experimental ensemble. Overall, the performance with the alternate discretization is comparable to that of Example 7.2.

7.2 Colored Process Noise

The Kalman filter makes several assumptions about the process noise $\boldsymbol{w}(n)$ and measurement noise $\boldsymbol{v}(n)$, namely that both are white and uncorrelated with one another. In this section we modify the SMM to handle colored (i.e., non-white) process noise. Section 7.3 considers the case where $\boldsymbol{w}(n)$ and $\boldsymbol{v}(n)$ are correlated, and Section 7.4 handles colored measurement noise.

Suppose that $\boldsymbol{w}(n)$ is colored and stationary. Then it possesses a power spectrum $S_w(z)$, but $S_w(z)$ does not equal a constant. Suppose that we can perform a spectral factorization of $S_w(z)$ as in Section 4.4. Then

$$S_w(z) = S_w^+(z) S_w^+(z^{-1}),$$

where $S_w^+(z)$ has all poles and zeros inside the unit circle. Therefore $S_w^+(z)$ is the system function of a stable LTI system.

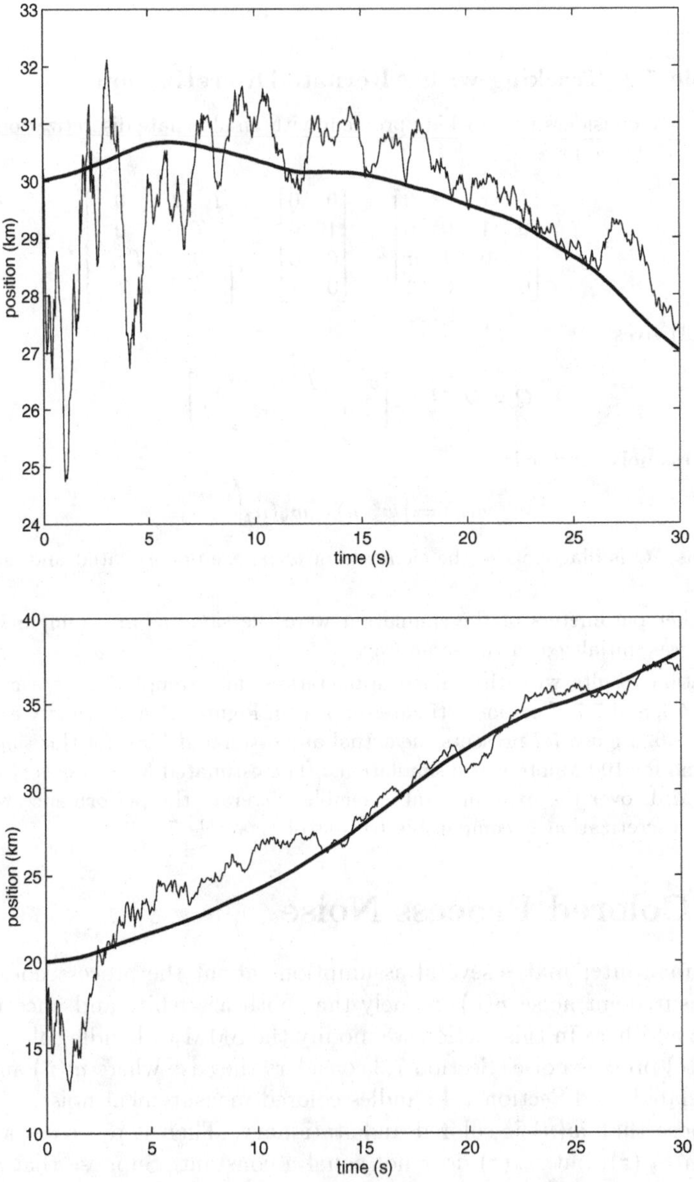

Figure 7.5. Target tracking results for Example 7.2. **Top:** Actual horizontal position (thick curve) and estimated position from the Kalman filter (thin curve). **Bottom:** Vertical position and estimates.

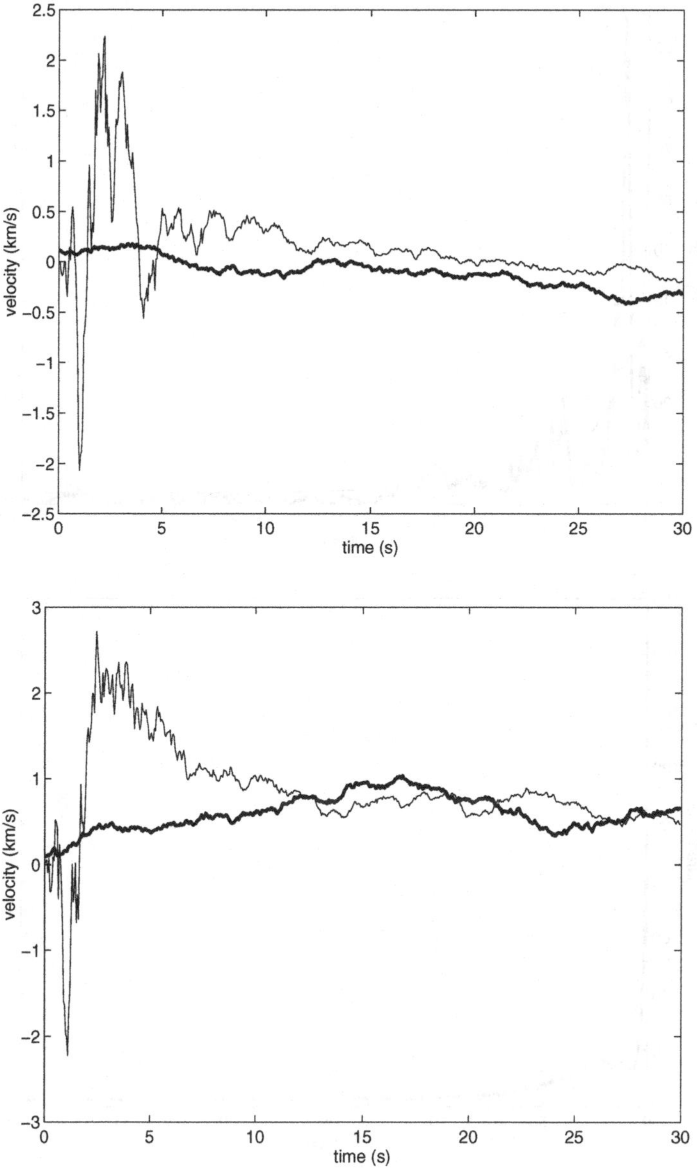

Figure 7.6. Target tracking results for Example 7.2, cont. **Top:** Actual horizontal velocity (thick curve) and Kalman estimate (thin curve). **Bottom:** Vertical velocity and Kalman estimate.

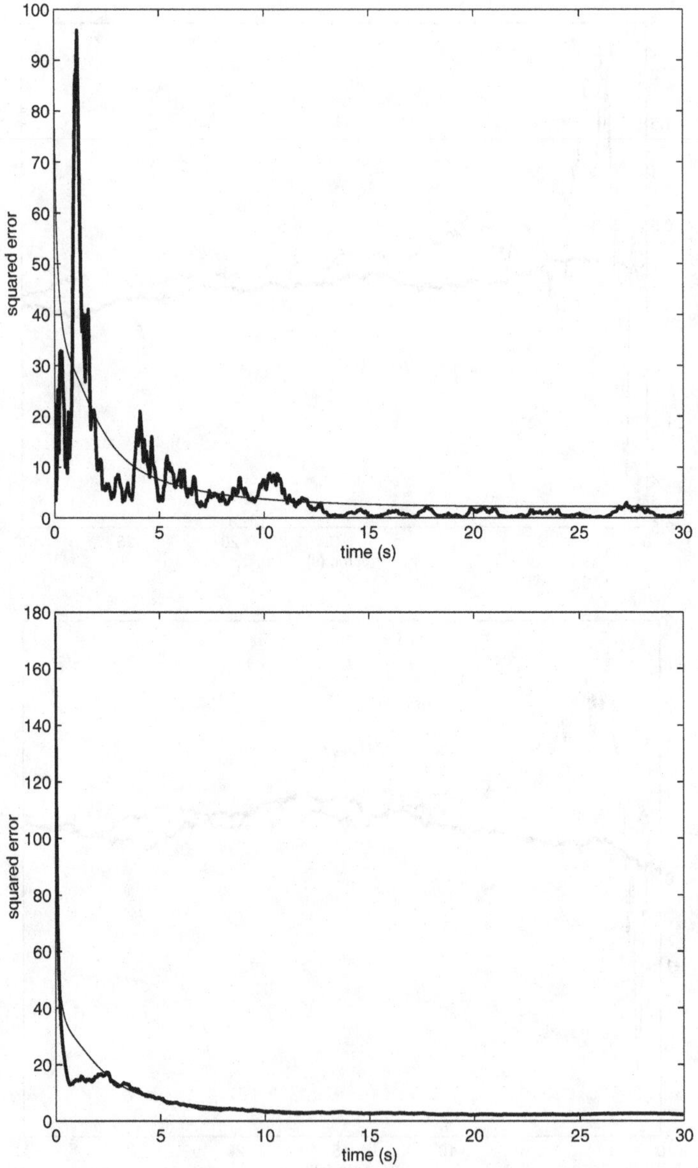

Figure 7.7. Target tracking results for Example 7.2, cont. **Top:** Actual MSE (thick curve) and estimated MSE ($\mathrm{tr}\,(P(n))$) of the Kalman filter (thin curve) for a single simulation. **Bottom:** Actual and Kalman-estimated MSEs averaged over 100 simulations.

We can use canonical forms from Appendix A to find a state model for $S_w^+(z)$. Let Φ_w, Γ_w, C_w, and D_w constitute such a state model, so

$$\boldsymbol{x}_w(n+1) = \Phi_w \boldsymbol{x}_w(n) + \Gamma_w \boldsymbol{w}'(n) \tag{7.13}$$
$$\boldsymbol{w}(n) = C_w \boldsymbol{x}_w(n) + D_w \boldsymbol{w}'(n), \tag{7.14}$$

and

$$S_w^+(z) = C_w(zI - \Phi_w)^{-1}\Gamma_w + D_w.$$

Let $\boldsymbol{w}'(n)$ be zero-mean white noise with unit variance ($\sigma_{w'}^2 = 1$), and suppose $\boldsymbol{w}'(n)$ is the input to $S_w^+(z)$. Then the output process will have power spectrum $S_w(z)$. Figure 7.2 demonstrates this relationship. As a result, we may view $\boldsymbol{w}(n)$ as the output of a LTI system driven by white noise, and $S_w^+(z)$ may be considered a *noise-shaping filter.*

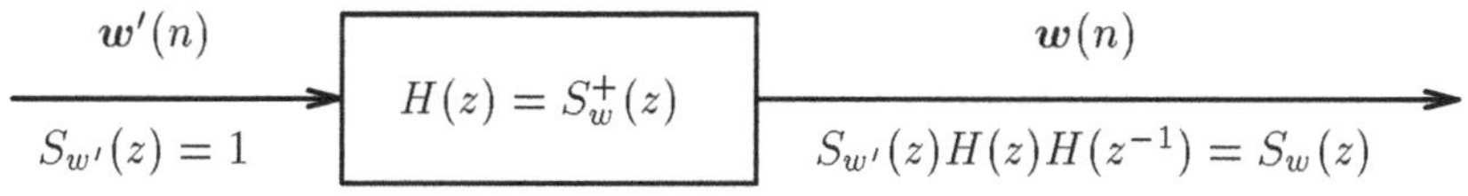

Figure 7.8. A noise-shaping filter. White noise $\boldsymbol{w}'(n)$ is input to the filter $H(z)$, which shapes the flat power spectrum of $\boldsymbol{w}'(n)$ into that of a colored process $\boldsymbol{w}(n)$.

Now we have replaced the colored process noise by a system driven by white noise. Next we modify the SMM with colored process noise to create a new SMM that conforms to the assumptions in the usual Kalman filter. Define a new state vector $\boldsymbol{x}'(n)$ by

$$\boldsymbol{x}'(n) \triangleq \begin{bmatrix} \boldsymbol{x}(n) \\ \boldsymbol{x}_w(n) \end{bmatrix}.$$

Substituing (7.14) into (5.36), we have

$$\boldsymbol{x}(n+1) = \Phi \boldsymbol{x}(n) + \Gamma C_w \boldsymbol{x}_w(n) + \Gamma D_w \boldsymbol{w}'(n).$$

Then our new SMM becomes

$$\boldsymbol{x}'(n+1) = \begin{bmatrix} \Phi & \Gamma C_w \\ 0 & \Phi_w \end{bmatrix} \boldsymbol{x}'(n) + \begin{bmatrix} \Gamma D_w \\ \Gamma_w \end{bmatrix} \boldsymbol{w}'(n) \tag{7.15}$$
$$\boldsymbol{z}(n) = \begin{bmatrix} C & 0 \end{bmatrix} \boldsymbol{x}'(n) + \boldsymbol{v}(n), \tag{7.16}$$

where $\boldsymbol{w}'(n)$ is zero-mean white noise with unit variance, and $\boldsymbol{w}'(n)$ and $\boldsymbol{v}(n)$ are uncorrelated.

Equations (7.15) and (7.16) define a new SMM that has *white* process noise. Notice that the Kalman filter itself does not change, but the SMM has been augmented. We may implement the Kalman filter using this new SMM.

Remarks

We have only considered the case of spectral factorization for a scalar process $\boldsymbol{w}(n)$. Factorization of a vector process $\boldsymbol{w}(n)$ can also be performed, although it is more complicated and requires a knowledge of MIMO system function matrices. See, for example, Brogan [8]. Additionally, Anderson and Moore [15] consider state models for generating nonstationary colored processes from white noise.

Example 7.3 Colored Process Noise

Let us return to the target tracking problem in Example 7.2. Recall that the process noise is $\boldsymbol{w}(n) = \begin{bmatrix} w_x(n) & w_y(n) \end{bmatrix}^{\mathrm{T}}$. Suppose that the process noise components $w_x(n)$ and $w_y(n)$ are colored. Specifically, let

$$w_x(n) = \alpha_x w_x(n-1) + \beta_x w'_x(n),$$

and

$$w_y(n) = \alpha_y w_y(n-1) + \beta_y w'_y(n),$$

where $w'_x(n)$ and $w'_y(n)$ are white Gaussian noise process with unit variance and are uncorrelated with each other ($Q' = \mathrm{E}\left[\boldsymbol{w}'(n)(\boldsymbol{w}'(n))^{\mathrm{T}}\right] = I_2$).

Now we may describe the process noise by its own state model after (7.13) and (7.14):

$$\begin{bmatrix} x_{w_x}(n+1) \\ x_{w_y}(n+1) \end{bmatrix} = \begin{bmatrix} \alpha_x & 0 \\ 0 & \alpha_y \end{bmatrix} \begin{bmatrix} x_{w_x}(n) \\ x_{w_y}(n) \end{bmatrix} + \begin{bmatrix} \beta_x & 0 \\ 0 & \beta_y \end{bmatrix} \begin{bmatrix} w'_x(n) \\ w'_y(n) \end{bmatrix}$$
$$\begin{bmatrix} w_x(n) \\ w_y(n) \end{bmatrix} = \begin{bmatrix} 1 & 0 \\ 0 & 1 \end{bmatrix} \begin{bmatrix} x_{w_x}(n) \\ x_{w_y}(n) \end{bmatrix}.$$

To incorporate the dynamics of $\boldsymbol{w}(n)$ in the SMM, we define a new state vector

$$\vec{\boldsymbol{x}}'(n) = \begin{bmatrix} x(n) \\ \dot{x}(n) \\ y(n) \\ \dot{y}(n) \\ w_x(n) \\ w_y(n) \end{bmatrix}.$$

According to (7.15) and (7.16), the new SMM becomes

$$\vec{x}'(n+1) = \begin{bmatrix} 1 & T & 0 & 0 & T^2/2 & 0 \\ 0 & 1 & 0 & 0 & T & 0 \\ 0 & 0 & 1 & T & 0 & T^2/2 \\ 0 & 0 & 0 & 1 & 0 & T \\ 0 & 0 & 0 & 0 & \alpha_x & 0 \\ 0 & 0 & 0 & 0 & 0 & \alpha_y \end{bmatrix} \vec{x}'(n) + \begin{bmatrix} 0 & 0 \\ 0 & 0 \\ 0 & 0 \\ 0 & 0 \\ \beta_x & 0 \\ 0 & \beta_y \end{bmatrix} \begin{bmatrix} w'_x(n) \\ w'_y(n) \end{bmatrix}$$

$$z(n) = \begin{bmatrix} 1 & 0 & 0 & 0 & 0 & 0 \\ 0 & 0 & 1 & 0 & 0 & 0 \end{bmatrix} \vec{x}'(n) + v(n).$$

The measurement noise is not affected by the new SMM, so R remains the same.

During simulations, the parameters for generating $w(n)$ were $\alpha_x = 0.985$, $\alpha_y = 0.95$, and

$$\beta_i = \sqrt{(1-\alpha_i^2)\sigma_{w_c}^2/T}, \quad i = x, y.$$

The reader may verify that this choice of parameters yields

$$\sigma_{w_i}^2 = \sigma_{w_{c_i}}^2/T, \quad i = x, y,$$

so that $Q = \mathrm{E}\left[w(n)w^{\mathrm{T}}(n)\right] = Q_c/T$ as in Example 7.2.

The Kalman filter was initialized with

$$\hat{\vec{x}}^-(0) = \begin{bmatrix} z_1(0) & 0 & z_2(0) & 0 & 0 & 0 \end{bmatrix}^{\mathrm{T}},$$

and

$$P^-(0) = \mathrm{diag}\left(\begin{bmatrix} 50 & 10 & 50 & 10 & 1 & 1 \end{bmatrix}\right).$$

All other parameters were identical to those in Example 7.2.

Simulation results of the observations and estimates appear in Figure 7.9 and the top of Figure 7.10. Results of applying a regular Kalman filter that does not consider the colored property of $w(n)$ are also shown. The regular filter is identical to the filter employed in Example 7.2. Not surprisingly, the Kalman filter that accounts for colored process noise produced more accurate estimates than the regular filter. As time passes, the regular filter lost accuracy.

To examine the ensemble performance, 100 Monte Carlo simulations were conducted. The bottom of Figure 7.10 shows the actual MSE and the estimated MSE for the colored-noise filter; the estimated MSE closely matched the true MSE. Figure 7.11 shows the actual and estimated MSE for the regular filter. As time progressed, the actual MSE decreased briefly but then increased. Yet the estimated MSE became small, which means that the filter believed it was producing an accurate estimate.

Finally, the bottom plot of Figure 7.11 shows the relative improvement in true MSE that results from using the colored-noise filter rather than the regular filter. Although the filters initially perform comparably, after 30 seconds the colored-noise filter has reduced the actual MSE 7 dB over the regular filter.

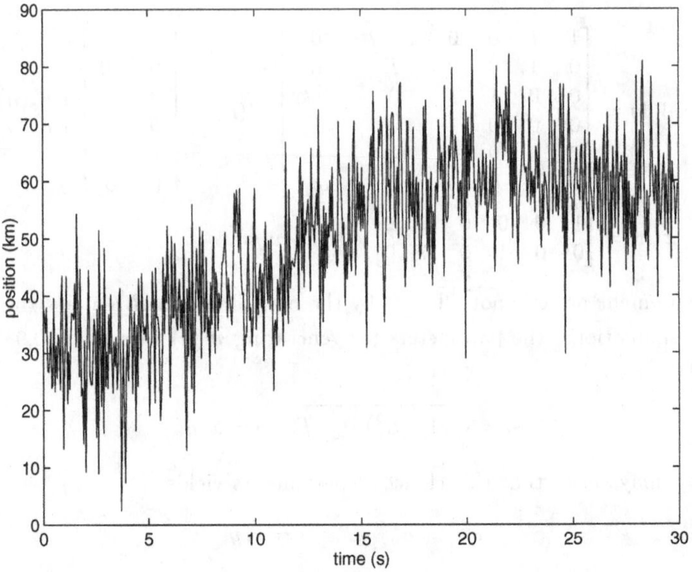

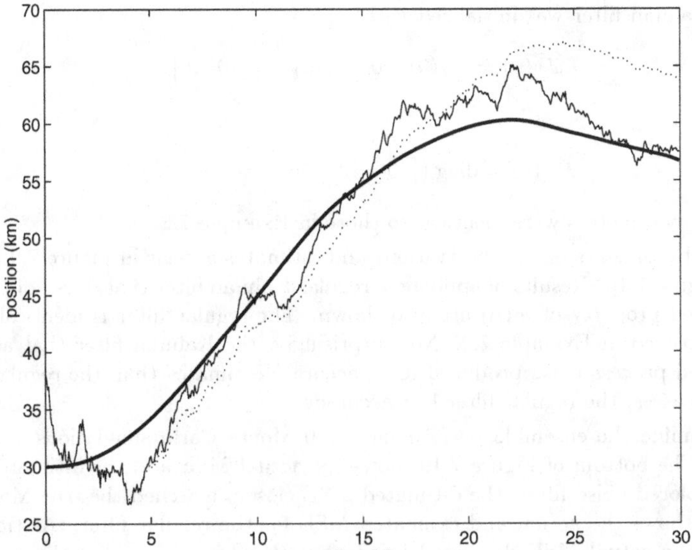

Figure 7.9. Colored process noise results for Example 7.3. **Top:** Noisy observations of the target's horizontal position. **Bottom:** Actual horizontal position (thick curve), Kalman estimate without considering the colored nature of the process noise (dotted curve), and Kalman estimate incorporating colored noise (thin curve).

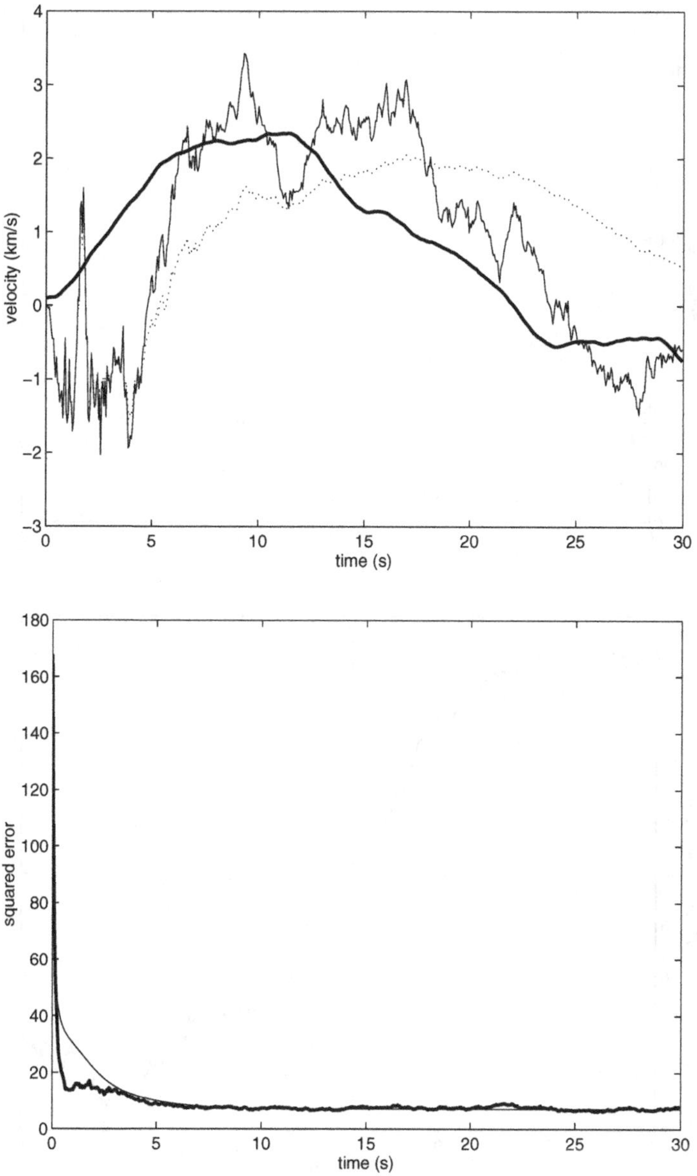

Figure 7.10. Colored process noise results for Example 7.3, cont. **Top:** Actual horizontal velocity (thick curve), estimated velocity with regular Kalman filter (dotted curve), and estimated velocity when colored noise is incorporated. Results are for a single simulation. **Bottom:** Actual MSE (thick curve) and colored-noise Kalman filter estimated MSE (thin curve). Results are for 100 Monte Carlo simulations.

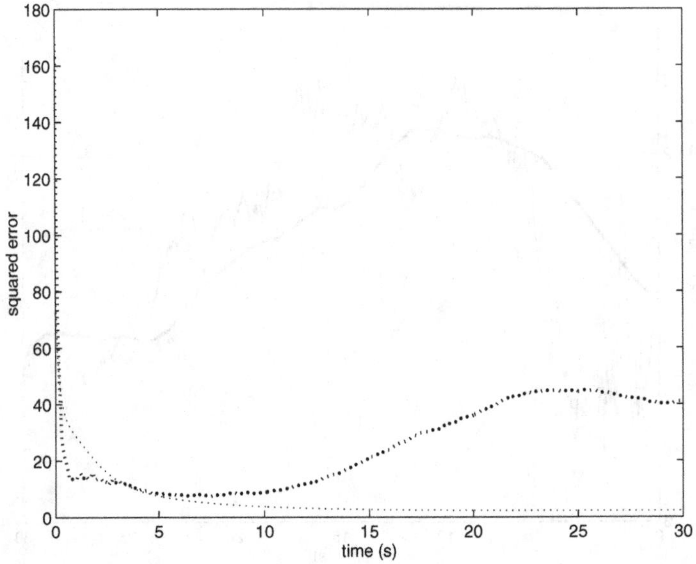

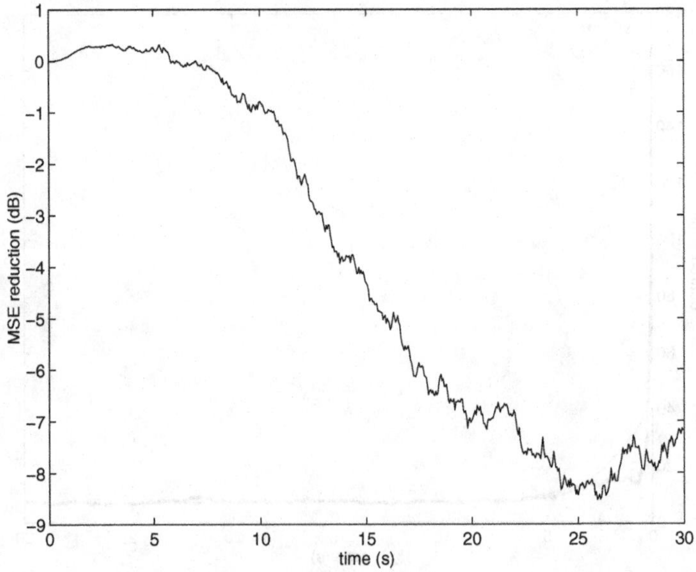

Figure 7.11. Colored process noise results for Example 7.3, cont. Results from 100 Monte Carlo simulations. **Top:** Actual MSE (thick dotted curve) and estimated MSE with regular Kalman filter (thin dotted curve). **Bottom:** Average reduction in actual MSE due to using the colored-noise Kalman filter instead of the regular Kalman filter.

7.3 Correlated Noises

In many applications, the assumption that $\boldsymbol{w}(n)$ and $\boldsymbol{v}(n)$ are independent is not valid. By exploiting the correlation between them, we expect to improve our estimates of $\boldsymbol{x}(n)$ and $\boldsymbol{s}(n)$. In this case, the SMM of Equations (5.36–5.40) applies, but rather than $\mathrm{E}\left[\boldsymbol{w}(i)\boldsymbol{v}^{\mathrm{T}}(j)\right] = \mathbf{0}$, we have

$$\mathrm{E}\left[\boldsymbol{w}(i)\boldsymbol{v}^{\mathrm{T}}(j)\right] = S\delta(i-j). \tag{7.17}$$

Since $\boldsymbol{w}(n)$ and $\boldsymbol{v}(n)$ are white, they are *not* correlated over time, but at any specific time index n they are correlated with cross-correlation S.

Notice that $\boldsymbol{x}(n)$ and $\boldsymbol{v}(n)$ remain uncorrelated. Hence, Equations (5.80), (5.81), and (5.82) still hold, namely,

$$\begin{aligned}
K(n) &= P^-(n)C^{\mathrm{T}}\left[CP^-(n)C^{\mathrm{T}} + R\right]^{-1}, \\
\hat{x}(n) &= \hat{x}^-(n) + K(n)\left[z(n) - C\hat{x}^-(n)\right], \\
P(n) &= P^-(n) - K(n)CP^-(n).
\end{aligned}$$

Estimating the Process Noise

By analogy with (5.83) it can be shown that

$$\hat{\boldsymbol{x}}^-(n+1) = \Phi\hat{\boldsymbol{x}}(n) + \Gamma\hat{\boldsymbol{w}}(n), \tag{7.18}$$

but due to the correlation between $\boldsymbol{w}(n)$ and $\boldsymbol{v}(n)$, $\hat{\boldsymbol{w}}(n)$ is generally nonzero. We must determine $\hat{\boldsymbol{w}}(n)$.

Just as we used the innovations (Sections 6.1 and 6.2) to estimate $\boldsymbol{x}(n)$, we may use them to estimate $\boldsymbol{w}(n)$. Since $\boldsymbol{v}(n)$ is white noise, $\hat{\boldsymbol{v}}^-(n) = 0$, so the innovation $\boldsymbol{\varepsilon}(n)$ satisfies

$$\boldsymbol{\varepsilon}(n) = \boldsymbol{z}(n) - C\hat{\boldsymbol{x}}^-(n) - \hat{\boldsymbol{v}}^-(n) = \boldsymbol{z}(n) - C\hat{\boldsymbol{x}}^-(n).$$

To estimate $\boldsymbol{w}(n)$ from the innovations, we apply (6.22) to find

$$\hat{\boldsymbol{w}}(n) = \hat{\boldsymbol{w}}^-(n) + \mathrm{E}\left[\boldsymbol{w}(n)\boldsymbol{\varepsilon}^{\mathrm{T}}(n)\right]\left(\mathrm{Cov}\left[\boldsymbol{\varepsilon}(n)\right]\right)^{-1}\boldsymbol{\varepsilon}(n). \tag{7.19}$$

Because $\boldsymbol{w}(n)$ is white noise, $\hat{\boldsymbol{w}}^-(n) = 0$.

Now we must find the two matrices that multiply $\boldsymbol{\varepsilon}(n)$ in (7.19). Since $\boldsymbol{x}(n)$ and $\hat{\boldsymbol{x}}^-(n)$ depend only on $\boldsymbol{w}(n-1)$, $\boldsymbol{w}(n-2), \ldots$, the first matrix is

$$\begin{aligned}
\mathrm{E}\left[\boldsymbol{w}(n)\boldsymbol{\varepsilon}^{\mathrm{T}}(n)\right] &= \mathrm{E}\left[\boldsymbol{w}(n)\left(C\boldsymbol{x}(n) + \boldsymbol{v}(n) - C\hat{\boldsymbol{x}}^-(n)\right)^{\mathrm{T}}\right] \\
&= \mathrm{E}\left[\boldsymbol{w}(n)\boldsymbol{v}^{\mathrm{T}}(n)\right] = S.
\end{aligned}$$

We determined the second matrix $(\mathrm{Cov}\left[\varepsilon(n)\right])^{-1}$ in Section 6.2; it is given by Equation (6.25), which we restate here:

$$\mathrm{E}\left[\varepsilon(n)\varepsilon^{\mathrm{T}}(n)\right] = CP^{-}(n)C^{\mathrm{T}} + R.$$

Then we conclude that (7.19) becomes

$$\hat{\boldsymbol{w}}(n) = S\left[CP^{-}(n)C^{\mathrm{T}} + R\right]^{-1}\left[\boldsymbol{z}(n) - C\hat{\boldsymbol{x}}^{-}(n)\right]. \tag{7.20}$$

The *a priori* estimate

Now we seek a recursive expression for computing $\hat{\boldsymbol{x}}^{-}(n)$. Substituting (7.20), (5.80), and (5.81) into (7.18), we have

$$\begin{aligned}\hat{\boldsymbol{x}}^{-}(n+1) &= \Phi\hat{\boldsymbol{x}}^{-}(n) \\ &\quad + \left[\Phi P^{-}(n)C^{\mathrm{T}} + \Gamma S\right]\left[CP^{-}(n)C^{\mathrm{T}} + R\right]^{-1}\left[\boldsymbol{z}(n) - C\hat{\boldsymbol{x}}^{-}(n)\right].\end{aligned}$$

Define a *Kalman gain for correlated noises* $K_c(n)$ as

$$\boxed{K_c(n) = \left[\Phi P^{-}(n)C^{\mathrm{T}} + \Gamma S\right]\left[CP^{-}(n)C^{\mathrm{T}} + R\right]^{-1},} \tag{7.21}$$

so the desired *a priori* estimate is given by

$$\boxed{\hat{\boldsymbol{x}}^{-}(n+1) = \Phi\hat{\boldsymbol{x}}^{-}(n) + K_c(n)\left[\boldsymbol{z}(n) - C\hat{\boldsymbol{x}}^{-}(n)\right].} \tag{7.22}$$

The *a priori* Error Covariance

We still need a recursion for updating $P^{-}(n+1)$ so that we can update $K_c(n)$ correctly. We begin by noting that

$$\tilde{\boldsymbol{x}}^{-}(n+1) = \left[\Phi - K_c(n)C\right]\tilde{\boldsymbol{x}}^{-}(n) + \Gamma\boldsymbol{w}(n) - K_c(n)\boldsymbol{v}(n).$$

Then

$$\begin{aligned}P^{-}(n+1) = \left[\Phi - K_c(n)C\right] & P^{-}(n)\left[\Phi - K_c(n)C\right]^{\mathrm{T}} \\ &+ \Gamma Q\Gamma^{\mathrm{T}} + K_c(n)RK_c^{\mathrm{T}}(n) \\ &- \Gamma SK_c^{\mathrm{T}}(n) - K_c(n)S^{\mathrm{T}}\Gamma^{\mathrm{T}}.\end{aligned}$$

Direct substitution of (7.21) for $K_c(n)$ produces the desired result:

$$\boxed{P^{-}(n+1) = \Phi P^{-}(n)\Phi^{\mathrm{T}} + \Gamma Q\Gamma^{\mathrm{T}} - K_c(n)\left[CP^{-}(n)C^{\mathrm{T}} + R\right]K_c^{\mathrm{T}}(n).} \tag{7.23}$$

Recursion

We can now combine the equations to obtain the filter. Since $\boldsymbol{w}(n)$ and $\boldsymbol{v}(n)$ are correlated at the same time index n, the time update has changed. Since they are not correlated over time, however, the measurement update remains the same as in the regular Kalman filter. Initialization is the same as for the regular Kalman filter (5.75) and (5.76).

- *Measurement update.* Identical to regular Kalman filter:

$$K(n) = P^{-}(n)C^{\mathrm{T}}\left[CP^{-}(n)C^{\mathrm{T}} + R\right]^{-1} \tag{7.24}$$

$$\hat{x}(n) = \hat{x}^{-}(n) + K(n)\left[z(n) - C\hat{x}^{-}(n)\right] \tag{7.25}$$

$$P(n) = P^{-}(n) - K(n)CP^{-}(n) \tag{7.26}$$

- *Time update.* Modified for correlated noises:

$$K_c(n) = \left[\Phi P^{-}(n)C^{\mathrm{T}} + \Gamma S\right]\left[CP^{-}(n)C^{\mathrm{T}} + R\right]^{-1} \tag{7.27}$$

$$\hat{x}^{-}(n+1) = \Phi\hat{x}^{-}(n) + K_c(n)\left[z(n) - C\hat{x}^{-}(n)\right] \tag{7.28}$$

$$\begin{aligned} P^{-}(n+1) &= \Phi P^{-}(n)\Phi^{\mathrm{T}} + \Gamma Q \Gamma^{\mathrm{T}} \\ &\quad - K_c(n)\left[CP^{-}(n)C^{\mathrm{T}} + R\right]K_c^{\mathrm{T}}(n) \end{aligned} \tag{7.29}$$

- *Time increment.* Increment n and repeat.

From (7.27) and (7.29), we see that S affects the *a priori* estimation error covariance $P^{-}(n)$ to account for the correlation between $\boldsymbol{w}(n)$ and $\boldsymbol{v}(n)$. Note that if $S = 0$, equation (7.29) reduces to the Riccati equation (5.89), and (7.27) reduces to the usual Kalman gain pre-multiplied by Φ.

In the case of nonstationary correlation between $\boldsymbol{w}(n)$ and $\boldsymbol{v}(n)$, we have

$$\mathrm{E}\left[\boldsymbol{w}(i)\boldsymbol{v}^{\mathrm{T}}(j)\right] = S(i)\delta(i-j). \tag{7.30}$$

Then S is replaced by $S(n)$ in the preceding derivation and recursion equation (7.27).

Example 7.4 Correlated Noise

We again consider the target tracking situation in Example 7.2. We have $\boldsymbol{w}(n) = \left[\boldsymbol{w}_x(n) \quad \boldsymbol{w}_y(n)\right]^{\mathrm{T}}$ and $\boldsymbol{v}(n) = \left[\boldsymbol{v}_x(n) \quad \boldsymbol{v}_y(n)\right]^{\mathrm{T}}$.

Suppose that these noise processes are related to one another via

$$\boldsymbol{v}(n) = A(n)\boldsymbol{w}(n-1) + B(n)\boldsymbol{d}(n),$$

where $\boldsymbol{d}(n) = \begin{bmatrix} d_x(n) & d_y(n) \end{bmatrix}^{\mathrm{T}}$ is white Gaussian noise with covariance I, and $\boldsymbol{d}(n)$ is uncorrelated with $\boldsymbol{w}(n)$. It follows that $\boldsymbol{w}(n)$ and $\boldsymbol{v}(n)$ are correlated, with

$$S(n) = \mathrm{E}\left[\boldsymbol{w}(n)\boldsymbol{v}^{\mathrm{T}}(n)\right] = Q(n)A^{\mathrm{T}}(n).$$

The covariance of $\boldsymbol{v}(n)$ becomes

$$R(n) = A(n)Q(n)A^{\mathrm{T}}(n) + B(n)B^{\mathrm{T}}(n).$$

For the case of constant dynamics, $A(n) = A$, $B(n) = B$, $Q(n) = Q$, $R(n) = R$, and $S(n) = S$.

In this example, we used

$$A = \begin{bmatrix} 39.75 & 0 \\ 0 & 28.00 \end{bmatrix} \quad \text{and} \quad B = \begin{bmatrix} 0.9984 & 0 \\ 0 & 0.9476 \end{bmatrix}.$$

This choice of parameters ensured that the discretized noise covariances were the same as in Example 7.2:

$$Q = \begin{bmatrix} 0.050 & 0 \\ 0 & 0.128 \end{bmatrix} \quad \text{and} \quad R = \begin{bmatrix} 80.00 & 0 \\ 0 & 101.25 \end{bmatrix}.$$

Then

$$S = \begin{bmatrix} 1.9875 & 0 \\ 0 & 3.5840 \end{bmatrix}.$$

By analogy with the correlation coefficient (3.54), the normalized cross-correlation matrix is given by

$$\begin{bmatrix} \dfrac{\mathrm{E}[w_x(n)v_x(n)]}{\sigma_{w_x}\sigma_{v_x}} & \dfrac{\mathrm{E}[w_x(n)v_y(n)]}{\sigma_{w_x}\sigma_{v_y}} \\ \dfrac{\mathrm{E}[w_y(n)v_x(n)]}{\sigma_{w_y}\sigma_{v_x}} & \dfrac{\mathrm{E}[w_y(n)v_y(n)]}{\sigma_{w_y}\sigma_{v_y}} \end{bmatrix} = \begin{bmatrix} 0.9938 & 0 \\ 0 & 0.9956 \end{bmatrix},$$

which shows that $w_x(n)$ and $v_x(n)$ are highly correlated, as are $w_y(n)$ and $v_y(n)$.

Tracking results for a single simulation appear in Figures 7.12 and 7.13 (top). The results are compared with a regular Kalman filter. Both filters have comparable tracking performance for this simulation.

To compare the ensemble performance, 100 Monte Carlo simulations were conducted. Figure 7.13 (bottom) shows the true MSE and estimated MSE for the Kalman filter with correlated noise; the estimated MSE was very close to the actual MSE. Figure 7.14 (top) compares the true and estimated MSE for the regular Kalman filter. Here, the true MSE was actually smaller than the MSE that the filter estimated. Since the estimated MSE was incorrect, the filter did not adjust the estimates as well as possible. As the bottom of Figure 7.14 shows, the correlated-noise filter eventually reduced its true MSE to about one-third of the true MSE of the regular filter.

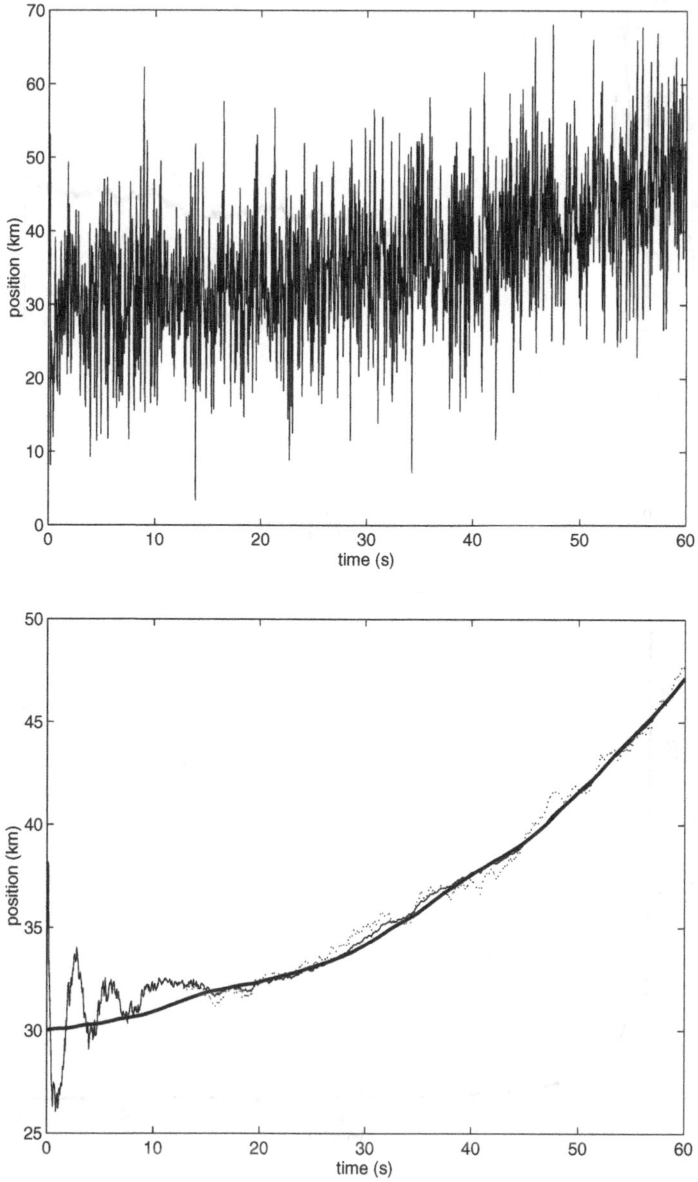

Figure 7.12. Correlated noise results for Example 7.4. **Top:** Noisy observations of the target's horizontal position. **Bottom:** Actual horizontal position (thick curve), Kalman estimate without considering correlated noise (dotted curve), and Kalman estimate incorporating correlated noise (thin curve).

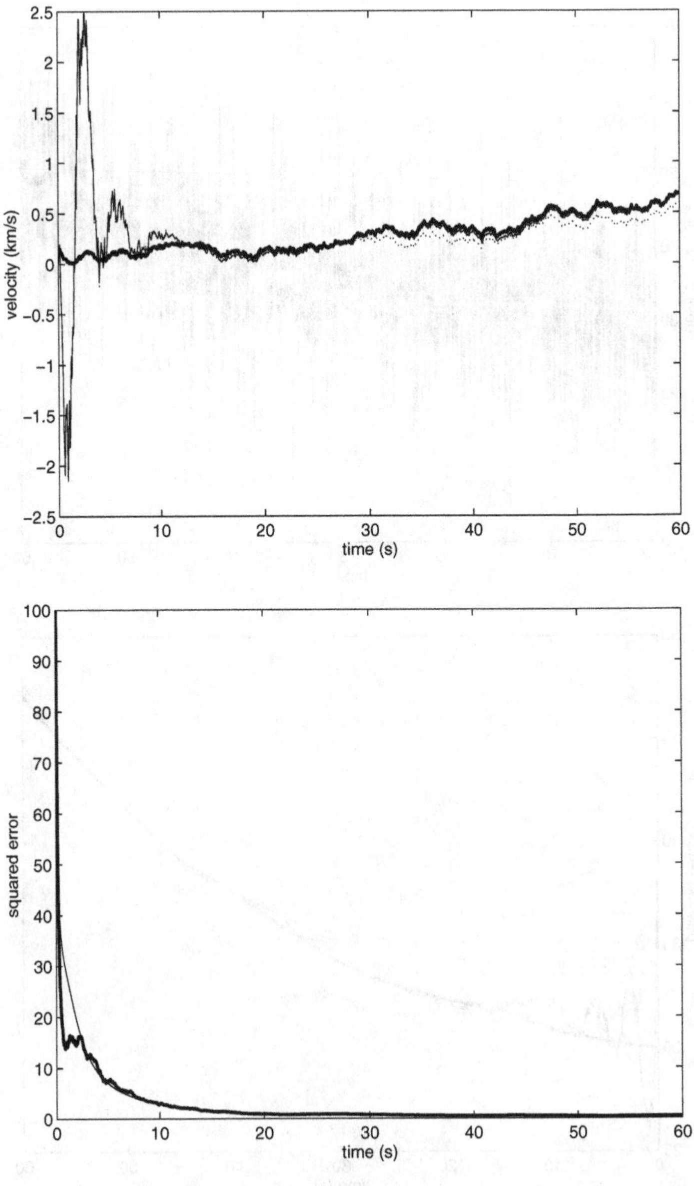

Figure 7.13. Correlated noise results for Example 7.4, cont. **Top:** Actual horizontal velocity (thick curve), estimated velocity with regular Kalman filter (dotted curve), and estimated velocity with the correlated-noise filter (thin curve). Results are for a single simulation. **Bottom:** Actual MSE (thick curve) and correlated-noise Kalman filter estimated MSE (thin curve). Results are for 100 Monte Carlo simulations.

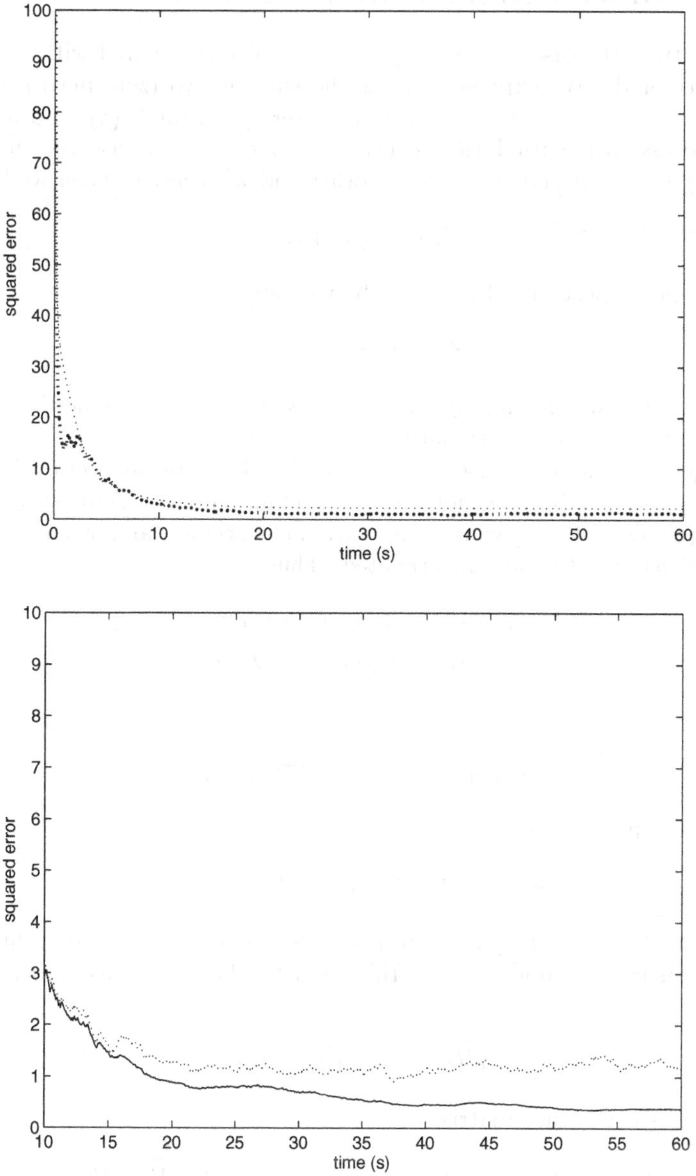

Figure 7.14. Correlated noise results for Example 7.4, cont. Results from 100 Monte Carlo simulations. **Top:** Actual MSE (thick dotted curve) and estimated MSE with regular Kalman filter (thin dotted curve). **Bottom:** Averaged true MSE of the correlated-noise Kalman filter (thin curve) and of the regular Kalman filter (dotted curve).

7.4 Colored Measurement Noise

Now we study the case where $\boldsymbol{w}(n)$ is white and uncorrelated with $\boldsymbol{v}(n)$, but $\boldsymbol{v}(n)$ is colored. We express $\boldsymbol{v}(n)$ as the sum of two (zero-mean) random processes: a colored process $\boldsymbol{c}(n)$ with power spectrum $S_c(z)$, and a white-noise process $\boldsymbol{d}(n)$ with $\mathrm{E}\left[\boldsymbol{d}(i)\boldsymbol{d}^{\mathrm{T}}(j)\right] = R_d\delta(i-j)$. We assume that $\boldsymbol{c}(n)$ and $\boldsymbol{d}(n)$ are uncorrelated with each other and with $\boldsymbol{w}(n)$. Then we have

$$\boldsymbol{v}(n) = \boldsymbol{c}(n) + \boldsymbol{d}(n). \tag{7.31}$$

The power spectrum of $\boldsymbol{v}(n)$ can be written as

$$S_v(z) = S_c(z) + R_d. \tag{7.32}$$

Note that since we are employing power spectra, we have implicitly assumed that $\boldsymbol{v}(n)$ and $\boldsymbol{c}(n)$ are stationary.

To generate $\boldsymbol{c}(n)$, we use arguments just like those in Section 7.2. We perform a spectral factorization on $S_c(z)$ to find a noise-shaping system with input $\boldsymbol{c}'(n)$, (zero-mean) white noise with unit variance, and output $\boldsymbol{c}(n)$. Of course, $\boldsymbol{c}'(n)$ and $\boldsymbol{d}(n)$ are uncorrelated. Thus,

$$\begin{aligned} \boldsymbol{x}_c(n+1) &= \Phi_c\boldsymbol{x}_c(n) + \Gamma_c\boldsymbol{c}'(n) \\ \boldsymbol{c}(n) &= C_c\boldsymbol{x}_c(n) + D_c\boldsymbol{c}'(n), \end{aligned}$$

and

$$S_c^+(z) = C_c(zI - \Phi_c)^{-1}\Gamma_c + D_c.$$

As a result, $\boldsymbol{v}(n)$ is given by

$$\boldsymbol{v}(n) = C_c\boldsymbol{x}_c(n) + D_c\boldsymbol{c}'(n) + \boldsymbol{d}(n).$$

The term $D_c\boldsymbol{c}'(n) + \boldsymbol{d}(n)$ is white noise. Since the usual Kalman filter has white measurement noise, we let this term be the new measurement noise. Define

$$\boldsymbol{v}'(n) = D_c\boldsymbol{c}'(n) + \boldsymbol{d}(n),$$

so $\boldsymbol{v}'(n)$ has covariance matrix

$$R' = \mathrm{Cov}\left[\boldsymbol{v}'(n)\right] = \mathrm{Cov}\left[D_c\boldsymbol{c}'(n) + \boldsymbol{d}(n)\right] = D_cD_c^{\mathrm{T}} + R_d. \tag{7.33}$$

Now we can create a new SMM that accounts for the colored measurement noise. Define a new state vector

$$\boldsymbol{x}'(n) = \begin{bmatrix} \boldsymbol{x}(n) \\ \boldsymbol{x}_c(n) \end{bmatrix}.$$

Then the new SMM becomes

$$\boldsymbol{x}'(n+1) = \begin{bmatrix} \Phi & 0 \\ 0 & \Phi_c \end{bmatrix} \boldsymbol{x}'(n) + \begin{bmatrix} \Gamma & 0 \\ 0 & \Gamma_c \end{bmatrix} \begin{bmatrix} \boldsymbol{w}(n) \\ \boldsymbol{c}'(n) \end{bmatrix} \tag{7.34}$$

$$\boldsymbol{z}(n) = \begin{bmatrix} C & C_c \end{bmatrix} \boldsymbol{x}'(n) + \boldsymbol{v}'(n). \tag{7.35}$$

In the course of generating the new SMM, we have modified the process noise. The new process noise,

$$\boldsymbol{w}'(n) = \begin{bmatrix} \boldsymbol{w}(n) \\ \boldsymbol{c}'(n) \end{bmatrix},$$

is white with covariance matrix

$$Q'(n) = \text{Cov}\left[\boldsymbol{w}'(n)\right] = \text{Cov}\left[\begin{bmatrix} \boldsymbol{w}(n) \\ \boldsymbol{c}'(n) \end{bmatrix}\right] = \begin{bmatrix} Q(n) & 0 \\ 0 & I \end{bmatrix}. \tag{7.36}$$

As a result, $\boldsymbol{w}'(n)$ and $\boldsymbol{v}'(n)$ both contain a $\boldsymbol{c}'(n)$ component, so that they are correlated. For the cross-correlation matrix we find

$$\begin{aligned} S(n) &= \text{E}\left[\boldsymbol{w}'(n)(\boldsymbol{v}'(n))^{\text{T}}\right] = \text{E}\left[\begin{bmatrix} \boldsymbol{w}(n) \\ \boldsymbol{c}'(n) \end{bmatrix} \left[D_c\boldsymbol{c}'(n) + \boldsymbol{d}(n)\right]^{\text{T}}\right] \\ &= \begin{bmatrix} 0 \\ D_c^{\text{T}} \end{bmatrix}. \end{aligned} \tag{7.37}$$

Equations (7.34–7.35) describe the new SMM. It is driven by white process noise $\boldsymbol{w}'(n)$ with covariance matrix given by (7.36), and includes white measurement noise $\boldsymbol{v}'(n)$ with covariance matrix given by (7.33). $\boldsymbol{w}'(n)$ and $\boldsymbol{v}'(n)$ are *correlated* according to (7.37), so we must use the modified Kalman filter in Section 7.3.

It is possible for $\boldsymbol{v}(n)$ to be a vector process, although we have not discussed this case. Also, $\boldsymbol{v}(n)$ can sometimes be nonstationary. References for both these possibilities are given in the Remarks of Section 7.2.

7.5 Target Tracking with Polar Measurements

In Section 7.1 we demonstrated the use of the Kalman filter for tracking a target moving with constant velocity. For simplicity we assumed the measuring device, e.g., a radar, returned measurements in Cartesian coordinates. Actual radar tracking systems usually employ spherical coordinates: range, elevation, and azimuth (r, θ, ϕ).

We consider a simplified version of the target tracking problem with polar coordinates: range and angle (r, θ). This example demonstrates the key concepts in implementing the Kalman filter without introducing an excessive amount of additional mathematics.

The actual Cartesian and polar coordinates are related via

$$x(n) = r(n)\cos\theta(n),$$
$$y(n) = r(n)\sin\theta(n),$$

and

$$r(n) = \sqrt{x^2(n) + y^2(n)},$$
$$\theta(n) = \tan_\theta^{-1}(x(n), y(n)),$$

where $\tan_\theta^{-1}(x, y)$ is defined by

$$\tan_\theta^{-1}(x,y) \triangleq \begin{cases} \tan^{-1}(y/x), & x > 0, \\ \pi/2, & x = 0,\ y > 0, \\ -\pi/2, & x = 0,\ y < 0, \\ \tan^{-1}(y/x) + \pi, & x < 0,\ y \geq 0, \\ \tan^{-1}(y/x) - \pi, & x < 0,\ y < 0. \end{cases} \tag{7.38}$$

Since $-\pi/2 < \tan^{-1}\omega < \pi/2$, we have defined $\tan_\theta^{-1}(x, y)$ to allow for $-\pi < \theta \leq \pi$.

The measurements are given by

$$\boldsymbol{r}(n) = r(n) + \boldsymbol{v}_r(n) \tag{7.39}$$
$$\boldsymbol{\theta}(n) = \theta(n) + \boldsymbol{v}_\theta(n), \tag{7.40}$$

where $\boldsymbol{v}_r(n)$ and $\boldsymbol{v}_\theta(n)$ are uncorrelated, zero-mean white noise processes with respective variances σ_r^2 and σ_θ^2. We use boldface to indicate the measured position of the target, which is a *random process*, and normal text to indicate the actual position of the target, which is deterministic.

We may address the problem in several ways:

1. Continue to use the Kalman filter for Cartesian measurements (Section 7.1) but use Relations (7.39) and (7.40) to relate $\boldsymbol{v}_r(n)$ and $\boldsymbol{v}_\theta(n)$ to Cartesian processes $\boldsymbol{v}_x(n)$ and $\boldsymbol{v}_y(n)$.

2. Continue to use the Cartesian state vector and use a linear approximation to the measurement equation.

3. Derive a new state model based on polar observations.

Converting Coordinates

We already have the Kalman filter for a Cartesian state model and Cartesian observations. Let us try the first option[1] and convert the polar measurements

[1] The second possibility is considered in Section 8.5. We do not treat the third option.

to fit this estimator. We begin by relating the Cartesian and polar coordinates for the measured values, $\boldsymbol{x}(n)$ and $\boldsymbol{y}(n)$. We have

$$\boldsymbol{x}(n) = \boldsymbol{r}(n)\cos\boldsymbol{\theta}(n) = f_x(\boldsymbol{r}(n), \boldsymbol{\theta}(n)),$$

and

$$\boldsymbol{y}(n) = \boldsymbol{r}(n)\sin\boldsymbol{\theta}(n) = f_y(\boldsymbol{r}(n), \boldsymbol{\theta}(n)).$$

We write $\boldsymbol{x}(n)$ by performing a Taylor series expansion of f_x about the true position $(r(n), \theta(n))$:

$$\begin{aligned}\boldsymbol{x}(n) = f_x(r,\theta)\Big|_{\substack{r=r(n)\\ \theta=\theta(n)}} &+ \frac{\partial f_x}{\partial r}\Big|_{\substack{r=r(n)\\ \theta=\theta(n)}} [\boldsymbol{r}(n) - r(n)] \\ &+ \frac{\partial f_x}{\partial \theta}\Big|_{\substack{r=r(n)\\ \theta=\theta(n)}} [\boldsymbol{\theta}(n) - \theta(n)] + \cdots .\end{aligned}$$

Similarly, we have the following expansion for $\boldsymbol{y}(n)$,

$$\begin{aligned}\boldsymbol{y}(n) = f_y(r,\theta)\Big|_{\substack{r=r(n)\\ \theta=\theta(n)}} &+ \frac{\partial f_y}{\partial r}\Big|_{\substack{r=r(n)\\ \theta=\theta(n)}} [\boldsymbol{r}(n) - r(n)] \\ &+ \frac{\partial f_y}{\partial \theta}\Big|_{\substack{r=r(n)\\ \theta=\theta(n)}} [\boldsymbol{\theta}(n) - \theta(n)] + \cdots .\end{aligned}$$

We then *approximate* these expressions by truncating them after the first partial derivatives. Then

$$\begin{aligned}\boldsymbol{x}(n) &\approx r(n)\cos\theta(n) + [\boldsymbol{r}(n) - r(n)]\cos\theta(n) \\ &\quad + [\boldsymbol{\theta}(n) - \theta(n)]r(n)(-\sin\theta(n)) \\ &\approx x(n) + [r(n) + \boldsymbol{v}_r(n) - r(n)]\cos\theta(n) \\ &\quad - [\theta(n) + \boldsymbol{v}_\theta(n) - \theta(n)]r(n)\sin\theta(n) \\ &\approx x(n) + \boldsymbol{v}_r(n)\cos\theta(n) - \boldsymbol{v}_\theta(n)r(n)\sin\theta(n),\end{aligned}$$

and

$$\begin{aligned}\boldsymbol{y}(n) &\approx r(n)\sin\theta(n) + [\boldsymbol{r}(n) - r(n)]\sin\theta(n) \\ &\quad + [\boldsymbol{\theta}(n) - \theta(n)]r(n)\cos\theta(n) \\ &\approx y(n) + \boldsymbol{v}_r(n)\sin\theta(n) + \boldsymbol{v}_\theta(n)r(n)\cos\theta(n).\end{aligned}$$

Define the *Cartesian* measurement noise $\boldsymbol{v}(n)$ by

$$\begin{aligned}\boldsymbol{v}(n) &\triangleq \begin{bmatrix}\boldsymbol{v}_r(n)\cos\theta(n) - \boldsymbol{v}_\theta(n)r(n)\sin\theta(n) \\ \boldsymbol{v}_r(n)\sin\theta(n) + \boldsymbol{v}_\theta(n)r(n)\cos\theta(n)\end{bmatrix} \\ &= \begin{bmatrix}\boldsymbol{v}_x(n) \\ \boldsymbol{v}_y(n)\end{bmatrix}. \end{aligned} \tag{7.41}$$

Note that $\boldsymbol{v}(n)$ and its components are not the same as the measurement noise in Section 7.1. Then we model the *Cartesian* states by

$$\boldsymbol{x}(n) = x(n) + \boldsymbol{v}_x(n), \quad \text{and} \quad \boldsymbol{y}(n) = y(n) + \boldsymbol{v}_y(n).$$

The observation vector in Section 7.1 may be written as

$$\boldsymbol{z}(n) = \begin{bmatrix} x(n) & y(n) \end{bmatrix}^{\mathrm{T}} + \begin{bmatrix} \boldsymbol{v}_x(n) & \boldsymbol{v}_y(n) \end{bmatrix}^{\mathrm{T}}.$$

As a result, we can use the same Kalman filter as in Section 7.1. We only need to find the new covariance matrix associated with $\boldsymbol{v}(n)$.

Measurement Noise Covariance

To find $R(n)$, we write

$$\begin{aligned} R(n) &= \mathrm{E}\left[\boldsymbol{v}(n)\boldsymbol{v}^{\mathrm{T}}(n)\right] \\ &= \mathrm{E}\left[\begin{bmatrix} \boldsymbol{v}_x^2(n) & \boldsymbol{v}_x(n)\boldsymbol{v}_y(n) \\ \boldsymbol{v}_x(n)\boldsymbol{v}_y(n) & \boldsymbol{v}_y^2(n) \end{bmatrix}\right] \\ &= \begin{bmatrix} \mathrm{E}\left[\boldsymbol{v}_x^2(n)\right] & \mathrm{E}\left[\boldsymbol{v}_x(n)\boldsymbol{v}_y(n)\right] \\ \mathrm{E}\left[\boldsymbol{v}_x(n)\boldsymbol{v}_y(n)\right] & \mathrm{E}\left[\boldsymbol{v}_y^2(n)\right] \end{bmatrix}. \end{aligned}$$

Substituting (7.41) into this expression, we find

$$R(n) = \begin{bmatrix} \sigma_r^2 \cos^2\theta(n) + \sigma_\theta^2 r^2(n)\sin^2\theta(n) & \frac{1}{2}\left[\sigma_r^2 - \sigma_\theta^2 r^2(n)\right]\sin 2\theta(n) \\ \frac{1}{2}\left[\sigma_r^2 - \sigma_\theta^2 r^2(n)\right]\sin 2\theta(n) & \sigma_r^2 \sin^2\theta(n) + \sigma_\theta^2 r^2(n)\cos^2\theta(n) \end{bmatrix}. \tag{7.42}$$

Note that because the measurement noise is considered to be in discrete-time form in (7.39) and (7.40), there is no normalization due to discretization. Also, $R(n)$ varies with time, so $\boldsymbol{v}(n)$ is nonstationary. At each time n, $\boldsymbol{v}(n)$ displays correlation between its components. However, the change from Cartesian to polar coordinates does not affect the temporal behavior of the measurement noise in either coordinate system. Hence $\boldsymbol{v}(n)$ remains a white noise sequence because it does not possess any *temporal* correlation.

We also see from (5.80) and (5.81) that we need $R(n)$ before we can estimate $\hat{\boldsymbol{x}}(n)$. Moreover, $R(n)$ depends upon $r(n)$ and $\theta(n)$, but these are effectively the quantities that we want to estimate! At time $n-1$, $\hat{\boldsymbol{x}}^-(n)$ provides the best estimate of the true position $(x(n), y(n))$, and we can use these values to estimate $(r(n), \theta(n))$. As a result, the Kalman filter depends upon the measurements $\boldsymbol{z}(n)$ and cannot be tested off-line.

Implementation

The SMM is the same as in Section 7.1, Equations (7.6) and (7.7). We initialize the Kalman filter as usual, with the addition of σ_r^2 and σ_θ^2. The filter recursion begins with predicted estimates of $r(n)$ and $\theta(n)$,

$$\hat{\boldsymbol{r}}^-(n) = \sqrt{(\hat{\boldsymbol{x}}^-(n))^2 + (\hat{\boldsymbol{y}}^-(n))^2}$$
$$\hat{\boldsymbol{\theta}}^-(n) = \tan_\theta^{-1}(\hat{\boldsymbol{x}}^-(n), \hat{\boldsymbol{y}}^-(n)).$$

Next, $R(n)$ is approximated via (7.42) with $r(n) = \hat{r}^-(n)$ and $\theta(n) = \hat{\boldsymbol{\theta}}^-(n)$. The regular Kalman filter recursion of Section 5.6 completes the estimator.

Example 7.5 Target Tracking with Polar Measurements

This filter has been simulated with MATLAB. The initial conditions were similar to those in Example 7.1. The sampling rate was $T = 50$ ms, $\sigma_{w_{cx}} = 50$ m/s, and $\sigma_{w_{cy}} = 80$ m/s. Of course, the measurement noise had a polar form, with $\sigma_{v_r} = 4$ km, and $\sigma_{v_\theta} = 15$ degrees.

We initialized the Kalman filter with

$$\hat{\boldsymbol{x}}^-(0) = \begin{bmatrix} z_1(0) & 0 & z_2(0) & 0 \end{bmatrix}^{\mathrm{T}},$$

and $P^-(0) = \mathrm{diag}\left(\begin{bmatrix} 50 & 10 & 50 & 10 \end{bmatrix}\right)$.

Simulation results appear in Figures 7.15 and 7.16. Only the tracking performance for the horizontal coordinate is shown, since the results for the vertical coordinate are similar. Some tracking fluctations occurred because of the truncated Taylor series approximation. The most notable fluctuation occurred in the velocity estimate, which overshot the true velocity before tracking it correctly.

The MSE of the estimates appears at the bottom of Figure 7.16. Here the effect of the truncated Taylor series is also visible. However, the filter was still able to track the target accurately after about 15 seconds.

7.6 System Identification

We generally use the Kalman filter to estimate the states of a system, and these states frequently correspond to physical quantities described by the signal model. However, the Kalman filter is just an estimator, so by appropriately defining $\boldsymbol{x}(n)$, we can use the Kalman filter to estimate virtually any random quantity.

Suppose we have a system with an unknown model. By measuring its input-output behavior, we wish to find a mathematical model for the system. Let $u(n)$ denote the input and $y(n)$ the output. We assume the system is LTI

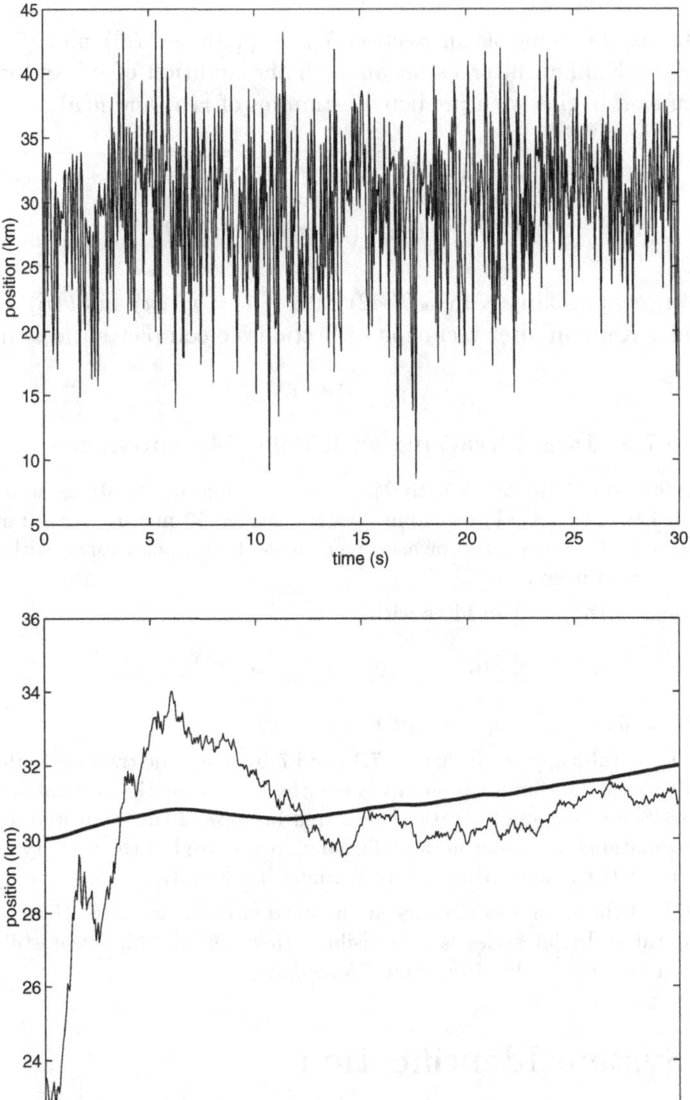

Figure 7.15. Target tracking results for Example 7.5. **Top:** Noisy observations $z_1(n)$ of the target's horizontal position. **Bottom:** Actual horizontal position $\vec{x}_1(n)$ or $x(n)$ (thick curve) and estimated position $\hat{\vec{x}}_1(n)$ or $\hat{x}(n)$ from the Kalman filter (thin curve).

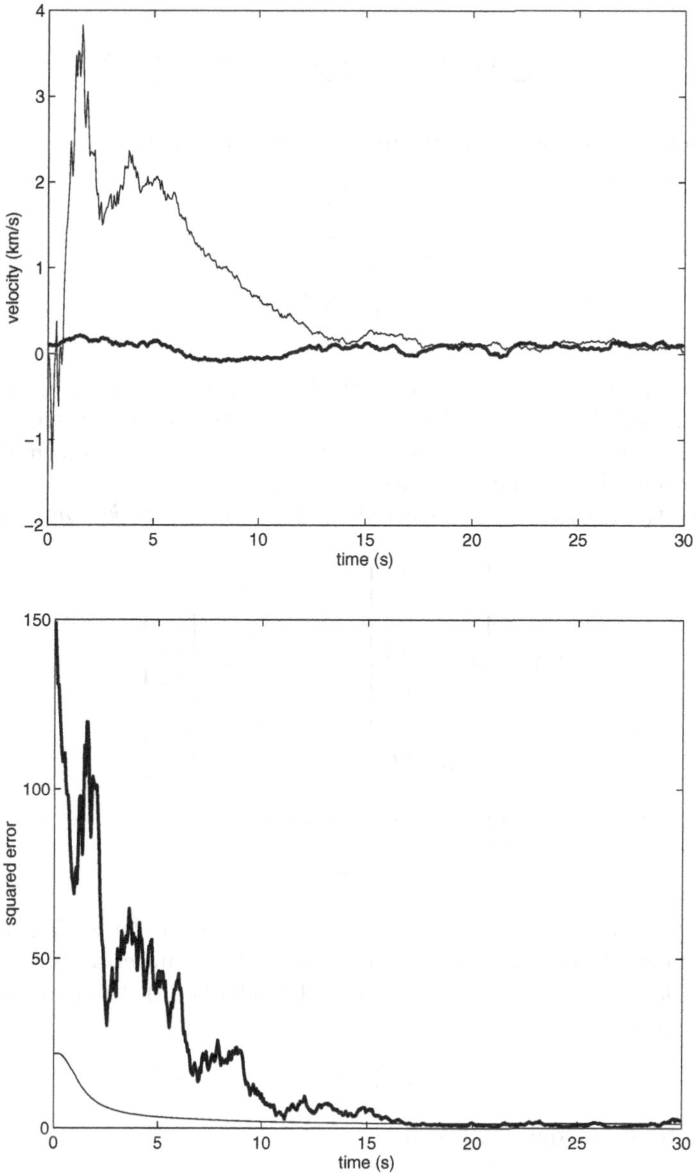

Figure 7.16. Target tracking results for Example 7.5, cont. **Top:** Actual horizontal velocity $\vec{x}_2(n)$ or $\dot{x}(n)$ (thick curve) and Kalman estimate (thin curve). **Bottom:** Actual MSE (thick curve) and estimated MSE ($\mathrm{tr}\,(P(n))$) of the Kalman filter (thin curve).

and can be modeled by the difference equation

$$y(n) = \sum_{i=1}^{N} a_i y(n-i) + \sum_{i=0}^{M} b_i u(n-i), \quad M < N. \tag{7.43}$$

This difference equation corresponds to the system function

$$H(z) = \frac{\sum_{i=0}^{M} b_i z^{-i}}{1 - \sum_{i=1}^{N} a_i z^{-i}}.$$

We attempt to determine the coefficients $a_1, \ldots, a_N$, $b_0, \ldots, b_M$. This problem is called *system identification* or *parameter estimation* since the unknowns $\{a_i\}$ and $\{b_i\}$ are the system parameters. We need to formulate the problem to fit the Kalman filter assumptions.

Define the *input-output vector* $r(n)$ and the *parameter vector* x by

$$r(n) \triangleq \begin{bmatrix} y(n-1) \\ \vdots \\ y(n-N) \\ u(n) \\ \vdots \\ u(n-M) \end{bmatrix}, \quad \text{and} \quad x \triangleq \begin{bmatrix} a_1 \\ \vdots \\ a_N \\ b_0 \\ \vdots \\ b_M \end{bmatrix}.$$

Then the input-output equation (7.43) becomes

$$y(n) = r^{\mathrm{T}}(n) x.$$

Since we estimate the parameters in x at each time n, let $\boldsymbol{x}(n)$ denote this time-varying estimate. It is a random quantity due to uncertainty in the true values of a_i, b_i. Since we assumed an LTI model (7.43), the parameters do not change with time. Thus,

$$\boldsymbol{x}(n+1) = I\boldsymbol{x}(n). \tag{7.44}$$

Our input-output equation becomes

$$\boldsymbol{y}(n) = r^{\mathrm{T}}(n)\boldsymbol{x}(n) + \boldsymbol{v}(n), \tag{7.45}$$

where $\boldsymbol{v}(n)$ is a white noise process that accounts for modeling errors such as small nonlinearities. Since $\boldsymbol{x}(n)$ and $\boldsymbol{v}(n)$ are random, $\boldsymbol{y}(n)$ becomes random as well. Note that (7.45) is a time-varying equation, so we are using a time-varying state model.

Equations (7.44) and (7.45) fit the SMM of Section 5.4 with $\boldsymbol{w}(n) \equiv 0$, $\Phi = I$, and $C(n) = \boldsymbol{r}^{\mathrm{T}}(n)$. Now we may apply a Kalman filter to identify the system parameters.

Example 7.6 Identification of a Second-Order System

Suppose we must estimate the parameters of a system assumed to have second order. We assume a second-order LTI signal model,

$$y(n) = a_1 y(n-1) + a_2 y(n-2) + b_0 u(n) + b_1 u(n-1).$$

Then based on the preceding development, we define $C(n)$ and $\boldsymbol{x}(n)$ by

$$C(n) \triangleq \begin{bmatrix} y(n-1) \\ y(n-2) \\ u(n) \\ u(n-1) \end{bmatrix}^{\mathrm{T}}, \quad \text{and} \quad \boldsymbol{x}(n) \triangleq \begin{bmatrix} a_1 \\ a_2 \\ b_0 \\ b_1 \end{bmatrix}.$$

It follows that $\Phi = I_4$, the 4×4 identity matrix, and $\boldsymbol{w}(n) \equiv 0$.

We have implemented this estimator in MATLAB. The actual system was

$$H(z) = \frac{2 - 3z^{-1}}{1 - 0.7z^{-1} + 0.1z^{-2}},$$

$u(n)$ was random white Gaussian noise with $\sigma_u^2 = 100$, and the modeling noise $v(n)$ was white Gaussian noise with $\sigma_v^2 = 36$. Our initial value for $P^-(n)$ was $P^-(0) = 100I_4$.

After 100 iterations, the Kalman filter returned an estimated system

$$\hat{H}(z) = \frac{2.0885 - 2.9899z^{-1}}{1 - 0.6904z^{-1} + 0.1132z^{-2}}.$$

Example 7.7 Identification of a Fourth-Order LTI System

Suppose we have a fourth-order system that, unknown to us, has system function

$$\begin{aligned} H(z) &= \frac{(1 + 0.2z^{-1})(1 - 2z^{-1})(1 + 4.5z^{-1})}{(1 - 0.6z^{-1})(1 + 0.95z^{-1})(1 + j0.8z^{-1})(1 - j0.8z^{-1})} \\ &= \frac{1 + 2.7z^{-1} - 8.5z^{-2} - 1.8z^{-3}}{1 + 0.35z^{-1} + 0.07z^{-2} - 0.2240z^{-3} - 0.3648z^{-4}}, \end{aligned}$$

and we wish to estimate $H(z)$.

Let us assume a fourth-order model,

$$y(n) = \sum_{i=1}^{4} a_i y(n-i) + \sum_{i=0}^{3} b_i u(n-i).$$

Then

$$C(n) \triangleq \begin{bmatrix} y(n-1) \\ y(n-2) \\ y(n-3) \\ y(n-4) \\ u(n) \\ u(n-1) \\ u(n-2) \\ u(n-3) \end{bmatrix}^{\mathrm{T}}, \quad \text{and} \quad x(n) \triangleq \begin{bmatrix} a_1 \\ a_2 \\ a_3 \\ a_4 \\ b_0 \\ b_1 \\ b_2 \\ b_3 \end{bmatrix}.$$

We set $\Phi = I_8$, and $w(n) \equiv 0$.

The MATLAB simulation of this estimator used Gaussian noise for $u(n)$ and $v(n)$ with $\sigma_u^2 = 121$ and $\sigma_v^2 = 25$. The initial value for $P^-(n)$ was $P^-(0) = 100I_8$.

After 100 iterations, the estimated system is

$$\begin{aligned} \hat{H}(z) &= \frac{1.0173 + 2.6684z^{-1} - 8.5989z^{-2} - 1.7488z^{-3}}{1 + 0.3386z^{-1} + 0.0577z{-2} + 0.2154z^{-3} - 0.3706z^{-4}} \\ &= 1.0173 \times \frac{(1 + 0.1927z^{-1})(1 - 2.0094z^{-1})}{(1 - 0.6116z^{-1})(1 + 0.9496z^{-1})} \\ &\quad \times \frac{(1 + 4.4398z^{-1})}{(1 + (0.0003 + j0.7989)z^{-1})(1 + (0.0003 - j0.7989)z^{-1})}. \end{aligned}$$

Remarks on System Identification

The two examples just presented demonstrate the utility of the Kalman filter for identifying system parameters. Nonetheless the examples present highly ideal circumstances for several reasons:

- The systems being identified fit the difference equation model.
- The assumed model order is exactly correct.
- The system poles and zeros are sufficiently separated so that they are not confused. That is, the poles are far enough apart that two poles (or zeros) do not behave like a single pole (or zero).

These conditions are not always exactly satisfied in practice. However, often only a few poles and zeros dominate the behavior of the system, so a low-order model is usually adequate. Alternative techniques for system identification exist [16, 17, 4], and depending upon the application, they may have properties that make them preferable to the Kalman filter.

Perfect Identification

Finally, suppose that the system is known to be an Nth-order LTI system. Then we can set $v(n) \equiv 0$ since the model is exactly correct. Now $R \equiv 0$, and since we have no random inputs, we expect the Kalman filter to return the exact system parameters. In fact, it does so after $2N$ iterations (note that we have $2N$ unknowns in $\{a_i\}$ and $\{b_i\}$).

We must exercise care when applying the Kalman filter, however. When the system parameters are estimated exactly, $P^-(n)$ becomes the zero matrix, and then the term $[C(n)P^-(n)C^{\mathrm{T}}(n) + R]$ becomes singular. According to (5.80), the Kalman gain $K(n)$ is undefined. Of course, as soon as $P^-(n) = 0$, we have completely determined the system and can stop the estimation process.

Problems

7.1. Section 7.5 demonstrated a method that permits polar-coordinate observations to be used with a Cartesian-coordinate Kalman filter. In actual radar systems, positional measurements are provided in spherical coordinates, (r, θ, ϕ). r is the range of the object, θ is the angle of elevation, and ϕ is the angle of azimuth.

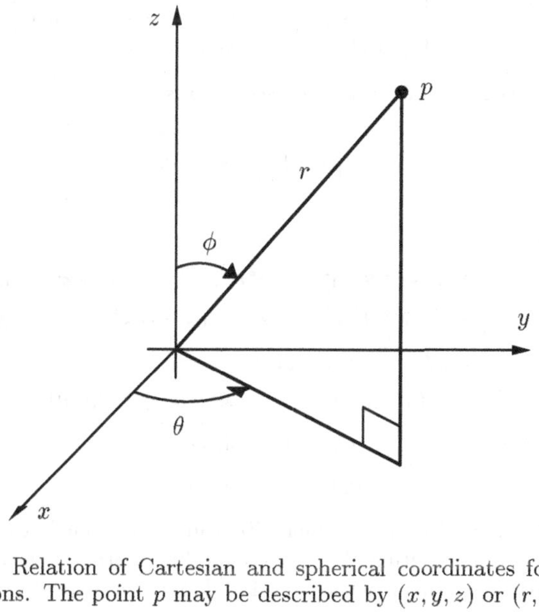

Figure 7.17. Relation of Cartesian and spherical coordinates for a point in three dimensions. The point p may be described by (x, y, z) or (r, θ, ϕ).

Suppose the noisy measurement of an object's location is modeled by the random vector

$$[\mathbf{r} \quad \boldsymbol{\theta} \quad \boldsymbol{\phi}]^{\mathrm{T}} = [r \quad \theta \quad \phi]^{\mathrm{T}} + [\mathbf{v}_r \quad \mathbf{v}_\theta \quad \mathbf{v}_\phi]^{\mathrm{T}},$$

where v_r, v_θ, and v_ϕ are zero-mean, independent random variables with variances σ_r^2, σ_θ^2, and σ_ϕ^2, respectively.

A point in spherical coordinates may be converted into Cartesian coordinates via the relations

$$x = r\sin\theta\sin\phi, \quad y = r\sin\theta\cos\phi, \quad \text{and} \quad z = r\cos\theta.$$

The point (x, y, z) is related to (r, θ, ϕ) by

$$r = \sqrt{x^2+y^2+z^2}, \quad \theta = \tan_\theta^{-1}\left(\frac{y}{x}\right),$$

and

$$\phi = \cos^{-1}\left(\frac{z}{\sqrt{x^2+y^2+z^2}}\right).$$

(a) Use the truncated expansion technique of Section 7.5 to find an approximate relationship

$$\begin{bmatrix} \boldsymbol{x} & \boldsymbol{y} & \boldsymbol{z} \end{bmatrix}^{\mathrm{T}} \approx \begin{bmatrix} x & y & z \end{bmatrix}^{\mathrm{T}} + \begin{bmatrix} \boldsymbol{v}_x & \boldsymbol{v}_y & \boldsymbol{v}_z \end{bmatrix}^{\mathrm{T}}.$$

(b) Let $\boldsymbol{v} = \begin{bmatrix} \boldsymbol{v}_x & \boldsymbol{v}_y & \boldsymbol{v}_z \end{bmatrix}^{\mathrm{T}}$. Show that

$$\boldsymbol{v} = \begin{bmatrix} \sin\theta\sin\phi & r\cos\theta\sin\phi & r\sin\theta\cos\phi \\ \sin\theta\cos\phi & r\cos\theta\cos\phi & -r\sin\theta\sin\phi \\ \cos\theta & -r\sin\theta & 0 \end{bmatrix} \begin{bmatrix} \boldsymbol{v}_r \\ \boldsymbol{v}_\theta \\ \boldsymbol{v}_\phi \end{bmatrix}.$$

(c) Show that the covariance matrix of $\boldsymbol{v}$ is

$$\mathrm{E}\left[\boldsymbol{v}\boldsymbol{v}^{\mathrm{T}}\right] = \begin{bmatrix} \sigma_x^2 & \sigma_{xy}^2 & \sigma_{xz}^2 \\ \sigma_{xy}^2 & \sigma_y^2 & \sigma_{yz}^2 \\ \sigma_{xz}^2 & \sigma_{yz}^2 & \sigma_z^2 \end{bmatrix}.$$

with

$$\begin{aligned}
\sigma_x^2 &= \sigma_r^2\sin^2\theta\sin^2\phi + \sigma_\theta^2 r^2\cos^2\theta\sin^2\phi + \sigma_\phi^2 r^2\sin^2\theta\cos^2\phi \\
\sigma_y^2 &= \sigma_r^2\sin^2\theta\cos^2\phi + \sigma_\theta^2 r^2\cos^2\theta\cos^2\phi + \sigma_\phi^2 r^2\sin^2\theta\sin^2\phi \\
\sigma_z^2 &= \sigma_r^2\sin^2 + \sigma_\theta^2 r^2\sin^2\theta \\
\sigma_{xy}^2 &= \tfrac{1}{2}\left(\sigma_r^2\sin^2\theta + \sigma_\theta^2 r^2\cos^2\theta - \sigma_\phi^2 r^2\sin^2\theta\right)\sin 2\phi \\
\sigma_{xz}^2 &= \tfrac{1}{2}\left(\sigma_r^2 - \sigma_\theta^2 r^2\right)\sin 2\theta\sin\phi \\
\sigma_{yz}^2 &= \tfrac{1}{2}\left(\sigma_r^2 - \sigma_\theta^2 r^2\right)\sin 2\theta\cos\phi.
\end{aligned}$$

7.2. The performance of an equipment item in some manufacturing operation is monitored by computing the expected value $\mathrm{E}[\boldsymbol{z}(n)]$ of measurements $\boldsymbol{z}(n)$ taken on the equipment, where $n = 1, 2,$ ldots. The expected value is modeled by

$$\begin{aligned}
\mathrm{E}[\boldsymbol{z}(n+1)] &= \mathrm{E}[\boldsymbol{z}(n)] + \boldsymbol{w}(n) \\
\boldsymbol{z}(n) &= \mathrm{E}[\boldsymbol{z}(n)] + \boldsymbol{v}(n),
\end{aligned}$$

where $\boldsymbol{w}(n)$ and $\boldsymbol{v}(n)$ are independent zero-mean white noise terms with variances σ_w^2 and σ_v^2, respectively.

(a) Based on the above model, give the equations for the Kalman filter for estimating the expected value of $z(n)$.

(b) Based on the above model, give the equations for the steady-state Kalman filter for estimating the expected value of $z(n)$.

(c) Suppose that $z(n) = 3 + v(n)$ with the variance of $v(n)$ equal to 1. Generate a sample realization of $z(n)$ for $n = 1, 2, \ldots, 100$. Use MATLAB to plot the sample realization.

(d) Apply the sample realization in Part (c) to the filters found in Parts (a) and (b). Use MATLAB to plot the estimates on the same graph and compare the results.

7.3. The performance of an equipment item is also sometimes monitored by computing the mean square $E[z^2(n)]$ of measurements $z(n)$ taken on the equipment, where $n = 1, 2, \ldots$. The mean square is modeled by

$$E[z^2(n+1)] = E[z^2(n)] + w(n)$$
$$z^2(n) = E[z^2(n)] + v(n),$$

where $w(n)$ and $v(n)$ are independent zero-mean white noise terms with variances σ_w^2 and σ_v^2, respectively.

(a) Based on the above model, give the equations for the Kalman filter for estimating the mean square of $z(n)$.

(b) Based on the above model, give the equations for the steady-state Kalman filter for estimating the mean square of $z(n)$.

(c) Apply the sample realization in Part (c) of Problem 7.2 to the filters found in Parts (a) and (b) of this problem. Use MATLAB to plot the estimates on the same graph and compare the results.

7.4. Two cars are moving down an expressway with constant velocities $V_1(t)$ and $V_2(t)$, respectively. A radar provides noisy measurements $z(n) = D(t) + v(t)$ of the distance $D(t)$ between the two cars at time t, where $v(t)$ is zero- mean white noise with variance σ_v^2. Assume that the velocities of the two cars are known exactly.

(a) Set up a linear time-invariant state model for estimating $D(n)$ at time $t = n$, $n = 1, 2, \ldots$. Express your answer by giving the definition of the state $x(n)$ and the coefficient matrices of the state model and the covariances of the noise terms.

(b) Based on the model found in Part (a), give the equations for the Kalman filter for estimating $D(n)$.

(c) Suppose that at time $t = 0$, one car is ahead of the other by 100 ft, the velocity of the car in front is 65 miles per hour, and the velocity of the car in the rear is 70 miles per hour. Assuming that the variance of $v(n)$ is equal to 10. Generate a sample realization of $z(n)$ for $n = 1, 2, \ldots, 100$. Use MATLAB to plot the sample realization.

(d) Apply the sample realization in Part (c) to the filter found in Part (b). Use MATLAB to plot the estimates. On the same plot, show the actual distance between the cars.

7.5. An object is moving with constant velocity. The x-coordinate position of the object at time nT is $p(n)$, where T is the sampling interval. Measurements $z(n)$ of $p(n)$ are given by $z(n) = p(n) + v(n)$, where $v(n)$ is zero-mean white noise with variance 3, and $v(n)$ is independent of $p(n)$.

(a) Suppose that $p(0) = 15$ and $p(100) = 5$. Generate a sample realization of the measurements $z(n)$ for $n = 1$ to 100 and give a MATLAB plot of the result.

(b) Apply the sample realization in Part (a) to a Kalman filter and use MATLAB to plot the estimates of the position. On the same plot, show the actual object trajectory.

7.6. An object is moving with constant acceleration. The x-coordinate position of the object at time nT is $p(n)$, where T is the sampling interval. Measurements $z(n)$ of $p(n)$ are given by $z(n) = p(n) + v(n)$, where $v(n)$ is zero-mean white noise with variance 10, and $v(n)$ is independent of $p(n)$.

(a) Generate a three-dimensional state model for the x-coordinate motion of the object with the state variables equal to $p(n)$, $\dot{p}(n)$, and $\ddot{p}(n)$.

(b) Suppose that $p(0) = 15$, $\dot{p}(0) = 1$, $p(100) = 605$, and $\dot{p}(100) = 11$. Generate a sample realization of the measurements $z(n)$ for $n = 1$ to 100 and give a MATLAB plot of the result.

(c) Apply the sample realization in Part (b) to a three-dimensional Kalman filter based on the state model found in Part (a). Use MATLAB to plot the estimates of the position and on the same plot, show the actual object trajectory.

7.7. For the object in Problem 7.6, use a two-dimensional state model with process noise to estimate the object position $p(n)$ where $p(0) = 15$, $\dot{p}(0) = 1$, $p(100) = 605$, and $\dot{p}(100) = 11$. Apply the sample realization in Part (b) of Problem 7.6 to a two-dimensional Kalman filter based on the two- dimensional state model. Use MATLAB to plot the estimates of the position and on the same plot, show the actual object trajectory and the estimates computed in Problem 7.6. Does the filter perform better than the one in Problem 7.6?

7.8. An object has x-coordinate position $p(t) = \theta_1(t + 10) + \theta_2 t^2$, where θ_1 and θ_2 are unknown constants. Measurements are given by $z(n) = p(n) + v(n)$ where the sampling interval $T = 1$ and $v(n)$ is zero-mean white noise with variance 2.

(a) Set up a two-dimensional state model for estimating $p(n)$.

(b) Using your model in Part (a), determine the equations for the Kalman filter that produces an estimate of $p(n)$ at time n based on the measurements $p(i)$, $i = 1, 2, \dots, n$.

7.9. The motion of an object with x-coordinate position $p(t)$ is given by $\ddot{p}(t) = 0$. A sensor provides noisy measurements $z(n)$ of the sum of x-coordinate position $p(0.1n)$ and the x-coordinate velocity $\dot{p}(0.1n)$; that is

$$z(n) = p(0.1n) + \dot{p}(0.1n) + v(n),$$

where $v(n)$ is zero-mean white noise with variance 5.

(a) Give the equations for the Kalman filter that provides estimates of the position $p(0.1n)$ and velocity $\dot{p}(0.1n)$.

(b) Generate a sample realization of $z(n)$ for $n = 1, 2, \ldots, 100$, in the case when $p(0) = 5$ and $p(10) = 30$. Use MATLAB to plot the realization.

(c) Apply the sample realization in Part (b) to the Kalman filter found in Part (a). Use MATLAB to plot the estimates of the position and velocity, and on the same plots, show the actual object position and velocity.

7.10. A LTI discrete-time system is given by

$$y(n) = -y(n-1) + 0.5y(n-2) + u(n),$$

where $u(n)$ is the input and $y(n)$ is the output.

(a) Compute $y(n)$ for $0 \leq n \leq 100$ when $u(n) = 1$ for $n \geq 0$, and $y(-2) = y(-1) = 0$.

(b) Using the data in Part (a) and the Kalman filter approach, identify a two-dimensional model.

(c) Plot the output of the model found in Part (b) when $u(n) = 1$ for $n \geq 0$, and $y(-2) = y(-1) = 0$, and on the same plot, show the output of the actual system. Compare the results.

7.11. Repeat Problem 7.10, but now take $u(n)$ to be a sample realization of white Gaussian noise with variance 1.

Chapter 8

Nonlinear Estimation

The Kalman filter is based on the assumption of a linear SMM. However, many physical systems are described by nonlinear equations. In this chapter, we consider modifying the Kalman filter to address nonlinear systems. We derive the resulting estimator, known as the *extended Kalman filter* (EKF).

We then present a number of example applications for the EKF. We begin by demonstrating how the EKF can be used to estimate the parameters of a neural network for nonlinear system identification. Next, the EKF is applied to frequency demodulation and target tracking with polar measurements. The chapter concludes by introducing the problem of tracking multiple targets.

8.1 The Extended Kalman Filter

The Nonlinear State/Measurement Model

Let us consider a general nonlinear, time-invariant state model of the form

$$\boldsymbol{x}(n+1) = \phi(\boldsymbol{x}(n)) + \Gamma \boldsymbol{w}(n) \tag{8.1}$$

$$\boldsymbol{z}(n) = \gamma(\boldsymbol{x}(n)) + \boldsymbol{v}(n), \tag{8.2}$$

where $\boldsymbol{w}(n)$ and $\boldsymbol{v}(n)$ are uncorrelated, zero-mean, white noise processes with covariance matrices $Q(n)$ and $R(n)$, respectively.[1] In this model, Γ is a matrix, so the input dynamics remain linear and independent of $\boldsymbol{x}(n)$.

The operators $\phi(x)$ and $\gamma(x)$ represent nonlinear vector-valued functions

[1] Note that in the nonlinear SMM, ϕ denotes the nonlinear state dynamics and should not be confused with a PDF. Likewise, γ represents the nonlinear output function, not the impulse response of an LTI system.

of x and n. Let x consist of m states; then ϕ has the form

$$\phi(x) = \begin{bmatrix} \phi_1(x) \\ \phi_2(x) \\ \vdots \\ \phi_m(x) \end{bmatrix},$$

where each $\phi_i(x)$, is a nonlinear scalar-valued function of x. Since the measurements $\boldsymbol{z}(n)$ are p-vectors, γ consists of p nonlinear scalar-valued function $\gamma_i(x)$, i.e.,

$$\gamma(x) = \begin{bmatrix} \gamma_1(x) \\ \gamma_2(x) \\ \vdots \\ \gamma_p(x) \end{bmatrix}.$$

Linearization of the SMM

The Kalman filter assumes a linear SMM, so the next step involves *linearization* of the original SMM. We assume ϕ and γ are sufficiently smooth in x so that each has a valid Taylor series expansion. Then, given the realization $\hat{\boldsymbol{x}}(n) = \hat{x}(n)$, we expand ϕ into a Taylor series about $\hat{x}(n)$:

$$\phi(\boldsymbol{x}(n)) = \phi(\hat{x}(n)) + J_\phi(\hat{x}(n))\,[\boldsymbol{x}(n) - \hat{x}(n)] + \cdots, \tag{8.3}$$

where $J_\phi(x)$ is the Jacobian of ϕ evaluated at x. Recall that if $\beta(x)$ is a vector-valued function consisting of k scalar-valued functions $\beta_i(x)$ and x is an N-vector, i.e.,

$$\beta(x) = \begin{bmatrix} \beta_1(x) \\ \beta_2(x) \\ \vdots \\ \beta_k(x) \end{bmatrix} \quad \text{and} \quad x = \begin{bmatrix} x_1 \\ x_2 \\ \vdots \\ x_N \end{bmatrix},$$

then $J_\beta(x) = \partial\beta/\partial x$ denotes the $k \times N$ Jacobian matrix of $\beta(x)$ with respect to x:

$$J_\beta(x) = \frac{\partial \beta}{\partial x} = \begin{bmatrix} \frac{\partial \beta_1}{\partial x_1} & \frac{\partial \beta_1}{\partial x_2} & \cdots & \frac{\partial \beta_1}{\partial x_N} \\ \frac{\partial \beta_2}{\partial x_1} & \frac{\partial \beta_2}{\partial x_2} & \cdots & \frac{\partial \beta_2}{\partial x_N} \\ \vdots & \vdots & \ddots & \vdots \\ \frac{\partial \beta_k}{\partial x_1} & \frac{\partial \beta_k}{\partial x_2} & \cdots & \frac{\partial \beta_k}{\partial x_N} \end{bmatrix}. \tag{8.4}$$

Likewise, we expand γ about the realization $\hat{x}^-(n)$:

$$\gamma(\boldsymbol{x}(n)) = \gamma(\hat{x}^-(n)) + J_\gamma(\hat{x}^-(n))\left[\boldsymbol{x}(n) - \hat{x}^-(n)\right] + \cdots. \tag{8.5}$$

We keep only the first two terms in the expansions (8.3) and (8.5). The resulting expressions create first-order approximations of ϕ and γ and provide *linear* functions of $\boldsymbol{x}(n)$ The input dynamics remain linear since Γ is a matrix.

We now have a *linearized SMM* that is given by

$$\boldsymbol{x}(n+1) = \phi(\hat{x}(n)) + J_\phi(\hat{x}(n))\left[\boldsymbol{x}(n) - \hat{x}(n)\right] + \Gamma\boldsymbol{w}(n) \tag{8.6}$$

$$\boldsymbol{z}(n) = \gamma(\hat{x}^-(n)) + J_\gamma(\hat{x}^-(n))\left[\boldsymbol{x}(n) - \hat{x}^-(n)\right] + \boldsymbol{v}(n) \tag{8.7}$$

and we use this model for implementing the EKF. Observe that this SMM is dependent upon the realizations $\hat{x}(n)$ and $\hat{x}^-(n)$.

Time Update

We now derive the filter recursion for the EKF. We first consider the *time update.* We assume that $\hat{\boldsymbol{x}}(n-1)$ is unbiased and seek the *a priori* estimate $\hat{x}^-(n)$. We want $\hat{x}^-(n)$ to be unbiased, so it must satisfy

$$\mathrm{E}\left[\boldsymbol{x}(n) - \hat{x}^-(n)\middle|\, \boldsymbol{Z}^- = Z^-\right] = \boldsymbol{0},$$

which produces the condition:

$$\hat{x}^-(n) = \mathrm{E}\left[\boldsymbol{x}(n)\middle|\, \boldsymbol{Z}^- = Z^-\right] \tag{8.8}$$

$$= \mathrm{E}\left[\phi(\hat{\boldsymbol{x}}(n-1)) + J_\phi(\hat{\boldsymbol{x}}(n-1))\left[\boldsymbol{x}(n-1) - \hat{\boldsymbol{x}}(n-1)\right] + \Gamma\boldsymbol{w}(n-1)\middle|\, \boldsymbol{Z}^- = Z^-\right]. \tag{8.9}$$

Given that $\boldsymbol{Z}^- = Z^-$, $\hat{\boldsymbol{x}}(n-1)$ is a constant vector $\hat{x}(n-1)$. Then (8.9) becomes

$$\hat{x}^-(n) = \phi(\hat{x}(n-1)) + J_\phi(\hat{x}(n-1))\mathrm{E}\left[\boldsymbol{x}(n-1) - \hat{\boldsymbol{x}}(n-1)\middle|\, \boldsymbol{Z}^- = Z^-\right] + \Gamma\mathrm{E}\left[\boldsymbol{w}(n-1)\middle|\, \boldsymbol{Z}^- = Z^-\right].$$

Since $\hat{\boldsymbol{x}}(n-1)$ is unbiased,

$$\mathrm{E}\left[\boldsymbol{x}(n-1) - \hat{\boldsymbol{x}}(n-1)\middle|\, \boldsymbol{Z}^- = Z^-\right] = \boldsymbol{0}.$$

Also, $\boldsymbol{w}(n-1)$ is independent of $\boldsymbol{z}(1), \ldots, \boldsymbol{z}(n-1)$, so

$$\mathrm{E}\left[\boldsymbol{w}(n-1)\middle|\, \boldsymbol{Z}^- = Z^-\right] = \boldsymbol{0}.$$

Then the *a priori* estimate is given by

$$\boxed{\hat{x}^-(n) = \phi(\hat{x}(n-1)),} \tag{8.10}$$

where $\hat{x}(n-1)$ is the realization of $\hat{\boldsymbol{x}}(n-1)$. Note that by (8.8), $\hat{x}^-(n)$ is equal to the conditional expectation of $\boldsymbol{x}(n)$ given $\boldsymbol{Z}^- = Z^-$. Ideally, $\hat{x}^-(n)$ is the MMSE estimate of $\boldsymbol{x}(n)$, but because of the linearization, $\hat{x}^-(n)$ is usually suboptimal.

We also seek a recursion for the *a priori conditional* error covariance $P^-(n)$. $P^-(n)$ is a *conditional* covariance matrix because the SMM has been linearized about $\hat{x}(n-1)$, which depends upon the measurements in Z^-. Therefore, $P^-(n)$ must also be conditioned on Z^-, so it is defined by

$$P^-(n) = \mathrm{Cov}\left[\boldsymbol{x}(n) - \hat{\boldsymbol{x}}^-(n) \middle| \boldsymbol{Z}^- = Z^-\right] \tag{8.11}$$

$$= \mathrm{E}\left[\left(\boldsymbol{x}(n) - \hat{\boldsymbol{x}}^-(n)\right)\left(\boldsymbol{x}(n) - \hat{\boldsymbol{x}}^-(n)\right)^{\mathrm{T}} \middle| \boldsymbol{Z}^- = Z^-\right]. \tag{8.12}$$

Substituting (8.6) into (8.11) and recalling that $\tilde{\boldsymbol{x}}(n-1) = \boldsymbol{x}(n-1) - \hat{\boldsymbol{x}}(n-1)$, we have

$$P^-(n) = \mathrm{Cov}\left[\phi(\hat{\boldsymbol{x}}(n-1)) + J_\phi(\hat{\boldsymbol{x}}(n-1))\tilde{\boldsymbol{x}}(n-1) + \Gamma\boldsymbol{w}(n-1) - \hat{\boldsymbol{x}}^-(n) \middle| \boldsymbol{Z}^- = Z^-\right].$$

From (8.10), $\phi(\hat{\boldsymbol{x}}(n-1)) = \hat{\boldsymbol{x}}^-(n)$, so this expression becomes

$$P^-(n) = \mathrm{Cov}\left[J_\phi(\hat{\boldsymbol{x}}(n-1))\tilde{\boldsymbol{x}}(n-1) + \Gamma\boldsymbol{w}(n-1) \middle| \boldsymbol{Z}^- = Z^-\right] \tag{8.13}$$

Because $\boldsymbol{Z}^-$ is given ($\boldsymbol{Z}^- = Z^-$), $\hat{\boldsymbol{x}}(n-1)$ is a constant vector $\hat{x}(n-1)$. Also, $\tilde{\boldsymbol{x}}(n-1)$ is independent of $\boldsymbol{w}(n-1)$. Then (8.13) reduces to

$$P^-(n) = J_\phi(\hat{x}(n-1))\mathrm{Cov}\left[\tilde{\boldsymbol{x}}(n-1) \middle| \boldsymbol{Z}^- = Z^-\right] J_\phi^{\mathrm{T}}(\hat{x}(n-1)) + \Gamma Q(n-1)\Gamma^{\mathrm{T}}. \tag{8.14}$$

Define the *a posteriori conditional* error covariance $P(n)$ as

$$P(n) = \mathrm{Cov}\left[\tilde{\boldsymbol{x}}(n) \middle| \boldsymbol{Z} = Z\right] = \mathrm{Cov}\left[\boldsymbol{x}(n) - \hat{\boldsymbol{x}}(n) \middle| \boldsymbol{Z} = Z\right] \tag{8.15}$$

Then, from (8.14), the desired recursion for $P^-(n)$ is

$$\boxed{P^-(n) = J_\phi(\hat{x}(n-1))P(n-1)J_\phi^{\mathrm{T}}(\hat{x}(n-1)) + \Gamma Q(n-1)\Gamma^{\mathrm{T}}.} \tag{8.16}$$

Measurement Update

We now consider the measurement update. Assume that $\hat{\boldsymbol{x}}^-(n)$ is unbiased and the error $\tilde{\boldsymbol{x}}^-(n)$ is orthogonal to the measurements $\boldsymbol{z}(1), \ldots, \boldsymbol{z}(n-1)$. Given the new measurement $\boldsymbol{z}(n)$, we seek the LMMSE estimate $\hat{x}(n)$. We assume that $\hat{x}(n)$ has the form

$$\hat{x}(n) = b(n) + K(n)z(n), \tag{8.17}$$

where $b(n)$ is an N-vector to be determined, $K(n)$ is an $N \times p$ matrix to be determined, and $z(n)$ is the realization of measurement $\boldsymbol{z}(n)$.

Since $\hat{x}(n)$ should be unbiased, we have

$$\mathrm{E}\left[\boldsymbol{x}(n) - \hat{x}(n) \middle| \boldsymbol{Z} = Z\right] = \mathbf{0}$$

We substitute (8.7) and (8.17) into this expression to obtain

$$\mathrm{E}\left[\boldsymbol{x}(n) - b(n) - K(n)\left[\gamma(\hat{\boldsymbol{x}}^-(n)) + J_\gamma(\hat{\boldsymbol{x}}^-(n))\tilde{\boldsymbol{x}}^-(n) + \boldsymbol{v}(n)\right] \middle| \boldsymbol{Z} = Z\right] = \mathbf{0}.$$

Now $\hat{\boldsymbol{x}}^-(n)$ is constant because $\boldsymbol{Z} = Z$ (note that $\hat{\boldsymbol{x}}^-(n)$ does not depend on $\boldsymbol{z}(n) = z(n)$ but only on $\boldsymbol{Z}^- = Z^-$). Then

$$\begin{aligned} b(n) &= \mathrm{E}\left[b(n) \middle| \boldsymbol{Z} = Z\right] \\ &= -K(n)\gamma(\hat{x}^-(n)) - K(n)J_\gamma(\hat{x}^-(n))\mathrm{E}\left[\tilde{\boldsymbol{x}}^-(n) \middle| \boldsymbol{Z} = Z\right] \\ &\quad + \mathrm{E}\left[\boldsymbol{x}(n) \middle| \boldsymbol{Z} = Z\right] - K(n)\mathrm{E}\left[\boldsymbol{v}(n) \middle| \boldsymbol{Z} = Z\right]. \end{aligned} \tag{8.18}$$

Because $\hat{\boldsymbol{x}}^-(n)$ is unbiased and $\boldsymbol{v}(n)$ is independent of the measurements in $\boldsymbol{Z}$, (8.18) becomes

$$b(n) = -K(n)\gamma(\hat{x}^-(n)) + \mathrm{E}\left[\boldsymbol{x}(n) \middle| \boldsymbol{Z} = Z\right]$$

Next, we use (8.8) to write this expression as

$$b(n) = -K(n)\gamma(\hat{x}^-(n)) + \hat{x}^-(n).$$

Therefore,

$$b(n) = \hat{x}^-(n) - K(n)\gamma(\hat{x}^-(n)),$$

so that (8.17) becomes

$$\boxed{\hat{x}(n) = \hat{x}^-(n) + K(n)\left[z(n) - \gamma(\hat{x}^-(n))\right].} \tag{8.19}$$

We must still find the matrix $K(n)$. By the orthogonality principle (3.52), we must satisfy

$$\mathrm{E}\left[(\boldsymbol{x}(n) - \hat{\boldsymbol{x}}(n))\,\boldsymbol{z}^{\mathrm{T}}(i) \middle| \boldsymbol{Z} = Z\right] = \mathbf{0}, \quad i = 1, \ldots, n. \tag{8.20}$$

We substitute (8.19) into (8.20) and get

$$\mathrm{E}\left[\left\{\boldsymbol{x}(n) - \hat{\boldsymbol{x}}^-(n) - K(n)\left[\boldsymbol{z}(n) - \gamma(\hat{\boldsymbol{x}}^-(n))\right]\right\}\boldsymbol{z}^{\mathrm{T}}(i) \middle| \boldsymbol{Z} = Z\right] = \mathbf{0},$$

which via (8.7) becomes

$$\mathrm{E}\left[\left\{\tilde{\boldsymbol{x}}^-(n) - K(n)\left[J_\gamma(\hat{\boldsymbol{x}}^-(n))\tilde{\boldsymbol{x}}^-(n) + \boldsymbol{v}(n)\right]\right\}\boldsymbol{z}^{\mathrm{T}}(i) \middle| \boldsymbol{Z} = Z\right] = \mathbf{0}, \qquad i = 1, \ldots, n. \tag{8.21}$$

Now, for $i = 1, 2, \ldots, n-1$, we have assumed that

$$\mathrm{E}\left[\tilde{\boldsymbol{x}}^-(n)\boldsymbol{z}^{\mathrm{T}}(i)\middle|\,\boldsymbol{Z}=Z\right] = \mathbf{0},$$

and $\boldsymbol{v}(n)$ is independent of the measurements in $\boldsymbol{Z}^-$. Thus, Condition (8.21) is already satisfied for $i = 1, 2, \ldots, n-1$. We must only consider the remaining case: $i = n$.

We set $i = n$ and substitute (8.7) into (8.21):

$$\begin{aligned}\mathrm{E}\Big[&\big(\tilde{\boldsymbol{x}}^-(n) - K(n)J_\gamma(\hat{\boldsymbol{x}}^-(n))\tilde{\boldsymbol{x}}^-(n) - K(n)\boldsymbol{v}(n)\big)\\ &\times \big(\gamma(\hat{\boldsymbol{x}}^-(n)) + J_\gamma(\hat{\boldsymbol{x}}^-(n))\tilde{\boldsymbol{x}}^-(n) + \boldsymbol{v}(n)\big)^{\mathrm{T}}\Big|\,\boldsymbol{Z}=Z\Big] = \mathbf{0}. \qquad (8.22)\end{aligned}$$

Given $\boldsymbol{Z}=Z$, $\hat{\boldsymbol{x}}^-(n) = \hat{x}^-(n)$, a constant vector. Also, $\hat{\boldsymbol{x}}^-(n)$ is unbiased and independent of the last measurement $\boldsymbol{z}(n)$, so

$$\mathrm{E}\left[\tilde{\boldsymbol{x}}^-(n)\middle|\,\boldsymbol{Z}=Z\right] = \mathrm{E}\left[\tilde{\boldsymbol{x}}^-(n)\middle|\,\boldsymbol{Z}^-=Z^-\right] = \mathbf{0}.$$

Then (8.22) can be written as

$$P^-(n)J_\gamma^{\mathrm{T}}(\hat{x}^-(n)) - K(n)J_\gamma(\hat{x}^-(n))P^-(n)J_\gamma^{\mathrm{T}}(\hat{x}^-(n)) - K(n)R(n) = \mathbf{0}.$$

Solving this equation for $K(n)$, we have the *Kalman gain for the EKF*:

$$\boxed{K(n) = P^-(n)J_\gamma^{\mathrm{T}}(\hat{x}^-(n))\left[J_\gamma(\hat{x}^-(n))P^-(n)J_\gamma^{\mathrm{T}}(\hat{x}^-(n)) + R(n)\right]^{-1}.} \qquad (8.23)$$

Finally, we need an expression for $P(n)$, the *a posteriori conditional* error covariance matrix. From (8.15) and (8.7)

$$\begin{aligned}P(n) &= \mathrm{Cov}\left[\boldsymbol{x}(n) - \hat{\boldsymbol{x}}(n) - K(n)\left[J_\gamma(\hat{\boldsymbol{x}}^-(n))\tilde{\boldsymbol{x}}^-(n) + \boldsymbol{v}(n)\right]\middle|\,\boldsymbol{Z}=Z\right]\\ &= P^-(n) - P^-(n)J_\gamma^{\mathrm{T}}(\hat{x}^-(n))K^{\mathrm{T}}(n) - K(n)J_\gamma(\hat{x}^-(n))P^-(n)\\ &\quad + K(n)\left[J_\gamma(\hat{x}^-(n))P^-(n)J_\gamma^{\mathrm{T}}(\hat{x}^-(n)) + R(n)\right]K^{\mathrm{T}}(n) \qquad (8.24)\end{aligned}$$

Direct substitution for $K(n)$ from (8.23) in the last term shows that the last term in (8.24) is just $P^-(n)J_\gamma^{\mathrm{T}}(\hat{x}^-(n))K^{\mathrm{T}}(n)$. Then $P(n)$ is given by

$$\boxed{P(n) = P^-(n) - K(n)J_\gamma(\hat{x}^-(n))P^-(n).} \qquad (8.25)$$

Summary

The EKF begins with a nonlinear SMM of the form (8.1–8.2). This model is linearized to create a new, approximate SMM given by (8.6–8.7).

Given initial conditions $\hat{x}^-(0)$ and $P^-(0)$, the recursion proceeds as

- *Measurement update.* Acquire $z(n)$ and compute:

$$K(n) = P^-(n) J_\gamma^{\mathrm{T}}(\hat{x}^-(n)) \left[J_\gamma(\hat{x}^-(n)) P^-(n) J_\gamma^{\mathrm{T}}(\hat{x}^-(n)) + R(n) \right]^{-1} \tag{8.26}$$

$$\hat{x}(n) = \hat{x}^-(n) + K(n) \left[z(n) - \gamma(\hat{x}^-(n)) \right] \tag{8.27}$$

$$P(n) = P^-(n) - K(n) J_\gamma(\hat{x}^-(n)) P^-(n) \tag{8.28}$$

- *Time update.* Compute:

$$P^-(n+1) = J_\phi(\hat{x}(n)) P(n) J_\phi^{\mathrm{T}}(\hat{x}(n)) + \Gamma Q(n) \Gamma^{\mathrm{T}} \tag{8.29}$$

$$\hat{x}^-(n+1) = \phi(\hat{x}(n)) \tag{8.30}$$

- *Time increment.* Increment n and repeat.

Remarks

The key to the EKF lies in the linearization of the original SMM. Generally the EKF is no longer optimal in the LMMSE sense. Indeed, if the linearization does not provide a reasonably accurate description of the system dynamics, the state estimates may diverge. Note that if ϕ and γ are linear functions of x, then the EKF reduces to the regular Kalman filter.

Observe that the Jacobians $J_\phi(\hat{\boldsymbol{x}}(n))$ and $J_\gamma(\hat{\boldsymbol{x}}^-(n))$ and the covariance matrices $P^-(n)$ and $P(n)$ require the estimates $\hat{x}(n)$ and $\hat{x}^-(n)$. As a result, the EKF cannot be tested off-line; it requires real or simulated data.

8.2 An Alternate Measurement Update

The EKF measurement update given by (8.26)–(8.28) may not yield accurate estimates if the linearized measurement equation (8.7) is not a good approximation of the given measurement equation (8.2). In this section we shall derive an alternate measurement update developed by Bellaire [18] that can yield much better estimates in general than those provided by the EKF formulation. The first step is to derive an expression for the posterior conditional density function $f_{\boldsymbol{x}(n)}(x(n)|\boldsymbol{Z}_n = Z_n)$ in terms of the prior density function $f_{\boldsymbol{x}(n)}(x(n)|\boldsymbol{Z}_{n-1} = Z_{n-1})$, where $\boldsymbol{Z}_i = \{\boldsymbol{z}(0), \boldsymbol{z}(1), \ldots \boldsymbol{z}(i)\}$ for any $i \geq 0$. We follow the derivation given by Bellaire [18]. To simplify the notation, in the following development we drop the subscript $\boldsymbol{x}(n)$ on the conditional density functions $f_{\boldsymbol{x}(n)}(x(n)|\boldsymbol{Z}_n = Z_n)$ and $f_{\boldsymbol{x}(n)}(x(n)|\boldsymbol{Z}_{n-1} = Z_{n-1})$.

By definition of $f_{\boldsymbol{x}(n)}\left(x(n)|\boldsymbol{Z}_n = Z_n\right)$, we have

$$f\left(x(n)|\boldsymbol{Z}_n = Z_n\right) = \frac{f\left(x(n), Z_n\right)}{f\left(Z_n\right)} = \frac{f\left(x(n), z(n), Z_{n-1}\right)}{f\left(Z_n\right)}. \tag{8.31}$$

Using Bayes' formula, (8.31) can be rewritten as

$$f\left(x(n)|\boldsymbol{Z}_n = Z_n\right) = \frac{f\left(z(n)|\boldsymbol{x}(n) = x(n), \boldsymbol{Z}_{n-1} = Z_{n-1}\right) f\left(x(n), Z_{n-1}\right)}{f\left(z(n)|\boldsymbol{Z}_{n-1} = Z_{n-1}\right) f\left(Z_{n-1}\right)}. \tag{8.32}$$

Now since the measurement noise $\boldsymbol{v}(n)$ is white and $\boldsymbol{z}(n)$ is independent of Z_{n-1} when $\boldsymbol{x}(n)$ is given, we have

$$f\left(z(n)|\boldsymbol{x}(n) = x(n), \boldsymbol{Z}_{n-1} = Z_{n-1}\right) = f\left(z(n)|\boldsymbol{x}(n) = x(n)\right). \tag{8.33}$$

Also

$$f\left(x(n)|\boldsymbol{Z}_{n-1} = Z_{n-1}\right) = \frac{f\left(x(n), Z_{n-1}\right)}{f\left(Z_{n-1}\right)}. \tag{8.34}$$

Then inserting (8.33) and (8.34) into (8.32) gives

$$f\left(x(n)|\boldsymbol{Z}_n = Z_n\right) = \frac{f\left(z(n)|\boldsymbol{x}(n) = x(n)\right) f\left(x(n)|\boldsymbol{Z}_{n-1} = Z_{n-1}\right)}{f\left(z(n)|\boldsymbol{Z}_{n-1} = Z_{n-1}\right)}. \tag{8.35}$$

It follows from (8.2) that

$$f\left(z(n)|\boldsymbol{x}(n) = x(n)\right) = f_{\boldsymbol{v}}\left(z(n) - \gamma(x(n))\right), \tag{8.36}$$

where $f_{\boldsymbol{v}}$ is the density function of the measurement noise $\boldsymbol{v}(n)$. If $\boldsymbol{v}(n)$ is Gaussian, then

$$f_{\boldsymbol{v}}\left(z(n) - \gamma(x(n))\right) = c_{\boldsymbol{v}} \exp\left[-\frac{1}{2}\left[z(n) - \gamma(x(n))\right]^{\mathrm{T}} R^{-1}\left[z(n) - \gamma(x(n))\right]\right], \tag{8.37}$$

where $c_{\boldsymbol{v}}$ is a constant and $R = \mathrm{Cov}[\boldsymbol{v}(n)]$.

We assume that $f_{\boldsymbol{x}(n)}\left(x(n)|\boldsymbol{Z}_{n-1} = Z_{n-1}\right)$ is Gaussian with mean

$$\hat{x}^{-}(n) = \mathrm{E}\left[\boldsymbol{x}(n)|\boldsymbol{Z}_{n-1} = Z_{n-1}\right]$$

and covariance

$$P^{-}(n) = \mathrm{Cov}\left[\boldsymbol{x}(n) - \hat{\boldsymbol{x}}^{-}(n)|\boldsymbol{Z}_{n-1} = Z_{n-1}\right].$$

Then

$$f\left(x(n)|\boldsymbol{Z}_{n-1}=Z_{n-1}\right)=$$
$$c_{\text{prior}}(n)\exp\left[-\frac{1}{2}\left[x(n)-\hat{x}^{-}(n)\right]^{\mathrm{T}}\left(P^{-}(n)\right)^{-1}\left[x(n)-\hat{x}^{-}(n)\right]\right], \quad (8.38)$$

and using (8.36)–(8.38) in (8.35) gives

$$f\left(x(n)|\boldsymbol{Z}_{n}=Z_{n}\right)=$$
$$c(n)\exp\left[-\frac{1}{2}\left[x(n)-\hat{x}^{-}(n)\right]^{\mathrm{T}}\left(P^{-}(n)\right)^{-1}\left[x(n)-\hat{x}^{-}(n)\right]\right.$$
$$\left.-\frac{1}{2}\left[z(n)-\gamma(x(n))\right]^{\mathrm{T}}R^{-1}\left[z(n)-\gamma(x(n))\right]\right], \quad (8.39)$$

where $c(n)$ is a function of $z(n)$ and n.

The conditional density function $f\left(x(n)|\boldsymbol{Z}_{n}=Z_{n}\right)$ given by (8.39) is not Gaussian in general due to the nonlinear function $\gamma(x(n))$ in the exponential function. However, if $f\left(x(n)|\boldsymbol{Z}_{n}=Z_{n}\right)$ is unimodal (i.e., has a single peak), we can take the *a posteriori* estimate $\hat{x}(n)$ to be

$$\hat{x}(n)=\text{ value of } x(n) \text{ for which } f\left(x(n)|\boldsymbol{Z}_{n}=Z_{n}\right) \text{ is maximum.} \quad (8.40)$$

It follows directly from (8.39) that $\hat{x}(n)$ given by (8.40) is the value of $x(n)$ that minimizes

$$\frac{1}{2}\left[x(n)-\hat{x}^{-}(n)\right]^{\mathrm{T}}\left(P^{-}(n)\right)^{-1}\left[x(n)-\hat{x}^{-}(n)\right]$$
$$+\frac{1}{2}\left[z(n)-\gamma(x(n))\right]^{\mathrm{T}}R^{-1}\left[z(n)-\gamma(x(n))\right] \quad (8.41)$$

Hence the problem of computing the measurement update $\hat{x}(n)$ reduces to finding the value of $x(n)$ that minimizes the function in (8.41). If the function γ in the measurement equation (8.2) is linear, there is a closed-form expression for the minimizer of (8.41), and this expression is exactly the same as the measurement update in the Kalman filter. If γ is nonlinear which is the case of interest here, in general there is no closed-form expression for the minimizer, and in this case, the minimizer of (8.41) must be computed numerically using an iterative procedure.

To compute the minimizer of (8.41), we first write it in the form

$$\frac{1}{2}\Psi^{\mathrm{T}}(x(n))S^{-1}(n)\Psi(x(n)), \quad (8.42)$$

where $\Psi(x(n))$ is the $(N+p)$-vector function of $x(n)$ given by

$$\Psi(x(n))=\begin{bmatrix}x(n)-\hat{x}^{-}(n)\\ z(n)-\gamma(x(n))\end{bmatrix}, \quad (8.43)$$

and $S(n)$ is the $(N+p)\times(N+p)$ symmetric matrix given by

$$S(n)=\begin{bmatrix}P^{-}(n) & 0\\ 0 & R\end{bmatrix}.$$

A very popular numerical method for finding the minimizer of (8.42) is the Gauss-Newton iterative procedure given by

$$x_{i+1}=x_i-\left[J_{\Psi}^{\mathrm{T}}(x_i)S^{-1}(n)J_{\Psi}(x_i)\right]^{-1}J_{\Psi}^{\mathrm{T}}(x_i)S^{-1}(n)\Psi(x_i),\qquad i=0,1,2,\ldots \tag{8.44}$$

where x_i is the ith iterate and $J_{\Psi}(x_i)$ is the Jacobian of $\Psi(x(n))$ evaluated at x_i. The Gauss-Newton iteration (8.44) can be started with the initialization $x_0=\hat{x}^{-}(n)$. The use of (8.44) to generate the measurement update results in the iterated extended Kalman filter. However, we shall not pursue this here. It is also worth noting that if the *a posteriori* estimate $\hat{x}(n)$ is taken to be the first iterate with $x_0=\hat{x}^{-}(n)$; that is,

$$\hat{x}(n)=x_1=\hat{x}^{-}(n)-\left[J_{\Psi}^{\mathrm{T}}\left(\hat{x}^{-}(n)\right)S^{-1}(n)J_{\Psi}\left(\hat{x}^{-}(n)\right)\right]^{-1}\times J_{\Psi}^{\mathrm{T}}\left(\hat{x}^{-}(n)\right)S^{-1}(n)\Psi\left(\hat{x}^{-}(n)\right), \tag{8.45}$$

and we obtain the measurement update for the extended Kalman filter. In other words, (8.26)–(8.28) follow from (8.45). The derivation follows as a special case of the constructions given below.

In general, a better numerical procedure than the Gauss-Newton method for computing the minimizer of (8.41) or (8.42) is the Levenberg-Marquardt algorithm given by

$$x_{i+1}=x_i-\left[J_{\Psi}^{\mathrm{T}}(x_i)S^{-1}(n)J_{\Psi}(x_i)+\mu I\right]^{-1}J_{\Psi}^{\mathrm{T}}(x_i)S^{-1}(n)\Psi(x_i),\qquad i=0,1,2,\ldots, \tag{8.46}$$

where I is the $N\times N$ identity matrix and μ? is a positive constant called the Levenberg- Marquardt parameter. Note that if $\mu=0$, the Levenberg-Marquardt algorithm (8.46) reduces to the Gauss-Newton algorithm (8.44). By a proper choice of μ, the Levenberg- Marquardt algorithm will converge to a local minimizer of (8.42), whereas such may not the case for the Gauss-Newton algorithm (8.44). Hence, the Levenberg-Marquardt algorithm can result in a much improved performance over the Gauss-Newton method.

We will take the *a posteriori* estimate $\hat{x}(n)$ to be the first iterate of the Levenberg-Marquardt algorithm (8.46) with $x_0=\hat{x}^{-}(n)$. Hence

$$\hat{x}(n)=x_1=\hat{x}^{-}(n)-\left[J_{\Psi}^{\mathrm{T}}\left(\hat{x}^{-}(n)\right)S^{-1}(n)J_{\Psi}\left(\hat{x}^{-}(n)\right)+\mu I\right]^{-1}\times J_{\Psi}^{\mathrm{T}}\left(\hat{x}^{-}(n)\right)S^{-1}(n)\Psi\left(\hat{x}^{-}(n)\right). \tag{8.47}$$

We will show that (8.47) can be written in the form

$$\hat{x}(n) = \hat{x}^-(n) + K(n)\left[z(n) - \gamma\left(\hat{x}^-(n)\right)\right] \tag{8.48}$$

for some $N \times p$ matrix $K(n)$.

First, from (8.43) we have that

$$J_\Psi\left(\hat{x}^-(n)\right) = \begin{bmatrix} I \\ -J_\gamma\left(\hat{x}^-(n)\right)\end{bmatrix},$$

and using the definitions of $\Psi(x(n))$ and $S(n)$ yields

$$\begin{aligned} J_\Psi^{\mathrm{T}}\left(\hat{x}^-(n)\right) S^{-1}(n) J_\Psi\left(\hat{x}^-(n)\right) \\ = \left(P^-(n)\right)^{-1} + J_\Psi^{\mathrm{T}}\left(\hat{x}^-(n)\right) R^{-1} J_\Psi\left(\hat{x}^-(n)\right), \end{aligned} \tag{8.49}$$

and

$$J_\Psi^{\mathrm{T}}\left(\hat{x}^-(n)\right) S^{-1}(n)\Psi\left(\hat{x}^-(n)\right) = -J_\gamma^{\mathrm{T}}\left(\hat{x}^-(n)\right) R^{-1}. \tag{8.50}$$

Inserting (8.49) and (8.50) into (8.44) results in (8.48) with

$$K(n) = \left[\left(P^-(n)\right)^{-1} + J_\Psi^{\mathrm{T}}\left(\hat{x}^-(n)\right) R^{-1} J_\Psi\left(\hat{x}^-(n)\right) + \mu I\right]^{-1} J_\gamma^{\mathrm{T}}\left(\hat{x}^-(n)\right) R^{-1}. \tag{8.51}$$

Defining

$$\left[\tilde{P}(n)\right]^{-1} = \left(P^-(n)\right)^{-1} + \mu I \tag{8.52}$$

and using the matrix inversion lemma (see [9]), we can rewrite $K(n)$ in the form

$$K(n) = \tilde{P}(n) J_\Psi^{\mathrm{T}}\left(\hat{x}^-(n)\right)\left[J_\gamma^{\mathrm{T}}\left(\hat{x}^-(n)\right)\tilde{P}(n) J_\gamma\left(\hat{x}^-(n)\right) + R\right]^{-1}. \tag{8.53}$$

Note that the expression (8.53) for $K(n)$ is the same as the expression (8.26) for $K(n)$ in the EKF measurement update except that $P^-(n)$ is replaced by $\tilde{P}(n)$ in (8.53).

The last step to complete the measurement update is the determination of the error covariance matrix update $P(n)$ given by

$$P(n) = \mathrm{Cov}\left[\boldsymbol{x}(n) - \hat{\boldsymbol{x}}(n) | \boldsymbol{Z}_n = Z_n\right], \tag{8.54}$$

where

$$\hat{\boldsymbol{x}}(n) = \hat{\boldsymbol{x}}^-(n) + K(n)\left[\boldsymbol{z}(n) - \gamma\left(\hat{\boldsymbol{x}}^-(n)\right)\right], \tag{8.55}$$

with $K(n)$ given by (8.53). Note that the error covariance given by (8.54) is a conditional covariance which corresponds to the definition of $P^-(n)$ given above. When the state model is linear, the error covariance matrix of the linear MMSE estimator (i.e., the Kalman filter) is the same as the conditional error covariance. However, this result is not true in the nonlinear formulation under consideration here due in part to the dependence of $K(n)$ on $\hat{x}^-(n)$ [see (8.53)]

To determine $P(n)$, insert (8.55) into (8.54), which gives

$$P(n) = \mathrm{Cov}\left[\boldsymbol{x}(n) - \hat{\boldsymbol{x}}^-(n) - K(n)\left[\boldsymbol{z}(n) - \gamma\left(\hat{\boldsymbol{x}}^-(n)\right)\right] | \boldsymbol{Z}_n = Z_n\right] \quad (8.56)$$

Inserting $\boldsymbol{z}(n) = \gamma(\boldsymbol{x}(n)) + \boldsymbol{v}(n)$ into (8.56) and defining $\tilde{\boldsymbol{x}}^-(n) = \boldsymbol{x}(n) - \hat{\boldsymbol{x}}^-(n)$ gives

$$P(n) = \mathrm{Cov}\left[\tilde{\boldsymbol{x}}(n) - K(n)\left[\gamma(\boldsymbol{x}(n)) - \gamma\left(\hat{\boldsymbol{x}}^-(n)\right)\right] - K(n)\boldsymbol{v}(n) | \boldsymbol{Z}_n = Z_n\right]. \quad (8.57)$$

We can approximate $\gamma(\boldsymbol{x}(n)) - \gamma\left(\hat{\boldsymbol{x}}^-(n)\right)$ by

$$\gamma(\boldsymbol{x}(n)) - \gamma\left(\hat{\boldsymbol{x}}^-(n)\right) = J_\gamma\left(\hat{\boldsymbol{x}}^-(n)\right)\tilde{\boldsymbol{x}}^-(n),$$

and inserting this into (8.57) gives

$$P(n) = \mathrm{Cov}\left[\left[I - K(n)J_\gamma\left(\hat{\boldsymbol{x}}^-(n)\right)\right]\tilde{\boldsymbol{x}}^-(n) - K(n)\boldsymbol{v}(n) | \boldsymbol{Z}_n = Z_n\right]. \quad (8.58)$$

Now $\hat{\boldsymbol{x}}^-(n)$ and $\tilde{\boldsymbol{x}}^-(n)$ are independent of $\boldsymbol{v}(n)$, and thus (8.58) can be rewritten as

$$P(n) = \mathrm{Cov}\left[\left[I - K(n)J_\gamma\left(\hat{\boldsymbol{x}}^-(n)\right)\right]\tilde{\boldsymbol{x}}^-(n) | \boldsymbol{Z}_n = Z_n\right] + \mathrm{Cov}\left[K(n)\boldsymbol{v}(n) | \boldsymbol{Z}_n = Z_n\right] \quad (8.59)$$

The covariance of $\boldsymbol{v}(n)$ is independent of $\boldsymbol{Z}_n$, and thus

$$\mathrm{Cov}\left[K(n)\boldsymbol{v}(n) | \boldsymbol{Z}_n = Z_n\right] = \mathrm{Cov}\left[K(n)\boldsymbol{v}(n)\right] = K(n)RK^{\mathrm{T}}(n). \quad (8.60)$$

The first term on the right-hand side of (8.59) can be approximated by

$$\mathrm{Cov}\left[\left[I - K(n)J_\gamma\left(\hat{\boldsymbol{x}}^-(n)\right)\right]\tilde{\boldsymbol{x}}^-(n) | \boldsymbol{Z}_n = Z_n\right] = \left[I - K(n)J_\gamma\left(\hat{\boldsymbol{x}}^-(n)\right)\right]P^-(n)\left[I - K(n)J_\gamma\left(\hat{\boldsymbol{x}}^-(n)\right)\right]^{\mathrm{T}}. \quad (8.61)$$

Inserting (8.60) and (8.61) into (8.59) yields

$$P(n) = \left[I - K(n)J_\gamma\left(\hat{\boldsymbol{x}}^-(n)\right)\right]P^-(n)\left[I - K(n)J_\gamma\left(\hat{\boldsymbol{x}}^-(n)\right)\right]^{\mathrm{T}} + K(n)RK^{\mathrm{T}}(n). \quad (8.62)$$

Equation (8.62) is the error covariance measurement update. It does not appear to be possible to simplify (8.62), although when $\mu = 0$ (so that $P^-(n) = \tilde{P}(n)$), (8.62) can be written in the form

$$P(n) = \left[I - K(n)J_\gamma\left(\hat{\boldsymbol{x}}^-(n)\right)\right]P^-(n),$$

which is the expression for $P(n)$ in the EKF measurement update [see (8.28)].

To summarize, the equations for the Levenberg-Marquardt measurement update are:

$$\hat{x}(n) = \hat{x}^-(n) + K(n)\left[z(n) - \gamma\left(\hat{x}^-(n)\right)\right] \tag{8.63}$$

$$K(n) = \tilde{P}(n)\left(J_\Psi\left(\hat{x}^-(n)\right)\right)^{\mathrm{T}}\left[J_\gamma^{\mathrm{T}}\left(\hat{x}^-(n)\right)\tilde{P}(n)J_\gamma\left(\hat{x}^-(n)\right) + R\right]^{-1} \tag{8.64}$$

$$\left[\tilde{P}(n)\right]^{-1} = \left(P^-(n)\right)^{-1} + \mu I \tag{8.65}$$

$$\begin{aligned} P(n) = {} & \left[I - K(n)J_\gamma\left(\hat{x}^-(n)\right)\right]P^-(n)\left[I - K(n)J_\gamma\left(\hat{x}^-(n)\right)\right]^{\mathrm{T}} \\ & + K(n)RK^{\mathrm{T}}(n). \end{aligned} \tag{8.66}$$

The equations for the time update are the same as in the EKF formulation [see (8.29)–(8.30)]. Again, we note that the expressions (8.63)–(8.66) reduce to the EKF measurement update in the case when $\mu = 0$.

8.3 Nonlinear System Identification Using Neural Networks

Consider a single-input single-output discrete-time system given by the difference equation

$$y(n) = f\left(y(n-1), y(n-2), \ldots, y(n-N), w(n), w(n-1), \ldots, w(n-M)\right), \tag{8.67}$$

where N is a positive integer, M is a nonnegative integer, $w(n)$ is the input, and $y(n)$ is the output. In general, the function f in (8.67) is nonlinear function of $y(i)$ for $i = 1, 2, \ldots, N$, and $u(i)$ for $i = 0, 1{,}2, \ldots, M$. The integers N and M in (8.67) are assumed to be known. If they are not known, one can use the modeling procedure given below for various choices of N and M until an acceptable result is obtained.

We could approximate (8.67) by a linear relationship given by

$$y(n) = \sum_{i=1}^{N} a_i y(n-i) + \sum_{i=0}^{M} b_i w(n-i) + v(n), \tag{8.68}$$

where $v(n)$ is the model error term. As discussed in Section 7.6, the system given by (8.68) has the state model

$$x(n+1) = Ix(n) \tag{8.69}$$
$$y(n) = c(n)x(n) + v(n), \tag{8.70}$$

where the state $x(n)$ is the $(N+M+1)$-dimensional vector given by

$$x(n) = \begin{bmatrix} a_1 \\ \vdots \\ a_N \\ b_0 \\ \vdots \\ b_M \end{bmatrix},$$

and $c(n)$ is the $(N+M+1)$-element row vector given by

$$c(n) = \begin{bmatrix} y(n-1) & \cdots & y(n-N) & w(n) & \cdots & w(n-M) \end{bmatrix}.$$

The parameters of the system (i.e., the a_i and the b_i) can then be estimated by using a Kalman filter based on the linear state model given by (8.69)–(8.70). The details are carried out in Section 7.6.

If the function f in (8.67) is nonlinear, the linear model defined by (8.68) is not likely to result in an accurate model. In the nonlinear case, an accurate model can be constructed by using a neural network that is defined in terms of neurons. A neuron with output ν and inputs α_1 , $\alpha_2, \ldots, \alpha_q$ is given by

$$\nu = \text{act}\left[\sum_{i=1}^{q} \alpha_i\right],$$

where act is the activation function. There are various possible choices for the activation function. One common example is the arctan function given by

$$\text{act}(\eta) = \frac{e^{\eta} - e^{-\eta}}{e^{\eta} + e^{-\eta}}. \tag{8.71}$$

The arctan function "compresses" all input values into the range from 0 to 1.

We can model the system given by (8.67) in terms of neurons. For example, when $N = 1$ and $M = 0$, a two neuron neural network model is given by

$$y(n) = c_1\chi_1(n) + c_2\chi_2(n) + v(n) \tag{8.72}$$
$$\chi_j(n) = \text{act}\left[a_{1j}y(n-1) + b_{1j}w(n)\right], \qquad j = 1, 2. \tag{8.73}$$

The signals $\chi_1(n)$ and $\chi_2(n)$ defined by (8.73) are signals located at the hidden nodes of the neural network. The term "hidden" means that these signals cannot be directly determined from either the inputs or the output of the neural network. The neural network (NN) given by (8.72)-(8.73) is said to be a feedforward NN with three layers (the input layer, hidden layer, and output layer).

It is known that any system given by (8.67) can be modeled arbitrarily closely by a three-layer feedforward NN if a sufficient number of nodes in the hidden layer are used. Unfortunately, there is no result that states how many hidden nodes must be used - one must determine this by trial-and-error.

The equations for a feedforward NN model with q hidden nodes are

$$y(n) = \sum_{j=1}^{q} c_j \chi_j(n) + v(n) \tag{8.74}$$

$$\chi_j(n) = \text{act}\left[\sum_{i=1}^{N} a_{ij} y(n-i) + \sum_{i=0}^{M} b_{ij} w(n-i)\right], \quad j = 1, 2, \ldots, q. \tag{8.75}$$

We can construct a state model for (8.74)–(8.75) as follows. First, define

$$\theta_j = \begin{bmatrix} a_{1j} \\ \vdots \\ a_{Nj} \\ b_{0j} \\ \vdots \\ b_{Mj} \end{bmatrix} \quad \text{and} \quad c = \begin{bmatrix} c_1 \\ c_2 \\ \vdots \\ c_q \end{bmatrix}.$$

Note that for each value of j, θ_j is a $(N + M + 1)$-element column vector. Letting $\phi(n)$ denote the $(N + M + 1)$-element row vector

$$\phi(n) = \begin{bmatrix} y(n-1) & \ldots & y(n-N) & w(n) & \ldots & w(n-m) \end{bmatrix},$$

we can then combine (8.74) and (8.75) into a single equation given by

$$y(n) = \sum_{j=1}^{q} c_j \text{act}\left[\phi(n)\theta_j\right] + v(n). \tag{8.76}$$

Finally, let $x(n)$ denote the $q(N + M + 2)$-element state vector defined by

$$x(n) = \begin{bmatrix} \theta_1 \\ \theta_2 \\ \vdots \\ \theta_q \\ c \end{bmatrix}.$$

Then the system defined by (8.74)–(8.75) has the state model

$$x(n+1) = Ix(n) \tag{8.77}$$

$$y(n) = \gamma\,(x(n)) + v(n), \tag{8.78}$$

where

$$\gamma\,(x(n)) = \sum_{j=1}^{q} c_j \text{act}\,[\phi(n)\theta_j]. \tag{8.79}$$

The problem of identifying the parameters of the system (8.74)–(8.75) then reduces to the problem of estimating the state $x(n)$ of the system given by (8.77)–(8.79). Note that since (8.77) is linear, the time update is given by

$$\hat{x}^-(n) = \hat{x}(n) \tag{8.80}$$

$$P^-(n+1) = P(n) + Q, \tag{8.81}$$

where $Q = \text{Cov}[\boldsymbol{v}(n)]$.

Since the function γ in (8.79) is nonlinear, to compute the measurement update we need to use the EKF or the Levenberg-Marquardt measurement update. Either approach requires that we compute the Jacobian J_γ of $\gamma\,(x(n))$, where $\gamma\,(x(n))$ is given by (8.79). By definition of $x(n)$,

$$\begin{aligned} J_\gamma\,(x(n)) &= \frac{\partial\gamma\,(x(n))}{\partial x(n)} \\ &= \begin{bmatrix} \frac{\partial\gamma(x(n))}{\partial\theta_1} & \frac{\partial\gamma(x(n))}{\partial\theta_1} & \cdots & \frac{\partial\gamma(x(n))}{\partial\theta_q} & \frac{\partial\gamma(x(n))}{\partial c} \end{bmatrix}. \end{aligned} \tag{8.82}$$

It follows from (8.79) that

$$\frac{\partial\gamma\,(x(n))}{\partial\theta_j} = c_j \left[\frac{\partial \text{act}(\eta)}{\partial\eta}\right] \theta_j^{\text{T}}, \quad j = 1, 2, \ldots, q, \; \eta = \phi(n)\theta_j, \tag{8.83}$$

and

$$\frac{\partial\gamma\,(x(n))}{\partial c} = \left[\text{act}\,(\phi(n)\theta_1) \quad \text{act}\,(\phi(n)\theta_2) \quad \ldots \quad \text{act}\,(\phi(n)\theta_q)\right]. \tag{8.84}$$

Note that if act is the arctan function defined by (8.71), then (8.83) becomes

$$\frac{\partial\gamma\,(x(n))}{\partial\theta_j} = c_j \left[\frac{4}{e^{\phi(n)\theta_j} + e^{-\phi(n)\theta_j}}\right] \theta_j^{\text{T}}, \quad j = 1, 2, \ldots, q. \tag{8.85}$$

Using (8.82)–(8.85), we can compute the Jacobian $J_\gamma\,(x(n))$, and then the measurement update can be computed using either the EKF or the Levenberg-Marquardt formulation.

8.4 Frequency Demodulation

Having introduced the EKF, let us now consider a simple example: the demodulation of a frequency-modulated signal.

Frequency Modulation

Frequency modulation (FM) is a well-known technique for transmitting analog waveforms. A continuous-time message $\boldsymbol{m}_c(t)$ modulates the angle of a sinusoidal carrier $\boldsymbol{c}_c(t)$ [19]. Let A_0 denote the carrier amplitude and Ω_0 the carrier frequency.

The carrier and message are related by

$$\begin{aligned} \boldsymbol{c}_c(t) &= A_0 \cos\left(\Omega_0 t + \beta_0 \int_0^t \boldsymbol{m}_c(\tau) d\tau\right) \\ &= A_0 \cos\left(\Omega_0 t + \boldsymbol{\theta}_c(t)\right), \end{aligned} \tag{8.86}$$

where

$$\boldsymbol{\theta}_c(t) = \beta_0 \int_0^t \boldsymbol{m}_c(\tau) d\tau, \tag{8.87}$$

where β_0 is the modulation index. The problem of interest is to estimate the message $\boldsymbol{m}_c(t)$ from noisy measurements of the form $\boldsymbol{c}_c(t) + \boldsymbol{v}_c(t)$. This process is known as frequency demodulation.

A State Model

In order to apply the EKF, we must derive a state model. The message is a bandlimited signal in the frequency range $-\Omega_m < \Omega < \Omega_m$. We therefore model $\boldsymbol{m}_c(t)$ as the output of a lowpass filter that has cutoff frequency Ω_m and is excited by white noise $\boldsymbol{w}_c(t)$ with variance $\sigma^2_{w_c}$.

We employ a first-order Butterworth filter. This filter does not provide exceptional frequency selectivity, but it is adequate to illustrate the use of the EKF. The Laplace transform $H_c(s)$ representation of the filter is [20]

$$H_c(s) = \frac{M_c(s)}{W_c(s)} = \frac{\Omega_m}{s + \Omega_m}. \tag{8.88}$$

As a result, the differential equation relating $\boldsymbol{m}_c(t)$ and $\boldsymbol{w}_c(t)$ is

$$\dot{\boldsymbol{m}}_c(t) = -\Omega_m \boldsymbol{m}_c(t) + \Omega_m \boldsymbol{w}_c(t).$$

We can easily relate $\boldsymbol{\theta}_c(t)$ and $\boldsymbol{m}_c(t)$. From (8.87), the derivative of $\boldsymbol{\theta}_c(t)$ is

$$\dot{\boldsymbol{\theta}}_c(t) = \beta_0 \boldsymbol{m}_c(t).$$

Defining a continuous-time state vector by $\boldsymbol{x}_c(t) \triangleq \begin{bmatrix} \boldsymbol{m}_c(t) & \boldsymbol{\theta}_c(t) \end{bmatrix}^{\mathrm{T}}$, we obtain the following continous-time state model.

$$\dot{\boldsymbol{x}}_c(t) = \begin{bmatrix} -\Omega_m & 0 \\ \beta_0 & 0 \end{bmatrix} \boldsymbol{x}_c(t) + \begin{bmatrix} \Omega_m \\ 0 \end{bmatrix} \boldsymbol{w}_c(t) \tag{8.89}$$

$$\boldsymbol{z}_c(t) = A_0 \cos\left(\Omega_0 t + \begin{bmatrix} 0 & 1 \end{bmatrix} \boldsymbol{x}_c(t)\right) + \boldsymbol{v}_c(t). \tag{8.90}$$

Discretization

We sample this system with sampling period T. When we discretize the system, the discrete-time state vector is

$$\boldsymbol{x}(n) = \boldsymbol{x}_c(nT) = \begin{bmatrix} \boldsymbol{m}(n) & \boldsymbol{\theta}(n) \end{bmatrix}^{\mathrm{T}} = \begin{bmatrix} \boldsymbol{m}_c(nT) & \boldsymbol{\theta}_c(nT) \end{bmatrix}^{\mathrm{T}}. \tag{8.91}$$

To discretize Equation (8.89), we use the Laplace transform relationship demonstrated in Appendix A

$$\Phi = \mathcal{L}^{-1}\left\{\left(sI - \begin{bmatrix} -\Omega_m & 0 \\ \beta_0 & 0 \end{bmatrix}\right)^{-1}\right\}\Bigg|_{t=T} = \mathcal{L}^{-1}\left\{\begin{bmatrix} \frac{1}{s+\Omega_m} & 0 \\ \frac{\beta_0}{s(s+\Omega_m)} & \frac{1}{s} \end{bmatrix}\right\}\Bigg|_{t=T}$$
$$= \begin{bmatrix} e^{-\Omega_m T} & 0 \\ \frac{\beta_0}{\Omega_m}\left(1 - e^{-\Omega_m T}\right) & 1 \end{bmatrix}. \tag{8.92}$$

Next, $\Gamma = I_2$, and Q is given by (A.45):

$$Q = \sigma_{w_c} \int_0^T e^{A\tau} B B^{\mathrm{T}} e^{A^{\mathrm{T}}\tau} d\tau,$$

where

$$e^{A\tau} = \begin{bmatrix} e^{-\Omega_m \tau} & 0 \\ \frac{\beta_0}{\Omega_m}\left(1 - e^{-\Omega_m \tau}\right) & 1 \end{bmatrix},$$

and $B = \begin{bmatrix} \Omega_m & 0 \end{bmatrix}^{\mathrm{T}}$. Direct substitution gives

$$Q = \sigma_{w_c}^2 \int_0^T \begin{bmatrix} \Omega_m^2 e^{-2\Omega_m \tau} & \beta_0 \Omega_m \left(e^{-\Omega_m \tau} - e^{-2\Omega_m \tau}\right) \\ \beta_0 \Omega_m \left(e^{-\Omega_m \tau} - e^{-2\Omega_m \tau}\right) & \beta_0^2 \left(1 - e^{-\Omega_m \tau}\right)^2 \end{bmatrix}.$$

We compute each integral in the above equation separately. Ultimately we obtain

$$Q = \begin{bmatrix} Q_{11} & Q_{12} \\ Q_{21} & Q_{22} \end{bmatrix}, \tag{8.93}$$

where

$$Q_{11} = \frac{\Omega_m}{2}\left(1 - e^{-2\Omega_m T}\right) \tag{8.94}$$

$$Q_{12} = Q_{21} = \frac{\beta_0}{2}\left(1 - 2e^{-\Omega_m T} + e^{-2\Omega_m T}\right) \tag{8.95}$$

$$Q_{22} = \frac{\beta_0^2}{2\Omega_m}\left(-3 + 2\Omega_m T + 4e^{-\Omega_m T} - e^{-2\Omega_m T}\right) \tag{8.96}$$

So the state model is

$$\boldsymbol{x}(n+1) = \begin{bmatrix} e^{-\Omega_m T} & 0 \\ \frac{\beta_0}{\Omega_m}\left(1-e^{-\Omega_m T}\right) & 1 \end{bmatrix} \boldsymbol{x}(n) + I_2 \boldsymbol{w}(n), \tag{8.97}$$

where $\boldsymbol{x}(n)$ is defined by (8.91) and Q is given by (8.93) through (8.96).

The carrier signal is then

$$c(n) = \gamma(\boldsymbol{x}(n)) = A_0 \cos\left(\Omega_0 nT + \begin{bmatrix} 0 & 1 \end{bmatrix} \boldsymbol{x}(n)\right). \tag{8.98}$$

Finally, the output equation (8.90) becomes

$$\boldsymbol{z}(n) = \gamma(\boldsymbol{x}(n)) + \boldsymbol{v}(n), \tag{8.99}$$

and $R = \sigma_{v_c}^2/T$. Note that the state and input dynamics of (8.97) are linear, but the output equation (8.99) is nonlinear. Hence, we only need to linearize (8.99).

Application of the EKF

Using the discrete-time SMM of (8.97) and (8.99), we can now apply the EKF to the demodulation problem. Since the state dynamics are linear, we only need to handle the output equation (8.99). Applying the EKF approximation to (8.98), we require

$$\begin{aligned} J_\gamma\left(\hat{x}^-(n)\right) &= \begin{bmatrix} 0 & -A_0 \sin\left(\Omega_0 nT + \begin{bmatrix} 0 & 1 \end{bmatrix} \hat{x}^-(n)\right) \end{bmatrix} \\ &= \left[0 \quad -A_0 \sin\left(\Omega_0 nT + \hat{\theta}^-(n)\right)\right] \end{aligned}$$

The *a posteriori* state estimate (8.27) becomes

$$\begin{aligned} \hat{x}(n) &= \hat{x}^-(n) + K(n)\left[z(n) - A_0 \cos\left(\Omega_0 nT + \begin{bmatrix} 0 & 1 \end{bmatrix} \hat{x}^-(n)\right)\right] \\ &= \hat{x}^-(n) + K(n)\left[z(n) - A_0 \cos\left(\Omega_0 nT + \hat{\theta}^-(n)\right)\right] \end{aligned}$$

and the demodulated message is then $\hat{m}(n) = \begin{bmatrix} 1 & 0 \end{bmatrix} \hat{x}(n)$.

With no *a priori* information, we may initialize the EKF with $\hat{x}^-(0) = \begin{bmatrix} 0 & 0 \end{bmatrix}^{\mathrm{T}}$ and $P^-(0) = \alpha I$, $\alpha > 0$.

Example 8.1 Frequency Demodulation

The EKF was simulated to perform frequency demodulation for commerical FM radio. The carrier signal had amplitude $A_0 = 1$ and frequency $f_0 = 100$ MHz. The message bandwidth $f_m = 15$ kHz modeled the typical maximum frequency of audio broadcasts [19]. (Note that $\Omega_0 = 2\pi f_0$ and $\Omega_m = 2\pi f_m$.) The frequency modulation index β_0 was 5.

In the simulations, the sampling frequency f_s was set to 250 MHz, so the sampling period is $T = 1/f_s = 9 \times 10^{-9} = 9$ ns. For the process noise, $\sigma_{w_c}^2 = 0.01$. The measurement noise had variance $\sigma_{v_c}^2 = 4 \times 10^{-12}$. Although this value may seem small, after discretization, $\sigma_v^2 = 0.001$. Then the carrier signal-to-noise ratio was

$$\text{SNR}_C = 10 \log_{10} \left(\frac{A_0^2}{2\sigma_v^2} \right) \approx 27 \text{ dB}.$$

Figure 8.1 shows the estimate $\hat{m}(n)$ of the message $m(n)$. Due to the small sampling period T, a short segment of the results is also displayed. The estimated message $\hat{m}(n)$ tracked the actual message $m(n)$ but did not match it exactly. This behavior is not surprising because the first-order lowpass filter of (8.88) does not bandlimit the message $m(n)$ particularly well. In addition, note that $m_c(t)$ is equal to the derivative of the angle $\theta_c(t)$, so estimates of the message $m(n)$ are highly sensitive to estimation errors in $\theta(n)$.

The EKF also estimates the angle $\theta(n)$, and results appear in Figure 8.2. The angle estimates were somewhat more accurate than the message estimates. The true MSE and the EKF's estimated MSE are shown in Figure 8.3.

8.5 Target Tracking Using the EKF

In this section we study the application of the EKF to the target tracking problem of Sections 7.1 and 7.5. We use the same premise as in Section 7.5: we model the target motion in Cartesian coordinates but obtain measurements in polar coordinates. The nonlinear relationship between these coordinate systems suggests that the EKF may be suitable for this problem.

The Nonlinear SMM

We maintain the system dynamics of (7.6), and the state vector is

$$\vec{\boldsymbol{x}}(n) = \begin{bmatrix} \boldsymbol{x}(n) & \dot{\boldsymbol{x}}(n) & \boldsymbol{y}(n) & \dot{\boldsymbol{y}}(n) \end{bmatrix}^{\mathrm{T}}.$$

However, we let the measurements be given in polar coordinates,

$$\boldsymbol{s}(n) = \begin{bmatrix} \boldsymbol{r}(n) \\ \boldsymbol{\theta}(n) \end{bmatrix} = \gamma(\vec{\boldsymbol{x}}(n)) = \begin{bmatrix} \sqrt{\boldsymbol{x}^2(n) + \boldsymbol{y}^2(n)} \\ \tan_\theta^{-1}(\boldsymbol{x}(n), \boldsymbol{y}(n)) \end{bmatrix},$$

where $\tan_\theta^{-1}(x, y)$ is defined by (7.38). So the observations are given by

$$\boldsymbol{z}(n) = \begin{bmatrix} \boldsymbol{r}(n) \\ \boldsymbol{\theta}(n) \end{bmatrix} + \begin{bmatrix} \boldsymbol{v}_r(n) \\ \boldsymbol{v}_\theta(n) \end{bmatrix}. \tag{8.100}$$

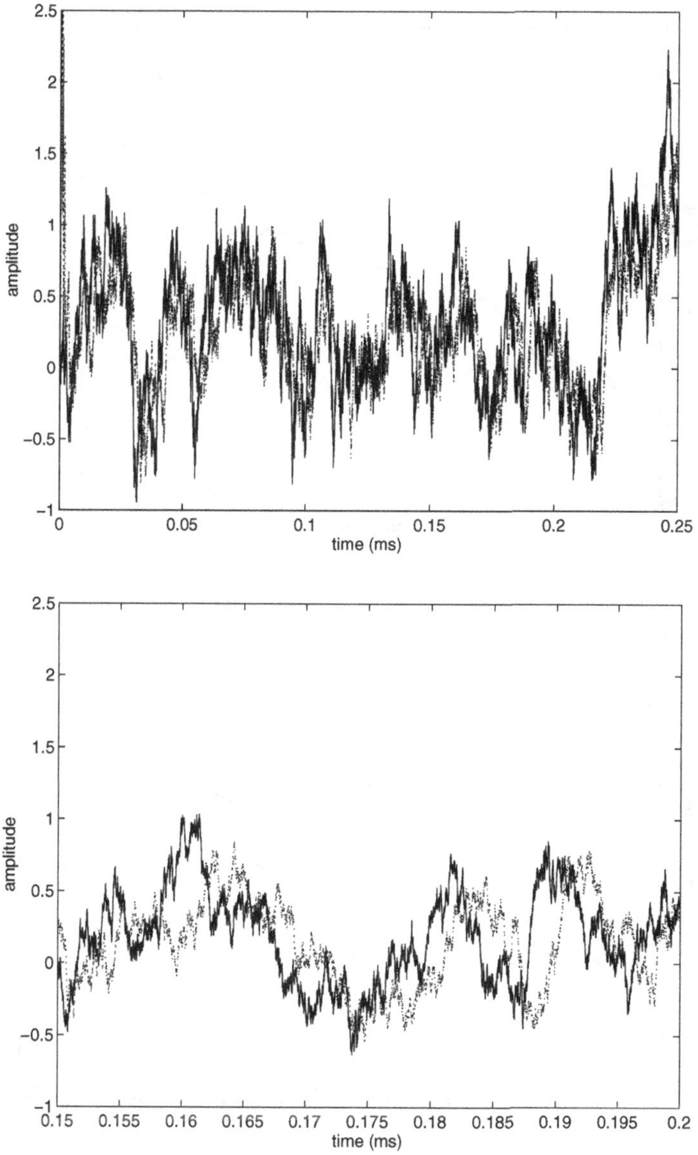

Figure 8.1. Frequency demodulation results for Example 8.1. **Top:** Actual message $m(n)$ (solid curve) and EKF estimate of message $\hat{m}(n)$ (dotted curve). **Bottom:** Detail of message.

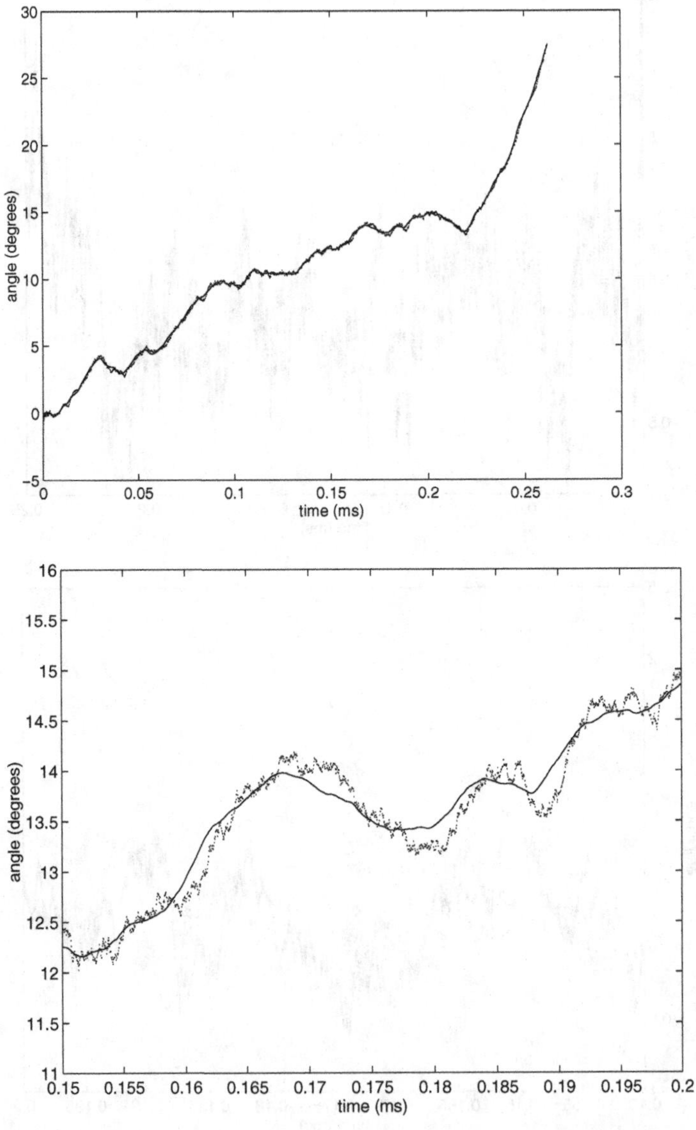

Figure 8.2. Frequency demodulation results for Example 8.1. **Top:** Actual angle $\theta(n)$ (solid curve) and EKF estimate $\hat{\theta}(n)$ (dotted curve). **Bottom:** Detail of angle and estimate.

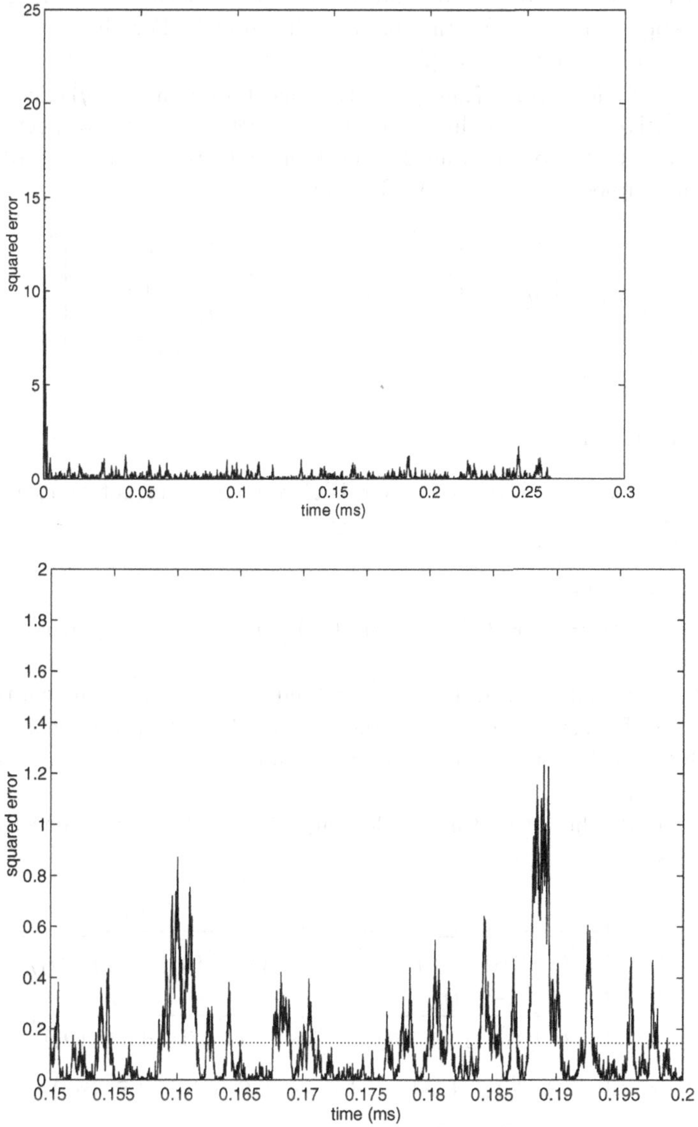

Figure 8.3. Frequency demodulation results for Example 8.1, cont. **Top:** True MSE (solid curve) and EKF estimate of the MSE ($\mathrm{tr}\,(P(n))$, dotted curve). **Bottom:** Detail of MSE.

Linearization

The system dynamics are linear; that is, $\phi(\vec{x}(n)) \equiv \Phi\vec{x}(n)$. so we do not need to make approximations for this part of the model. For the measurement equation we must calculate $J_\gamma(\vec{x})$.

$\tan_\theta^{-1}(x, y)$, defined in (7.38), is continuous at all points (x, y) except along the nonpositive x-axis. Furthermore, it is differentiable with respect to x at all points except the origin, and differentiable with respect to y at all points except the nonpositive x-axis. We find that

$$J_\gamma(\vec{x}) = \begin{bmatrix} \frac{x(n)}{\sqrt{x^2(n)+y^2(n)}} & 0 & \frac{x(n)}{\sqrt{x^2(n)+y^2(n)}} & 0 \\ \frac{-y(n)}{x^2(n)+y^2(n)} & 0 & \frac{x(n)}{x^2(n)+y^2(n)} & 0 \end{bmatrix}.$$

Implementation

We are now ready to implement our EKF for single target tracking. The linearized SMM is

$$\begin{aligned} \boldsymbol{x}(n+1) &= \Phi\boldsymbol{x}(n) + \Gamma\boldsymbol{w}(n) \\ \boldsymbol{z}(n) &= \gamma(\hat{x}^-(n)) + J_\gamma(\hat{x}^-(n))\left[\boldsymbol{x}(n) - \hat{x}^-(n)\right] + \boldsymbol{v}(n), \end{aligned}$$

where Φ is given by (7.3), $\Gamma = I_4$, and $\boldsymbol{w}(n)$ is a zero-mean white noise of the form of (7.4) and has covariance matrix Q as in (7.5). $\boldsymbol{v}(n)$ is zero-mean white noise, uncorrelated with $\boldsymbol{w}(n)$, and has covariance matrix $R = \mathrm{diag}\left(\left[\sigma_{v_r}^2 \quad \sigma_{v_\theta}^2\right]\right)$.

We initialize the estimator by selecting $\hat{x}^-(0)$ and $P^-(0)$. Then the filter recursion is

$$J_\gamma\left(\hat{\vec{x}}^-(n)\right) = \begin{bmatrix} \frac{\hat{x}^-(n)}{\sqrt{\left(\hat{x}^-(n)\right)^2+\left(\hat{y}^-(n)\right)^2}} & 0 & \frac{\hat{x}^-(n)}{\sqrt{\left(\hat{x}^-(n)\right)^2+\left(\hat{y}^-(n)\right)^2}} & 0 \\ \frac{-\hat{y}^-(n)}{\left(\hat{x}^-(n)\right)^2+\left(\hat{y}^-(n)\right)^2} & 0 & \frac{\hat{x}^-(n)}{\left(\hat{x}^-(n)\right)^2+\left(\hat{y}^-(n)\right)^2} & 0 \end{bmatrix}$$

$$K(n) = P^-(n)J_\gamma^{\mathrm{T}}\left(\hat{\vec{x}}^-(n)\right)\left[J_\gamma\left(\hat{\vec{x}}^-(n)\right)P^-(n)J_\gamma^{\mathrm{T}}\left(\hat{\vec{x}}^-(n)\right)+R(n)\right]^{-1}$$

$$\gamma\left(\hat{\vec{x}}^-(n)\right) = \begin{bmatrix} \sqrt{\left(\hat{x}^-(n)\right)^2+\left(\hat{y}^-(n)\right)^2} \\ \tan_\theta^{-1}(\hat{x}^-(n), \hat{y}^-(n)) \end{bmatrix}$$

$$\hat{\vec{x}}(n) = \hat{\vec{x}}^-(n) + K(n)\left[z(n) - \gamma\left(\hat{\vec{x}}^-(n)\right)\right]$$

$$P(n) = P^-(n) - K(n)J_\gamma\left(\hat{\vec{x}}^-(n)\right)P^-(n)$$

$$P^-(n+1) = \Phi P(n)\Phi^{\mathrm{T}} + \Gamma(n)Q(n)\Gamma^{\mathrm{T}}(n)$$

$$\hat{\vec{x}}^-(n+1) = \Phi\hat{\vec{x}}(n).$$

Example 8.2 Target Tracking Using the EKF

We simulated the EKF for this problem using the same parameters as in Example 7.5. Since measurements are now provided in polar form but our state vector employs Cartesian coordinates, we initialized the Kalman filter with measurement $z(n)$ and

$$\hat{\bar{x}}^-(0) = \begin{bmatrix} z_1(0)\cos z_2(0) & 0 & z_1(0)\sin z_2(0) & 0 \end{bmatrix}^{\mathrm{T}},$$

and $P^-(0) = \mathrm{diag}\left(\begin{bmatrix} 50 & 10 & 50 & 10 \end{bmatrix}\right)$.

Tracking results for the horizontal component appear in Figure 8.4. As the EKF tracked the target, a noticeable decaying sinusoidal pattern was evident, due to the linearization of $\tan_\theta^{-1}(x, y)$ in the EKF. Figure 8.5 (top) shows the velocity track and compares the true and estimated MSE. Both these graphs show the sinusoidal behavior in the estimates. The estimated MSE did not have this pattern because the EKF used the linearized state model to estimate the MSE.

8.6 Multiple Target Tracking

We have examined the application of the Kalman filter and EKF to track an object in several of the preceding sections. Now let us turn our attention to the problem of tracking several objects at once, which forms the *multiple target tracking* (MTT) problem.

Overview

At a first glance, the MTT problem appears very similar to the single target tracking problem. However, MTT presents several difficulties. We briefly touch upon some of the issues involved.

Data Association At each time n, a group of measurements becomes available. Most existing tracking systems must determine which measurement corresponds to which target. In this section we present an estimation scheme that does not require knowledge of the target-measurement associations.

Track Initiation The tracking system must be able to determine how many targets are present. Properly choosing the number of targets remains an open problem. We do not consider this problem here.

Appearance and Disappearance of Targets The number of objects of interest may change during tracking. The system should be able to identify when targets are no longer present and when new targets appear and adjust accordingly. We do not concern ourselves with this aspect.

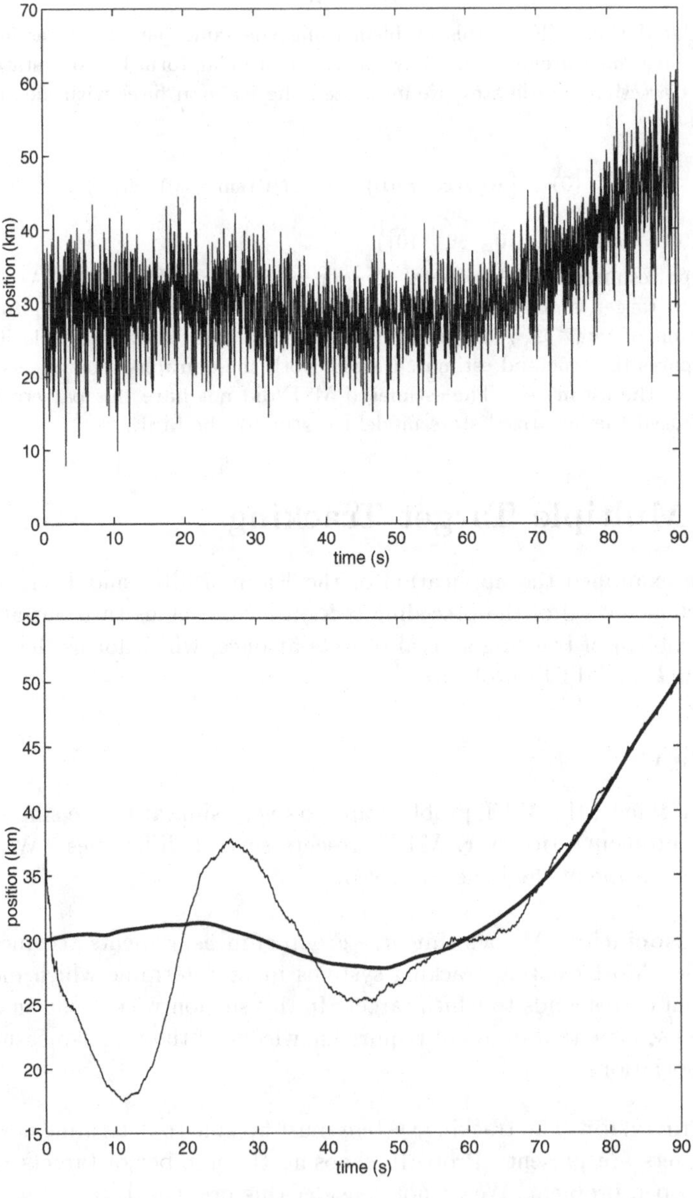

Figure 8.4. Target tracking results for Example 8.2. **Top:** Noisy observations $z_1(n)$ of the target's horizontal position. **Bottom:** Actual horizontal position $\vec{x}_1(n)$ or $x(n)$ (thick curve) and estimated position $\hat{\vec{x}}_1(n)$ or $\hat{x}(n)$ from the Kalman filter (thin curve).

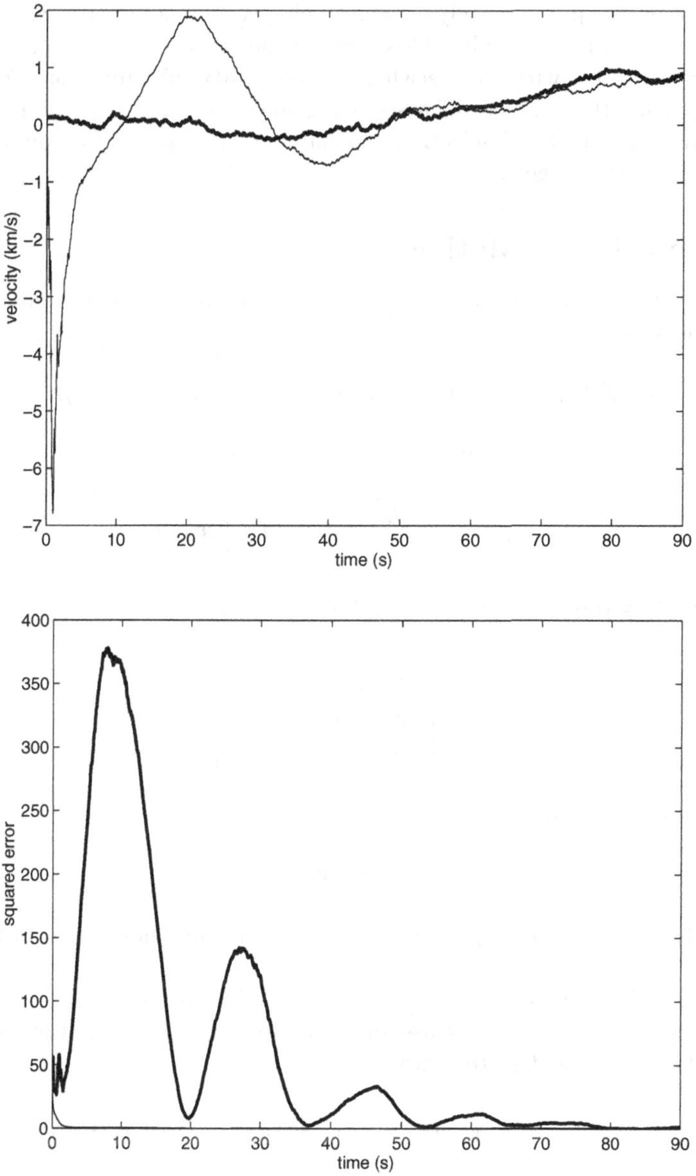

Figure 8.5. Target tracking results for Example 8.2, cont. **Top:** Actual horizontal velocity $\vec{x}_2(n)$ or $\dot{x}(n)$ (thick curve) and Kalman estimate (thin curve). **Bottom:** Actual MSE (thick curve) and estimated MSE $(\mathrm{tr}\,(P(n)))$ of the Kalman filter (thin curve).

Let us consider the following scenario: We wish to track three targets that travel with approximately constant velocity. For simplicity, we consider one-dimensional motion only. This scenario demonstrates an application of the EKF for MTT without overwhelming difficulty. We must only concern ourselves with the problem of associating measurements with targets. We do not have the tasks of selecting the number of targets or accounting for obsolete and new targets.

Modeling Target Motion

Let $x_{ci}(t)$ denote the position of the ith target at time t. Then we define the continuous-time state vector

$$\vec{x}_c(t) \triangleq \begin{bmatrix} x_{1c}(t) & x_{2c}(t) & x_{3c}(t) & \dot{x}_{1c}(t) & \dot{x}_{2c}(t) & \dot{x}_{3c}(t) \end{bmatrix}^{\mathrm{T}}.$$

Then the constant-velocity state model becomes

$$\dot{\vec{\boldsymbol{x}}}_c(t) = \begin{bmatrix} 0_3 & I_3 \\ 0_3 & 0_3 \end{bmatrix} \vec{\boldsymbol{x}}_c(t) + \begin{bmatrix} 0_3 \\ I_3 \end{bmatrix} \boldsymbol{w}_c(t),$$

where $\boldsymbol{w}_c(t) = [\boldsymbol{w}_{1c}(t)\ \boldsymbol{w}_{2c}(t)\ \boldsymbol{w}_{3c}(t)]^{\mathrm{T}}$ is zero-mean white noise with covariance

$$Q_c = \begin{bmatrix} \sigma^2_{w_{1c}} & 0 & 0 \\ 0 & \sigma^2_{w_{2c}} & 0 \\ 0 & 0 & \sigma^2_{w_{3c}} \end{bmatrix}. \tag{8.101}$$

The measurements are produced via

$$\boldsymbol{z}_c(t) = \begin{bmatrix} I_3 & 0_3 \end{bmatrix} \vec{\boldsymbol{x}}_c(t) + \boldsymbol{v}'_c(t),$$

where $\boldsymbol{v}'_c(t) = [\boldsymbol{v}'_{1c}(t)\ \boldsymbol{v}'_{2c}(t)\ \boldsymbol{v}'_{3c}(t)]^{\mathrm{T}}$ is zero-mean white noise with covariance $\sigma^2_{v'_c} I_3$.

We use the results of Appendix A to obtain the discretized version of this system. For simplicity we assume a zero-order hold on the process noise $\boldsymbol{w}_c(t)$. The discretized state vector is

$$\vec{x}(n) \triangleq \begin{bmatrix} x_1(n) & x_2(n) & x_3(n) & \dot{x}_1(n) & \dot{x}_2(n) & \dot{x}_3(n) \end{bmatrix}^{\mathrm{T}},$$

and the discretized signal model becomes

$$\vec{\boldsymbol{x}}(n+1) = \begin{bmatrix} I_3 & TI_3 \\ 0_3 & I_3 \end{bmatrix} \vec{\boldsymbol{x}}(n) + \begin{bmatrix} (T^2/2)I_3 \\ TI_3 \end{bmatrix} \boldsymbol{w}(n), \tag{8.102}$$

where $\boldsymbol{w}(n)$ is a 3-vector zero-mean white noise process and $Q = Q_c/T$.

Based on our assumptions about the sensors, the initial measurement equation is

$$\boldsymbol{z}'(n) = \begin{bmatrix} I_3 & 0_3 \end{bmatrix} \vec{\boldsymbol{x}}(n) + \boldsymbol{v}'(n), \tag{8.103}$$

where $\boldsymbol{v}'(n) = \boldsymbol{v}'_c(nT)$ is the measurement noise. Each $\boldsymbol{v}'_i(n)$ is zero-mean and white with covariance $\sigma^2_{v'_c}/T$. Then the $\boldsymbol{v}'_i(n)$ are independent and $\boldsymbol{v}'(n)$ is zero-mean white noise with covariance matrix $\sigma^2_{v'_c} I_3/T$. (We adopt the prime notation because later we will introduce new processes $\boldsymbol{z}(n)$ and $\boldsymbol{v}(n)$.)

Symmetric Measurement Equations

As stated at the beginning of this section, we must address the problem of associating the observations $\boldsymbol{z}'(n)$ with the targets. The simplest approach would be to assign observation $\boldsymbol{z}'_i(n)$ with the ith target. Clearly this technique suffers from flaws since the sensors do not distinguish between targets.

Another method suggests assigning the observation $\boldsymbol{z}_i(n)$ to the estimate $\hat{\boldsymbol{x}}_j^-(n-1)$ to which $\boldsymbol{z}_i(n)$ is closest. However, if two targets pass each other, then the tracker may assign observations incorrectly and exchange the targets.

One popular technique for MTT is the interacting multiple model (IMM) approach [21]. The IMM approach uses several state models for the system dynamics and switches between them by deciding which model is most appropriate. The different state models may be a collection of Kalman filters or EKFs. However, a complete discussion of IMM would take us beyond the scope of this text. We instead consider another approach that avoids having to consider model switching. This method is based on the *symmetric measurement equation* (SME) [22].

The SME method is based on the following idea: we can combine the observations $\boldsymbol{z}'_i(n)$ to create *new observations* $\boldsymbol{z}_j(n)$ that are independent of the possible permutations of the $\boldsymbol{z}'_i(n)$. In this way we avoid data association based upon indexing or proximity.

We say that a scalar-valued function (called a *functional*) with this property is *symmetric with respect to its arguments.* In this section we use *symmetric* in this sense. For example, the functional $f(w, x, y, z) = wxyz$ produces the same result for any ordering of the arguments; hence f is symmetric. Additional examples of symmetric functionals are $f(w, x, y, z) = wxy + wxz + wyz + xyz$ and $g(x, y, z) = (xy)^2 + (xz)^2 + (yz)^2$.

In general the new observations $\boldsymbol{z}(n)$ may be expressed as

$$\boldsymbol{z}(n) = \gamma(\boldsymbol{z}'(n)) = \begin{bmatrix} \gamma_1(\boldsymbol{z}'(n)) \\ \vdots \\ \gamma_N(\boldsymbol{z}'(n)) \end{bmatrix}, \tag{8.104}$$

where each $\gamma_i(\boldsymbol{z}'(n))$ is a symmetric functional. The tracking filter resulting from such a formulation is called the SME filter.

The example symmetric functionals suggest the use of sums of products for the γ_i, although other functionals are available[2]. Define

$$\gamma(z') = \begin{bmatrix} z'_1 + z'_2 + z'_3 \\ z'_1 z'_2 + z'_1 z'_3 + z'_2 z'_3 \\ z'_1 z'_2 z'_3 \end{bmatrix}, \tag{8.105}$$

so that the new observations are given by

$$\boldsymbol{z}(n) = \begin{bmatrix} \boldsymbol{z}'_1(n) + \boldsymbol{z}'_2(n) + \boldsymbol{z}'_3(n) \\ \boldsymbol{z}'_1(n)\boldsymbol{z}'_2(n) + \boldsymbol{z}'_1(n)\boldsymbol{z}'_3(n) + \boldsymbol{z}'_2(n)\boldsymbol{z}'_3(n) \\ \boldsymbol{z}'_1(n)\boldsymbol{z}'_2(n)\boldsymbol{z}'_3(n) \end{bmatrix}. \tag{8.106}$$

New Measurement Noise

We intend to apply the EKF, but the system dynamics remain linear and do not require approximation. Hence, we only need to consider the measurement noise that results after transformation by the SME. For the measurements, Equation (8.2) requires that $\boldsymbol{z}(n)$ have the form

$$\boldsymbol{z}(n) = \gamma(\vec{\boldsymbol{x}}(n)) + \boldsymbol{v}(n), \tag{8.107}$$

so each $\boldsymbol{z}_i(n)$ should be separable in this fashion.

Based on Assumption (8.103), we have $\boldsymbol{z}'_i(n) = \boldsymbol{x}_i(n) + \boldsymbol{v}_i(n)$. We substitute this relation into (8.106) and separate the terms to fit the form of (8.107). After some algebraic manipulations we find

$$\boldsymbol{z}(n) = \begin{bmatrix} \boldsymbol{x}_1 + \boldsymbol{x}_2 + \boldsymbol{x}_3 \\ \boldsymbol{x}_1\boldsymbol{x}_2 + \boldsymbol{x}_1\boldsymbol{x}_3 + \boldsymbol{x}_2\boldsymbol{x}_3 \\ \boldsymbol{x}_1\boldsymbol{x}_2\boldsymbol{x}_3 \end{bmatrix} + \begin{bmatrix} \boldsymbol{v}'_1 + \boldsymbol{v}'_2 + \boldsymbol{v}'_3 \\ (\boldsymbol{x}_1\boldsymbol{v}'_2 + \boldsymbol{v}'_1\boldsymbol{x}_2 + \boldsymbol{v}'_1\boldsymbol{v}'_2 + \boldsymbol{x}_1\boldsymbol{v}'_3 + \boldsymbol{v}'_1\boldsymbol{x}_3 \\ +\boldsymbol{v}'_1\boldsymbol{v}'_3 + \boldsymbol{x}_2\boldsymbol{v}'_3 + \boldsymbol{v}'_2\boldsymbol{x}_3 + \boldsymbol{v}'_2\boldsymbol{v}'_3) \\ (\boldsymbol{x}_1\boldsymbol{x}_2\boldsymbol{v}'_3 + \boldsymbol{x}_1\boldsymbol{v}'_2\boldsymbol{x}_3 + \boldsymbol{v}'_1\boldsymbol{x}_2\boldsymbol{x}_3 + \boldsymbol{x}_1\boldsymbol{v}'_2\boldsymbol{v}'_3 \\ +\boldsymbol{v}'_1\boldsymbol{x}_2\boldsymbol{v}'_3 + \boldsymbol{v}'_1\boldsymbol{v}'_2\boldsymbol{x}_3 + \boldsymbol{v}'_1\boldsymbol{v}'_2\boldsymbol{v}'_3) \end{bmatrix},$$

where dependence on n is omitted to conserve space.

[2] The γ_i must also satisfy an observability condition [23], and the symmetric functionals we have chosen do so.

Write $\boldsymbol{v}(n)$ as

$$\boldsymbol{v}(n) = \overbrace{\begin{bmatrix} 1 & 1 & 1 \\ \boldsymbol{x}_2+\boldsymbol{x}_3 & \boldsymbol{x}_1+\boldsymbol{x}_3 & \boldsymbol{x}_1+\boldsymbol{x}_2 \\ \boldsymbol{x}_2\boldsymbol{x}_3 & \boldsymbol{x}_1\boldsymbol{x}_3 & \boldsymbol{x}_1\boldsymbol{x}_2 \end{bmatrix}}^{V_1} \overbrace{\begin{bmatrix} \boldsymbol{v}_1' \\ \boldsymbol{v}_2' \\ \boldsymbol{v}_3' \end{bmatrix}}^{q_1}$$
$$+ \underbrace{\begin{bmatrix} 0 & 0 & 0 \\ 1 & 1 & 1 \\ \boldsymbol{x}_3 & \boldsymbol{x}_2 & \boldsymbol{x}_1 \end{bmatrix}}_{V_2} \underbrace{\begin{bmatrix} \boldsymbol{v}_1'\boldsymbol{v}_2' \\ \boldsymbol{v}_1'\boldsymbol{v}_3' \\ \boldsymbol{v}_2'\boldsymbol{v}_3' \end{bmatrix}}_{q_2} + \underbrace{\begin{bmatrix} 0 \\ 0 \\ 1 \end{bmatrix}}_{V_3} \underbrace{\begin{bmatrix} \boldsymbol{v}_1'\boldsymbol{v}_2'\boldsymbol{v}_3' \end{bmatrix}}_{q_3} \tag{8.108}$$

$$= V_1(\vec{\boldsymbol{x}}(n))q_1(\boldsymbol{v}'(n)) + V_2(\vec{\boldsymbol{x}}(n))q_2(\boldsymbol{v}'(n)) + V_3(\vec{\boldsymbol{x}}(n))q_3(\boldsymbol{v}'(n)). \tag{8.109}$$

The elements of $V_i(\boldsymbol{x}(n))$ and $q_i(\boldsymbol{v}'(n))$ are independent, and the $\boldsymbol{v}_i'(n)$ are zero-mean and independent. As a result, each $\boldsymbol{v}_i(n)$ is also zero-mean and independent, and therefore $\boldsymbol{v}(n)$ is zero-mean.

As a result of the independence between V_i and q_i,

$$\mathrm{E}\left[\boldsymbol{v}(n)\boldsymbol{v}^{\mathrm{T}}(\ell)\right] = \sum_{i=1}^{3} \mathrm{E}\left[V_i(\vec{\boldsymbol{x}}(n))q_i(\boldsymbol{v}'(n))q_i^{\mathrm{T}}(\boldsymbol{v}'(\ell))V_i^{\mathrm{T}}(\vec{\boldsymbol{x}}(\ell))\right].$$

Since the $\boldsymbol{v}_i'(n)$ are white, it follows that

$$\mathrm{E}\left[\boldsymbol{v}(n)\boldsymbol{v}^{\mathrm{T}}(\ell)\right] = 0, \quad n \neq \ell,$$

so that $\boldsymbol{v}(n)$ is white noise. When $n = \ell$ we have

$$R(n) = \sum_{i=1}^{3} \mathrm{Cov}\left[V_i(\vec{\boldsymbol{x}}(n))q_i(\boldsymbol{v}'(n))\right].$$

Since the $\boldsymbol{v}_i'(n)$ are zero-mean and independent, we find that

$$\mathrm{E}\left[q_i(\boldsymbol{v}'(n))q_j^{\mathrm{T}}(\boldsymbol{v}'(n))\right] = \begin{cases} \sigma_{q_i}^2 I_3, & i = j; \\ 0_3, & i \neq j. \end{cases}$$

We can use the definitions of $q_i(\boldsymbol{v}'(n))$ in (8.108) to find the $\sigma_{q_i}^2$ for $i = 1$, 2, 3. We find

$$\sigma_{q_1}^2 = \sigma_{v_1'}^2 + \sigma_{v_2'}^2 + \sigma_{v_3'}^2, \tag{8.110}$$
$$\sigma_{q_2}^2 = \sigma_{v_1'}^2\sigma_{v_2'}^2 + \sigma_{v_1'}^2\sigma_{v_3'}^2 + \sigma_{v_2'}^2\sigma_{v_3'}^2, \tag{8.111}$$
$$\sigma_{q_3}^2 = \sigma_{v_1'}^2\sigma_{v_2'}^2\sigma_{v_3'}^2. \tag{8.112}$$

Finally, we have

$$R(n) = \sum_{i=1}^{3} \sigma_{q_i}^2 \operatorname{Cov}\left[V_i(\vec{x}(n))\right].$$

In practice we drop the expected value and use our best estimate $\hat{\vec{x}}^-(n)$. Hence, we have

$$R(n) = \sum_{i=1}^{3} \sigma_{q_i}^2 V_i\left(\hat{\vec{x}}^-(n)\right) V_i^{\mathrm{T}}\left(\hat{\vec{x}}^-(n)\right). \tag{8.113}$$

Linearization

Next we determine $J_\gamma(\vec{x})$ to approximate $\gamma(\vec{x})$ in (8.105). Since each functional γ_i does not depend upon $\dot{x}_i(n)$, we find

$$J_\gamma(\vec{x}) = \frac{\partial \gamma}{\partial \vec{x}} = \begin{bmatrix} 1 & 1 & 1 & 0 & 0 & 0 \\ x_2 + x_3 & x_1 + x_3 & x_1 + x_2 & 0 & 0 & 0 \\ x_2 x_3 & x_1 x_3 & x_1 x_2 & 0 & 0 & 0 \end{bmatrix}.$$

Integral Feedback

At this point, we have enough information to set up and apply the EKF to the SME problem. However, the SME EKF suffers from instabilities when targets cross [24]. By adding integral feedback to the EKF, we may be able to stabilize the filter [25, 26]. From the regular EKF measurement update (8.27), the usual EKF employs

$$\begin{aligned} \hat{\vec{x}}(n) &= \hat{\vec{x}}^-(n) + K(n)\left[z(n) - \gamma\left(\hat{\vec{x}}^-(n)\right)\right] \\ &= \hat{\vec{x}}^-(n) + K(n)\left(z(n) - \hat{z}^-(n)\right), \end{aligned} \tag{8.114}$$

where we note from (8.107) that $\hat{z}^-(n) = \gamma\left(\hat{\vec{x}}^-(n)\right)$.

We simply add a feedback term from the past measurements,

$$\hat{\vec{x}}(n) = \hat{\vec{x}}^-(n) + K(n)\left(z(n) - \hat{z}^-(n)\right) + B_I \sum_{m=1}^{n-1} K(m) K_I \left(z(m) - \hat{z}^-(m)\right), \tag{8.115}$$

where K_I is a 3×3 weighting matrix that determines the time constant of the integral term, and B_I is the integral input matrix. We use $K_I = k_I I_3$, where k_I is a scalar, and $B_I^{\mathrm{T}} = \begin{bmatrix} I_3 & 0_3 \end{bmatrix}^{\mathrm{T}}$. Note that the integral term is actually a summation because we operate in discrete-time. The terminology carries over from the classical *proportional-integral* (PI) controller in continuous-time control systems. Hence, we have the *proportional-integral extended Kalman filter* (PI-EKF).

We can rewrite (8.115) as a pair of coupled equations, one that performs the measurement update and one that computes the integral term. Define

$$\theta(n) = \sum_{m=1}^{n-1} K(m) K_I \left(z(m) - \hat{z}^-(m) \right), \tag{8.116}$$

so that (8.115) becomes

$$\hat{\vec{x}}(n) = \hat{\vec{x}}^-(n) + K(n) \left(z(n) - \hat{z}^-(n) \right) + B_I \theta(n) \tag{8.117}$$

$$\theta(n+1) = \theta(n) + K(n) K_I \left(z(n) - \hat{z}^-(n) \right). \tag{8.118}$$

Tracking System Recursion

We are now prepared to implement the tracking recursion for the PI-EKF.

- *Measurement update.* Acquire measurement $z'(n)$ and compute:

$$R(n) = \sum_{i=1}^{3} \sigma_{q_i}^2 V_i \left(\hat{\vec{x}}^-(n) \right) V_i^{\mathrm{T}} \left(\hat{\vec{x}}^-(n) \right) \tag{8.119}$$

$$J_\gamma(n) = \left. \begin{bmatrix} 1 & 1 & 1 & 0 & 0 & 0 \\ x_2 + x_3 & x_1 + x_3 & x_1 + x_2 & 0 & 0 & 0 \\ x_2 x_3 & x_1 x_3 & x_1 x_2 & 0 & 0 & 0 \end{bmatrix} \right|_{\vec{x} = \hat{\vec{x}}^-(n)} \tag{8.120}$$

$$K(n) = P^-(n) J_\gamma^{\mathrm{T}}(n) \left[J_\gamma(n) P^-(n) J_\gamma^{\mathrm{T}}(n) + R(n) \right]^{-1} \tag{8.121}$$

$$\hat{z}^-(n) = \gamma \left(\hat{\vec{x}}^-(n) \right) = \left. \begin{bmatrix} x_1 + x_2 + x_3 \\ x_1 x_2 + x_1 x_3 + x_2 x_3 \\ x_1 x_2 x_3 \end{bmatrix} \right|_{\vec{x} = \hat{\vec{x}}^-(n)} \tag{8.122}$$

$$z(n) = \begin{bmatrix} z_1'(n) + z_2'(n) + z_3'(n) \\ z_1'(n) z_2'(n) + z_1'(n) z_3'(n) + z_2'(n) z_3'(n) \\ z_1'(n) z_2'(n) z_3'(n) \end{bmatrix} \tag{8.123}$$

$$\hat{\vec{x}}(n) = \hat{\vec{x}}^-(n) + K(n) \left(z(n) - \hat{z}^-(n) \right) + B_I \theta(n) \tag{8.124}$$

$$\theta(n) = \theta(n-1) + K(n-1) K_I \left(z(n) - \hat{z}^-(n) \right) \tag{8.125}$$

$$P(n) = P^-(n) + K(n) J_\gamma(n) P^-(n) \tag{8.126}$$

Observe that (8.123) generates the symmetric measurements to avoid the problem of data association. The updates (8.124) and (8.125) incorporate the integration term.

- *Time update.* Compute:

$$P^-(n+1) = \Phi P(n)\Phi^{\mathrm{T}} + \Gamma Q \Gamma^{\mathrm{T}} \tag{8.127}$$

$$\hat{\vec{x}}^-(n+1) = \Phi \hat{\vec{x}}(n) \tag{8.128}$$

Note that the time-update dynamics are linear. Also, recall that $Q = Q_c/T$, where Q_c is given by (8.101).

- *Time increment.* Increment n and repeat.

Example 8.3 Multiple Target Tracking

We now apply the PI-EKF to the problem of tracking three targets using the SME of (8.105). The matrices Φ and Γ are as given in (8.102). The sampling period T was 0.1 second. For the continuous-time process noise, we used a diagonal matrix Q_c with $\mathrm{diag}(Q_c) = \begin{bmatrix} 0.1 & 0.2 & 0.13 \end{bmatrix}$. For the continuous-time measurement noise, we set $\sigma^2_{v'_c} = 4$.

The integration weight k_I was experimentally chosen to be 0.025. The true initial positions of the targets were 20, 0, and −25 units (e.g., meters), respectively. The respective initial velocities for the targets were −33, 10, and 25 units/second.

50 Monte Carlo simulations were conducted using both the EKF and PI-EKF. Note that the EKF recursion is the same as the PI-EKF, except that the EKF sets $k_I = 0$ to disable the integral term. Example target paths and noisy observations for a single trial appear in Figure 8.6. The corresponding EKF and PI-EKF tracks are shown in Figure 8.7. Not until the last crossing at 12 seconds was the EKF able to track the targets reliably. However, the PI-EKF acquired the targets before the first crossing at 4 seconds, although it displayed some oscillation before the initial transient vanished.

Figure 8.8 displays the target positions and tracks, averaged over all simulations. Both the EKF and PI-EKF demonstrated an initial transient due to their incorrect estimates of the initial state vector $\vec{x}(0)$. The transient was much shorter for the PI-EKF, which acquired the tracks after about 3 seconds; the EKF did not acquire tracks until about 11 seconds had elapsed. At target crossings (e.g., at 5, 8, and 14 seconds), the EKF failed to maintain accurate tracks, while the PI-EKF maintained its tracks at each crossing.

Finally, Figure 8.9 shows the mean-square position and velocity errors— summed over all three targets— for the EKF and PI-EKF. The PI-EKF position error dropped steadily and fell below 1 after 4 seconds. In contrast, the EKF error grew and only fell below 1 after 9 seconds. The last track crossing, at 14 seconds, and the proximity of two targets caused the EKF position error to increase again. Both filters demonstrated comparable behavior for estimating the velocities of the targets.

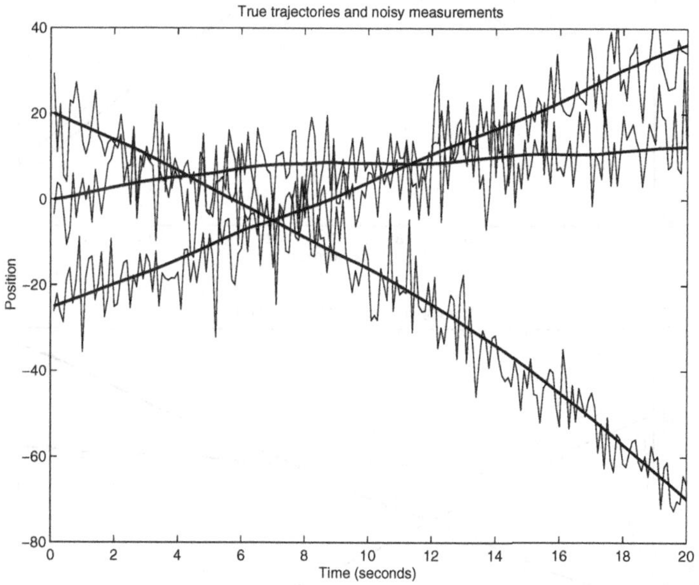

Figure 8.6. A sample realization for multiple target tracking in Example 8.3. Thick lines show actual target trajectories. Thin lines show noisy observations.

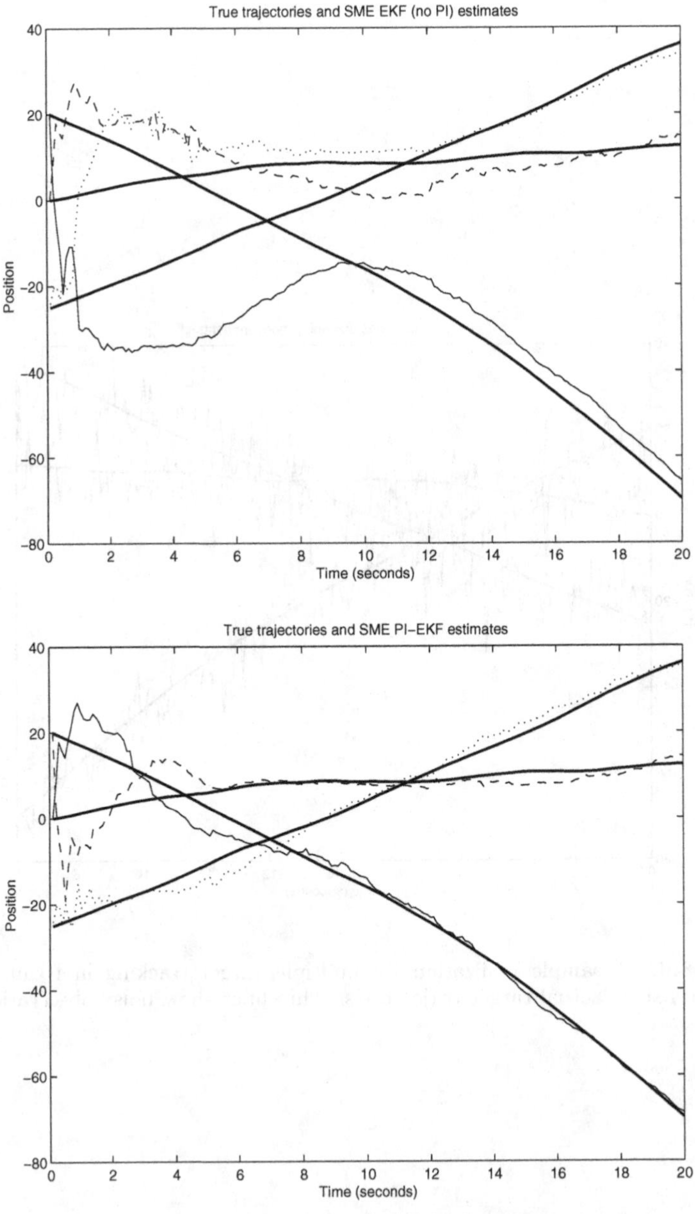

Figure 8.7. Example estimated trajectories for multiple target tracking in Example 8.3. **Top:** Estimated tracks for EKF (no integration). **Bottom:** Estimated tracks for PI-EKF.

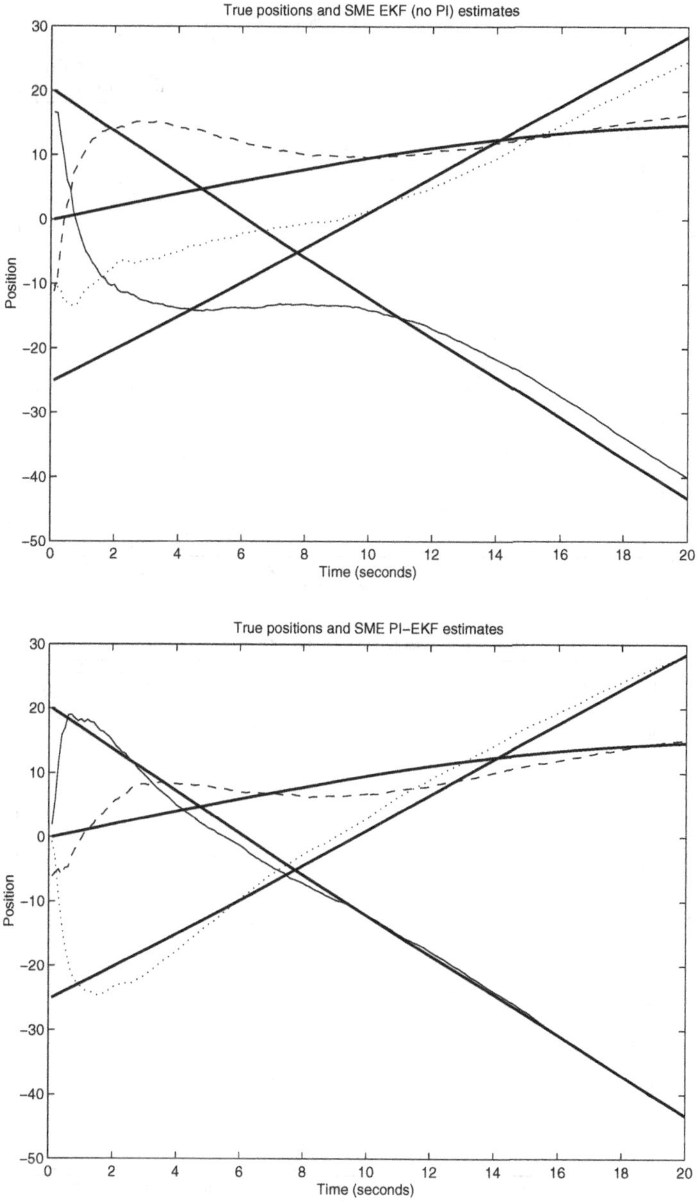

Figure 8.8. Average tracks for 50 Monte Carlo simulations of MTT. Thick, solid curves indicate true target positions. Thin curves show average tracks. **Top:** Average tracks for EKF (no integration). **Bottom:** Average tracks for PI-EKF.

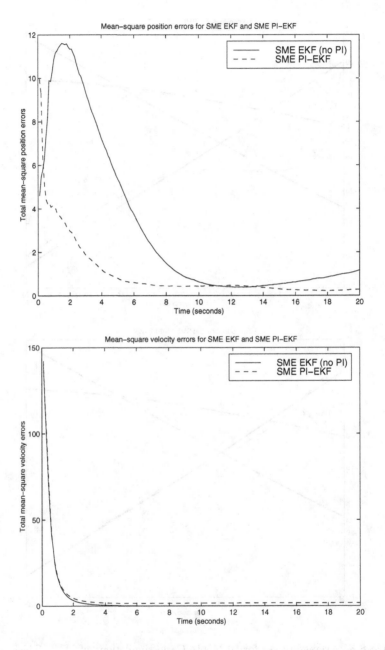

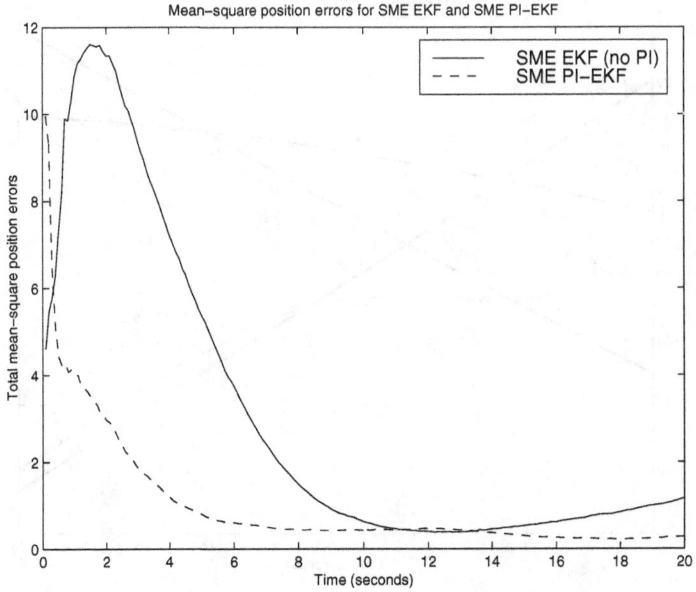

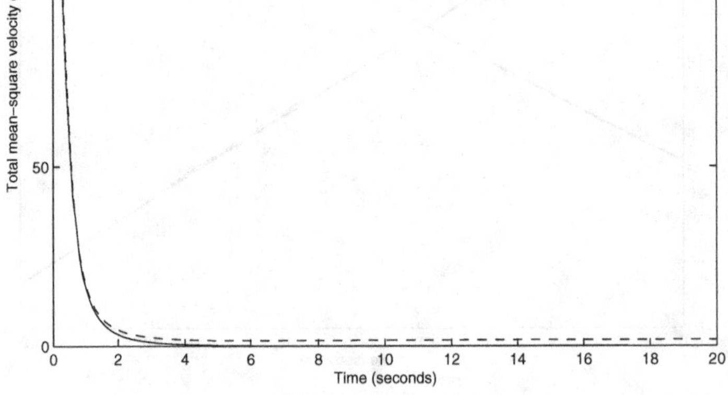

Figure 8.9. Mean-square errors for 50 Monte Carlo simulations of MTT. **Top:** Position errors. **Bottom:** Velocity errors.

Problems

8.1. In the derivation of the EKF (Section 8.1), the *a posteriori* update (8.27) for $\hat{\boldsymbol{x}}(n)$ given $\hat{\boldsymbol{x}}^-(n)$ and $\boldsymbol{z}(n)$ was derived by applying the orthogonality principle (3.52). The update can also be derived by using the innovations (Section 6.1).

(a) Show that $\hat{\boldsymbol{z}}^-(n) = \gamma(\hat{\boldsymbol{x}}^-(n))$. Then the innovation $\varepsilon(n)$ is given by $\varepsilon(n) = \boldsymbol{z}(n) - \gamma(\hat{\boldsymbol{x}}^-(n))$.

(b) Show that (8.27) can be derived directly from Equation (6.7).

8.2. Measurements $\boldsymbol{z}(n)$ of an unknown constant s are given by $\boldsymbol{z}(n) = h(s)+\boldsymbol{v}(n)$ where h is an invertible function and $\boldsymbol{v}(n)$ is zero-mean white noise with variance σ_v^2. Let $\hat{s}_{\mathrm{ML}}(n)$ denote the ML estimate of s at time n based on the measurements $\boldsymbol{z}(i) = z(i)$ for $i = 1, 2, \ldots, n$.

(a) Is it possible to express $\hat{s}_{\mathrm{ML}}(n)$ in the form

$$\hat{s}_{\mathrm{ML}}(n) = h^{-1}\left[\frac{1}{n}\sum_{i=1}^{n} z(i)\right]?$$

Justify your answer.

(b) Is it possible to express the estimate given in Part (a) in recursive form? If so, derive the recursive form. If not, show that there is no recursive form.

8.3. In Problem 8.2, suppose that $h(s) = e^s$. Give the equations for the EKF that provides an estimate of s based on the measurements $\boldsymbol{z}(i) = z(i)$ for $i = 1, 2, \ldots, n$.

8.4. A random signal $\boldsymbol{s}(n)$ has the state model

$$\begin{aligned}\boldsymbol{x}_1(n+1) &= \mathrm{act}[\boldsymbol{x}_1(n) + \boldsymbol{x}_2(n)]\\ \boldsymbol{x}_2(n+1) &= \boldsymbol{x}_1(n) + \gamma(n)\\ \boldsymbol{s}(n) &= \boldsymbol{x}_1(n)\boldsymbol{x}_2(n),\end{aligned}$$

where act is the arctan activation function, $\gamma(n)$ is zero-mean white noise with variance 5, and $\gamma(n)$ is independent of $\boldsymbol{x}_1(0)$ and $\boldsymbol{x}_2(0)$. Measurements $\boldsymbol{z}(n)$ are given by $\boldsymbol{z}(n) = \boldsymbol{s}(n) + \boldsymbol{v}(n)$, where $\boldsymbol{v}(n)$ is zero-mean white noise with variance 2 and $\boldsymbol{v}(n)$ is independent of $\boldsymbol{x}_1(0)$, $\boldsymbol{x}_2(0)$, and $\gamma(n)$. Give the equations for the EKF that provides an estimate of $\boldsymbol{x}(n) = [\boldsymbol{x}_1(n) \quad \boldsymbol{x}_2(n)]^{\mathrm{T}}$ based on the measurements $\boldsymbol{z}(i) = z(i)$ for $i = 1, 2, \ldots, n$.

8.5. Consider the continuous-time state model

$$\dot{x}_1(t) = 0, \qquad \dot{x}_2(t) = x_1(t)x_2(t)$$

with discrete measurements $\boldsymbol{z}(n) = \boldsymbol{x}_1(n)+\boldsymbol{x}_2(n)+\boldsymbol{v}(n)$, where the sampling interval $T = 1$, $\boldsymbol{v}(n)$ is zero-mean white noise with variance σ_v^2, and $n = 1, 2, \ldots$. Give the equations for the EKF that provides an estimate of $\boldsymbol{x}(n) = [\boldsymbol{x}_1(n) \quad \boldsymbol{x}_2(n)]^{\mathrm{T}}$ based on the measurements $\boldsymbol{z}(i) = z(i)$ for $i = 1, 2, \ldots, n$.

8.6. Measurements of two constants c_1 and c_2 are given by

$$z_1(n) = c_i + v_1(n), \quad i = 1 \text{ or } 2,$$
$$z_2(n) = c_j + v_2(n), \quad j = 1 \text{ or } 2, j \neq i,$$

where $v_1(n)$ and $v_2(n)$ are independent zero-mean white noise terms with variance σ_v^2. The measurement equations reveal that there is a data association problem, that is, we do not know if $z_1(n)$ is a measurement of c_1 or c_2. By considering the sum and the product of $z_1(n)$ and $z_2(n)$, give the equations for the SME EKF that provides an estimate of $c = \begin{bmatrix} c_1 & c_2 \end{bmatrix}^{\mathrm{T}}$ based on the measurements $z_1(i) = z_1(i)$, $z_2(i) = z_2(i)$, for $i = 1, 2, \ldots, n$.

8.7. Again consider the estimation of the two constants c_1 and c_2 in Problem 8.6 with the data association problem. Instead of using a SME EKF, generate a nonlinear recursive filter for estimating c_1 and c_2 using the inverse of the function

$$f(c_1, c_2) = \begin{bmatrix} c_1 + c_2 \\ c_1 c_2 \end{bmatrix}.$$

Give the equation for the filter with all coefficients evaluated and determine the inverse of the function f.

8.8. A pendulum hanging from a ceiling is given by the differential equation

$$\frac{d^2\theta(t)}{dt^2} + t \sin\theta(t) = 0,$$

where $\theta(t)$ is the angular position of the pendulum at time t. Discrete measurements of $\theta(t)$ are given by $z(n) = \theta(n) + v(n)$, where the sampling interval $T = 1$, $v(n)$ is zero-mean white noise with variance σ_v^2, and $n = 1, 2, \ldots$. Give the equations for the EKF that provides an estimate of $\theta(n)$ based on the measurements $z(i) = z(i)$ for $i = 1, 2, \ldots, n$.

8.9. Three targets have x-coordinate positions given by

$$p_1(t) = t, \quad p_2(t) = 100 - t, \quad p_3(t) = 50.$$

(a) A radar provides noisy measurements of the positions with the variance of the noise equal to 2. Generate a sample realization of the three target trajectories for $t = n$, $n = 1, 2, \ldots, 100$.

(b) Apply the sample realization in Part (a) to a EKF SME filter and plot the results using MATLAB. Is tracking maintained when all three targets cross? Increase the measurement noise variance until loss of tracking occurs.

8.10. A nonlinear discrete-time system is given by

$$y(n) = 0.9y(n-1)u(n) + u(n),$$

where $u(n)$ is the input and $y(n)$ is the output.

(a) Compute $y(n)$ for $0 \le n \le 100$ when $u(n) = 1$ for $n \ge 0$, and $y(-1) = 0$.

(b) Using the data in Part (a) and the EKF, identify a feedforward neural network model with one hidden node.

(c) Plot the output of the model found in Part (b) when $u(n) = 1$ for $n \geq 0$, and $y(-2) = y(-1) = 0$, and on the same plot, show the output of the actual system. Compare the results.

8.11. Repeat Problem 8.10, but now take $u(n)$ to be a sample realization of white Gaussian noise with variance 1.

8.12. Section 8.6 presented the SME filter for tracking three targets in one dimension. This problem considers MTT in two dimensions.

Describe the (x, y) position of the ith target by

$$\boldsymbol{p}_{ic}(t) = \begin{bmatrix} \boldsymbol{x}_{ic}(t) & \boldsymbol{y}_{ic}(t) \end{bmatrix}.$$

Then define the continuous-time state vector by

$$\vec{\boldsymbol{x}}_c^{\mathrm{T}}(t) \triangleq \begin{bmatrix} \boldsymbol{p}_{1c}(t) & \boldsymbol{p}_{2c}(t) & \boldsymbol{p}_{3c}(t) & \dot{\boldsymbol{p}}_{1c}(t) & \dot{\boldsymbol{p}}_{2c}(t) & \dot{\boldsymbol{p}}_{3c}(t) \end{bmatrix}.$$

Assume the targets move with approximately constant velocity, so the discretized model becomes

$$\vec{\boldsymbol{x}}(n+1) = \begin{bmatrix} I_6 & TI_6 \\ 0_6 & I_6 \end{bmatrix} \vec{\boldsymbol{x}}(n) + \begin{bmatrix} (T^2/2)_6 \\ TI_6 \end{bmatrix} \boldsymbol{w}(n)$$

$$\boldsymbol{z}'(n) = \begin{bmatrix} I_6 & 0_6 \end{bmatrix} \vec{\boldsymbol{x}}(n) + \boldsymbol{v}'(n),$$

where $Q = \sigma_w^2 I_6$, $\boldsymbol{v}'(n)$ has the form

$$\boldsymbol{v}(n) = \begin{bmatrix} v'_{x_1}(n) & v'_{y_1}(n) & v'_{x_2}(n) & v'_{y_2}(n) & v'_{x_3}(n) & v'_{y_3}(n) \end{bmatrix}^{\mathrm{T}},$$

and $R = \sigma_{v'}^2 I_6$.

(a) Express the position of the ith target as a complex number,

$$\boldsymbol{p}_i(n) = \boldsymbol{x}_i(n) + j\boldsymbol{y}_i(n),$$

with $j = \sqrt{-1}$. Then the symmetric functionals are chosen as the real and imaginary parts of the sums of products in (8.105),

$$\gamma(p(n)) = \begin{bmatrix} \mathrm{Re}\,(p_1(n) + p_2(n) + p_3(n)) \\ \mathrm{Im}\,(p_1(n) + p_2(n) + p_3(n)) \\ \mathrm{Re}\,(p_1(n)p_2(n) + p_1(n)p_3(n) + p_2(n)p_3(n)) \\ \mathrm{Im}\,(p_1(n)p_2(n) + p_1(n)p_3(n) + p_2(n)p_3(n)) \\ \mathrm{Re}\,(p_1(n)p_2(n)p_3(n)) \\ \mathrm{Im}\,(p_1(n)p_2(n)p_3(n)) \end{bmatrix}.$$

Show that

$$\gamma(p) = \begin{bmatrix} x_1 + x_2 + x_3 \\ y_1 + y_2 + y_3 \\ x_1x_2 + x_1x_3 + x_2x_3 - (y_1y_2 + y_1y_3 + y_2y_3) \\ x_1y_2 + x_2y_1 + x_1y_3 + x_3y_1 + x_2y_3 + x_3y_2 \\ x_1x_2x_3 - (y_1y_2x_3 + y_1y_3x_2 + y_2y_3x_1) \\ x_1x_2y_3 + x_1x_3y_2 + x_2x_3y_1 - y_1y_2y_3 \end{bmatrix}.$$

(b) Show that the new observations $z(n)$ may be written in the form (8.107),

$$z(n) = \gamma(\vec{x}(n)) + v(n).$$

(c) Let $p(n) = [x_1(n)\ y_1(n)\ x_2(n)\ y_2(n)\ x_3(n)\ y_3(n)]^{\mathrm{T}}$. That is, $p(n)$ is the positional part of $\vec{x}(n)$. Show that $v(n)$ may be expressed as

$$v(n) = \sum_{i=1}^{3} V_i(p(n))q_i(v'(n)),$$

where the V_i are given by

$$V_1(p) = J_\gamma(p) = \frac{\partial \gamma}{\partial p}$$

$$= \left[\begin{array}{cccc} 1 & 0 & 1 & \\ 0 & 1 & 0 & \\ x_2+x_3 & -(y_2+y_3) & x_1+x_3 & \cdots \\ y_2+y_3 & x_2+x_3 & y_1+y_3 & \\ x_2x_3-y_2y_3 & -(y_2x_3+y_3x_2) & x_1x_3-y_1y_3 & \\ y_2x_3+y_3x_2 & x_2x_3-y_2y_3 & y_1x_3+y_3x_1 & \end{array}\right.$$

$$\left.\begin{array}{cccc} & 0 & 1 & 0 \\ & 1 & 0 & 1 \\ \cdots & -(y_1+y_3) & x_1+x_2 & -(y_1+y_2) \\ & x_1+x_3 & y_1+y_2 & x_1+x_2 \\ & -(y_1x_3+y_3x_1) & x_1x_2-y_1y_2 & -(y_1x_2+y_2x_1) \\ & x_1x_3-y_1y_3 & y_1x_2+y_2x_1 & x_1x_2-y_1y_2 \end{array}\right],$$

$$V_2(p) = \left[\begin{array}{ccccccc} 0 & & & \cdots & & 0 & \\ 0 & & & \cdots & & 0 & \\ 1 & 1 & 1 & -1 & -1 & -1 & \cdots \\ 0 & & & \cdots & & 0 & \\ x_3 & x_2 & x_1 & -x_3 & -x_2 & -x_1 & \\ y_3 & y_2 & y_1 & y_3 & y_2 & y_1 & \end{array}\right.$$

$$\left.\begin{array}{ccccccc} & 0 & & & \cdots & & 0 \\ & 0 & & & \cdots & & 0 \\ & 0 & & & \cdots & & 0 \\ \cdots & 1 & & & \cdots & & 1 \\ & -y_3 & -y_3 & -y_2 & -y_2 & -y_1 & -y_1 \\ & x_3 & x_3 & x_2 & x_2 & x_1 & x_1 \end{array}\right],$$

and

$$V_3(p) = \begin{bmatrix} 0 & & & \cdots & & & & 0 \\ 0 & & & \cdots & & & & 0 \\ 0 & & & \cdots & & & & 0 \\ 0 & & & \cdots & & & & 0 \\ 1 & 0 & 0 & 0 & 0 & 1 & 1 & 1 \\ 0 & -1 & 1 & 1 & 1 & 0 & 0 & 0 \end{bmatrix},$$

and the q_i are given by

$$q_1(v') = \begin{bmatrix} v'_{x_1} \\ v'_{y_1} \\ v'_{x_2} \\ v'_{y_2} \\ v'_{x_3} \\ v'_{y_3} \end{bmatrix}, \quad q_2(v') = \begin{bmatrix} v'_{x_1} v'_{x_2} \\ v'_{x_1} v'_{x_3} \\ v'_{x_2} v'_{x_3} \\ v'_{y_1} v'_{y_2} \\ v'_{y_1} v'_{y_3} \\ v'_{y_2} v'_{y_3} \\ v'_{x_1} v'_{y_2} \\ v'_{x_2} v'_{y_1} \\ v'_{x_1} v'_{y_3} \\ v'_{x_3} v'_{y_1} \\ v'_{x_2} v'_{y_3} \\ v'_{x_3} v'_{y_2} \end{bmatrix},$$

and

$$q_3(v') = \begin{bmatrix} v'_{x_1} v'_{x_2} v'_{x_3} \\ v'_{y_1} v'_{y_2} v'_{y_3} \\ v'_{x_1} v'_{x_2} v'_{y_3} \\ v'_{x_1} v'_{x_3} v'_{y_3} \\ v'_{x_2} v'_{x_3} v'_{y_1} \\ v'_{y_1} v'_{y_2} v'_{x_3} \\ v'_{y_1} v'_{y_3} v'_{x_2} \\ v'_{y_2} v'_{y_3} v'_{x_1} \end{bmatrix}.$$

(d) By the same arguments as in Section 8.6, $\boldsymbol{v}(n)$ is zero-mean white noise. Show that the covariance $R(n)$ associated with $\boldsymbol{v}(n)$ is equal to

$$R(n) = \sum_{i=1}^{3} \sigma^2_{v'_i} \mathrm{Cov}\left[V_i(\boldsymbol{p}(n))\right].$$

Appendix A

The State Representation

Various aspects of the state representation that are needed in the book are given in this appendix. An in-depth treatment of the state formulation can be found in [8], [9], or [27].

A linear time-invariant m-input p-output N-dimensional continuous-time system can be represented by the state model

$$\dot{x}(t) = Ax(t) + Bw(t) \tag{A.1}$$

$$y(t) = Cx(t), \tag{A.2}$$

where (A.1) is the state equation and (A.2) is the output equation. In (A.1) and (A.2), A is the $N \times N$ **system matrix**, B is the $N \times m$ **input matrix**, C is the $p \times N$ **output matrix**, $w(t)$ is the m-dimensional input vector, $x(t)$ is the N-dimensional state vector, and $y(t)$ is the p-dimensional output vector. If the system has a single output so that $p = 1$, in this book we shall sometimes use lower case "c" for the output matrix, in which case (A.2) becomes $y(t) = cx(t)$.

The complete solution to the state equation (A.1) starting with initial state $x(t_0)$ at initial time t_0 is

$$x(t) = e^{A(t-t_0)}x(t_0) + \int_{t_0}^{t} e^{A(t-\tau)}Bw(\tau)\,d\tau, \quad t > t_0. \tag{A.3}$$

The system is said to be *asymptotically stable* if the system state $x(t)$ converges to 0 as $t \to \infty$ for any initial state $x(t_0)$ at any initial time t_0 and with $w(t) = 0$ for all $t > t_0$. It follows from (A.3) that the system is asymptotically stable if and only if all the eigenvalues of A have real parts that are strictly less than zero (i.e., the eigenvalues of A lie in the open left half plane).

Using (A.2) and (A.3), we have that the complete output response of the system is given by

$$y(t) = Ce^{A(t-t_0)}x(t_0) + \int_{t_0}^{t} Ce^{A(t-\tau)}Bw(\tau)\,d\tau, \quad t > t_0. \tag{A.4}$$

If the initial state $x(t_0)$ is zero and the initial time t_0 is taken to be $-\infty$, (A.4) reduces to the input/output representation

$$y(t) = \int_{-\infty}^{t} H(t-\tau)w(\tau)\,d\tau, \tag{A.5}$$

where $H(t)$ is the $p \times m$ impulse response function matrix given by

$$H(t) = \begin{cases} 0, & t < 0; \\ Ce^{At}B, & t \geq 0. \end{cases} \tag{A.6}$$

Taking the Laplace transform of both sides of (A.5) results in the transfer function representation

$$Y(s) = H(s)W(s),$$

where $Y(s)$ is the transform of $y(t)$, $W(s)$ is the transform of $w(t)$, and $H(s)$ is the transfer function matrix equal to the transform of $H(t)$. The Laplace transform of the matrix exponential $e^{At}, t \geq 0$ is equal to $(sI - A)^{-1}$ where I is the $N \times N$ identity matrix. Thus, from (A.6) it is seen that the transfer function matrix $H(s)$ is given by $H(s) = C(sI - A)^{-1}B$.

A.1 Discrete-Time Case

In this book we are primarily interested in the state model for discrete-time systems. We can generate a discrete-time state model by discretizing the continuous-time state model given by (A.1) and (A.2). To accomplish this, let T be the sampling interval, and set $t_0 = nT$ and $t = nT + T$ in (A.3), where n is the discrete-time index. This yields

$$x(nT+T) = e^{AT}x(nT) + \int_{nT}^{nT+T} e^{A(nT+T-\tau)}Bw(\tau)\,d\tau. \tag{A.7}$$

Note that e^{AT} can be determined by inverse transforming $(sI - A)^{-1}$ and setting $t = T$ in the result.

If $w(\tau)$ is approximately constant over each interval $nT \leq \tau < nT + T$, then $w(\tau) \approx w(nT)$ for $nT \leq \tau < nT + T$, and (A.7) can be rewritten as

$$x(nT+T) = x^{AT}x(nT) + \left[\int_{nT}^{nT+T} e^{A(nT+T-\tau)}B\,d\tau\right] w(nT). \tag{A.8}$$

Defining

$$\Gamma = \int_{nT}^{nT+T} e^{A(nT+T-\tau)} B \, d\tau,$$

via a change of variables it follows that Γ can be expressed in the form

$$\gamma = \int_0^T e^{A\tau} B \, d\tau. \tag{A.9}$$

Finally, defining $\Phi = e^{AT}$, from (A.8) we obtain the following discrete-time state equation:

$$x(nT+T) = \Phi x(nT) + \Gamma w(nT), \tag{A.10}$$

and setting $t = nT$ in both sides of (A.2), we obtain the following discrete-time output equation:

$$y(nT) = Cx(nT). \tag{A.11}$$

We shall drop the notation "T" for the sampling interval in (A.10) and (A.11), in which case the general form of the linear time-invariant finite-dimensional state model is

$$x(n+1) = \Phi x(n) + \Gamma w(n) \tag{A.12}$$

$$y(n) = Cx(n). \tag{A.13}$$

In some cases of interest, the output $y(n)$ at time n may depend on the input $w(n)$ at time n, in which case (A.13) becomes

$$y(n) = Cx(n) + Dw(n) \tag{A.14}$$

where D is a $p \times p$ matrix, called the *direct-feed matrix.*

The complete solution to the state equation (A.12) starting with initial state $x(n_0)$ at initial time n_0 is

$$x(n) = \Phi^{n-n_0} x(n_0) + \sum_{i=n_0}^{n-1} \Phi^{n-i-1} \Gamma w(n), \quad n > n_0. \tag{A.15}$$

The system is said to be asymptotically stable if $x(n)$ converges to 0 as $n \to \infty$ for any initial state $x(n_0)$ at any initial time n_0 and with $w(n) = 0$ for all $n \geq n_0$. It follows from (A.15) that the system is asymptotically stable if and only if all the eigenvalues of Φ have magnitudes that are strictly less than one (i.e., all eigenvalues of Φ lie within the unit circle of the complex plane). Using (A.14) and (A.15), we have that the complete output response is

$$y(n) = C\Phi^{n-n_0} x(n_0) + \sum_{i=n_0}^{n-1} C\Phi^{n-i-1} \Gamma w(n) + Dw(n), \quad n > n_0. \tag{A.16}$$

If the initial state $x(n_0)$ is zero and the initial time n_0 is taken to be $-\infty$, (A.16) reduces to the input/output representation

$$y(n) = \sum_{i=-\infty}^{n} H(n-i)w(i), \tag{A.17}$$

where $H(n)$ is the $p \times m$ unit-pulse response function matrix given by

$$H(n) = \begin{cases} 0, & n < 0; \\ D, & n = 0; \\ CA^n B, & n \geq 1. \end{cases} \tag{A.18}$$

Taking the z-transform of both sides of (A.17) results in the transfer function representation

$$Y(z) = H(z)W(z),$$

where Y(z) is the z-transform of $y(n)$, $W(z)$ is the z-transform of $w(n)$, and $H(z)$ is the transfer function matrix equal to the z-transform of $H(n)$. The z-transform of A^n, $n \geq 1$, is equal to $(zI - A)^{-1}$. It follows from (A.18) that the transfer function matrix $H(z)$ is given by

$$H(z) = C(zI - A)^{-1}B + D. \tag{A.19}$$

A.2 Construction of State Models

In the single-input single-output (SISO) case (i.e., when $m = p = 1$), the transfer function $H(z)$ given by (A.19) is a rational function of z having the form

$$H(z) = \frac{B(z)}{A(z)}, \tag{A.20}$$

where $B(z)$ and $A(z)$ are polynomials in z given by

$$B(z) = \sum_{i=0}^{M} b_i z^i, \tag{A.21}$$

$$A(z) = z^N + \sum_{i=0}^{N-1} a_i z^i, \tag{A.22}$$

with the degree M of $B(z)$ less than or equal to the degree N of $A(z)$. Conversely, if a SISO linear time-invariant N-dimensional discrete-time system with input $w(n)$ and output $y(n)$ is given by the transfer function $H(z)$ defined by (A.20)-(A.22) with $M < N$, the system has the state model

$$x(n+1) = \Phi x(n) + \Gamma w(n) \tag{A.23}$$
$$y(n) = Cx(n), \tag{A.24}$$

where

$$\Phi = \begin{bmatrix} 0 & 1 & 0 & \cdots & 0 \\ 0 & 0 & 1 & \cdots & 0 \\ \vdots & \vdots & \vdots & \vdots & \vdots \\ -a_0 & -a_1 & -a_2 & \cdots & -a_{N-1} \end{bmatrix}, \tag{A.25}$$

$$\Gamma = \begin{bmatrix} 0 \\ 0 \\ \vdots \\ 1 \end{bmatrix}, \quad C = \begin{bmatrix} b0 & b1 & \cdots & bM & 0 & \cdots & 0 \end{bmatrix}. \tag{A.26}$$

The verification that this is a state model of the system can be carried out by showing that $C(zI - A)^{-1}B$ is equal to $H(z)$ given by (A.20)-(A.22). The details are left to the reader.

If the degree M of $B(z)$ in (A.20) is equal to N, $H(z)$ can be expressed in the form

$$H(z) = D + \frac{B(z)}{A(z)}, \tag{A.27}$$

where D is a constant and $B(z)$ and $A(z)$ can be expressed in the form (A.21) and (A.22) with M less than N. The system with transfer function $H(z)$ given by (A.27) has the state model

$$x(n+1) = \Phi x(n) + \Gamma w(n) \tag{A.28}$$
$$y(n) = Cx(n) + Dw(n), \tag{A.29}$$

where Φ, Γ, and C are given by (A.25) and (A.26). Note that if $D = 0$, this state model reduces to the one given above in the case when $H(z) = B(z)/A(z)$.

The state model with Φ, Γ, and C given by (A.25) and (A.26) is only one among an infinite number of possible state representations of the system with transfer function $H(z) = B(z)/A(z)$. Due to the special form of Φ and Γ, the state model given by (A.23)–(A.24) or (A.28)–(A.29) is an example of a canonical form, called the *control canonical form.* The reason for this terminology follows from results in state feedback control which are not considered in this book. Another special state model for the system with transfer function $H(z) = B(z)/A(z)$ is the *observable canonical form* given by (when $\deg B(z) < \deg A(z)$)

$$x(n+1) = \Phi^{\mathrm{T}} x(n) + C^{\mathrm{T}} w(n) \tag{A.30}$$

$$y(n) = \Gamma^{\mathrm{T}} x(n), \tag{A.31}$$

where Φ, Γ, and C are again given by (A.25) and (A.26), and the superscript "T" denotes matrix transposition.

A.3 Dynamical Properties

Given the m-input p-output N-dimensional discrete-time system with state model

$$x(n+1) = \Phi x(n) + \Gamma w(n)$$
$$y(n) = Cx(n),$$

a state x is said to be *observable* if x can be uniquely determined from the response values $y(0), y(1), \ldots, y(N-1)$, resulting from initial state $x(0) = x$ with $w(n) = 0$ for $n \geq 0$. The system or the pair (Φ, C) is said to be *observable* if all states (ranging over the set of all N-element column vectors) are observable. To determine a condition for observability, first note that when $x(0) = x$ and $w(n) = 0$ for $n \geq 0$, the output response values are given by

$$y(i) = C\Phi^{i-1} x, \quad i = 1, 2, \ldots, N-1. \tag{A.32}$$

Writing (A.32) in matrix form gives

$$\begin{bmatrix} y(0) \\ y(1) \\ \vdots \\ y(N-1) \end{bmatrix} = \begin{bmatrix} C \\ C\Phi \\ \vdots \\ C\Phi^{N-1} \end{bmatrix} x. \tag{A.33}$$

This equation can be solved for any x if and only if the $pN \times N$ matrix

$$O_N = \begin{bmatrix} C \\ C\Phi \\ \vdots \\ C\Phi^{N-1} \end{bmatrix} \tag{A.34}$$

has rank N. The matrix O_N is called the *N-step observability matrix*, and thus the system or the pair (Φ, C) is observable if and only if the rank of the N-step observability matrix O_N is equal to the system dimension N.

Suppose that the pair (Φ, C) is not observable and let Ψ denote the set of all unobservable states; that is,

$$\Psi = \{x : x \text{ cannot be determined from } y(0), y(1), \ldots, y(N-1), \\ \text{when } x(0) = x \text{ and } w(n) = 0 \text{ for } n \geq 0\}$$

The system or the pair (Φ, C) is said to be *detectable* if the system state $x(n)$ converges to 0 as $n \to \infty$ whenever $x(0) = x$ for any $x \in \Psi$ with $w(n) = 0$ for $n \geq 0$. Clearly, detectability is a weaker condition than observability; that is, if (Φ, C) is observable, it is detectable, but the converse is not true in general.

A test for detectability can be generated in terms of the eigenvalues λ_1, λ_2, ..., λ_N, of the system matrix Φ: The pair (Φ, C) is detectable if and only if

$$\operatorname{rank} \begin{bmatrix} \lambda_i I - \Phi \\ C \end{bmatrix} = N \quad \text{for all } i \text{ such that } |\lambda_i| \geq 1. \tag{A.35}$$

The proof of the detectability condition (A.35) is beyond the scope of this book.

The dual of detectability is called *stabilizability*. A necessary and sufficient condition for the pair (Φ, Γ) to be stabilizable is that

$$\operatorname{rank} \begin{bmatrix} \lambda_i I - \Phi & \Gamma \end{bmatrix} = N \quad \text{for all } i \text{ such that } |\lambda_i| \geq 1. \tag{A.36}$$

Stabilizability arises in the study of state feedback control which as noted before is not considered in this book. The condition also appears in a result in Chapter 5 involving the properties of the Kalman filter which is the reason why the condition for stabilizability is given here.

A.4 Discretization of Noise Covariance Matrices

For the Kalman filter, the discrete-time SMM is often a discretized representation of a continuous-time state model of the form:

$$\dot{\boldsymbol{x}}_c(t) = A\boldsymbol{x}_c(t) + B\boldsymbol{w}_c(t) \tag{A.37}$$

$$\boldsymbol{z}_c(t) = C\boldsymbol{x}_c(t) + \boldsymbol{v}_c(t), \tag{A.38}$$

and initial state $\boldsymbol{x}_c(t_0) = \boldsymbol{x}_{c0}$. $\boldsymbol{w}_c(t)$ and $\boldsymbol{v}_c(t)$ are uncorrelated, white-noise, continuous-time random signals with covariances

$$\mathrm{E}\left[\boldsymbol{w}_c(t)\boldsymbol{w}_c^{\mathrm{T}}(t-\tau)\right] = Q_c(t)\delta_c(t-\tau), \tag{A.39}$$

and

$$\mathrm{E}\left[\boldsymbol{v}_c(t)\boldsymbol{v}_c^{\mathrm{T}}(t-\tau)\right] = R_c(t)\delta_c(t-\tau). \tag{A.40}$$

If $\boldsymbol{w}_c(t)$ and $\boldsymbol{v}_c(t)$ are stationary, as is often the case, $Q_c(t) = Q_c$ and $R_c(t) = R_c$ for all t. $\delta_c(t)$ is the continuous-time impulse function:

$$\delta_c(t) = 0, \quad t \neq 0, \quad \text{and} \quad \int_{-\varepsilon}^{\varepsilon} \delta_c(\tau)\, d\tau \quad = 1, \quad \text{for any } \varepsilon > 0. \qquad \text{(A.41)}$$

Then the discretized SMM with sampling period T has the form of (5.36) and (5.38)

$$\boldsymbol{x}(n+1) = \Phi \boldsymbol{x}(n) + \Gamma \boldsymbol{w}(n)$$
$$\boldsymbol{z}(n) = C\boldsymbol{x}(n) + \boldsymbol{v}(n),$$

where $\Phi = e^{AT}$, Γ is given by (A.9) (or (A.46), see below), and $\boldsymbol{w}(n)$ and $\boldsymbol{v}(n)$ are uncorrelated, white-noise, discrete-time random signals with covariance matrices given by (6.32) and (6.33), respectively:

$$\mathrm{E}\left[\boldsymbol{w}(i)\boldsymbol{w}^{\mathrm{T}}(j)\right] = Q(i)\delta(i-j),$$

and

$$\mathrm{E}\left[\boldsymbol{v}(i)\boldsymbol{v}^{\mathrm{T}}(j)\right] = R(i)\delta(i-j).$$

In the case of stationary noises, $Q(n) = Q$ and $R(n) = R$ for all n.

It is therefore necessary to relate the discrete-time covariance matrices $Q(n)$ and $R(n)$ to their continuous-time counterparts $Q_c(t)$ and $R_c(t)$. First consider discretization of $R_c(t)$. It may appear that

$$R(n)\delta(n) = R_c(nT)\delta_c(nT).$$

However, the impulses $\delta(n)$ and $\delta_c(t)$ are not equivalent. The discrete-time impulse $\delta(n)$ remains finite for all n, but the continuous-time impulse $\delta_c(t)$ is unbounded at $t = 0$.

To remedy this incompatibility, Lewis [7] suggests the following method. Define the continuous-time unit rectangle by

$$b(t) = \begin{cases} 1, & -1/2 \leq t \leq 1/2; \\ 0, & \text{otherwise.} \end{cases}$$

Then $\delta_c(t)$ can be written as

$$\delta_c(t) = \lim_{T \to 0} \frac{1}{T} b(t/T)$$

and the continuous-time covariance becomes

$$R_c(t)\delta_c(t)\big|_{t=nT} = \left[\lim_{T \to 0} R(n)T \times \frac{1}{T} b(t/T)\right]\Bigg|_{t=nT}.$$

(The extra factor of T ensures that the area of the rectangle remains unity.) In order for the right-hand side of this equation to equal the left-hand side, it follows that

$$R(n) = R_c(nT)/T. \tag{A.42}$$

This expression gives the proper relationship between $R(n)$ and $R_c(t)$.

We may discretize B_w in two ways. First, we may assume a zero-order hold, i.e., we have $\boldsymbol{w}(n) = \boldsymbol{w}_c(nT)$. Note that if $\boldsymbol{w}_c(t)$ is a p-vector, then so is $\boldsymbol{w}(n)$. Γ is given by (A.9), and by the same argument that produced (A.42), $Q(n)$ is given by

$$Q(n) = Q_c(nT)/T. \tag{A.43}$$

Second, it is possible to incorporate the system dynamics from time $t = nT$ to time $t = nT + T$. In this case, $\boldsymbol{w}(n)$ is given by

$$\boldsymbol{w}(n) = \int_{nT}^{nT+T} e^{A(nT+T-\tau)} B\boldsymbol{w}_c(\tau)\, d\tau. \tag{A.44}$$

The intervals of integration for $\boldsymbol{w}(n)$ and $\boldsymbol{w}(n+1)$ do not overlap, so $\boldsymbol{w}(n)$ is white noise. As a result of (A.44), $\boldsymbol{w}(n)$ is a vector that is the same length as $\boldsymbol{x}(n)$ (an N-vector). To find the covariance associated with $\boldsymbol{w}(n)$, write

$$\begin{aligned} Q(n) &= \mathrm{E}\left[\boldsymbol{w}(n)\boldsymbol{w}^{\mathrm{T}}(n)\right] \\ &= \mathrm{E}\left[\int_{nT}^{nT+T} e^{A(nT+T-\tau)} B\boldsymbol{w}_c(\tau)\, d\tau \right. \\ &\qquad \left. \times \int_{nT}^{nT+T} \boldsymbol{w}_c^{\mathrm{T}}(\rho) B^{\mathrm{T}} e^{A^{\mathrm{T}}(nT+T-\rho)}\, d\rho\right] \\ &= \int_{nT}^{nT+T}\int_{nT}^{nT+T} e^{A(nT+T-\tau)} B\mathrm{E}\left[\boldsymbol{w}_c(\tau)\boldsymbol{w}_c^{\mathrm{T}}(\rho)\right] \\ &\qquad \times B^{\mathrm{T}} e^{A^{\mathrm{T}}(nT+T-\rho)}\, d\tau\, d\rho. \end{aligned}$$

Now $\mathrm{E}\left[\boldsymbol{w}_c(\tau)\boldsymbol{w}_c^{\mathrm{T}}(\rho)\right] = Q_c(\tau)\delta_c(\tau - \rho)$, so

$$Q(n) = \int_{nT}^{nT+T} e^{A(nT+T-\tau)} BQ_c(\tau)B^{\mathrm{T}} e^{A^{\mathrm{T}}(nT+T-\tau)}\, d\tau.$$

By a change of variables, this equation becomes

$$Q(n) = \int_0^T e^{A\tau} BQ_c(\tau)B^{\mathrm{T}} e^{A^{\mathrm{T}}\tau}\, d\tau. \tag{A.45}$$

Because the dynamics of $e^{At}B$ are directly incorporated into $\boldsymbol{w}(n)$ and $Q(n)$ in (A.45), this method uses

$$\Gamma = I, \tag{A.46}$$

where I is the $N \times N$ identity matrix.

Appendix B

The z-transform

This appendix provides a brief description of the bilateral z-transform. Excellent discussions of the z-transform and all its properties appear in [20, 6].

Given a deterministic sequence $x(n)$, its *two-sided* or *bilateral z-transform* $X(z)$ is defined as the power series

$$X(z) = \mathcal{Z}\{x(n)\} = \sum_{n=-\infty}^{\infty} x(n)z^{-n}, \tag{B.1}$$

where z is a complex number and $\mathcal{Z}\{\cdot\}$ denotes the transform operation. Hence, the signal values $x(n)$ are the coefficients of the power series in z^{-n}. When discussing the z-transform, the complex plane will sometimes be referred to as the *z-plane*. Denote the association between a sequence and its z-transform by

$$x(n) \overset{\mathcal{Z}}{\longleftrightarrow} X(z). \tag{B.2}$$

In some applications, the *one-sided* or *unilateral z-transform* is of interest; it is defined by

$$X^{+}(z) = \sum_{n=0}^{\infty} x(n)z^{-n}.$$

Of course, if $x(n) = 0$ for $n < 0$, then $X(z) = X^{+}(z)$. However, we will concern ourselves primarily with the bilateral z-transform. See, for example, [20, 28] for a discussion of the unilateral z-transform.

Since z is a complex number, it may be expressed in polar form as $z = re^{j\omega}$, where $r = |z|$, $\omega = \angle z$, and $j = \sqrt{-1}$. See Figure B.1. Then (B.1) becomes

$$X(z)|_{z=re^{j\omega}} = X(re^{j\omega}) = \sum_{n=-\infty}^{\infty} x(n)r^{-n}e^{-j\omega n}. \tag{B.3}$$

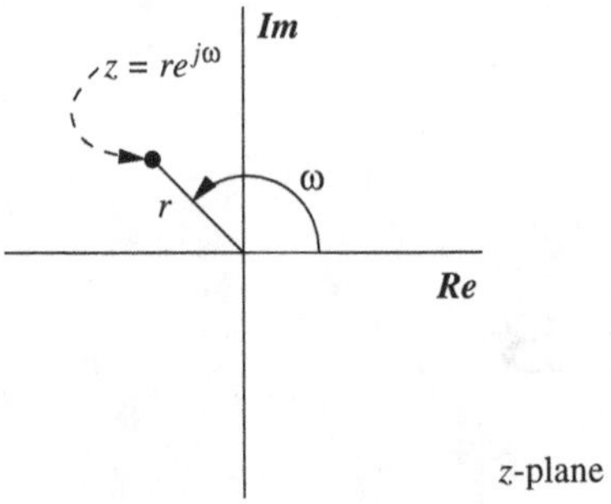

Figure B.1. Representation of a point in the complex plane or z-plane.

For most purposes, it is useful to express the z-transform in closed form. We restrict our attention to the case where the z-transform $X(z)$ is in *rational form*, i.e., $X(z)$ can be written as the ratio of two polynomials in z^{-1}:

$$X(z) = \frac{\sum_{m=0}^{M} b_m z^{-m}}{\sum_{n=0}^{N} a_n z^{-n}} \tag{B.4}$$

The term *zero* of $X(z)$ refers to a value of z for which $X(z) = 0$. Similarly, a *pole* of $X(z)$ is a value of z for which $X(z)$ is infinite. The poles of $X(z)$ are the roots of the denominator polynomial $\sum_{n=0}^{N} a_n z^{-n}$, as well as possibly $z = 0$ and $z = \infty$. When $M < N$, the rational form $X(z)$ is said to be *proper*; otherwise, it is *improper*.

B.1 Region of Convergence

An infinite sum is said to *converge* if it is finite. The infinite sum in the definition of the z-transfom (B.1) depends upon both the signal $x(n)$ and the value of z. Given the signal $x(n)$, the *region of convergence* (ROC) is the set of values of z for which (B.1) converges. It is important to realize that *the z-transform of a signal is not complete unless its associated ROC is also specified.*

Example B.1 **z-transform of an Exponential Signal**

Given signal $x(n) = a^n 1(n)$, we wish to find the z-transform $X(z)$ and its ROC. From (B.1), we have

$$X(z) = \sum_{n=-\infty}^{\infty} x(n)z^{-n} = \sum_{n=0}^{\infty} a^n z^{-n} = \sum_{n=0}^{\infty} (az^{-1})^n .$$

Now we employ the infinite sum

$$\sum_{n=0}^{\infty} \lambda^n = \frac{1}{1-\lambda}, \quad |\lambda| < 1,$$

and substitute $\lambda = az^{-1}$ into our expression for $X(z)$. The ROC is determined by the condition $|az^{-1}| = |\lambda| < 1$, which translates into $|z| > |a|$.

Hence, the z-transform of $x(n)$ is

$$X(z) = \frac{1}{1-az^{-1}}, \quad \mathrm{ROC}_x = \{z : |z| > |a|\}.$$

A sufficient condition for convergence of (B.1) follows. From (B.3),

$$\begin{aligned} |X(z)| &= \left| \sum_{n=-\infty}^{\infty} x(n) r^{-n} e^{-j\omega n} \right| \\ &\leq \sum_{n=-\infty}^{\infty} \left| x(n) r^{-n} \right| \left| e^{-j\omega n} \right| \\ &\leq \sum_{n=-\infty}^{\infty} \left| x(n) r^{-n} \right|. \end{aligned}$$

Therefore, if

$$\sum_{n=-\infty}^{\infty} \left| x(n) r^{-n} \right| < \infty, \tag{B.5}$$

then $|X(z)| < \infty$ and (B.1) converges. In other words, if $x(n)r^{-n}$ is *absolutely summable* (meaning (B.5) holds), then the z-transform converges.

Observe that when $r = 1$, (B.3) reduces to the discrete-time Fourier transform of $x(n)$. In addition, with $r = 1$, (B.5) reduces to the condition that $x(n)$ must be absolutely summable for the Fourier transform $X(e^{j\omega})$ of $x(n)$ to exist. Observe that $|z| = 1$ describes a circle with unit radius and centered at the origin of the z-plane; this circle is called the *unit circle*. Then the Fourier transform can be interpreted as the z-transform evaluated on the unit circle. Thus, the z-transform generalizes the Fourier transform $X(e^{j\omega})$:

$$X(e^{j\omega}) = X(z)|_{z=e^{j\omega}}.$$

Figure B.2 shows the unit circle. Observe that the frequency variable ω corresponds to the angle of $z = e^{j\omega}$. Hence, sweeping ω from 0 to $\pi/2$ corresponds to sweeping z in a counterclockwise direction on the unit circle from $z = 1$ to $z = j$.

In the z-plane, the graph of $|z| = r$ is a circle of radius r centered at the origin. For every value of r for which (B.5) holds, the z-transform converges,

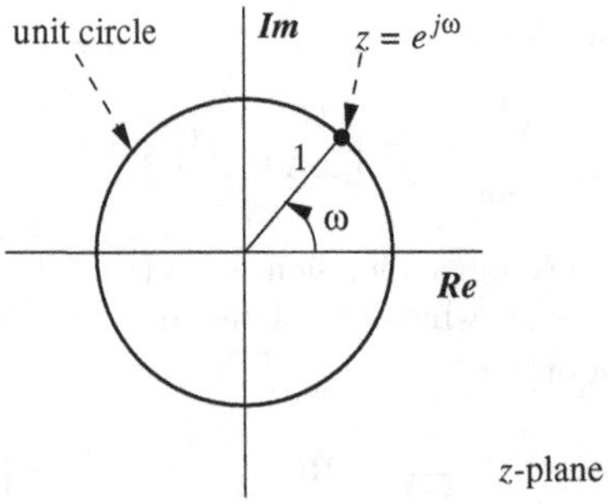

Figure B.2. The unit circle in the z-plane.

so the ROC is the union of concentric circles of different radii. As a result, the ROC forms a ring-shaped region, whose inner and outer boundaries may include the origin and infinity, respectively.

For z-transforms in rational form (B.4), the ROC has several important and useful properties, which we state without proof. Detailed explanations of these properties appear in [6].

- The ROC is a ring centered at the origin in the z-plane and whose inner and outer boundaries may include the origin and infinity, respectively.
- The Fourier transform of $x(n)$ converges absolutely if and only if the unit circle lies within the ROC of $X(z)$.
- The ROC is a connected region and does not contain any poles of $X(z)$.
- Suppose that $x(n)$ has finite support; that is, $x(n)$ is zero for $n < N_1$ and $n > N_2$ ($N_1 \leq N_2$). Then the ROC is the entire z-plane, with the possible exceptions of the origin and infinity.
- Suppose that $x(n)$ is left-sided; that is, $x(n)$ is zero for $n > N > -\infty$. Then the ROC is a disk, centered at the origin with radius equal to the magnitude of the smallest-magnitude non-zero pole. The ROC may also include the origin itself. See Figure B.3a.
- Conversely, if $x(n)$ is right-sided ($x(n) = 0$ for $n < N < \infty$), then the ROC is the entire z-plane except for a disk, centered at the origin with radius equal to the magnitude of the largest-magnitude finite pole. The ROC may also include infinity. See Figure B.3b.
- If $x(n)$ is a two-sided signal (i.e., $x(n)$ is of infinite duration but is neither right-sided nor left-sided), then the ROC is a ring, bounded by the smallest-magnitude non-zero pole and the largest-magnitude finite pole of $X(z)$. See Figure B.3c.

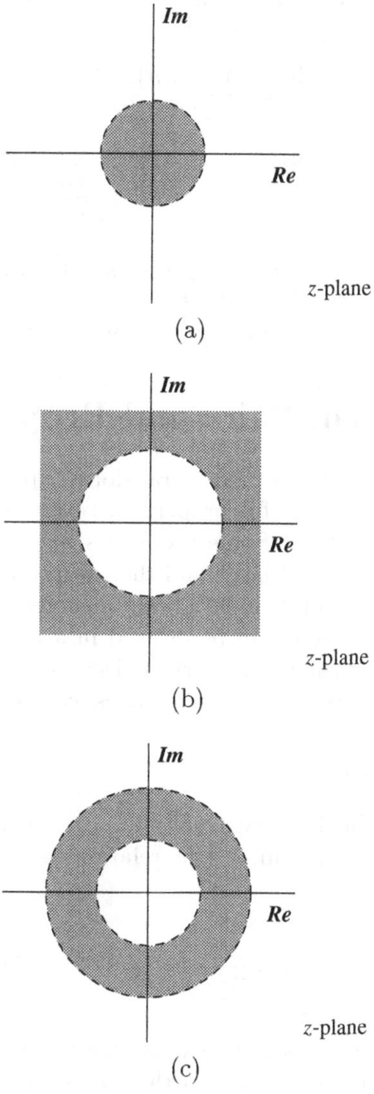

Figure B.3. Possible forms of the ROC for a rational z-transform. (a) ROC corresponding to a left-sided signal. (b) ROC corresponding to a right-sided signal. (c) ROC corresponding to a two-sided signal.

Note that if $X(z)$ is finite at $z=\infty$, then the signal $x(n)=\mathcal{Z}^{-1}\{X(z)\}$ is causal ($x(n)=0$ for $n<0$). This result emerges from the power series (B.1). If $x(n)$ was not causal, then for at least one $n_0>0$, $x(-n_0)\neq 0$. Then

$$\begin{aligned}\lim_{z\to\infty} X(z) &= \lim_{z\to\infty} \sum_{n=-\infty}^{\infty} x(n)z^{-n} \\ &= \lim_{z\to\infty} x(-n_0)z^{n_0} + \sum_{\substack{n=-\infty \\ n\neq -n_0}}^{\infty} x(n)z^{-n},\end{aligned}$$

and since $\lim_{z\to\infty} x(-n_0)z^{n_0}=\infty$, $X(z)$ does not converge at $z=\infty$. By a similar argument, if $X(z)$ is finite at z 0, then the signal $x(n)=\mathcal{Z}^{-1}\{X(z)\}$ has the property $x(n)=0$ for $n>0$ (but note that $x(0)$ may be non-zero).

B.2 z-transform Pairs and Properties

A number of useful signals and their z-transforms appear in Table B.1. The z-transform also has many useful properties, several of which are given in Table B.2 without derivation. Some properties involve the relationship between two sequences, $x(n)$ and $y(n)$, and their respective z-transforms, $X(z)$ and $Y(z)$. The ROCs of $X(z)$ and $Y(z)$ are denoted by ROC_x and ROC_y, respectively. These properties can be derived in a manner analogous to the derivations of Fourier transform properties. Derivations of these z-transform pairs and properties, as well as additional ones, can be found in [20, 6].

Parseval's Relation

Given two (complex-valued) signals $x(n)$ and $y(n)$, their z-transforms possess an energy-preserving relationship. This relationship is known as *Parseval's relation* and is given by

$$\sum_{n=-\infty}^{\infty} x(n)y^*(n) = \frac{1}{2\pi j}\oint_C X(v)Y^*(1/v^*)v^{-1}\,dv, \tag{B.6}$$

where the asterisk (*) denotes complex conjugation and C is a counterclockwise contour of integration that lies in the intersection of the ROC of $X(z)$ and the ROC of $Y^*(1/z^*)$.

If ROC_x and ROC_y both include the unit circle, then we may set $v=e^{j\omega}$ in (B.6), which becomes

$$\sum_{n=-\infty}^{\infty} x(n)y^*(n) = \frac{1}{2\pi}\int_{2\pi} X(e^{j\omega})Y^*(e^{j\omega})\,d\omega. \tag{B.7}$$

Signal	z-transform	ROC
$\delta(n)$	1	Entire z-plane
$1(n)$	$\frac{1}{1-z^{-1}}$	$\|z\| < 1$
$-1(-n-1)$	$\frac{1}{1-z^{-1}}$	$\|z\| > 1$
$\delta(n-n_0)$	z^{-n_0}	Entire z-plane except $z=0$ (if $n_0 > 0$) or $z=\infty$ (if $n_0 < 0$)
$a^n 1(n)$	$\frac{1}{1-az^{-1}}$	$\|z\| > \|a\|$
$-a^n 1(-n-1)$	$\frac{1}{1-az^{-1}}$	$\|z\| < \|a\|$

Table B.1. Table of z-transform pairs.

Signal	z-transform	ROC
$ax(n)+by(n)$	$aX(z)+bY(z)$	Contains $\text{ROC}_x \cap \text{ROC}_y$
$x(n-n_0)$	$z^{-n_0}X(z)$	ROC_x, possibly with the addition or removal of $z=0$ or $z=\infty$
$a^n x(n)$	$X(z/a)$	$\|a\| \times \text{ROC}_x$
$nx(n)$	$-z\frac{dX(z)}{dz}$	ROC_x, possibly with the addition or removal of $z=0$ or $z=\infty$
$x(-n)$	$X(z^{-1})$	$1/\text{ROC}_x$
$x(n)*y(n)$	$X(z)Y(z)$	Contains $\text{ROC}_x \cap \text{ROC}_y$

Table B.2. Table of z-transform properties.

Equation (B.7) is Parseval's relation for the Fourier transform. In many cases, however, it is simpler to evaluate (B.6) using the Cauchy residue theorem, rather than to evaluate (B.7).

B.3 The Inverse z-transform

Given a z-transform $X(z)$ and its ROC, the *inverse z-transform* of $X(z)$ is the corresponding signal $x(n)$, that is,

$$x(n) = \mathcal{Z}^{-1}\{X(z)\}.$$

Formally, the inverse z-transform of $X(z)$ can be determined via a contour integral of $X(z)$ in the z-plane. This method is presented later. However, it is usually simpler to determine $x(n)$ by other means.

Inspection

The simplest manner for finding the inverse z-transform is by *inspection.* That is, we find $x(n)$ by identifying a known z-transform pair that matches $X(z)$ (including the ROC of $X(z)$) and possibly exploiting some properties of the z-transform.

Example B.2 Inverse z-transform by Inspection

Suppose we wish to find the inverse z-transform of

$$X(z) = \frac{0.64z^{-2}}{1 - 0.8z^{-1}}, \quad \mathrm{ROC}_x = \{z : |z| > 0.8\}.$$

We write $X(z) = 0.64z^{-2}Y(z)$, where

$$Y(z) = \frac{1}{1 - 0.8z^{-1}}, \quad \mathrm{ROC}_y = \{z : |z| > 0.8\},$$

and the inverse z-transform of $Y(z)$ is known:

$$Y(z) \stackrel{\mathcal{Z}}{\longleftrightarrow} y(n) = (0.8)^n 1(n).$$

Then the z-transform properties show that the inverse z-transform of $X(z)$ is given by

$$\begin{aligned} X(z) \stackrel{\mathcal{Z}}{\longleftrightarrow} x(n) &= (0.64)y(n-2) = (0.64)(0.8)^{n-2}1(n-2) \\ &= (0.8)^n 1(n-2). \end{aligned}$$

Power Series Expansion

As the z-transform is defined (B.1) as a power series in z^{-n} (note that n may be negative), $X(z)$ may also be converted into a power series in z^{-n} to find $x(n)$. This inversion method is thus known as the *power series expansion* method.

Example B.3 Inverse z-transform by Power Series Expansion

Suppose we want to find the signal $x(n)$ whose z-transform is

$$X(z) = \left(1 - \tfrac{1}{2}z^{-1}\right)\left(1 + z^{-1}\right)^2\left(\tfrac{1}{2} + 3z\right), \quad \mathrm{ROC}_x = \{z : 0 < |z| < \infty\}.$$

Multiplying out this expression, we obtain the power series

$$X(z) = 3z + 5 + \tfrac{3}{4}z^{-1} - \tfrac{3}{4}z^{-2} - \tfrac{1}{4}z^{-3}.$$

Therefore, the signal $x(n)$ is

$$x(n) = 3\delta(n+1) + 5\delta(n) + \tfrac{3}{4}\delta(n-1) - \tfrac{3}{4}\delta(n-2) - \tfrac{1}{4}\delta(n-3).$$

In some cases, the power series expansion can be obtained by long division in z and/or z^{-1}.

Example B.4 Inverse z-transform by Long Division

Given

$$X(z) = \frac{1}{1 + 0.5z^{-1}}, \quad \mathrm{ROC}_x = \{z : |z| < 0.5\},$$

we want to find $x(n)\mathcal{Z}^{-1}\{X(z)\}$. Although we could use the inspection method, in this example we use long division.

Since ROC_x is a disk, $x(n)$ is a left-sided signal. Also, $X(z)$ may be written as

$$X(z) = \frac{1}{1 + 0.5z^{-1}} = \frac{z}{z + 0.5},$$

and thus $X(z)$ is finite at $z = 0$, which means that $x(n) = 0$ for $n > 0$. Therefore, we seek a power series expansion in z, which is obtained via long division below.

$$\begin{array}{r|lllllll}
 & 2z & - & 4z^2 & + & 8z^3 & - & \cdots \\
\hline
0.5 + z & z & & & & & & \\
 & z & + & 2z^2 & & & & \\
\cline{2-4}
 & & & -2z^2 & & & & \\
 & & & -2z^2 & - & 4z^3 & & \\
\cline{4-6}
 & & & & & 4z^3 & & \\
 & & & & & \vdots & &
\end{array}$$

Ultimately, we have

$$\begin{aligned} X(z) &= 2z - 4z^2 + 8z^3 - 16z^4 + \ldots \\ &= -(-0.5)^{-1}z - (-0.5)^2z^2 - (-0.5)^3z^3 - (0.5)^4z^4 - \ldots \\ &= -\sum_{n=1}^{\infty}(-0.5)^n z^n, \end{aligned}$$

so that $x(n) = -(-0.5)^n 1(-n-1)$.

Method of Partial Fractions

When an analytical solution for the inverse z-transform is desired, neither the inspection method nor the power series method may be adequate. A third technique for determining the inverse z-transform of a rational $X(z)$ is the *method of partial fractions*, which decomposes $X(z)$ into a sum of simpler z-transforms.

With $X(z)$ in the form of (B.4), $X(z)$ can be rewritten as

$$X(z) = \frac{z^N \sum_{m=0}^{M} b_m z^{M-m}}{z^M \sum_{n=0}^{N} a_n z^{N-n}}.$$

Thus, $X(z)$ has M nonzero, finite zeros and N nonzero, finite poles. If $M > N$, then $X(z)$ has $M-N$ poles at $z=0$; conversely, if $M < N$, then $X(z)$ has $N-M$ zeros at $z=0$. Lastly, $\lim_{z\to\infty} X(z) = b_0/a_0$, so $X(z)$ has no poles or zeros at infinity. Let p_n denote the nth pole ($n = 0, 1, \ldots, N$) of $X(z)$.

Suppose that all poles are of first order; that is, no pole p_n is repeated. Then $X(z)$ can be written in the form

$$X(z) = \sum_{k=0}^{M-N} c_k z^{-k} + \sum_{n=1}^{N} \frac{\lambda_n}{1 - p_n z^{-1}}, \tag{B.8}$$

where the p_n are the poles of $X(z)$ and

$$\lambda_n = (1 - p_n z^{-1})X(z)\big|_{z=p_n}. \tag{B.9}$$

If $M < N$, then the first summation in (B.8) is zero, i.e., non-existent. Otherwise ($M \geq N$), it can be obtained by long division.

More generally, if d_n denotes the order of pole p_n, then $X(z)$ can be expressed as

$$X(z) = \sum_{k=0}^{M-N} c_k z^{-k} + \sum_{n=1}^{N}\left[\sum_{\ell=1}^{d_n} \frac{\mu_{n,\ell}}{(1 - p_n z^{-1})^\ell}\right], \tag{B.10}$$

where

$$\mu_{n,\ell} = \frac{1}{(d_n - \ell)!(-p_n)^{d_n-\ell}} \left[\frac{d^{d_n-\ell}}{dz^{d_n-\ell}} (1 - p_n z)^{d_n} X(z^{-1}) \right] \Bigg|_{z=p_n^{-1}} . \tag{B.11}$$

Note that (B.11) is expressed in z rather than z^{-1}. Also, when $\ell = d_n$, (B.11) reduces to (B.9), i.e.,

$$\mu_{n,n} = (1 - p_n z^{-1}) X(z)\big|_{z=p_n} . \tag{B.12}$$

It is sometimes more convenient to compute the $\mu_{n,\ell}$ in another manner. First, convert (B.10) into rational form by placing all terms over a common denominator. Second, set the numerator of this result equal to the numerator in (B.4). Finally, solve a set of linear equations that relate $\mu_{n,\ell}$ and c_k to b_m.

Example B.5 Inverse z-transform by Partial Fractions

We seek the inverse z-transform $x(n)$ of

$$X(z) = \frac{2 - 1.1z^{-1} - 5.56z^{-2} + 6.08z^{-3} - 1.92z^{-4}}{1 - 3.1z^{-1} + 3.04z^{-2} - 0.96z^{-3}},$$

$$\mathrm{ROC}_x = \{z : 0.8 < |z| < 1.5\}.$$

Hence, $M = 4$ and $N = 3$.

Let $N(z)$ and $D(z)$ denote the numerator and denominator of $X(z) = N(z)/D(z)$, respectively:

$$N(z) = 2 - 1.1z^{-1} - 5.56z^{-2} + 6.08z^{-3} - 1.92z^{-4},$$

and

$$D(z) = 1 - 3.1z^{-1} + 3.04z^{-2} - 0.96z^{-3} = (1 - 1.5z^{-1})(1 - 0.8z^{-1})^2.$$

The factorization of $D(z)$ shows that $X(z)$ has a single pole at $z = 1.5$ and a double pole at $z = 0.8$. Thus, we set $p_1 = 1.5$, $d_1 = 1$, and $p_2 = 0.8$, $d_2 = 2$.

Since $M > N$, we perform long division in z^{-1} to put $X(z)$ into the form of (B.10):

$$X(z) = 2z^{-1} + \frac{\mu_{1,1}}{1 - 1.5z^{-1}} + \frac{\mu_{2,1}}{1 - 0.8z^{-1}} + \frac{\mu_{2,2}}{(1 - 0.8z^{-1})^2}. \tag{B.13}$$

Then the coefficients $\mu_{1,1}$ and $\mu_{2,2}$ are given by (B.12):

$$\mu_{1,1} = (1 - 1.5z^{-1}) X(z)\big|_{z=1.5} = \frac{N(z)}{(1 - 0.8z^{-1})^2}\bigg|_{z=1.5} = 1,$$

and

$$\mu_{2,2} = (1 - 0.8z^{-1})^2 X(z)\big|_{z=0.8} = \frac{N(z)}{1 - 1.5z^{-1}}\bigg|_{z=0.8} = 1.$$

To find $\mu_{2,1}$, we use (B.11):

$$\begin{aligned}\mu_{2,1} &= \frac{1}{1!(-0.8)^1}\left[\frac{d}{dz}(1-0.8z)^2 X(z^{-1})\right]\Bigg|_{z=(0.8)^{-1}} \\ &= (-1.25)\left[\frac{d}{dz}\frac{N(z^{-1})}{1-1.5z}\right]\Bigg|_{z=(0.8)^{-1}},\end{aligned}$$

and

$$\frac{d}{dz}\frac{N(z^{-1})}{1-1.5z} = \frac{(1-1.5z)N(z^{-1}) - (-1.5)\left[\frac{d}{dz}N(z^{-1})\right]}{(1-1.5z)^2}.$$

We find that $\mu_{2,1} = 0$.

As an alternate method for finding the $\mu_{n,\ell}$, we convert (B.13) into rational form:

$$\begin{aligned}X(z) = {} & \left[2z^{-1} - 6.2z^{-2} + 6.08z^{-3} - 1.92z^{-4} + \mu_{1,1}(1-1.6z^{-1}+0.64z^{-2})\right. \\ & \left. + \mu_{2,1}(1-2.3z^{-1}+1.2z^{-2}) + \mu_{2,2}(1-1.5z^{-1})\right] \\ & \div \left[(1-1.5z^{-1})(1-0.8z^{-1})^2\right].\end{aligned}$$

The denominator of this expression is just $D(z)$. Let $P(z)$ be the numerator, i.e.,

$$\begin{aligned}P(z) = {} & 2z^{-1} - 6.2z^{-2} + 6.08z^{-3} - 1.92z^{-4} + \mu_{1,1}(1-1.6z^{-1}+0.64z^{-2}) \\ & + \mu_{2,1}(1-2.3z^{-1}+1.2z^{-2}) + \mu_{2,2}(1-1.5z^{-1}).\end{aligned}$$

Then $X(z) = P(z)/D(z) = N(z)/D(z)$, so $P(z) = N(z)$. From $P(z) = N(z)$, we collect the coefficients of z^0, z^{-1}, and z^{-2} (none of the $\mu_{n,\ell}$ multiply z^{-3} or z^{-4}). Doing so produces a system of linear equations:

$$\begin{bmatrix} 1.00 & 1.00 & 1.00 \\ -1.60 & -2.30 & -1.50 \\ 0.64 & 1.20 & 0.00 \end{bmatrix}\begin{bmatrix}\mu_{1,1} \\ \mu_{2,1} \\ \mu_{2,2}\end{bmatrix} + \begin{bmatrix} 0.00 \\ 2.00 \\ -6.20 \end{bmatrix} = \begin{bmatrix} 2.00 \\ -1.10 \\ -5.56 \end{bmatrix}.$$

Solving this system gives

$$\begin{bmatrix}\mu_{1,1} & \mu_{2,1} & \mu_{2,2}\end{bmatrix}^{\mathrm{T}} = \begin{bmatrix}1 & 0 & 1\end{bmatrix}^{\mathrm{T}}.$$

Both methods for finding the $\mu_{n,\ell}$ indicate that $X(z)$ can be written as

$$X(z) = 2z^{-1} + \frac{1}{1-1.5z^{-1}} + \frac{1}{(1-0.8z^{-1})^2}.$$

We can now use the inspection method to find the inverse z-transform; the desired signal $x(n)$ is

$$x(n) = 2\delta(n-1) - (1.5)^n 1(-n-1) + \left[(0.8)^n 1(n) * (0.8)^n 1(n)\right].$$

Contour Integration

As mentioned above, the formal means for computing the inverse z-transform is via a contour integral in the z-plane. The inverse z-transform is formally defined as

$$x(n) = \frac{1}{j2\pi} \oint_C X(z) z^{n-1} \, dz, \tag{B.14}$$

where C is a closed contour in the z-plane that is contained in the ROC and encircles the origin, and the integral on C is taken in the counterclockwise direction.

We do not make extensive use of (B.14) because the other methods for determining the inverse z-transform are adequate for our purposes. However, we mention it for completeness and for the benefit of readers who are familiar with complex variables [5].

When $X(z)$ is a rational z-transform, the Cauchy residue theorem allows (B.14) to be computed conveniently via

$$\begin{aligned} x(n) &= \frac{1}{j2\pi} \oint_C X(z) z^{n-1} \, dz \\ &= \sum \left[\text{residues of } X(z) z^{n-1} \text{ inside } C \right]. \end{aligned} \tag{B.15}$$

Additional discussion of the inverse z-transform via contour integration appears in [6]. An excellent discussion of complex variables, including the Cauchy residue theorem, can be found in [5].

As a final note, if the unit circle is contained within the ROC, then C may be chosen as the unit circle ($z = e^{j\omega}$), and (B.14) reduces to the definition of the inverse Fourier transform

$$x(n) = \frac{1}{2\pi} \int_{2\pi} X(e^{j\omega}) e^{j\omega n} \, d\omega. \tag{B.16}$$

Appendix C

Stability of the Kalman Filter

This appendix presents a proof of the stability of the Kalman filter for the general time-varying, nonstationary case and provides a supplement to Section 5.7. The proofs detail the work of Deyst and Price [29], Deyst [30], and Jazwinski [14].

We begin with the time-varying SMM of (6.29–6.33) with nonstationary noises (6.32) and (6.33) in Section 5.4. We also make assumptions (6.39–6.42) in Section 5.7. We also assume that all matrices in the SMM are bounded above in norm.

The notions of observability and controllability provide a set of sufficient conditions for the stability of the Kalman filter. Because the Kalman filter is time-varying, we present a brief explanation of these concepts for time-varying systems. More detail appears in, for example, [8, 9, 10].

C.1 Observability

A deterministic linear system is said to be *observable at* n_0 if the initial state $x(n_0)$ can be reconstructed from a *finite* number of observations $z(n_0), \ldots, z(n_0+N(x(n_0)))$. N is a positive integer that may depend on $x(n_0)$.

Let us develop this idea for systems excited by random processes. Now we want to obtain an estimate $\hat{\boldsymbol{x}}(n_0)$ from the observations $\boldsymbol{z}(n_0), \ldots, \boldsymbol{z}(n_0+N(\boldsymbol{x}(n_0)))$. For simplicity we assume no process noise, i.e., $\boldsymbol{w}(n) \equiv 0$ and $Q(n) \equiv 0$ for all n.

The Kalman filter equations may be written in the form,

$$P(n) = \Big[\Phi^{\mathrm{T}}(n_0,n)P^{-1}(n_0)\Phi(n_0,n) + \sum_{i=n_0+1}^{n} \Phi^{\mathrm{T}}(i,n)C^{\mathrm{T}}(i)R^{-1}(i)C(i)\Phi(i,n)\Big]^{-1} \tag{C.1}$$

$$\hat{\boldsymbol{x}}(n) = P(n)\Big[\Phi^{\mathrm{T}}(n_0,n)P^{-1}(n_0)\Phi(n_0,n)\hat{\boldsymbol{x}}^{-}(n_0) + \sum_{i=n_0+1}^{n} \Phi^{\mathrm{T}}(i,n)C^{\mathrm{T}}(i)R^{-1}(i)\boldsymbol{z}(i)\Big] \tag{C.2}$$

$$P^{-}(n+1) = \Phi(n+1,n)P(n)\Phi^{\mathrm{T}}(n+1,n). \tag{C.3}$$

The summation in (C.1) is called the *information matrix*,[1] which we define by

$$\mathcal{I}(m,n) \triangleq \sum_{i=m}^{n} \Phi^{\mathrm{T}}(i,n)C^{\mathrm{T}}(i)R^{-1}(i)C(i)\Phi(i,n). \tag{C.4}$$

We can use the Kalman filter to obtain $\hat{\boldsymbol{x}}(n_0{+}N)$ from $\boldsymbol{z}(n_0), \ldots, \boldsymbol{z}(n_0{+}N)$. Then we can use the recurrence relation below (with n running backwards from n_0+N to n_0)

$$\hat{\boldsymbol{x}}(n-1) = \Phi(n-1,n)P^{-}(n)\left[P^{-1}(n)\hat{\boldsymbol{x}}(n) - C^{\mathrm{T}}(n)R^{-1}(n)\boldsymbol{z}(n)\right] \tag{C.5}$$

to find $\hat{\boldsymbol{x}}(n_0)$. There is an implicit dependence on $\boldsymbol{x}(n_0)$ because of the $P^{-1}(n_0)$ term in (C.1) and (C.2)

In the worst case we have no knowledge of the initial state $\boldsymbol{x}(n_0)$, so let us assume our initial error covariance matrix $P^{-}(n_0)$ is unbounded. Equivalently, $(P^{-}(n_0))^{-1} = 0$. From (5.98),

$$P(n_0) = \left[C^{\mathrm{T}}(n_0)R^{-1}(n_0)C(n_0)\right]^{-1}$$

Then (C.1) becomes

$$P(n) = \left[\sum_{i=n_0}^{n} \Phi^{\mathrm{T}}(i,n)C^{\mathrm{T}}(i)R^{-1}(i)C(i)\Phi(i,n)\right]^{-1} = \mathcal{I}^{-1}(n_0,n).$$

From this expression we see that $P(n_0{+}N)$ will be singular unless

$$\mathcal{I}(n_0, n_0{+}N) > 0, \tag{C.6}$$

[1] The information matrix is *not* identical to the observability Gramian, usually defined (cf. [8, 9, 7]) as $\mathcal{O}(m,n) \triangleq \sum_{i=m}^{n} \Phi^{\mathrm{T}}(i,m)C^{\mathrm{T}}(i)R^{-1}(i)C(i)\Phi(i,m)$.

for some finite $N = N(\boldsymbol{x}(n_0)) > 0$.

Based upon (C.2) and (C.5), if (C.6) is not satisfied, then we cannot obtain $\hat{\boldsymbol{x}}(n_0)$. On the other hand, if (C.6) does hold, then we can find $\hat{\boldsymbol{x}}(n_0)$. As a result, we say that a linear system is observable at n_0 if and only if $\mathcal{I}(n_0, n_0+N)$ is nonsingular.

When we can generate $\hat{\boldsymbol{x}}(n_0)$ for all initial states $\boldsymbol{x}(n_0)$, we say that the system is *completely observable at* n_0. It follows that a linear system is completely observable at n_0 if and only if there is an integer $N > 0$, independent of $\boldsymbol{x}(n_0)$, such that

$$\mathcal{I}(n_0, n_0+N) > 0. \tag{C.7}$$

The information matrix is said to grow *uniformly* if its rate of growth remains within fixed bounds. That is, there exist an integer $N > 0$ and real numbers $0 < \beta_1 < \beta_2 < \infty$ such that

$$\beta_1 I \leq \mathcal{I}(n-N, n) \leq \beta_2 I, \quad \forall\, n \geq N. \tag{C.8}$$

Equation (C.8) says that over any time interval of length $N+1$, the information matrix will increase by at least β_1 but not by more than β_2. Equation (C.8) is a necessary and sufficient condition for *uniform complete observability* (UCO).

C.2 Controllability

Controllability is the dual of observability. A deterministic linear system is *controllable at* n_0 if there is a *finite-duration* input sequence $u(n_0), \ldots, u(n_0+N-1)$ that will drive the states from $x(n_0)$ to $x(n_0+N) = 0$.[2]

In a stochastic framework, the dual of the MMSE observer problem is the linear quadratic regulator (LQR) problem. Given a linear system with initial state $\boldsymbol{x}(n_0)$ (a random variable), we try to determine the minimum-energy input sequence that drives $\boldsymbol{x}(n)$ to zero. We will not address the LQR problem in greater detail here; the interested reader may consult [10]. We present only the appropriate controllability conditions.

By analogy with complete observability, a linear system is *completely controllable at* n_0 if and only if there is an integer $N > 0$ such that

$$\mathcal{C}(n_0, n_0+N) > 0,$$

where $\mathcal{C}(m, n)$ is the *controllability Gramian*,

$$\mathcal{C}(m, n) \triangleq \sum_{i=m}^{n-1} \Phi(n, i+1)\Gamma(i)Q(i)\Gamma^{\mathrm{T}}(i)\Phi^{\mathrm{T}}(n, i+1). \tag{C.9}$$

[2] $N-1$ appears instead of N because the input $u(n)$ affects $x(n+1)$, not $x(n)$.

In other words, there is an input sequence of length N that drives $\boldsymbol{x}(n)$ to zero for every initial state $\boldsymbol{x}(n_0)$.

A linear system is *uniformly completely controllable* (UCC) if and only if there exist an integer $N > 0$ and real numbers $0 < \alpha_1 < \alpha_2 < \infty$ such that

$$\alpha_1 I \leq \mathcal{C}(n-N, n) \leq \alpha_2 I, \quad \forall\, n \geq N. \tag{C.10}$$

C.3 Types of Stability

When speaking of the stability of a system, we generally refer to the internal stability of the system. That is, we are concerned with the behavior of the system without any forcing inputs ($u(n) = \boldsymbol{w}(n) \equiv 0$).

The norm of the state vector, $||x(n)||$, is often used to define types of stability. For all $x(n)$, $||x(n)|| \geq 0$, and $||x(n)|| = 0$ if and only if $x(n) = 0$. Thus, $||x(n)||$ may be viewed as a measure of energy of the states $x(n)$. When the state energy is bounded or tends to zero, we say the system is stable, i.e., the state energy does not grow without bound.

The unforced system is also called the *homogeneous system* and takes the form:

$$x(n+1) = \Psi(n+1, n)x(n). \tag{C.11}$$

$\Psi(n+1, n)$ is called the *state-transition matrix* because it describes the dynamics of the states with no forcing input. That is, $\Psi(n+1, n)$ completely characterizes the change in the states from time n to time $n+1$ due to the internal behavior of the system.

The state-transition matrix has several important properties. The semigroup property describes the manner in which state-transition matrices may be combined:

$$\Psi(n, m)\Psi(m, j) = \Psi(n, j), \quad j < m < n. \tag{C.12}$$

The identity property applies, namely,

$$\Psi(n, n) = I. \tag{C.13}$$

Lastly, if $\Psi(n, j)$ is nonsingular, then the time-reversal property holds:

$$\Psi^{-1}(n, j) = \Psi(j, n). \tag{C.14}$$

If $\Psi(n, j)$ is a discretized version of the state-transition matrix for a continuous system, then (C.14) will hold.

We now briefly present several types of stability. Although formal definitions are included, the emphasis lies in the conceptual notions of stability that precede each definition.

The first type of stability is *stability in the sense of Lyapunov* (SISL) or *Lyapunov stability.* Conceptually, it means that if the initial state energy is small enough, then the state energy will always be bounded. Formally, we have

Definition C.1 (Lyapunov Stability) *A system is* stable *if* $\forall\ \epsilon > 0$ *there exists* $\delta(\epsilon, n_0) > 0$ *such that if* $||x(n_0)|| < \delta$, *then* $||x(n)|| < \epsilon,\ \forall\ n > n_0$.

Another common form of stability is *asymptotic stability.* This type of stability means that if the initial state energy is small enough, then the state energy will approach zero as $n \to \infty$.

Definition C.2 (Asymptotic Stability) *A system is* asymptotically stable *if it is SISL and if there exists* $\delta(n_0) > 0$ *such that if* $||x(n_0)|| < \delta$, *then* $\lim_{n\to\infty} ||x(n)|| = 0$.

In Definitions C.1 and C.2, δ depends upon the initial time n_0, so the stability of the system depends upon the initial time. This property is important since the Kalman filter is time-varying. When system stability is independent of the initial time, we say the system is *uniformly stable* (in the sense of Lyapunov) or *uniformly asymptotically stable* (UAS). Also, note that uniform stability refers to initial-time independence while uniform complete observability and controllability refer to the existence of bounds on the matrices in (C.8) and (C.10), respectively.

If, for a particular n_0, the state energy goes to zero as $n \to \infty$ for *any* initial state $x(n_0)$, we say the system is *globally asymptotically stable* (GAS). Note that *uniform* asymptotic stability indicates independence of the initial *time* n_0, but *global* asymptotic stability refers to independence of the initial *state* $x(n_0)$.

We can combine these two types of independence to define *global uniform asymptotic stability* (GUAS), which means that, regardless of the initial state *and* the initial time, the state energy approaches zero as $n \to \infty$. Clearly GUAS is a stronger form of stability than either UAS or GAS. We define GUAS defined formally as

Definition C.3 (GUAS 1) *A system is* globally uniformly asymptotically stable *if for all* $x(n_0)$, $\lim_{n\to\infty} ||x(n)|| = 0$.

An equivalent definition is

Definition C.4 (GUAS 2) *A system is* globally uniformly asymptotically stable *if there exist real numbers* $a, b > 0$ *such that*

$$||\Psi(n, n_0)|| \leq ae^{-b(n-n_0)}, \quad n \geq n_0. \tag{C.15}$$

In this case, the norm of the state-transition matrix approaches zero as $n \to \infty$. As a result, the state vector will approach zero as well.

In the course of deriving a set of sufficient conditions that ensure the global uniform asymptotic stability of the Kalman filter, we will make use of the following theorem by Kalman and Bertram [31].

Theorem C.1 *A linear system with state vector $x(n)$ is globally uniformly asymptotically stable if there exist real-valued scalar functions V, γ_1, γ_2, and γ_3 such that*

$$0 < \gamma_1(\|x(n)\|) \leq V(x(n), n) \leq \gamma_2(\|x(n)\|),$$

$$n \geq N, \; x(n) \neq 0, \qquad \text{(C.16)}$$

and

$$V(x(n), n) - V(x(n-N), n-N) \leq \gamma_3(\|x(n)\|) < 0,$$

$$n \geq N, x(n) \neq 0, \qquad \text{(C.17)}$$

and

$$\gamma_1(0) = \gamma_2(0) = 0, \quad \lim_{a \to \infty} \gamma_1(a) = \infty. \qquad \text{(C.18)}$$

$V(x(n), n)$ is called a *Lyapunov function* of the system. Equations (C.16) and (C.18) state that γ_1 and γ_2 are zero if and only if the state energy is zero, i.e., $x(n) = 0$. Otherwise γ_1 and γ_2 are positive. If the state energy becomes infinite, then γ_1 and γ_2 become infinite. Hence, the functions γ_1 and γ_2 are alternative measures of the state energy. By (C.16) V is bounded by γ_1 and γ_2, so V may also be considered a measure of the state energy.

Equation (C.17) indicates that V is decreasing as long as $x(n) \neq 0$, and by (C.16) V is bounded below by zero. Therefore $V(x(n), n)$ must decrease to zero as n increases. In other words, the state energy decreases to zero asymptotically, and thus $x(n) \to 0$ as $n \to \infty$.

C.4 Positive-Definiteness of $P(n)$

Lemma C.1 *If the system (6.29–6.33), (6.39–6.42) is UCC and $P^-(0) \geq 0$, then $P(n) > 0$ and $P^-(n) > 0$ for all $n \geq N$.*

Proof. [14] From (5.98) and (5.84), it is clear that if $P(N) > 0$, then $P(n) > 0$ and $P^-(n) > 0$ for all $n \geq N$. As a result, we only need to prove that $P(N) > 0$. $P(N)$ is guaranteed to be positive-semidefinite, so if $P(N)$ is nonsingular, then $P(N) > 0$.

Let us assume that $P(N)$ is singular. Then there exists a vector $v \neq 0$ such that

$$v^{\mathrm{T}} P(N) v = 0. \tag{C.19}$$

By (5.83) and (5.81) the Kalman filter is given by

$$\hat{\boldsymbol{x}}(n) = [I - K(n)C(n)]\Phi(n,n-1)\hat{\boldsymbol{x}}(n-1) + K(n)\boldsymbol{z}(n).$$

Let Ψ be the state-transition matrix of the Kalman filter, so

$$\Psi(n,n-1) = [I - K(n)C(n)]\Phi(n,n-1).$$

Define

$$S(n) \triangleq \Psi(N,n)P(n)\Psi^{\mathrm{T}}(N,n), \tag{C.20}$$

and consider the difference

$$\begin{aligned} S(n) - S(n-1) &= \Psi(N,n)P(n)\Psi^{\mathrm{T}}(N,n) - \Psi(N,n-1)P(n-1)\Psi^{\mathrm{T}}(N,n-1) \\ &= \Psi(N,n)P(n)\Psi^{\mathrm{T}}(N,n) \\ &\quad - \Psi(N,n)\Psi(n,n-1)P(n-1)\Psi^{\mathrm{T}}(n,n-1)\Psi^{\mathrm{T}}(N,n) \\ &= \Psi(N,n)\left\{P(n) - [I - K(n)C(n)]\Phi(n,n-1)P(n-1)\right. \\ &\quad \left.\times\Phi^{\mathrm{T}}(n,n-1)[I - K(n)C(n)]^{\mathrm{T}}\right\}\Psi^{\mathrm{T}}(N,n). \end{aligned} \tag{C.21}$$

It is easy to verify that

$$P(n) = [I - K(n)C(n)]P^{-}(n)[I - K(n)C(n)]^{\mathrm{T}} + K(n)R(n)K^{\mathrm{T}}(n).$$

From (5.84), we have

$$\Phi(n,n-1)P(n-1)\Phi^{\mathrm{T}}(n,n-1) = P^{-}(n) - \Gamma(n-1)Q(n-1)\Gamma^{\mathrm{T}}(n-1).$$

Then (C.21) becomes

$$\begin{aligned} S(n) - S(n-1) = \Psi(N,n)&\left\{[I - K(n)C(n)]\Gamma(n-1)Q(n-1)\Gamma^{\mathrm{T}}(n-1)\right. \\ &\left.\times[I - K(n)C(n)]^{\mathrm{T}} + K(n)R(n)K^{\mathrm{T}}(n)\right\}\Psi^{\mathrm{T}}(N,n). \end{aligned} \tag{C.22}$$

From (C.20), $S(0) \geq 0$, and from (C.22), $S(n) - S(n-1) \geq 0$. By assumption (C.19) and the fact that $\Psi(n,n) = I$, we conclude that $v^{\mathrm{T}} S(N) v = 0$. Then we have

$$v^{\mathrm{T}}[S(N) - S(N-1)]v = v^{\mathrm{T}} S(N) v - v^{\mathrm{T}} S(N-1) v = -v^{\mathrm{T}} S(N-1) v \geq 0,$$

and hence $v^T S(N-1) v = 0$. Continuing in this manner, we conclude that

$$v^{\mathrm{T}} S(n) v = 0, \quad 0 \leq n \leq N, \tag{C.23}$$

which we write via (C.20) as

$$v^{\mathrm{T}}\Psi(N,n)P(n)\Psi^{\mathrm{T}}(N,n)v = 0, \quad 0 \leq n \leq N.$$

Therefore,

$$P(n)\Psi^{\mathrm{T}}(N,n)v = 0, \quad 0 \le n \le N. \tag{C.24}$$

Next define

$$w(n) \triangleq \Psi^{\mathrm{T}}(N,n)v, \quad \text{so} \quad w(N) = v.$$

It follows that

$$\begin{aligned} w(n-1) &= \Psi^{\mathrm{T}}(N,n-1)v = \Psi^{\mathrm{T}}(N,n-1)\Psi^{-\mathrm{T}}(N,n)w(n) \\ &= [\Psi(n,N)\Psi(N,n-1)]^{\mathrm{T}} w(n) \\ &= \Psi^{\mathrm{T}}(n,n-1)w(n), \end{aligned}$$

and

$$\begin{aligned} \Phi^{\mathrm{T}}(n-1,n)w(n-1) &= \Phi^{\mathrm{T}}(n-1,n)\Psi^{\mathrm{T}}(n,n-1)w(n) \\ &= \Phi^{\mathrm{T}}(n-1,n)\Psi^{\mathrm{T}}(n,n-1)\Psi^{\mathrm{T}}(N,n)v \\ &= \Phi^{\mathrm{T}}(n-1,n)\Psi^{\mathrm{T}}(N,n-1)v \\ &= \Phi^{\mathrm{T}}(n-1,n)\Phi^{\mathrm{T}}(n,n-1) \\ &\qquad \times [I-K(n)C(n)]^{\mathrm{T}}\Psi^{\mathrm{T}}(N,n)v \\ &= [I-K(n)C(n)]^{\mathrm{T}}\Psi^{\mathrm{T}}(N,n)v. \end{aligned} \tag{C.25}$$

Multiplying both sides of (C.25) on the left by $\Phi^{-\mathrm{T}}(n-1,n)$, we find

$$\begin{aligned} w(n-1) &= \Phi^{-\mathrm{T}}(n-1,n)[I-K(n)C(n)]^{\mathrm{T}}\Psi^{\mathrm{T}}(N,n)v \\ &= \Phi^{\mathrm{T}}(n,n-1)w(n) - \Phi^{\mathrm{T}}(n,n-1)C^{\mathrm{T}}(n)K^{\mathrm{T}}(n)w(n). \end{aligned} \tag{C.26}$$

Note that by (5.80),

$$K^{\mathrm{T}}(n)w(n) = [C(n)P^{-}(n)C^{\mathrm{T}}(n) + R(n)]^{-1}C(n)P^{-}(n)\Psi^{\mathrm{T}}(N,n)v,$$

and by (5.82) and (C.24), we have

$$[I-K(n)C(n)]P^{-}(n)\Psi^{\mathrm{T}}(N,n)v = 0,$$

and since $I - K(n)C(n)$ is generally not the zero matrix, we have

$$P^{-}(n)\Psi^{\mathrm{T}}(N,n)v = 0.$$

Thus (C.26) reduces to

$$w(n-1) = \Phi^{\mathrm{T}}(n,n-1)w(n). \tag{C.27}$$

Solving (C.27) for $w(n)$, we have

$$\begin{aligned} w(n) &= \Phi^{-\mathrm{T}}(n,n-1)w(n-1) \\ &= \Phi^{\mathrm{T}}(n-1,n)\Phi^{\mathrm{T}}(n-2,n-1)w(n-2) \\ &= \Phi^{\mathrm{T}}(n-1,n)\Phi^{\mathrm{T}}(n-2,n-1)\cdots\Phi^{\mathrm{T}}(N,N+1)w(N) \\ &= \Phi^{\mathrm{T}}(N,n)v. \end{aligned} \tag{C.28}$$

Returning to (C.23), we conclude that $v^{\mathrm{T}}[S(n) - S(n-1)]v = 0,\ 1 \leq n \leq N$. Then (C.22) gives

$$v^{\mathrm{T}}\Psi(N,n)\Big\{[I - K(n)C(n)]\Gamma(n-1)Q(n-1)\Gamma^{\mathrm{T}}(n-1) \\ \times [I - K(n)C(n)]^{\mathrm{T}} + K(n)R(n)K^{\mathrm{T}}(n)\Big\}\Psi^{\mathrm{T}}(N,n)v = 0, \quad 1 \leq n \leq N.$$

Extracting the first term from the braced sum produces

$$v^{\mathrm{T}}\Psi(N,n)[I - K(n)C(n)]\Gamma(n-1)Q(n-1) \\ \times \Gamma^{\mathrm{T}}(n-1)[I - K(n)C(n)]^{\mathrm{T}}\Psi^{\mathrm{T}}(N,n)v = 0, \quad 1 \leq n \leq N. \tag{C.29}$$

From (C.25), (C.27), and (C.28) we have

$$\begin{aligned}[I - K(n)C(n)]^{\mathrm{T}}\Psi^{\mathrm{T}}(N,n)v &= \Phi^{\mathrm{T}}(n-1,n)w(n-1) \\ &= \Phi^{\mathrm{T}}(n-1,n)\Phi^{\mathrm{T}}(n,n-1)w(n) \\ &= \Phi^{\mathrm{T}}(N,n)v.\end{aligned}$$

Then (C.29) becomes

$$v^{\mathrm{T}}\Phi(N,n)\Gamma(n-1)Q(n-1)\Gamma^{\mathrm{T}}(n-1)\Phi^{\mathrm{T}}(N,n)v = 0, \quad 1 \leq n \leq N.$$

Finally, we sum over $1 \leq n \leq N$ to obtain

$$v^{\mathrm{T}}\left[\sum_{n=1}^{N}\Phi(N,n)\Gamma(n-1)Q(n-1)\Gamma^{\mathrm{T}}(n-1)\Phi^{\mathrm{T}}(N,n)\right]v = 0. \tag{C.30}$$

A simple change of variables shows that the bracketed term in (C.30) to be equivalent to (C.10), the condition for UCC. However, this result contradicts the hypothesis that the system is UCC. Thus $P(N)$ cannot be singular, so $P(N) > 0$.

Q.E.D. ■

C.5 An Upper Bound for $P(n)$

Lemma C.2 *Let the system (6.29–6.33), (6.39–6.42) be UCO and UCC, and let $P^{-}(0) \geq 0$. Then $P(n)$ is uniformly bounded above,*

$$P(n) \leq \frac{\beta_1 + N\alpha_2\beta_2^2}{\beta_1^2}I, \quad \forall\ n \geq N.$$

Proof. [29, 30, 32, 14] It will be useful to relate $\boldsymbol{x}(i)$ to $\boldsymbol{x}(n)$ and $\boldsymbol{w}(i), \ldots, \boldsymbol{w}(n-1)$ for $i < n$. For any such i,

$$\begin{aligned}\boldsymbol{x}(i+1) &= \Phi(i+1,i)\boldsymbol{x}(i) + \Gamma(i)\boldsymbol{w}(i) \\ \boldsymbol{x}(i+2) &= \Phi(i+2,i+1)\boldsymbol{x}(i+1) + \Gamma(i+1)\boldsymbol{w}(i+1) \\ &\ \ \vdots \\ \boldsymbol{x}(n) &= \Phi(n,n-1)\boldsymbol{x}(n-1) + \Gamma(n-1)\boldsymbol{w}(n-1)\end{aligned}$$

Then

$$\begin{aligned} x(i) &= \Phi^{-1}(i+1,i)\,[x(i+1)-\Gamma(i)w(i)] \\ x(i+1) &= \Phi^{-1}(i+2,i+1)\,[x(i+2)-\Gamma(i+1)w(i+1)] \\ &\vdots \\ x(n-1) &= \Phi^{-1}(n,n-1)\,[x(n)-\Gamma(n-1)w(n-1)]\,. \end{aligned}$$

So

$$\begin{aligned} x(i) &= \Phi(i,i+1)\,\{\Phi(i+1,i+2)\,[x(i+2)-\Gamma(i+1)w(i+1)]-\Gamma(i)w(i)\} \\ &= \Phi(i,i+1)\cdots\Phi(n-1,n)x(n)-\Phi(i,i+1)\,[\Gamma(i)w(i) \\ &\quad +\Phi(i+1,i+2)\Gamma(i+1)w(i+1)+\cdots \\ &\quad +\Phi(i+1,i+2)\cdots\Phi(n-1,n)\Gamma(n-1)w(n-1)] \\ &= \Phi(i,n)x(n)-\Phi(i,n)\sum_{j=i}^{n-1}\Phi(n,j+1)\Gamma(j)w(j). \end{aligned} \tag{C.31}$$

Using (6.30), (6.31), and (C.31), the observation $z(i)$ can be expressed as

$$z(i)=C(i)\Phi(i,n)x(n)+v(i)-C(i)\Phi(i,n)\sum_{j=i}^{n-1}\Phi(n,j+1)\Gamma(j)w(j). \tag{C.32}$$

Let us consider a *suboptimal* linear estimate $\breve{x}(n)$ given by

$$\breve{x}(n)=\mathcal{I}^{-1}(n-N,n)\sum_{i=n-N}^{n}\Phi^{\mathrm{T}}(i,n)C^{\mathrm{T}}(i)R^{-1}(i)z(i),\quad n\geq N. \tag{C.33}$$

Since $\breve{x}(n)$ is suboptimal,

$$\mathrm{Cov}\,[x(n)-\breve{x}(n)]\triangleq \mathrm{E}\left[(x(n)-\breve{x}(n))\,(x(n)-\breve{x}(n))^{\mathrm{T}}\right]\geq P(n). \tag{C.34}$$

We substitute (C.33) into (C.34) and obtain

$$\begin{aligned} &\mathrm{Cov}\,[x(n)-\breve{x}(n)] \\ &\qquad = \mathrm{Cov}\left[x(n)-\mathcal{I}^{-1}(n-N,n)\sum_{i=n-N}^{n}\Phi^{\mathrm{T}}(i,n)C^{\mathrm{T}}(i)R^{-1}(i)z(i)\right]. \end{aligned}$$

Substituting (C.32) for $z(i)$ produces

$$\begin{aligned}
&\operatorname{Cov}[\boldsymbol{x}(n)-\check{\boldsymbol{x}}(n)] \\
&\quad= \operatorname{Cov}\Bigg[\boldsymbol{x}(n)-\mathcal{I}^{-1}(n-N,n)\sum_{i=n-N}^{n}\Phi^{\mathrm{T}}(i,n)C^{\mathrm{T}}(i)R^{-1}(i) \\
&\qquad\times\left(C(i)\Phi(i,n)\boldsymbol{x}(n)+\boldsymbol{v}(i)-C(i)\Phi(i,n)\sum_{j=i}^{n-1}\Phi(n,j+1)\Gamma(j)\boldsymbol{w}(j)\right)\Bigg] \\
&\quad= \operatorname{Cov}\Bigg[\boldsymbol{x}(n)-\mathcal{I}^{-1}(n-N,n)\mathcal{I}(n-N,n)\boldsymbol{x}(n) \\
&\qquad-\mathcal{I}^{-1}(n-N,n)\sum_{i=n-N}^{n}\Phi^{\mathrm{T}}(i,n)C^{\mathrm{T}}(i)R^{-1}(i) \\
&\qquad\times\left(\boldsymbol{v}(i)-C(i)\Phi(i,n)\sum_{j=i}^{n-1}\Phi(n,j+1)\Gamma(j)\boldsymbol{w}(j)\right)\Bigg] \\
&\quad= \operatorname{Cov}\Bigg[-\mathcal{I}^{-1}(n-N,n)\sum_{i=n-N}^{n}\Phi^{\mathrm{T}}(i,n)C^{\mathrm{T}}(i)R^{-1}(i) \\
&\qquad\times\left(\boldsymbol{v}(i)-C(i)\Phi(i,n)\sum_{j=i}^{n-1}\Phi(n,j+1)\Gamma(j)\boldsymbol{w}(j)\right)\Bigg] \\
&\quad= \operatorname{Cov}\Bigg[-\mathcal{I}^{-1}(n-N,n)\sum_{i=n-N}^{n}\Phi^{\mathrm{T}}(i,n)C^{\mathrm{T}}(i)R^{-1}(i)\boldsymbol{v}(i)\Bigg] \\
&\qquad+\operatorname{Cov}\Bigg[\mathcal{I}^{-1}(n-N,n)\sum_{i=n-N}^{n}\Phi^{\mathrm{T}}(i,n)C^{\mathrm{T}}(i)R^{-1}(i)C(i)\Phi(i,n) \\
&\qquad\times\sum_{j=i}^{n-1}\Phi(n,j+1)\Gamma(j)\boldsymbol{w}(j)\Bigg] \qquad\qquad (\mathrm{C.35})
\end{aligned}$$

The first term in (C.35) is

$$\begin{aligned}
&-\operatorname{Cov}\Bigg[\mathcal{I}^{-1}(n-N,n)\sum_{i=n-N}^{n}\Phi^{\mathrm{T}}(i,n)C^{\mathrm{T}}(i)R^{-1}(i)\boldsymbol{v}(i)\Bigg] \\
&\quad=\mathcal{I}^{-1}(n-N,n)\left(\sum_{i=n-N}^{n}\Phi^{\mathrm{T}}(i,n)C^{\mathrm{T}}(i)R^{-1}(i)R(i)R^{-1}(i)C(i)\Phi(i,n)\right) \\
&\qquad\times\mathcal{I}^{-1}(n-N,n) \\
&\quad=\mathcal{I}^{-1}(n-N,n)\mathcal{I}(n-N,n)\mathcal{I}^{-1}(n-N,n) \\
&\quad=\mathcal{I}^{-1}(n-N,n).
\end{aligned}$$

Next we reverse the summations in the second term in (C.35) to obtain

$$\begin{aligned}
&\mathrm{Cov}\left[\boldsymbol{x}(n)-\check{\boldsymbol{x}}(n)\right]\\
&= \mathcal{I}^{-1}(n-N,n)+\mathrm{Cov}\left[\mathcal{I}^{-1}(n-N,n)\sum_{j=n-N}^{n-1}\left(\sum_{i=n-N}^{j}\Phi^{\mathrm{T}}(i,n)C^{\mathrm{T}}(i)\right.\right.\\
&\qquad \left.\left.\times R^{-1}(i)C(i)\Phi(i,n)\right)\Phi(n,j+1)\Gamma(j)\boldsymbol{w}(j)\right]\\
&= \mathcal{I}^{-1}(n-N,n)\\
&\quad +\mathrm{Cov}\left[\mathcal{I}^{-1}(n-N,n)\sum_{j=n-N}^{n-1}\mathcal{I}(n-N,j)\Phi(n,j+1)\Gamma(j)\boldsymbol{w}(j)\right]\\
&= \mathcal{I}^{-1}(n-N,n)+\mathcal{I}^{-1}(n-N,n)\sum_{j=n-N}^{n-1}\mathcal{I}(n-N,j)\Phi(n,j+1)\Gamma(j)Q(j)\\
&\qquad \times\Gamma^{\mathrm{T}}(j)\Phi^{\mathrm{T}}(n,j+1)\mathcal{I}(n-N,j)\mathcal{I}^{-1}(n-N,n). \qquad (C.36)
\end{aligned}$$

Since the system is UCO (C.8),

$$\mathcal{I}^{-1}(n-N,n)\le\frac{1}{\beta_1}I,\quad \forall\, n\ge N,$$

and

$$\mathcal{I}(n-N,j)\le\beta_2 I,\quad \text{for } j\le n,\quad \forall\, n\ge N.$$

The system is also UCC (C.10), so

$$\Phi(n,j+1)\Gamma(j)Q(j)\Gamma^{\mathrm{T}}(j)\Phi^{\mathrm{T}}(n,j+1)\le\alpha_2 I,\quad \text{for } j\le n-1,\quad \forall\, n\ge N.$$

Then (C.36) gives the inequality

$$\begin{aligned}
&\mathrm{Cov}\left[\boldsymbol{x}(n)-\check{\boldsymbol{x}}(n)\right]\\
&\le\left(\frac{1}{\beta_1}+\frac{1}{\beta_1^2}\sum_{j=n-N}^{n-1}\alpha_2\beta_2^2\right)I=\left(\frac{1}{\beta_1}+\frac{1}{\beta_1^2}N\alpha_2\beta_2^2\right)I,\quad n\ge N.
\end{aligned}$$

Finally, from (C.34) we have the desired result,

$$P(n)\le\mathrm{Cov}\left[\boldsymbol{x}(n)-\check{\boldsymbol{x}}(n)\right]\le\frac{\beta_1+N\alpha_2\beta_2^2}{\beta_1^2}I,\quad \forall\, n\ge N.$$

Q.E.D. ■

C.6 A Lower Bound for $P(n)$

Lemma C.3 *Let the system (6.29–6.33), (6.39–6.42) be UCO and UCC, and let $P^-(0)\ge 0$. Then $P(n)$ is uniformly bounded below,*

$$P(n)\ge\frac{\alpha_1^2}{\alpha_1+N\alpha_2^2\beta_2}I,\quad \forall\, n\ge N.$$

Proof. [29, 30, 32, 14] By Lemma C.1 $P(n) > 0$ and $P^-(n) > 0$ for all $n \geq N$. Hence $P(n)$ is invertible, and define the inverse of $P(n)$ by

$$W(n) \triangleq P^{-1}(n). \tag{C.37}$$

By (5.98) $W(n)$ becomes

$$W(n) = (P^-(n))^{-1} + C^{\mathrm{T}}(n)R^{-1}(n)C(n). \tag{C.38}$$

Then define

$$T(n) \triangleq (P^-(n))^{-1} \tag{C.39}$$

$$= W(n) - C^{\mathrm{T}}(n)R^{-1}(n)C(n), \tag{C.40}$$

and

$$T^-(n+1) \triangleq \Phi^{-\mathrm{T}}(n+1,n)W(n)\Phi^{-1}(n+1,n). \tag{C.41}$$

We apply (5.84) to arrive at

$$T(n) = \Big[\Phi(n,n-1)P(n-1)\Phi^{\mathrm{T}}(n,n-1) + \Gamma(n-1)Q(n-1)\Gamma^{\mathrm{T}}(n-1)\Big]^{-1}. \tag{C.42}$$

Note that

$$\begin{aligned}
&\Phi(n,n-1)P(n-1)\Phi^{\mathrm{T}}(n,n-1) \\
&\quad = \Phi(n,n-1)W^{-1}(n-1)\Phi^{\mathrm{T}}(n,n-1) \\
&\quad = \Big[\Phi^{-\mathrm{T}}(n,n-1)W(n-1)\Phi^{-1}(n,n-1)\Big]^{-1} \\
&\quad = (T^-(n))^{-1},
\end{aligned}$$

so (C.42) becomes

$$T(n) = \Big[(T^-(n))^{-1} + \Gamma(n-1)Q(n-1)\Gamma^{\mathrm{T}}(n-1)\Big]^{-1}. \tag{C.43}$$

Using (C.38) and (C.39) in conjunction with (C.41), we find

$$\begin{aligned}
T^-(n+1) &= \Phi^{-\mathrm{T}}(n+1,n)\Big[T(n) + C^{\mathrm{T}}(n)R^{-1}(n)C(n)\Big]\Phi^{-1}(n+1,n) \\
&= \Phi^{-\mathrm{T}}(n+1,n)T(n)\Phi^{-1}(n+1,n) \\
&\quad + \Phi^{-\mathrm{T}}(n+1,n)C^{\mathrm{T}}(n)R^{-1}(n)C(n)\Phi^{-1}(n+1,n).
\end{aligned} \tag{C.44}$$

Notice the similarity between (5.98) and (C.43) and between (5.84) and (C.44). We conclude that $T(n)$ may be interpreted as the estimation error covariance matrix of a related system, which we will denote with an overbar. The bar system is

$$\begin{aligned}
\boldsymbol{x}(n+1) &= \bar{\Phi}(n+1,n)\boldsymbol{x}(n) + \bar{\Gamma}(n)\bar{\boldsymbol{w}}(n) \\
\boldsymbol{z}(n) &= \bar{C}(n)\boldsymbol{x}(n) + \bar{\boldsymbol{v}}(n-1),
\end{aligned}$$

with

$$\bar{\Phi}(n+1,n) = \Phi^{-\mathrm{T}}(n+1,n), \tag{C.45}$$

$$\bar{\Gamma}(n) = \Phi^{-\mathrm{T}}(n+1,n)C^{\mathrm{T}}(n), \tag{C.46}$$

$$\bar{C}(n) = \Gamma^{\mathrm{T}}(n-1), \tag{C.47}$$

$$\mathrm{E}\left[w(n)\right] \equiv 0, \quad \bar{Q}(i) = \mathrm{E}\left[w^{\mathrm{T}}(i)w(j)\right] = R^{-1}(i)\delta(i-j),$$
$$\mathrm{E}\left[v(n)\right] \equiv 0, \quad \bar{R}(i) = \mathrm{E}\left[v^{\mathrm{T}}(i)v(j)\right] = Q^{-1}(i)\delta(i-j),$$

and

$$\mathrm{E}\left[w^{\mathrm{T}}(i)v(j)\right] \equiv 0.$$

Also, $\bar{\Phi}^{\mathrm{T}}(n+1,n)$ is nonsingular for all n, and

$$\bar{\Phi}^{\mathrm{T}}(n+1,n)\bar{\Phi}(n+1,n) \geq b_{\Phi} I > 0,$$

and the noise covariances are positive-definite and bounded below.

The bar system is UCO and UCC. To see that it is UCO, we plug (C.45) and (C.47) into Definition (C.4), which gives

$$\begin{aligned}
&\bar{\mathcal{I}}(n-N+1,n)\\
&= \sum_{i=n-N+1}^{n} \left(\Phi^{-\mathrm{T}}(i,n)\right)^{\mathrm{T}} \left(\Gamma^{\mathrm{T}}(i-1)\right)^{\mathrm{T}} \left(Q^{-1}(i-1)\right)^{-1} \Gamma^{\mathrm{T}}(i-1)\Phi^{-\mathrm{T}}(i,n)\\
&= \sum_{i=n-N+1}^{n} \Phi(n,i)\Gamma(i-1)Q(i-1)\Gamma^{\mathrm{T}}(i-1)\Phi^{-\mathrm{T}}(i,n).
\end{aligned}$$

Letting $j = i-1$, we have

$$\begin{aligned}
\bar{\mathcal{I}}(n-N+1,n) &= \sum_{j=n-N}^{n-1} \Phi(n,j+1)\Gamma(j)Q(j)\Gamma^{\mathrm{T}}(j)\Phi^{\mathrm{T}}(n,j+1)\\
&= \mathcal{C}(n-N,n).
\end{aligned}$$

Since the original system is UCC (C.10),

$$\alpha_1 I \leq \bar{\mathcal{I}}(n-N+1,n) \leq \alpha_2 I, \quad \forall\, n \geq N, \tag{C.48}$$

so the bar system is UCO.

In a similar manner, we can show that the bar system is UCC. Substituting (C.45) and (C.46) into (C.9), we get

$$\begin{aligned}
&\bar{\mathcal{C}}(n-N,n+1)\\
&= \sum_{i=n-N}^{n} \Phi^{\mathrm{T}}(i+1,n)\Phi^{\mathrm{T}}(i,i+1)C^{\mathrm{T}}(i)R^{-1}(i)C(i)\Phi(i,i+1)\Phi(i+1,n)\\
&= \sum_{i=n-N}^{n-1} \Phi^{\mathrm{T}}(i,n)C^{\mathrm{T}}(i)R^{-1}(i)C(i)\Phi(i,n)\\
&= \mathcal{I}(n-N,n).
\end{aligned}$$

The original system is UCO (C.8), so

$$\beta_1 I \leq \bar{C}(n-N, n+1) \leq \beta_2 I, \quad \forall\, n \geq N, \tag{C.49}$$

making the bar system UCC.

Now we can follow the same procedure as in the proof of Lemma C.2. We choose a suboptimal estimate in the bar system,

$$\check{\boldsymbol{x}}(n) = \bar{\mathcal{I}}^{-1}(n-N+1, n) \sum_{i=n-N+1}^{n} \bar{\Phi}^{\mathrm{T}}(i,n)\bar{C}^{\mathrm{T}}(i)\bar{R}^{-1}(i)\bar{C}(i)\bar{\Phi}^{\mathrm{T}}(i,n)\boldsymbol{z}(i), \quad n \geq N.$$

By the same method as the derivation of (C.36), we find

$$\begin{aligned}
&\mathrm{Cov}\,[\boldsymbol{x}(n) - \check{\boldsymbol{x}}(n)] \\
&\quad = \bar{\mathcal{I}}^{-1}(n-N+1, n) + \bar{\mathcal{I}}^{-1}(n-N+1, n) \sum_{j=n-N+1}^{n-1} \bar{\mathcal{I}}(n-N+1, j) \\
&\qquad \times \Phi^{\mathrm{T}}(j,n)C^{\mathrm{T}}(j)R^{-1}(j)C(j)\Phi(j,n)\bar{\mathcal{I}}(n-N+1, j)\bar{\mathcal{I}}^{-1}(n-N+1, n).
\end{aligned}$$

Being suboptimal, $\check{\boldsymbol{x}}(n)$ satisfies

$$\mathrm{Cov}\,[\boldsymbol{x}(n) - \check{\boldsymbol{x}}(n)] \geq T(n).$$

Then from (C.40),

$$\begin{aligned}
W(n) &= T(n) + C^{\mathrm{T}}(n)R^{-1}(n)C(n) \\
&\leq \mathrm{Cov}\,[\boldsymbol{x}(n) - \check{\boldsymbol{x}}(n)] + C^{\mathrm{T}}(n)R^{-1}(n)C(n) \\
&= \bar{\mathcal{I}}^{-1}(n-N+1, n) + \bar{\mathcal{I}}^{-1}(n-N+1, n) \sum_{j=n-N+1}^{n} \bar{\mathcal{I}}(n-N+1, j) \\
&\quad \times \Phi^{\mathrm{T}}(j,n)C^{\mathrm{T}}(j)R^{-1}(j)C(j)\Phi(j,n)\bar{\mathcal{I}}(n-N+1, j) \\
&\quad \times \bar{\mathcal{I}}^{-1}(n-N+1, n).
\end{aligned}$$

From (C.48) and (C.49),

$$P^{-1}(n) = W(n) \leq \frac{1}{\alpha_1} I + \frac{1}{\alpha_1^2} \sum_{j=n-N+1}^{n} \alpha_2^2 \beta_2 I, \quad \forall\, n \geq N,$$

so

$$P(n) \geq \frac{\alpha_1^2}{\alpha_1 + N\alpha_2^2\beta_2} I, \quad \forall\, n \geq N.$$

Q.E.D. ■

C.7 A Useful Control Lemma

Lemma C.4 *Suppose the linear system*

$$x(n) = \Phi(n, n-1)x(n-1) + u(n)$$

with initial condition

$$x_0 = x(n-N-1)$$

is UCO. Let $u^*(n)$ denote the finite-duration input sequence $\{u(n-N), \ldots, u(n)\}$ that minimizes the cost function

$$J = \sum_{i=n-N}^{n} \left[x^T(i)C^T(i)R^{-1}(i)C(i)x(i) + u^T(i)(P^-(i))^{-1}u(i)\right]. \tag{C.50}$$

Then the minimum cost J^* is bounded below; there are real numbers β_3, $\beta_5 > 0$ such that

$$J^* \geq \frac{\beta_3^2}{\beta_5}||x(n-N-1)||^2. \tag{C.51}$$

Proof. [29] To find $u^*(n)$, we define the following vectors:

$$X \triangleq \begin{bmatrix} x(n) \\ x(n-1) \\ \vdots \\ x(n-N) \end{bmatrix}, \quad \text{and} \quad U \triangleq \begin{bmatrix} u(n) \\ u(n-1) \\ \vdots \\ u(n-N) \end{bmatrix},$$

and the matrices:

$$M \triangleq \begin{bmatrix} R(n) & & & 0 \\ & R(n-1) & & \\ & & \ddots & \\ 0 & & & R(n-N) \end{bmatrix},$$

$$B \triangleq \begin{bmatrix} P^-(n) & & & 0 \\ & P^-(n-1) & & \\ & & \ddots & \\ 0 & & & P^-(n-N) \end{bmatrix},$$

$$L \triangleq \begin{bmatrix} C(n) & & & 0 \\ & C(n-1) & & \\ & & \ddots & \\ 0 & & & C(n-N) \end{bmatrix},$$

$$C \triangleq \begin{bmatrix} \Phi(n,n-N-1) \\ \Phi(n-1,n-N-1) \\ \vdots \\ \Phi(n-N,n-N-1) \end{bmatrix},$$

and

$$D \triangleq \begin{bmatrix} I & \Phi(n,n-1) & \Phi(n,n-2) & \cdots & \Phi(n,n-N) \\ & I & \Phi(n-1,n-2) & \cdots & \Phi(n-1,n-N) \\ & & I & \cdots & \Phi(n-2,n-N) \\ & & & \ddots & \vdots \\ 0 & & & & I \end{bmatrix}.$$

It is easily verified that J may be written as

$$J = X^{\mathrm{T}} L^{\mathrm{T}} M^{-1} L X + U^{\mathrm{T}} B^{-1} U,$$

and the output may be written as

$$X = C x_0 + DU.$$

Hence,

$$J = \left(U^{\mathrm{T}} D^{\mathrm{T}} + x_0^{\mathrm{T}} C^{\mathrm{T}}\right) L^{\mathrm{T}} M^{-1} L \left(C x_0 + DU\right) + U^{\mathrm{T}} B^{-1} U. \tag{C.52}$$

Let U^* represent the vector form of $u^*(n)$. To find U^* we set

$$\frac{\partial J}{\partial U} = 2\left(U^{\mathrm{T}} D^{\mathrm{T}} + x_0^{\mathrm{T}} C^{\mathrm{T}}\right) L^{\mathrm{T}} M^{-1} L D + 2U^{\mathrm{T}} B^{-1} = 0, \tag{C.53}$$

and

$$\frac{\partial^2 J}{\partial U^2} = 2D^{\mathrm{T}} L^{\mathrm{T}} M^{-1} L D + 2B^{-1} > 0. \tag{C.54}$$

Solving (C.53) for U, we find the desired control U^*:

$$U^* = -\left(D^{\mathrm{T}} L^{\mathrm{T}} M^{-1} L D + B^{-1}\right)^{-1} D^{\mathrm{T}} L^{\mathrm{T}} M^{-1} L C x_0. \tag{C.55}$$

We observe that (C.54) is always satisfied.

Substituting (C.55) into (C.52), we find that the minimum cost J^* is

$$J^* = x_0^{\mathrm{T}} C^{\mathrm{T}} L^{\mathrm{T}} \left(M + LDBD^{\mathrm{T}} L^{\mathrm{T}}\right)^{-1} L C x_0. \tag{C.56}$$

Let us investigate the sign of J^*. We have

$$\begin{aligned}\Phi^{\mathrm{T}}(n-N-1,n)C^{\mathrm{T}}L^{\mathrm{T}}M^{-1}LC\Phi(n-N-1,n)\\ &= \sum_{i=n-N}^{n} \Phi^{\mathrm{T}}(i,n)C^{\mathrm{T}}(i)R^{-1}(i)C(i)\Phi(i,n)\\ &= \mathcal{I}(n-N,n).\end{aligned}$$

Since the system is UCO (C.8), it follows that

$$\beta_1 I \leq \Phi^{\mathrm{T}}(n-N-1,n)C^{\mathrm{T}}L^{\mathrm{T}}M^{-1}LC\Phi(n-N-1,n) \leq \beta_2 I.$$

Because $R(n)$ is bounded below, $R^{-1}(n)$ is bounded above, and there exists a real number $0 < \lambda_1 < \infty$ such that $M^{-1} \leq \lambda_1 I$. Also, there exists $0 < \lambda_2 < \infty$ such that $0 < \lambda_2 \Phi^{\mathrm{T}}(n-N-1,n)\Phi(n-N-1,n) \leq \beta_1 I$. Then we have

$$\begin{aligned}&\lambda_2 \Phi^{\mathrm{T}}(n-N-1,n)\Phi(n-N-1,n)\\ &\quad\leq \beta_1 I\\ &\quad\leq \Phi^{\mathrm{T}}(n-N-1,n)C^{\mathrm{T}}L^{\mathrm{T}}M^{-1}LC\Phi(n-N-1,n)\\ &\quad\leq \Phi^{\mathrm{T}}(n-N-1,n)C^{\mathrm{T}}L^{\mathrm{T}}\left(\lambda_1 I\right)LC\Phi(n-N-1,n).\end{aligned}$$

We multiply by $x^{\mathrm{T}}(n)$ and $x(n)$ to obtain

$$\begin{aligned}&\lambda_2\, x^{\mathrm{T}}(n)\Phi^{\mathrm{T}}(n-N-1,n)\Phi(n-N-1,n)x(n)\\ &\qquad\leq \lambda_1\, x^{\mathrm{T}}(n)\Phi^{\mathrm{T}}(n-N-1,n)C^{\mathrm{T}}L^{\mathrm{T}}LC\Phi(n-N-1,n)x(n),\end{aligned}$$

which is equivalent to

$$\lambda_2\|x_0\|^2 \leq \lambda_1 x_0^{\mathrm{T}}C^{\mathrm{T}}L^{\mathrm{T}}LCx_0.$$

Let $\beta_3 = \sqrt{\lambda_2/\lambda_1}$, so $0 < \beta_3 < \infty$ and

$$\beta_3\|x_0\| \leq \|LCx_0\|. \tag{C.57}$$

Next, we recall that $R(n)$, $\Phi(n,n-1)$, $P^-(n)$, and $C(n)$ are bounded above in norm. Then there exist real numbers $0 < \beta_4 < \beta_5 < \infty$ such that

$$\beta_4 I \leq M + LDBD^{\mathrm{T}}L^{\mathrm{T}} \leq \beta_5 I,$$

so

$$\beta_4^{-1}I \geq \left(M + LDBD^{\mathrm{T}}L^{\mathrm{T}}\right)^{-1} \geq \beta_5^{-1}I. \tag{C.58}$$

Then we combine (C.56), (C.57), and (C.58) to obtain

$$J^* \geq x_0^{\mathrm{T}}C^{\mathrm{T}}L^{\mathrm{T}}\left(\beta_5^{-1}I\right)LCx_0 = \beta_5^{-1}\|LCx_0\|^2 \geq \beta_5^{-1}\beta_3^2\|x_0\|^2.$$

Hence,

$$J^* \geq \frac{\beta_3^2}{\beta_5}\|x(n-N-1)\|^2.$$

Therefore, provided that the initial state $x(n-N-1) \neq 0$, the minimum cost J^* is positive. **Q.E.D.** ■

C.8 A Kalman Filter Stability Theorem

We are now prepared to consider the stability of the Kalman filter. The homogeneous or unforced part of (5.99) is

$$\hat{\boldsymbol{x}}(n+1) = P(n+1)(P^-(n+1))^{-1}\Phi(n+1,n)\hat{\boldsymbol{x}}(n).$$

It is this part of the filter that determines internal stability, so the state-transition matrix is

$$\Psi(n+1,n) = P(n+1)(P^-(n+1))^{-1}\Phi(n+1,n).$$

Theorem C.2 *Let the system (6.29–6.33), (6.39–6.42) be UCO and UCC, and let $P^-(0) \geq 0$. Then the Kalman filter is GUAS.*

Proof. [29, 14] From Lemma C.1 $P(n) > 0$ for all $n \geq N$. Define a Lyapunov function for the Kalman filter by

$$V(\hat{\boldsymbol{x}}(n),n) \triangleq \hat{\boldsymbol{x}}^{\mathrm{T}}(n)P^{-1}(n)\hat{\boldsymbol{x}}(n), \tag{C.59}$$

and define the following bounding functions on V:

$$\gamma_1(\|\hat{\boldsymbol{x}}(n)\|) \triangleq \frac{\beta_1^2}{\beta_1 + N\alpha_2\beta_2^2}\|\hat{\boldsymbol{x}}(n)\|^2, \tag{C.60}$$

$$\gamma_2(\|\hat{\boldsymbol{x}}(n)\|) \triangleq \frac{\alpha_1 + N\alpha_2^2\beta_2}{\alpha_1^2}\|\hat{\boldsymbol{x}}(n)\|^2. \tag{C.61}$$

From Lemma C.2 we have

$$V(\hat{\boldsymbol{x}}(n),n) \geq \hat{\boldsymbol{x}}^{\mathrm{T}}(n)\left(\frac{\beta_1 + N\alpha_2\beta_2^2}{\beta_1^2}\right)^{-1}\hat{\boldsymbol{x}}(n) = \frac{\beta_1^2}{\beta_1 + N\alpha_2\beta_2^2}\|\hat{\boldsymbol{x}}(n)\|^2,$$

so that

$$V(\hat{\boldsymbol{x}}(n),n) \geq \gamma_1(\|\hat{\boldsymbol{x}}(n)\|), \quad \forall\, n \geq N, \tag{C.62}$$

and clearly

$$\gamma_1(0) = 0, \quad \text{and} \quad \lim_{a\to\infty}\gamma_1(a) = \infty. \tag{C.63}$$

Also, from Lemma C.3 we have

$$V(\hat{\boldsymbol{x}}(n),n) \leq \hat{\boldsymbol{x}}^{\mathrm{T}}(n)\left(\frac{\alpha_1 + N\alpha_2^2\beta_2}{\alpha_1^2}\right)\hat{\boldsymbol{x}}(n) = \frac{\alpha_1 + N\alpha_2^2\beta_2}{\alpha_1^2}\|\hat{\boldsymbol{x}}(n)\|^2$$

so that

$$V(\hat{\boldsymbol{x}}(n),n) \leq \gamma_2(\|\hat{\boldsymbol{x}}(n)\|), \quad \forall\, n \geq N, \tag{C.64}$$

and

$$\gamma_2(0) = 0. \tag{C.65}$$

We see that Equations (C.59–C.65) satisfy Requirements (C.16) and (C.18) of Theorem C.1. However, we must still find γ_3 and satisfy (C.17).

Let us write the states as the sum of two equations:

$$\hat{\boldsymbol{x}}(n) = \hat{\boldsymbol{x}}'(n) + \boldsymbol{u}(n), \tag{C.66}$$

where

$$\hat{\boldsymbol{x}}'(n) = \Phi(n, n-1)\hat{\boldsymbol{x}}(n-1) \tag{C.67}$$

$$\boldsymbol{u}(n) = \left[P(n)(P^{-}(n))^{-1} - I\right]\hat{\boldsymbol{x}}'(n). \tag{C.68}$$

It is easily verified that (C.66–C.68) do form the homogeneous part of the Kalman filter equation (5.99). Then it follows that

$$\hat{\boldsymbol{x}}(n) = P(n)(P^{-}(n))^{-1}\hat{\boldsymbol{x}}'(n) = P(n)(P^{-}(n))^{-1}\Phi(n, n-1)\hat{\boldsymbol{x}}(n-1). \tag{C.69}$$

The Lyapunov function becomes

$$\begin{aligned}
&V(\hat{\boldsymbol{x}}(n), n)\\
&\quad = \hat{\boldsymbol{x}}^{\mathrm{T}}(n)P^{-1}(n)\hat{\boldsymbol{x}}(n)\\
&\quad = \hat{\boldsymbol{x}}^{\mathrm{T}}(n)\left[2P^{-1}(n) - P^{-1}(n)\right]\hat{\boldsymbol{x}}(n)\\
&\quad = \hat{\boldsymbol{x}}^{\mathrm{T}}(n)\left[2P^{-1}(n) - (P^{-}(n))^{-1} - C^{\mathrm{T}}(n)R^{-1}(n)C(n)\right]\hat{\boldsymbol{x}}(n)\\
&\qquad + \hat{\boldsymbol{x}}'^{\mathrm{T}}(n)(P^{-}(n))^{-1}\hat{\boldsymbol{x}}'(n) - \hat{\boldsymbol{x}}'^{\mathrm{T}}(n)(P^{-}(n))^{-1}\hat{\boldsymbol{x}}'(n)\\
&\quad = \hat{\boldsymbol{x}}'^{\mathrm{T}}(n)(P^{-}(n))^{-1}\hat{\boldsymbol{x}}'(n) - \hat{\boldsymbol{x}}^{\mathrm{T}}(n)C^{\mathrm{T}}(n)R^{-1}(n)C(n)\hat{\boldsymbol{x}}(n)\\
&\qquad - \hat{\boldsymbol{x}}'^{\mathrm{T}}(n)(P^{-}(n))^{-1}\hat{\boldsymbol{x}}'(n) + 2\hat{\boldsymbol{x}}^{\mathrm{T}}(n)P^{-1}(n)\hat{\boldsymbol{x}}(n)\\
&\qquad - \hat{\boldsymbol{x}}^{\mathrm{T}}(n)(P^{-}(n))^{-1}\hat{\boldsymbol{x}}(n)\\
&\quad = \hat{\boldsymbol{x}}'^{\mathrm{T}}(n)(P^{-}(n))^{-1}\hat{\boldsymbol{x}}'(n) - \hat{\boldsymbol{x}}^{\mathrm{T}}(n)C^{\mathrm{T}}(n)R^{-1}(n)C(n)\hat{\boldsymbol{x}}(n)\\
&\qquad - \hat{\boldsymbol{x}}'^{\mathrm{T}}(n)(P^{-}(n))^{-1}\hat{\boldsymbol{x}}'(n) + \hat{\boldsymbol{x}}^{\mathrm{T}}(n)P^{-1}(n)P^{-}(n)(P^{-}(n))^{-1}\hat{\boldsymbol{x}}(n)\\
&\qquad + \hat{\boldsymbol{x}}^{\mathrm{T}}(n)(P^{-}(n))^{-1}P^{-}(n)P^{-1}(n)\hat{\boldsymbol{x}}(n) - \hat{\boldsymbol{x}}^{\mathrm{T}}(n)(P^{-}(n))^{-1}\hat{\boldsymbol{x}}(n)\\
&\quad = \hat{\boldsymbol{x}}'^{\mathrm{T}}(n)(P^{-}(n))^{-1}\hat{\boldsymbol{x}}'(n) - \hat{\boldsymbol{x}}^{\mathrm{T}}(n)C^{\mathrm{T}}(n)R^{-1}(n)C(n)\hat{\boldsymbol{x}}(n)\\
&\qquad - \hat{\boldsymbol{x}}'^{\mathrm{T}}(n)(P^{-}(n))^{-1}\hat{\boldsymbol{x}}'(n) + \hat{\boldsymbol{x}}'^{\mathrm{T}}(n)(P^{-}(n))^{-1}\hat{\boldsymbol{x}}(n)\\
&\qquad + \hat{\boldsymbol{x}}^{\mathrm{T}}(n)(P^{-}(n))^{-1}\hat{\boldsymbol{x}}'(n) - \hat{\boldsymbol{x}}^{\mathrm{T}}(n)(P^{-}(n))^{-1}\hat{\boldsymbol{x}}(n)\\
&\quad = \hat{\boldsymbol{x}}'^{\mathrm{T}}(n)(P^{-}(n))^{-1}\hat{\boldsymbol{x}}'(n) - \hat{\boldsymbol{x}}^{\mathrm{T}}(n)C^{\mathrm{T}}(n)R^{-1}(n)C(n)\hat{\boldsymbol{x}}(n)\\
&\qquad - \left[\hat{\boldsymbol{x}}(n) - \hat{\boldsymbol{x}}'(n)\right]^{\mathrm{T}}(P^{-}(n))^{-1}\left[\hat{\boldsymbol{x}}(n) - \hat{\boldsymbol{x}}'(n)\right]\\
&\quad = \hat{\boldsymbol{x}}'^{\mathrm{T}}(n)\Big[\Phi(n, n-1)P(n-1)\Phi^{\mathrm{T}}(n, n-1)\\
&\qquad + \Gamma(n-1)Q(n-1)\Gamma^{\mathrm{T}}(n-1)\Big]^{-1}\hat{\boldsymbol{x}}'(n)\\
&\qquad - \hat{\boldsymbol{x}}^{\mathrm{T}}(n)C^{\mathrm{T}}(n)R^{-1}(n)C(n)\hat{\boldsymbol{x}}(n) - \boldsymbol{u}^{\mathrm{T}}(n)(P^{-}(n))^{-1}\boldsymbol{u}(n).
\end{aligned} \tag{C.70}$$

Let

$$F(n) = \hat{\boldsymbol{x}}^{\mathrm{T}}(n)C^{\mathrm{T}}(n)R^{-1}(n)C(n)\hat{\boldsymbol{x}}(n) + \boldsymbol{u}^{\mathrm{T}}(n)(P^{-}(n))^{-1}\boldsymbol{u}(n), \tag{C.71}$$

and note that $F(n) > 0$. Then apply (C.67) to (C.70) so that the Lyapunov function becomes

$$\begin{aligned}
&V(\hat{\boldsymbol{x}}(n), n) \\
&\quad = \hat{\boldsymbol{x}}^{\mathrm{T}}(n-1)\Phi^{\mathrm{T}}(n,n-1)\Big\{\Phi(n,n-1)P(n-1)\Phi^{\mathrm{T}}(n,n-1) \\
&\qquad +\Gamma(n-1)Q(n-1)\Gamma^{\mathrm{T}}(n-1)\Big\}^{-1}\Phi(n,n-1)\hat{\boldsymbol{x}}(n-1) - F(n) \\
&\quad = \hat{\boldsymbol{x}}^{\mathrm{T}}(n-1)\Phi^{\mathrm{T}}(n,n-1)\Big\{\Phi(n,n-1)\Big[P(n-1) + \Phi(n-1,n)\Gamma(n-1) \\
&\qquad \times Q(n-1)\Gamma^{\mathrm{T}}(n-1)\Phi^{\mathrm{T}}(n-1,n)\Big]\Phi^{\mathrm{T}}(n,n-1)\Big\}^{-1}\Phi(n,n-1)\hat{\boldsymbol{x}}(n-1) \\
&\qquad - F(n) \\
&\quad = \hat{\boldsymbol{x}}^{\mathrm{T}}(n-1)\Big\{P(n-1) + \Phi(n-1,n)\Gamma(n-1)Q(n-1)\Gamma^{\mathrm{T}}(n-1) \\
&\qquad \times\Phi^{\mathrm{T}}(n-1,n)\Big\}^{-1}\hat{\boldsymbol{x}}(n-1) - F(n) \\
&\quad \le \hat{\boldsymbol{x}}^{\mathrm{T}}(n-1)P^{-1}(n-1)\hat{\boldsymbol{x}}(n-1) - F(n).
\end{aligned}$$

Therefore, by (C.59) and (C.71) we have

$$\begin{aligned}
&V(\hat{\boldsymbol{x}}(n), n) - V(\hat{\boldsymbol{x}}(n-1), n-1) \\
&\qquad \le -\hat{\boldsymbol{x}}^{\mathrm{T}}(n)C^{\mathrm{T}}(n)R^{-1}(n)C(n)\hat{\boldsymbol{x}}(n) - \boldsymbol{u}^{\mathrm{T}}(n)(P^{-}(n))^{-1}\boldsymbol{u}(n). \qquad \text{(C.72)}
\end{aligned}$$

Summing from $n-N$ to n, we obtain from (C.72) the inequality

$$\begin{aligned}
&V(\hat{\boldsymbol{x}}(n), n) - V(\hat{\boldsymbol{x}}(n-N), n-N) \\
&\le -\sum_{i=n-N}^{n}\Big[\hat{\boldsymbol{x}}^{\mathrm{T}}(i)C^{\mathrm{T}}(i)R^{-1}(i)C(i)\hat{\boldsymbol{x}}(i) + \boldsymbol{u}^{\mathrm{T}}(i)(P^{-}(i))^{-1}\boldsymbol{u}(i)\Big].
\end{aligned}$$

We substitute $x(i) = \hat{\boldsymbol{x}}(i)$ and $u(i) = \boldsymbol{u}(i)$ into Lemma C.4 and recall the cost function J (C.50) and the minimum cost J^* (C.51). Then

$$\begin{aligned}
V(\hat{\boldsymbol{x}}(n), n) - V(\hat{\boldsymbol{x}}(n-N), n-N) &\le -J \le -J^* \\
&\le -\frac{\beta_3^2}{\beta_5}\|\hat{\boldsymbol{x}}(n-N-1)\|^2. \qquad \text{(C.73)}
\end{aligned}$$

Let us define a matrix transformation θ by

$$\begin{aligned}
\theta(n, n-N-1) \triangleq &\left[P(n)(P^{-}(n))^{-1}\Phi(n,n-1)\right] \\
&\times\left[P(n-1)(P^{-}(n-1))^{-1}\Phi(n-1,n-2)\right]\times\cdots \\
&\times\left[P(n-N)(P^{-}(n-N))^{-1}\Phi(n-N,n-N-1)\right]. \qquad \text{(C.74)}
\end{aligned}$$

Note that all matrices composing $\theta(n, n-N-1)$ are nonsingular and bounded below in norm, so θ is nonsingular and bounded below in norm. From (C.69) and (C.74) we find

$$\hat{\boldsymbol{x}}(n-N-1) = \theta^{-1}(n, n-N-1)\hat{\boldsymbol{x}}(n).$$

Since $\theta^{-1}(n, n-N-1)$ is nonsingular, there exists a real number $0 < \beta_6 < \infty$ such that $\beta_6 I \leq \theta^{-1}(n, n-N-1)$. Therefore

$$\begin{aligned} \|\hat{\boldsymbol{x}}(n-N-1)\| &= \|\theta^{-1}(n, n-N-1)\hat{\boldsymbol{x}}(n)\| \\ &\geq \|\beta_6 I \hat{\boldsymbol{x}}(n)\| = \beta_6 \|\hat{\boldsymbol{x}}(n)\|. \end{aligned} \tag{C.75}$$

We define

$$\gamma_3(\|\hat{\boldsymbol{x}}(n)\|) \triangleq -\frac{\beta_3^2 \beta_6^2}{\beta_5} \|\hat{\boldsymbol{x}}(n)\|^2, \tag{C.76}$$

and apply (C.75) to (C.73), which gives

$$V(\hat{\boldsymbol{x}}(n), n) - V(\hat{\boldsymbol{x}}(n-N), n-N) \leq \gamma_3(\|\hat{\boldsymbol{x}}(n)\|) < 0, \quad \hat{\boldsymbol{x}}(n) \neq 0. \tag{C.77}$$

Equations (C.76) and (C.77) satisfy Requirement (C.17) of Theorem C.1. Therefore the Kalman filter is globally uniformly asymptotically stable. **Q.E.D.** ■

C.9 Bounds for $P(n)$

Theorem C.3 *Let the system (6.29–6.33), (6.39–6.42) be UCO and UCC, and let $P^-(0) \geq 0$. Then $P(n)$ is uniformly bounded,*

$$\frac{\alpha_1^2}{\alpha_1 + N\alpha_2^2\beta_2} I \leq P(n) \leq \frac{\beta_1 + N\alpha_2\beta_2^2}{\beta_1^2} I, \quad \forall\, n \geq N.$$

Proof. From the hypotheses, Lemmas C.2 and C.3 are satisfied. The stated result follows immediately. **Q.E.D.** ■

C.10 Independence of $P^-(n)$

Theorem C.4 *Let the system (6.29–6.33), (6.39–6.42) be UCO and UCC. Let $P_1^-(n)$ and $P_2^-(n)$ correspond to solutions of (5.84) with initial conditions $P_1^-(0) \geq 0$ and $P_2^-(0) \geq 0$, respectively. Let $\Delta P^-(n) = P_1^-(n) - P_2^-(n)$. Then there exist real numbers $a, b > 0$ such that*

$$\lim_{n\to\infty} \|\Delta P^-(n)\| \leq \lim_{n\to\infty} a^2 e^{-2bn} \|P_1^-(0) - P_2^-(0)\| = 0.$$

Proof. [14] We begin with

$$\begin{aligned} P_1^-(n+1) = \Phi(n+1, n)\,[I - K_1(n)C(n)]\,P_1^-(n)\Phi^{\mathrm{T}}(n+1, n) \\ + \Gamma(n)Q(n)\Gamma^{\mathrm{T}}(n), \end{aligned}$$

and

$$P_2^-(n+1) = \Phi(n+1,n)P_2^-(n)\left[I - K_2(n)C(n)\right]^{\mathrm{T}}\Phi^{\mathrm{T}}(n+1,n) \\ + \Gamma(n)Q(n)\Gamma^{\mathrm{T}}(n),$$

where $K_1(n)$ and $K_2(n)$ are the corresponding Kalman gains from (5.80).

Then the difference between $P_1^-(n+1)$ and $P_2^-(n+1)$ is

$$\begin{aligned}
&\Delta P^-(n+1) \\
&\quad = P_1^-(n+1) - P_2^-(n+1) \\
&\quad = \Phi(n+1,n)\left[I - K_1(n)C(n)\right]P_1^-(n)\Phi^{\mathrm{T}}(n+1,n) \\
&\qquad - \Phi(n+1,n)P_2^-(n)\left[I - K_2(n)C(n)\right]^{\mathrm{T}}\Phi^{\mathrm{T}}(n+1,n) \\
&\qquad + \Gamma(n)Q(n)\Gamma^{\mathrm{T}}(n) - \Gamma(n)Q(n)\Gamma^{\mathrm{T}}(n) \\
&\quad = \Phi(n+1,n)\left[I - K_1(n)C(n)\right]\left[P_1^-(n) - P_2^-(n)\right] \\
&\qquad \times \left[I - K_2(n)C(n)\right]^{\mathrm{T}}\Phi^{\mathrm{T}}(n+1,n) \\
&\qquad + \Phi(n+1,n)\left[I - K_1(n)C(n)\right]P_1^-(n)C^{\mathrm{T}}(n)K_2^{\mathrm{T}}(n)\Phi^{\mathrm{T}}(n+1,n) \\
&\qquad - \Phi(n+1,n)K_1(n)C(n)P_2^-(n)\left[I - K_2(n)C(n)\right]^{\mathrm{T}}\Phi^{\mathrm{T}}(n+1,n). \\
&\quad = \Phi(n+1,n)\left[I - K_1(n)C(n)\right]\Delta P^-(n) \\
&\qquad \times \left[I - K_2(n)C(n)\right]^{\mathrm{T}}\Phi^{\mathrm{T}}(n+1,n) + A(n),
\end{aligned} \tag{C.78}$$

where we have chosen

$$A(n) = \Phi(n+1,n)\left[I - K_1(n)C(n)\right]P_1^-(n)C^{\mathrm{T}}(n)K_2^{\mathrm{T}}(n)\Phi^{\mathrm{T}}(n+1,n) \\ - \Phi(n+1,n)K_1(n)C(n)P_2^-(n)\left[I - K_2(n)C(n)\right]^{\mathrm{T}}\Phi^{\mathrm{T}}(n+1,n).$$

To simplify the calculations, let

$$D_1(n) = \left[C(n)P_1^-(n)C^{\mathrm{T}}(n) + R(n)\right]^{-1},$$

and

$$D_2(n) = \left[C(n)P_2^-(n)C^{\mathrm{T}}(n) + R(n)\right]^{-1},$$

and drop the explicit dependence on n.

When we substitute for $K_1(n)$ and $K_2(n)$ we find

$$\begin{aligned}
A(n) &= \Phi\left(I - P_1^-C^{\mathrm{T}}D_1C\right)P_1^-C^{\mathrm{T}}D_2CP_2^-\Phi^{\mathrm{T}} \\
&\quad - \Phi P_1^-C^{\mathrm{T}}D_1CP_2^-\left(I - P_2^-C^{\mathrm{T}}D_2C\right)^{\mathrm{T}}\Phi^{\mathrm{T}} \\
&= \Phi P_1^-C^{\mathrm{T}}D_2CP_2^-\Phi^{\mathrm{T}} - \Phi P_1^-C^{\mathrm{T}}D_1CP_1^-C^{\mathrm{T}}D_2CP_2^-\Phi^{\mathrm{T}} \\
&\quad - \Phi P_1^-C^{\mathrm{T}}D_1CP_2^-\Phi^{\mathrm{T}} + \Phi P_1^-C^{\mathrm{T}}D_1CP_2^-C^{\mathrm{T}}D_2CP_2^-\Phi^{\mathrm{T}}.
\end{aligned}$$

Next we let

$$B(n) = D_2 - D_1 C P_1^- C^T D_2 - D_1 + D_1 C P_2^- C^T D_2,$$

so that

$$A(n) = \Phi P_1^- C^T B(n) C P_2^- \Phi^T.$$

Then we write

$$\begin{aligned} B(n) &= D_2 - D_1 C P_1^- C^T D_2 - D_1 + D_1 C P_2^- C^T D_2 \\ &\quad - D_1 R D_2 + D_1 R D_2. \\ &= D_2 - D_1 D_1^{-1} D_2 - D_1 + D_1 D_1^{-1} D_2 \\ &= 0. \end{aligned}$$

Hence, $A(n) = 0$ and (C.78) reduces to

$$\begin{aligned} \Delta P^-(n+1) = \Phi(n+1, n)\, [I - K_1(n) C(n)]\, \Delta P^-(n) \\ \times [I - K_2(n) C(n)]^T\, \Phi^T(n+1, n). \end{aligned} \quad \text{(C.79)}$$

Denote the filter state-transition matrices by

$$\Psi_i(n, n-1) = \Phi(n, n-1)\, [I - K_i(n) C(n)]$$

for $i = 1, 2$. Then (C.79) becomes

$$\Delta P^-(n) = \Psi_1(n, 0) \left[P_1^-(0) - P_2^-(0)\right] \Psi_2^T(n, 0). \quad \text{(C.80)}$$

From Theorem C.2 the *a posteriori* ($\hat{x}(n)$) system of the Kalman filter is globally uniformly asymptotically stable. However, this result clearly applies to the *a priori* ($\hat{x}^-(n)$) system as well. We apply the property (C.15) of Definition C.4. and take norms in (C.80) to obtain

$$\begin{aligned} \|\Delta P^-(n)\| &\le \|\Psi_1(n, 0) \left[P_1^-(0) - P_2^-(0)\right] \Psi_2^T(n, 0)\| \\ &\le a^2 e^{-2bn} \|P_1^-(0) - P_2^-(0)\|. \end{aligned}$$

Since a and b are positive,

$$\lim_{n\to\infty} \|\Delta P^-(n)\| \le \lim_{n\to\infty} a^2 e^{-2bn} \|P_1^-(0) - P_2^-(0)\| = 0.$$

Q.E.D. ■

Appendix D

The Steady-State Kalman Filter

This appendix presents conditions that guarantee the existence and stability of the steady-state Kalman filter. The proofs are based on the work of Anderson and Moore [15] and Lewis [7].

D.1 An Upper Bound on $P^-(n)$

Lemma D.1 *Let (Φ, C) be detectable. Let $P^-(n)$ satisfy the Kalman filter equation (5.89). Then for all $P^-(0) \geq 0$, $P^-(n)$ is bounded above,*

$$0 \leq P^-(n) \leq S < \infty.$$

Proof. [15] The pair (Φ, C) is detectable, so there exists a constant matrix L such that $(\Phi - LC)$ is asymptotically stable. Define a suboptimal estimate $\hat{\boldsymbol{x}}_L^-(n)$ by

$$\hat{\boldsymbol{x}}_L^-(n+1) = (\Phi - LC)\hat{\boldsymbol{x}}_L^-(n) + L\boldsymbol{z}(n). \tag{D.1}$$

Then the *a priori* estimation error associated with $\hat{\boldsymbol{x}}_L^-(n+1)$ is

$$\begin{aligned}
\tilde{\boldsymbol{x}}_L^-(n+1) &= \boldsymbol{x}(n+1) - \hat{\boldsymbol{x}}_L^-(n+1) \\
&= \Phi\boldsymbol{x}(n) + \Gamma\boldsymbol{w}(n) - (\Phi - LC)\hat{\boldsymbol{x}}_L^-(n) - L\boldsymbol{z}(n) \\
&= \Phi\boldsymbol{x}(n) + \Gamma\boldsymbol{w}(n) - (\Phi - LC)\hat{\boldsymbol{x}}_L^-(n) - L\left[C\boldsymbol{x}(n) + \boldsymbol{v}(n)\right] \\
&= (\Phi - LC)\tilde{\boldsymbol{x}}_L^-(n) + \Gamma\boldsymbol{w}(n) - L\boldsymbol{v}(n).
\end{aligned} \tag{D.2}$$

Equations (D.1) and (D.2) have the same homogeneous part, namely $\boldsymbol{y}(n+1) = (\Phi - LC)\boldsymbol{y}(n)$, so (D.2) is also asymptotically stable.

The estimation error covariance matrix is

$$\begin{aligned} P_L^-(n+1) &= \text{Cov}\left[(\Phi - LC)\tilde{x}_L^-(n) + \Gamma w(n) - Lv(n)\right] \\ &= (\Phi - LC)P_L^-(n)(\Phi - LC)^{\mathrm{T}} + \Gamma Q \Gamma^{\mathrm{T}} + LRL^{\mathrm{T}}. \end{aligned} \tag{D.3}$$

Equation (D.3) is called a *Lyapunov equation*, and because $(\Phi - LC)$ is asymptotically stable, (D.3) has a bounded limiting solution. That is, there is a constant finite matrix $P_{L\infty}^-$ such that

$$\lim_{n\to\infty} P_L^-(n) = P_{L\infty}^-. \tag{D.4}$$

Note that $P_{L\infty}^-$ may depend on the initial choice $P_L^-(0)$; however, the existence of $P_{L\infty}^-$ is our only concern here.

By hypothesis, $P^-(0) \geq 0$. Equation (5.84) shows that $P^-(n)$ is the sum of positive-semidefinite terms. Hence $P^-(n) \geq 0$. Because $\hat{x}_L^-(n)$ is a suboptimal estimate,

$$0 \leq P^-(n) \leq P_L^-(n), \quad \forall\, n.$$

In light of (D.4), $P^-(n)$ is bounded above by a finite matrix. **Q.E.D.** ■

D.2 A Stabilizability Lemma

Lemma D.2 *Let $\sqrt{Q}$ be any matrix such that $\sqrt{Q}\sqrt{Q}^T = Q$, where Q is positive definite. If $(\Phi, \Gamma\sqrt{Q})$ is stabilizable, then the steady-state Kalman filter with K_∞ given by (5.104) is asymptotically stable.*

Proof. [15] Assume $(\Phi, \Gamma\sqrt{Q})$ is stabilizable, which implies that if $\alpha v = \Phi^{\mathrm{T}} v$ and $\sqrt{Q}^{\mathrm{T}}\Gamma^{\mathrm{T}} = 0$ for some constant α and vector v, then $|\alpha| < 1$ or $v = 0$.

Let us assume that the steady-state Kalman filter is *not* asymptotically stable, so $\Phi(I - K_\infty C)$ has at least one eigenvalue that lies on or outside the unit circle. Let λ denote such an eigenvalue. Then

$$[\Phi(I - K_\infty C)]^{\mathrm{T}}\, v = \lambda v, \tag{D.5}$$

for some vector $v \neq 0$.

The ARE (5.103) may be written as[1]

$$P_\infty^- = \Phi[I - K_\infty C]P_\infty^-[I - K_\infty C]^{\mathrm{T}}\Phi^{\mathrm{T}} + \Phi K_\infty R K_\infty^{\mathrm{T}}\Phi^{\mathrm{T}} + \Gamma Q \Gamma^{\mathrm{T}}. \tag{D.6}$$

Multiply each side of (D.6) by v and v^{H}, where v^{H} is the conjugate transpose of v: $v^{\mathrm{H}} = \left(v^{\mathrm{T}}\right)^* = (v^*)^{\mathrm{T}}$. Then

$$\begin{aligned} v^{\mathrm{H}} P_\infty^- v = {} & v^{\mathrm{H}}\Phi[I - K_\infty C]P_\infty^-[I - K_\infty C]^{\mathrm{T}}\Phi^{\mathrm{T}} v \\ & + v^{\mathrm{H}}\Phi K_\infty R K_\infty^{\mathrm{T}}\Phi^{\mathrm{T}} v + v^{\mathrm{H}}\Gamma Q \Gamma^{\mathrm{T}} v, \end{aligned}$$

[1] The derivation of (D.5) is analogous to that of (D.3).

which, by (D.5), becomes

$$\left(1-|\lambda|^2\right) v^{\mathrm{H}} P_\infty^- v = v^{\mathrm{H}} \Phi K_\infty R K_\infty^{\mathrm{T}} \Phi^{\mathrm{T}} v + v^{\mathrm{H}} \Gamma Q \Gamma^{\mathrm{T}} v. \tag{D.7}$$

The left-hand side of (D.7) is nonpositive, and the right-hand side of (D.7) is nonnegative. Both sides must be zero. This requirement means that $K_\infty^{\mathrm{T}} \Phi^{\mathrm{T}} = 0$, and

$$\sqrt{Q}^{\mathrm{T}} \Gamma^{\mathrm{T}} = 0. \tag{D.8}$$

Then (D.5) produces

$$\Phi^{\mathrm{T}} v = \lambda v. \tag{D.9}$$

Now (D.8) and (D.9) hold for $|\lambda| \geq 1$ and $v \neq 0$. But these properties contradict the hypothesis of stabilizability. Hence $\Phi\left[I - K_\infty C\right]$, the steady-state Kalman filter, must be asymptotically stable. **Q.E.D. ■**

D.3 Preservation of Ordering

Lemma D.3 *Let the Kalman filter be initialized by two positive-semidefinite initial error covariance matrices, $P_1^-(0)$ and $P_2^-(0)$, respectively. Let $P_1^-(n)$ and $P_2^-(n)$ denote the solutions of (5.89) associated with $P_1^-(0)$ and $P_2^-(0)$, respectively. If $P_1^-(0) > P_2^-(0)$, then $P_1^-(n) > P_2^-(n)$ for all $n \geq 0$.*

Proof. [15] Employ mathematical induction. By hypothesis, $P_1^-(0) > P_2^-(0)$. Assume

$$P_1^-(n-1) > P_2^-(n-1). \tag{D.10}$$

Then we have[2]

$$P_1^-(n) = \min_L \left[\Phi(I - LC) P_1^-(n-1)(I - LC)^{\mathrm{T}} \Phi^{\mathrm{T}} + \Gamma Q \Gamma^{\mathrm{T}} + L R L^{\mathrm{T}}\right].$$

The Kalman gain K *is* the matrix L that minimizes $P_1^-(n)$, so

$$P_1^-(n) = \Phi(I - KC) P_1^-(n-1)(I - KC)^{\mathrm{T}} \Phi^{\mathrm{T}} + \Gamma Q \Gamma^{\mathrm{T}} + K R K^{\mathrm{T}}.$$

Now apply assumption (D.10).

$$\begin{aligned} P_1^-(n) &> \Phi(I - KC) P_2^-(n-1)(I - KC)^{\mathrm{T}} \Phi^{\mathrm{T}} + \Gamma Q \Gamma^{\mathrm{T}} + K R K^{\mathrm{T}} \\ &\geq \min_M \left[\Phi(I - MC) P_2^-(n-1)(I - MC)^{\mathrm{T}} \Phi^{\mathrm{T}} + \Gamma Q \Gamma^{\mathrm{T}} + M R M^{\mathrm{T}}\right] \\ &= P_2^-(n). \end{aligned}$$

Q.E.D. ■

[2] This relation may be derived in a manner analogous to (D.3).

D.4 Convergence when $P^-(0) = 0$

Lemma D.4 *Let the Kalman filter be initialized with $P^-(0) = 0$. Then $P^-(n)$ converges to a finite, positive-semidefinite matrix P^-_∞,*

$$0 \leq \lim_{n\to\infty} P^-(n) = P^-_\infty, \tag{D.11}$$

and P^-_∞ is a solution to the ARE (5.103).

Proof. [15] Consider the Kalman filter with two different initial conditions, namely $P_0^-(0) = 0$, which produces $P_0^-(n)$, and $P_{-1}^-(-1) = 0$, which produces $P_{-1}^-(n)$. Note that $P_0^-(0)$ starts at time $n = 0$ while $P_{-1}^-(-1)$ starts at time $n = -1$.

Certainly,

$$P_{-1}^-(0) \geq P_0^-(0) = 0.$$

We follow the same argument as in Lemma D.3. Assume

$$P_{-1}^-(n-1) \geq P_0^-(n-1).$$

Then

$$\begin{aligned}
P_{-1}^-(n) &= \min_L \left[\Phi(I - LC)P_{-1}^-(n-1)(I - LC)^{\mathrm{T}}\Phi^{\mathrm{T}} + \Gamma Q \Gamma^{\mathrm{T}} + LRL^{\mathrm{T}}\right] \\
&= \Phi(I - KC)P_{-1}^-(n-1)(I - KC)^{\mathrm{T}}\Phi^{\mathrm{T}} + \Gamma Q \Gamma^{\mathrm{T}} + KRK^{\mathrm{T}} \\
&\geq \Phi(I - KC)P_0^-(n-1)(I - KC)^{\mathrm{T}}\Phi^{\mathrm{T}} + \Gamma Q \Gamma^{\mathrm{T}} + KRK^{\mathrm{T}} \\
&\geq \min_M \left[\Phi(I-MC)P_0^-(n-1)(I-MC)^{\mathrm{T}}\Phi^{\mathrm{T}} + \Gamma Q \Gamma^{\mathrm{T}} + MRM^{\mathrm{T}}\right] \\
&= P_0^-(n).
\end{aligned} \tag{D.12}$$

Observe that

$$P_0^-(1) = \min_L \left[0 + \Gamma Q \Gamma^{\mathrm{T}} + LRL^{\mathrm{T}}\right] = \Gamma Q \Gamma^{\mathrm{T}},$$

and

$$P_{-1}^-(0) = \min_M \left[0 + \Gamma Q \Gamma^{\mathrm{T}} + MRM^{\mathrm{T}}\right] = \Gamma Q \Gamma^{\mathrm{T}} = P_0^-(1).$$

It follows that

$$P_0^-(n+1) = P_{-1}^-(n), \quad n \geq 0. \tag{D.13}$$

Applying (D.12) and (D.13), we have

$$P_0^-(n+1) \geq P_0^-(n), \quad n \geq 0.$$

Hence, $P_0^-(n)$ is a monotone increasing sequence of matrices. Lemma D.1 states that $P_0^-(n)$ is bounded above. Therefore, $P_0^-(n)$ converges to a finite, positive-semidefinite matrix, which we denote by P^-_∞. We indicate this result by

$$\lim_{n\to\infty} P^-(n)\Big|_{P^-(0)=0} = P^-_\infty.$$

$P_0^-(n)$ is a solution to the Riccati equation (5.101), so P^-_∞ is a solution to the ARE (5.103). **Q.E.D.** ■

D.5 Existence and Stability

Theorem D.1 *Let $\sqrt{Q}$ be any matrix such that $\sqrt{Q}\sqrt{Q}^T = Q$, where Q is positive definite, and let $(\Phi, \Gamma\sqrt{Q})$ be stabilizable. Then (Φ, C) is detectable if and only if*

- P^-_∞ *is the* unique *(finite) positive-semidefinite solution to the ARE in (5.103),*
- P^-_∞ *is independent of $P^-(0)$, provided $P^-(0) \geq 0$, and*
- *the steady-state Kalman filter with Kalman gain given by (5.104) is asymptotically stable.*

Proof. [15, 7] We first prove necessity, so assume (Φ, C) is detectable. By analogy with (D.3), we have

$$P^-(n+1) = \Phi[I - K(n)C]P^-(n)[I - K(n)C]^{\mathrm{T}}\Phi^{\mathrm{T}} + \Phi K(n)RK^{\mathrm{T}}(n)\Phi^{\mathrm{T}} + \Gamma Q\Gamma^{\mathrm{T}}.$$

Let $\Psi(n+1, n)$ be the state-transition matrix associated with the *a priori* estimate of the Kalman filter, so

$$\Psi(n+1, n) = \Phi\,[I - K(n)C].$$

Then

$$\begin{aligned} P^-(n) &= \Psi(n,0)P^-(0)\Psi^{\mathrm{T}}(n,0) + \text{positive-semidefinite terms} \\ &\geq \Psi(n,0)P^-(0)\Psi^{\mathrm{T}}(n,0). \end{aligned}$$

By Lemma D.1, $P^-(n) \leq S$ for $n \geq 0$ and any fixed $P^-(0) \geq 0$. Choose $P^-(0) = \alpha I$, $0 < \alpha < \infty$. Then

$$0 \leq \alpha\Psi(n,0)\Psi^{\mathrm{T}}(n,0) \leq S, \quad n \geq 0,$$

and $\Psi(n,0)$ is bounded above for all $n \geq 0$. We will make use of this result shortly.

By a derivation similar to that of (C.79), we have

$$P^-(n+1) - P^-_\infty = \Phi\,[I - K_\infty C]\,[P^-(n) - P^-_\infty]\,[I - K(n)C]^{\mathrm{T}}\,\Phi^{\mathrm{T}},$$

where P^-_∞ is the matrix defined in Lemma D.4. Let $P^-(0) = \alpha I$, so $\Psi(n,0)$ is bounded above, and

$$P^-(n) - P^-_\infty = \{\Phi\,[I - K_\infty C]\}^n\,[P^-(0) - P^-_\infty]\,\Psi^{\mathrm{T}}(n,0).$$

We have a bound on $\Psi(n,0)$, and $\Phi\,[I - K_\infty C]$ is asymptotically stable due to Lemma D.2. Therefore,

$$\lim_{n\to\infty} P^-(n) - P^-_\infty = 0,$$

which we represent by

$$\lim_{n\to\infty} P^-(n)\Big|_{P^-(0)=\alpha I} = P^-_\infty. \tag{D.14}$$

Now let $P^-(0)$ be an *arbitrary* positive-semidefinite matrix, not necessarily of the form $P^-(0) = \alpha I$. Choose $0 < \alpha < \infty$ such that $\alpha I > P^-_0$. We apply Lemma D.3 and obtain

$$P^-(n)\big|_{P^-(0)=\alpha I} > P^-(n) \geq P^-(n)\big|_{P^-(0)=0}, \quad n \geq 0.$$

Hence,

$$\lim_{n\to\infty} P^-(n)\Big|_{P^-(0)=\alpha I} > \lim_{n\to\infty} P^-(n) \geq \lim_{n\to\infty} P^-(n)\Big|_{P^-(0)=0}.$$

By (D.14) and Lemma D.4, we have

$$P^-_\infty > \lim_{n\to\infty} P^-(n) \geq P^-_\infty.$$

Therefore,

$$\lim_{n\to\infty} P^-(n) = P^-_\infty,$$

which proves that P^-_∞ is unique. Since Lemmas D.4 and D.2 are satisfied, P^-_∞ is a solution to the ARE (5.103), and K_∞ is given by (5.104).

We now prove sufficiency. Assume the steady-state Kalman filter exists and is asymptotically stable, so $\Phi\,[I - K_\infty C]$ is asymptotically stable. Let $L = \Phi K_\infty$, and thus (Φ, C) is detectable. **Q.E.D.** ■

Appendix E

Modeling Errors

Included in this appendix are a few proofs related to the problem of modeling errors and their effect on the Kalman filter. Section 6.4 addresses this problem and this appendix employs the same notation as Section 6.4.

E.1 Inaccurate Initial Conditions

The proof of Corollary 6.1 follows.

Corollary E.1 *Let $P^-_{\text{true}}(0) \leq P^-_m(0)$, $Q_p(n) \leq Q_m(n)$, and $R_p(n) \leq R_m(n)$ for all n. Additionally let the system model (6.45–6.46) be UCO and UCC. Then there exists an integer $N > 0$ and a real number $\alpha > 0$ such that*

$$P_{\text{true}}(n) \leq \alpha I, \quad n \geq N.$$

Proof. [14] By Theorem 6.3 there exists

$$\alpha = \frac{\beta_1 + N\alpha_2\beta_2^2}{\beta_1^2} > 0,$$

and

$$P_m(n) \leq \alpha I, \quad n \geq N.$$

Theorem 6.5 implies that $P_{\text{true}}(n)$ is bounded above by $P_m(n)$. **Q.E.D.** ■

E.2 Nonlinearities and Neglected States

To set up the proof of Corollary 6.2, we substitute (6.59) into (6.60), which yields

$$\begin{aligned} P_{\text{true}}(n+1) = {} & [I - K(n+1)C(n+1)]\,\Phi(n+1,n)P_{\text{true}}(n)\Phi^{\mathrm{T}}(n+1,n) \\ & \times [I - K(n+1)C(n+1)]^{\mathrm{T}} + F(n), \end{aligned} \tag{E.1}$$

with

$$F(n) = [I - K(n+1)C(n+1)] \left[\Gamma(n)Q_p(n)\Gamma^{\mathrm{T}}(n) + \Delta\Theta(n)\Delta\Theta^{\mathrm{T}}(n) \right.$$
$$+ \Phi(n+1, n)\mathrm{E}\left[\tilde{\boldsymbol{x}}_{\text{true}}(n)\right]\Delta\Theta^{\mathrm{T}}(n)$$
$$\left. +\Delta\Theta(n)\mathrm{E}\left[\tilde{\boldsymbol{x}}^{\mathrm{T}}_{\text{true}}(n)\right]\Phi^{\mathrm{T}}(n+1, n)\right] [I - K(n+1)C(n+1)]^{\mathrm{T}}$$
$$+ K(n+1)R_p(n+1)K^{\mathrm{T}}(n+1). \tag{E.2}$$

We now have a theorem by Price [33].

Theorem E.1 *Let the model (6.45–6.46) be UCO and UCC, and let $F(n)$ be uniformly bounded in (E.2). Also let $P^-_{\text{true}}(0)$ be bounded. Then $P_{\text{true}}(n)$ is uniformly bounded for all n.*

Proof. [14, 33] Denote the state-transition matrix of $P_{\text{true}}(n)$ by

$$\Psi(n+1, n) = [I - K(n+1)C(n+1)]\,\Phi(n+1, n).$$

Then equation (E.1) becomes

$$P_{\text{true}}(n+1) = \Psi(n+1, n)P_{\text{true}}(n)\Psi^{\mathrm{T}}(n+1, n) + F(n),$$

which has solution

$$P_{\text{true}}(n+1) = \Psi(n+1, n)P_{\text{true}}(n)\Psi^{\mathrm{T}}(n+1, n) + \sum_{i=0}^{n} \Psi(n+1, i+1)F(i)\Psi^{\mathrm{T}}(n+1, i+1).$$

By Theorem 6.2 the Kalman filter is GUAS. According to Definition C.4, there exist real numbers $a > 0$ and $b > 0$ such that $\|\Psi(n, j)\| \leq ae^{-b(n-j)}$, for $n \geq j$. Also, by hypothesis $F(n) \leq \gamma I$ for all n and some real number $\gamma > 0$.

Then

$$\begin{aligned}
\|P_{\text{true}}(n+1)\| &= \left\| \Psi(n+1, 0)P_{\text{true}}(0)\Psi^{\mathrm{T}}(n+1, 0) + \sum_{i=0}^{n} \Psi(n+1, i+1)F(i)\Psi^{\mathrm{T}}(n+1, i+1) \right\| \\
&\leq \left\| \Psi(n+1, 0)P_{\text{true}}(0)\Psi^{\mathrm{T}}(n+1, 0) \right\| + \sum_{i=0}^{n} \left\| \Psi(n+1, i+1)F(i)\Psi^{\mathrm{T}}(n+1, i+1) \right\| \\
&\leq \|\Psi(n+1, 0)\|^2 \|P_{\text{true}}(0)\| + \sum_{i=0}^{n} \|\Psi(n+1, i+1)\|^2 \|F(i)\| \\
&\leq \|P_{\text{true}}(0)\|\, a^2 e^{-2b(n+1)} + \gamma \sum_{i=0}^{n} a^2 e^{-2b(n-i)}.
\end{aligned}$$

Let $\alpha = a^2 \max\{\|P_{\text{true}}(0)\|, \gamma\}$, and $\beta = 2b$. Then

$$\|P_{\text{true}}(n+1)\| \leq \alpha \sum_{i=0}^{n+1} e^{-\beta i} \leq \alpha \sum_{i=0}^{\infty} e^{-\beta i} < \infty.$$

Q.E.D. ■

We now prove Corollary 6.2.

Corollary E.2 *Suppose the model (6.45–6.46) is UCO and UCC, $\Delta\Theta(n)$ is uniformly bounded, and $P^-_{\text{true}}(0)$ is bounded. Then $P_{\text{true}}(n)$ is uniformly bounded for $n \geq N$.*

Proof. From equations (6.52) and (6.53),

$$\begin{aligned} \mathrm{E}\left[\tilde{\boldsymbol{x}}_{\text{true}}(n+1)\right] = {} & \left[I - K(n+1)C(n+1)\right]\Phi(n+1,n)\mathrm{E}\left[\tilde{\boldsymbol{x}}_{\text{true}}(n)\right] \\ & + \left[I - K(n+1)C(n+1)\right]\Delta\Theta(n). \end{aligned}$$

$\Delta\Theta(n)$ is uniformly bounded and $[I - K(n+1)C(n+1)]\,\Phi(n+1,n)$ is UAS. In a manner analogous to the proof of Theorem E.1, it follows that $\mathrm{E}[\tilde{\boldsymbol{x}}_{\text{true}}(n)]$ is uniformly bounded for $n \geq N$.

All matrices in the model (6.45–6.46) are bounded. Thus $F(n)$ is also bounded and Theorem E.1 applies. **Q.E.D. ■**

References

[1] A. Papoulis, *Probability, Random Variables, and Stochastic Processes.* New York: McGraw-Hill, 3rd ed., 1991.

[2] N. Levinson, "The Wiener RMS (root mean square) error criterion in filter design and prediction," *Journal of Mathematics and Physics*, vol. 25, pp. 261–278, 1946.

[3] J. Durbin, "The fitting of time series models," *Review of the International Statistical Institute*, vol. 28, pp. 233–244, 1960.

[4] M. H. Hayes, *Statistical Digital Signal Processing and Signal Modeling.* New York: John Wiley & Sons, 1996.

[5] R. V. Churchill and J. W. Brown, *Complex Variables and Applications.* New York, New York: McGraw-Hill, 1984.

[6] A. V. Oppenheim and R. W. Schafer, *Discrete-Time Signal Processing.* Englewood Cliffs, New Jersey: Prentice-Hall, 1989.

[7] F. L. Lewis, *Optimal Estimation with an Introduction to Stochastic Control.* New York: John Wiley & Sons, 1986.

[8] W. L. Brogan, *Modern Control Theory.* Englewood Cliffs, New Jersey: Prentice-Hall, 3rd ed., 1991.

[9] T. Kailath, *Linear Systems.* Englewood Cliffs, New Jersey: Prentice-Hall, 1980.

[10] R. E. Kalman, "Contributions to the theory of optimal control," *Boletin de la Sociedad Matematica Mexicana*, vol. 5, pp. 102–119, Apr. 1960.

[11] H. Heffes, "The effect of erroneous models on the Kalman filter response," *IEEE Trans. Automatic Control*, vol. AC-11, pp. 541–543, 1966.

[12] T. Nishimura, "On the a priori information in sequential estimation problems," *IEEE Trans. Automatic Control*, vol. AC-11, pp. 197–204, Apr. 1966.

[13] T. Nishimura, "Correction to and extension of 'On the a priori information in sequential estimation problems'," *IEEE Trans. Automatic Control*, vol. AC-12, p. 123, 1967.

[14] A. H. Jazwinski, *Stochastic Processes and Filtering Theory.* New York: Academic Press, 1970.

[15] B. D. O. Anderson and J. B. Moore, *Optimal Filtering.* Englewood Cliffs, New Jersey: Prentice-Hall, 1979.

[16] G. C. Goodwin and K. S. Sin, *Adaptive Filtering Prediction and Control.* Englewood Cliffs, New Jersey: Prentice-Hall, 1984.

[17] S. Haykin, *Adaptive Filter Theory.* Englewood Cliffs, New Jersey: Prentice-Hall, 1986.

[18] R. L. Bellaire, *Nonlinear Estimation with Applications to Target Tracking.* PhD thesis, Georgia Institute of Technology, Atlanta, GA, Jun. 1996.

[19] S. Haykin, *Communication Systems.* New York: John Wiley & Sons, 2nd ed., 1983.

[20] A. V. Oppenheim, A. S. Willsky, and I. T. Young, *Signals and Systems.* Englewood Cliffs, New Jersey: Prentice-Hall, 1983.

[21] E. Mazor, A. Averbuch, Y. Bar-Shalom, and J. Dayan, "Interacting multiple model methods in target tracking: A survey," *IEEE Trans. Aerospace and Electronic Systems*, vol. 34, pp. 103–123, Jan. 1998.

[22] E. W. Kamen, "Multiple target tracking based on symmetric measurements," in *Proc. American Control Conference*, (Pittsburgh, PA, USA), pp. 263–268, 1989.

[23] E. W. Kamen and C. R. Sastry, "Multiple target tracking using products of position measurements," *IEEE Trans. Aerospace and Electronic Systems*, vol. 29, pp. 476–493, Apr. 1993.

[24] R. L. Bellaire and E. W. Kamen, "A new implementation of the SME filter approach to multiple target tracking," in *Proc. SPIE Signal and Data Proc. of Small Targets*, vol. 2759, (Orlando, FL, USA), pp. 477–487, 1996.

[25] Ö. Y. Baş, M. Ho, B. Shafai, and S. P. Linder, "Improving stability of EKF filter used by the symmetrical measurement equation approach to multiple target tracking," in *Proc. SPIE Intl. Symp. on Optical Science, Engineering, Instrumentation*, Jul. 1999.

[26] S. P. Linder, B. Shafai, and Ö. Y. Baş, "Improving track maintenance of crossing and manuevering targets." Submitted to *AIAA Guidance and Control Conference*, Aug. 1999.

[27] W. J. Rugh, *Linear System Theory.* Englewood Cliffs, NJ: Prentice Hall, 2 ed., 1996.

[28] E. W. Kamen and B. S. Heck, *Fundamentals of Signals and Systems with MATLAB.* Upper Saddle River, New Jersey: Prentice-Hall, 1997.

[29] J. J. Deyst, Jr. and C. F. Price, "Conditions for asymptotic stability of the discrete minimum-variance linear estimator," *IEEE Trans. Automatic Control*, vol. AC-13, pp. 702–705, Dec. 1968.

[30] J. J. Deyst, Jr., "Correction to 'Conditions for asymptotic stability of the discrete minimum-variance linear estimator'," *IEEE Trans. Automatic Control*, vol. AC-18, pp. 562–563, Oct. 1973.

[31] R. E. Kalman and J. E. Bertram, "Control system analysis and design via the 'second method' of Lyapunov: II," *Trans. ASME, J. Basic Engineering*, vol. 82, pp. 394–400, Jun. 1960.

[32] K. L. Hitz, T. E. Fortmann, and B. D. O. Anderson, "A note on the bounds on solutions of the Riccati equation," *IEEE Trans. Automatic Control*, vol. AC-17, pp. 178–180, 1972.

[33] C. F. Price, "An analysis of the divergence problem in the Kalman filter," *IEEE Trans. Automatic Control*, vol. AC-13, pp. 699–702, Dec. 1968.

Index

9 781852 331337